G000164918

# Advanced Concrete Technology

## Advanced Concrete Technology



| | |
|---|---|
| *Constituent Materials* | ISBN 0 7506 5103 2 |
| *Concrete Properties* | ISBN 0 7506 5104 0 |
| *Processes* | ISBN 0 7506 5105 9 |
| *Testing and Quality* | ISBN 0 7506 5106 7 |

# Advanced Concrete Technology

## Processes

Edited by
**John Newman**
Department of Civil Engineering
Imperial College
London

**Ban Seng Choo**
School of the Built Environment
Napier University
Edinburgh

ELSEVIER
BUTTERWORTH
HEINEMANN

AMSTERDAM BOSTON HEIDELBERG LONDON NEW YORK OXFORD
PARIS SAN DIEGO SAN FRANCISCO SINGAPORE SYDNEY TOKYO

Butterworth-Heinemann
An imprint of Elsevier
Linacre House, Jordan Hill, Oxford OX2 8DP
200 Wheeler Road, Burlington MA 01803

First published 2003



**British Library Cataloguing in Publication Data**
A catalogue record for this book is available from the British Library

**Library of Congress Cataloguing in Publication Data**
A catalogue record for this book is available from the Library of Congress

ISBN 0 7506 5105 9

For information on all Butterworth-Heinemann publications visit our website at www.bh.com

Typeset by Replika Press Pvt Ltd, India
Printed and bound in Great Britain

# Contents

# Part 3 Special processes and technology for particular types of structure

# Part 5 Exposed concrete finishes



# Part 6 Formwork

## Part 7 Precast concrete

## Part 8 Concrete roads

## Part 9 Industrial floors



## Part 10 Reinforced and prestressed concrete

# Preface

The book is based on the syllabus and learning objectives devised by the Institute of Concrete Technology for the Advanced Concrete Technology (ACT) course. The first ACT course was held in 1968 at the Fulmer Grange Training Centre of the Cement and Concrete Association (now the British Cement Association). Following a re-organization of the BCA the course was presented at Imperial College London from 1982 to 1986 and at Nottingham University from 1996 to 2002. With advances in computer-based communications technology the traditional residential course has now been replaced in the UK by a web-based distance learning version to focus more on self-learning rather than teaching and to allow better access for participants outside the UK. This book, as well as being a reference document in its own right, provides the core material for the new ACT course and is divided into four volumes covering the following general areas:

- constituent materials
- properties and performance of concrete
- types of concrete and the associated processes, plant and techniques for its use in construction
- testing and quality control processes.

The aim is to provide readers with an in-depth knowledge of a wide variety of topics within the field of concrete technology at an advanced level. To this end, the chapters are written by acknowledged specialists in their fields.

The book has taken a relatively long time to assemble in view of the many authors so the contents are a snapshot of the world of concrete within this timescale. It is hoped that the book will be revised at regular intervals to reflect changes in materials, techniques and standards.

John Newman
Ban Seng Choo

# Preface

# List of contributors

**Phil Bamforth**
19 Pursewardens Close, Ealing, London W13 9PN, UK

**Neil Beningfield**
RMC Readymix Ltd, RMC House, Evreux Way, Rugby, Warwickshire CV21 2DT, UK

**Tony Binns**
Tony Binns Training Workshops, PO Box 5328, Slough, SL2 3FL, UK

**British Cement Association**
Century House, Telford Avenue, Crowthorne, Berkshire RG45 6YS

**Ban Seng Choo**
School of the Built Enviroment, Napier University, Edinburgh

**John L. Clarke**
The Concrete Society, Century House, Telford Avenue, Crowthorne, Berkshire RG45 6YS, UK

**Rod Collins**
Building research Establishment, Bucknalls Lane, Garston, Watford WD2 7JR, UK

**Steve Crompton**
RMC Readymix Ltd, RMC House, Rugby, Warwickshire CV21 2DT, UK

**Joe Dewar**
J.D. Dewar Consultancy, 32 Tolethorpe Close, Oakham, Rutland LE15 6GF, UK

**Noel Dixon**
RMC Readymix Ltd, RMC House, Evreux Way, Rugby, Warwickshire CV21 2DT, UK

**Rob Gaimster**
RMC Readymix Ltd, RMC House, Evreux Way, Rugby, Warwickshire CV21 2DT, UK

**Geoffrey Griffiths**
135 Gertrude Road, West Bridgford, Nottingham NG2 5DA

**D.J. Hannant**
Department of Civil Engineering, University of Surrey, Guildford, Surrey GU2 5XH, UK

**Frank Hawes**
Gulls, Cley next the Sea, Holt, Norfolk NR25 7RR, UK

**Reg Horne**
Slipform International, Westgate, Aldridge, West Midlands WS9 8BS, UK

**Shaun A. Hurley**
Taylor Woodrow, 345 Ruislip Road, Southall, Middlesex UB1 2QX, UK

**S.A. Jefferis**
Department of Civil Engineering, University of Surrey, Guildford, Surrey GU2 5XH

**Eric Miller**
British Nuclear Fuels plc, Sellafield, Seascale, Cumbria CA20 1PG, UK

**Ron Montgomery**
Lafarge Aluminates, 730 London Road, Grays, Essex RM20 3NJ, UK

**John Newman**
Department of Civil Engineering, Imperial College, London, SW7 2BU, UK

**Tony Newton**
Lafarge Aluminates, 730 London Road, Grays, Essex RM20 3NJ, UK

**Phil Owens**
Rosebank, Donkey Lane, Tring, Hertfordshire, HP23 4DY, UK

**P.F. Pallet**
Pallet Temporary*Works* Ltd, 38 Anson Avenue, Lichfield, Staffordshire WS13 7EU, UK

**Bill Price**
Lafarge Cement, Manor Court, Chilton, Oxon OX11 0RN, UK

**John Richardson**
71 Disraeli Crescent, Park Estate, High Wycombe, Buckinghamshire HP13 5EW, UK

**Graham Taylor**
British Cement Association, Century House, Telford Avenue, Crowthorne, Berkshire RG45 6YS, UK

**Tony Threlfall**
16 Chiltern Road, Marlow, Buckinghamshire SL7 2PP, UK

**Neil Williamson**
Monofloor Technology Ltd, PO Box 367, Swindon, Wiltshire SN26 7EQ, UK

**David York**
Sitebatch Technologies, Victoria Stables, Bourne, Lincolnshire PE10 9JZ, UK

# Part 1

# Mix design

# 1

# Concrete mix design

*Joe Dewar*

This chapter surveys the subject of concrete mix design, identifying the aspects to be considered and the range of methods available. Particular emphasis is placed on the comprehensive series of initial tests of concrete, their accurate computerized simulation and simple approximate methods of design.

Note – The introduction of BS EN 206–1/BS 8500 for concrete specifications has entailed changes to traditional definitions and terminology; Readers are advised to consult these publications together with guidance provided by Harrison (2003). Where earlier publications are quoted some of the original terminology has been retained in this chapter particularly in figures and examples.

The main changes in terminology are

mix → concrete or design
trial mix → initial test
workability → consistence and consistence class
free water → effective water
transgrade → compressive strength class

## 1.1 Introduction

Before considering alternative approaches to mix design, it is necessary to differentiate between the various purposes for which a design may be required as any one or more of the following:

1 Accurate design for safety and economy for large-scale use:
   (a) for a single specification
   (b) for multiple specifications
   (c) to meet quality assurance requirements.

2 Comparisons of performance and economy of different materials or designs.
3 Approximate design for:
   (a) estimating costs
   (b) preparing trial mixes
   (c) checking submitted designs.
4 Safe design for instant use.

The best method to suit any of these purposes may be found as one or as a combination of the following:

1 Laboratory tests of concrete
2 Full-scale tests of concrete
3 Previous production data bank
4 'Comprehensive' methods of design simulating initial tests of concrete
5 'Simplified' methods of design
6 'Ready-to-use' standard data

The mix design process may be required to include

1 Appraisal of the specification for any constraints, e.g. materials, production and construction conditions
2 Assessing availability of materials
3 Selection of materials
4 Obtaining data and/or testing of materials
5 Identifying influences of mix design on concrete production and construction methods
6 Economic analyses concerning material and/or concrete production costs
7 Obtaining approval for designs
8 Recommending changes to specifications

The aspects of the fresh and hardened concrete to be covered by the design may include a selection of the following either because they are specified or because they are needed for production or construction purposes:

1 Consistence
2 Rate of stiffening
3 Cohesion
4 Plastic density and yield
5 Mix proportions
6 Batch quantities for production
7 Strength
8 Durability parameters
9 Air content
10 Special types of concrete or special properties, e.g. surface finish.

Not surprisingly, with all the different purposes, methods and aspects that may need to be covered, a wide range of solutions needs to be available, some of which will be purpose-made to suit the requirements of specific sectors of industry or individual companies and for particular applications.

For example, a ready-mixed concrete producer needs to have a large number of accurate mix designs ready for immediate use to cater for 'same-day' orders. Because of the large amount of test data being accumulated, the producer can utilize this information to develop and update the designs routinely within a control system.

On a large construction site where concrete is produced over a long period, a similar situation may apply except that the number of designs can be restricted to those applying at that site. At the other extreme, on a small site, where small quantities of concrete are mixed by hand, general-purpose safe ready-to-use mix designs are usually more appropriate.

Examples of some of the available solutions are outlined in this chapter. In each case, readers are referred to the cited reference for full details. Before this is done, a number of general principles which are common to each solution need to be introduced.

## 1.1.1 Interpreting the specification

Specification requirements need careful examination because of the different ways in which different aspects may be treated. For example, concrete may be specified by prescription or by performance. When concrete is prescribed, some or all of the materials or their proportions may be fixed in advance by the specifier. When mixes are specified by performance the producer has the responsibility of selecting the materials and their proportions to meet the test requirements for performance. In practice most specifications are a combination of both prescription and performance requirements.

Specified values may be absolute limits, class limits characteristic values or target values. When absolute limits, class limits, or characteristic values are specified it is necessary to make allowance for variations due to materials, batching and testing in assessing the target value. Such allowance is usually termed the design margin or simply the margin. For example, on the basis of BS EN 206–1/BS 8500 the interpretations of the specification shown in Table 1.1 would apply.

**Table 1.1** Interpretations of a specification

| Aspect | Specified value | Interpretation |
|---|---|---|
| Compressive strength class | C25/30 | Characteristic values; not more than 5% of possible results below 25 N/mm$^2$ for cylinders or 30 N/mm$^2$ for cubes, absolute limit of 21 N/mm$^2$ for cylinders or 26 N/mm$^2$ for cubes (see section 1.1.3) |
| Workability (slump) | Class S2 | Target value; tolerance ±25 mm |
| Air entrainment | 4% min *total* air | Target value, tolerance ±2% |
| Minimum cement content | 300 kg/m$^3$ | Minimum target value; absolute limit of 290 kg/m$^3$ for individual observed values |

## 1.1.2 Selection of materials

The selection of materials may be constrained by

1 Availability and cost
2 Conformity with standards
3 Conformity with the contract specification, e.g. maximum aggregate size.

Additionally the materials properties and the specified requirements for concrete influence the mix proportions and concrete costs. There may be interdependencies which effectively prohibit the use of particular materials or can be resolved only by using certain materials.

## 1.1.3 Relation between target strength and specified characteristic strength

Compressive strength is normally specified in terms of a characteristic value, $f_{ck}$, below which a stated maximum proportion of results, usually 5 per cent in Europe, is permitted to occur. For mix design purposes, relationships between strength and water/cement ratio or cement content are usually based on average strength, so that information is required on how to estimate the average strength to be targeted for any given specified strength ($f_{ck}$).

Statistical theory and concreting practice have combined to confirm that the proportion of results expected to be below the characteristic strength is related to the margin when expressed as a multiple, $k$, of the standard deviation, as shown in Table 1.2.

**Table 1.2** Statistical margin factors for strength

| Maximum percentage of results below the characteristic strength level ($f_{ck}$) | Minimum value of $k$ |
|---|---|
| 10 | 1.28 |
| 5 | 1.64 |
| 2.5 | 2.00 |
| 1 | 2.33 |
| 0.5 | 2.58 |
| 0.1 | 3.00 |

Thus, the minimum margin is 1.64 × standard deviation when the specified maximum percentage is 5 per cent. However, taking into account the need for safety to cover time delay in obtaining strength results and other aspects introducing uncertainty, it would be very unwise to operate at this margin. Indeed, compliance requirements for strength are usually set at a level to discourage the use of low values for $k$. In UK practice values of $k$ of less than 2 would not normally be adopted.

The Quality Scheme for Ready Mixed Concrete, QSRMC (2001), for example in the UK, defines the Target Mean Strength ($T$) for cubes as follows:

$$T \geq f_{cu} + M \text{ for values of } T \text{ of 27 N/mm}^2 \text{ and above}$$

or

$$T \geq \frac{f_{cu}}{1 + M/27} \text{ for values of } T \text{ below 27 N/mm}^2$$

where $M$ is the minimum design margin, $k \times SD$
$k$ is the margin factor
$SD$ is the current standard deviation for the control mix.

These formulae assume that standard deviation increases uniformly with strength level until a mean of 27 N/mm$^2$ is reached and is constant above that level as shown in Figure 1.1(a).

The minimum design margin, $M$, is not normally permitted to be less than 7 N/mm$^2$ or $k$ less than 2. When test rates are very low and for new plants and new combinations of materials, more onerous constraints are applied by QSRMC (2001). For example, initially a standard deviation of 6 N/mm$^2$ is to be assumed. Typical values for standard deviation lie in the range 3.5–6.5 N/mm$^2$ so that design margins are usually in the range 7–13 N/mm$^2$ for adequate test rates.

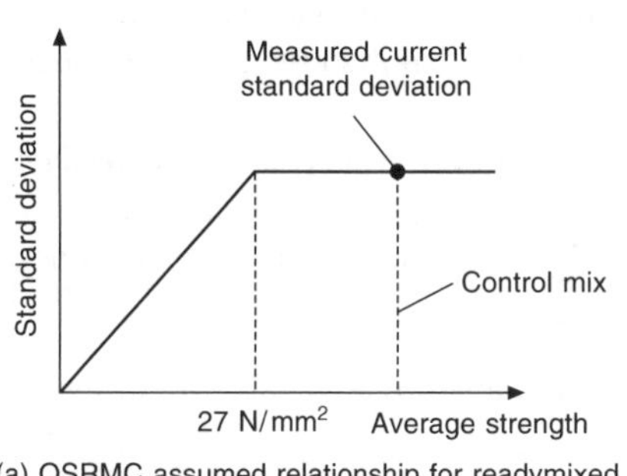


(a) QSRMC assumed relationship for readymixed concrete, based on *average* strength

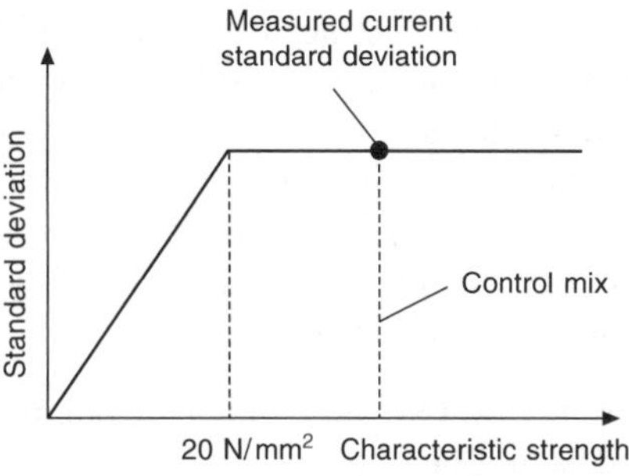


(b) DoE method of mix design assumed relationship for concrete, based on specified *characteristic* strength

**Figure 1.1** Assumed relationships between standard deviation and strength, enabling standard deviations to be estimated over the full range of strength when the current value is known at the control mix level.

The Department of the Environment, *Design of Normal Concrete Mixes*, (DoE, 1988), suggests that for values of $f_{ck}$ of 20 and above, standard deviations below 8 N/mm$^2$ should not be assumed unless more than 20 results have been obtained and values below 4 N/mm$^2$ should not be used even when more results are available. This is obviously good advice for site mixing on small and medium-sized contracts when control is limited and initially on larger contracts before control systems become fully operable. This publication also identifies that standard deviation can be assumed to be constant above a certain strength level but as a characteristic strength, in this case 20 N/mm$^2$ as shown in Figure 1.1(b).

Figures 1.1(a) and 1.1(b) referred to above illustrate the two commonly quoted assumptions concerning standard deviation, which are constant standard deviation above a certain level and constant coefficient of variation at lower levels. (*Note:* Constant coefficient of variation implies a constant ratio of standard deviation to average strength, i.e. a line of constant gradient, through the origin.)

ACI (1991) permits either assumption to be made so that standard deviations and margins to be used in design will differ depending on the assumption made, on the specified strength level and on the level at which standard deviation is measured for quality-control purposes.

## 1.1.4 Useful simplifying assumptions

There are a number of 'approximate' rules which can be used to simplify mix design but it needs to be recognized that they are approximate and that they may lead to unsafe or uneconomic mix designs. Some methods fall into the trap of attempting to be fully comprehensive while employing assumptions which lead to inaccuracies. The end result may be complexity and inaccuracy. Simplifying assumptions should be used only to make small adjustments to the results of trial batches or to produce further trial batches which will then be observed and adjusted before proceeding to full-scale production.

Examples of simplifying rules which are often adopted are:

1 *Water content is assumed constant over the range of cement content, for any given materials and slump.* This is a very convenient assumption but can be very misleading when applied over a wide range of cement contents when accuracy is paramount.

2 *Cement plus fine aggregate content is maintained constant.* This may lead to uncohesive lean concrete and over-cohesive rich concrete mixes if applied over a wide range of cement contents.
3 *Percentage of fine in total aggregate is maintained constant.* As (2) but greater problems.
4 *Single relationship is assumed between strength and w/c for a given cement type and for all aggregates.* If the relationship is nationally safe it will usually be uneconomic for large-scale production.

There are two simple rules which are completely valid and are applicable to all methods of mix design. These are:

5 *The sum of the weights of materials per cubic metre equals the plastic density.* The useful implication of this obvious rule is that if there are, say, three materials, e.g. cement ($C$), aggregate ($A$) and water ($W$), and if the quantities of any two of them, together with the plastic density ($PD$), are either known or assumed then the third can be calculated by difference:

$$(PD)\ \text{kg/m}^3 = (C) + (A) + (W)$$

*Note:* ($W$) is effective water and ($A$) is in the saturated and surface dried conditon (SSD).

6 *The sum of the volumes of materials per cubic metre is unity.* This is only of value if the relative density of each material is known, when

$$\frac{(C)}{RD_C \times 1000} + \frac{(FA)}{RD_{FA} \times 1000} + \frac{(CA)}{RD_{CA} \times 1000} + \frac{(W)}{1000} + \frac{\text{air}\,(\%)}{100} = 1$$

where ($W$) is the volume of effective free water in litres (= weight in kg), ($C$) is the weight of cement, ($FA$) and ($CA$) are respectively the weights of fine and coarse aggregates in the SSD condition, $RD_C$ is the relative density of the cement and $RD_{FA}$ and $RD_{CA}$ are the relative densities on a saturated and surface-dried basis (SSD) of the fine and coarse aggregates respectively.

*Note:* If admixtures are incorporated then their weights and volumes need to be taken into account.

## 1.1.5 Interaction between mix design, concrete production and construction

The number and types of materials may influence the storage requirements, e.g. the number of silos for powders, the speeds of batching, mixing and delivery, the necessity for special equipment and the training of staff. The cohesion of the concrete and its consistence and the rates of chemical reactions may affect the method and speed of construction and the time available for completing the process.

Thus, in extreme situations, the designer of a mix may have to consider several different solutions to ensure that total optimization is obtained. It is always preferable to consider the constraints, problems and possible solutions before commencing the design process rather than afterwards. Communication between interested parties is paramount.

## 1.2 Initial laboratory tests of concrete

There are a number of factors to keep in mind when preparing for, and making, initial tests.

1 All materials should be as representative as possible of those to be used for the construction.
2 Preferably materials should be used in the same condition as those to be used for the construction, i.e. usually wet, in the case of aggregates. Although greater accuracy may appear to be obtained by oven-drying, use of wet aggregates and allowing for moisture content may be preferable to making assumptions concerning absorption by dry aggregates of water from the concrete mix. Air expelled from the aggregates may remain in the concrete and may reduce the strength.
3 Cohesion should be checked at the upper tolerance level permitted for the specified consistence.
4 Small-scale batches will lose consistence due to evaporation (and absorption, if the aggregates are dried) at a much faster rate than large batches used in construction.
5 Interpolation between the results of two initial tests is preferable to extrapolation from a single initial test.
6 Results from an initial test are only representative of the qualities of the materials used in the test. It is necessary to consider making allowances for normal variations in larger-scale concrete production.

If two initial tests are required to enable interpolation then they should be made about 50 kg/m$^3$ apart in cement content, either by making two separate designs or by rule-of-thumb adjustments to the original design. For example, if it is expected that a design is over-safe at a cement content of 375 kg/m$^3$ a second test could be made at 325 kg/m$^3$, using (325/375) × original batch weight of cement, i.e. a reduction of 2.5 kg and with a corresponding increase in sand, as shown in Table 1.3.

**Table 1.3** Examples of designs for initial tests

| Materials | Quantities (kg) | |
|---|---|---|
| | Test 1 | Test 2 |
| Intended cement (kg/m$^3$) | 375 | 325 |
| Cement | 18.75 | 16.25 |
| Sand (SSD) | 30 | 32.50 |
| Gravel (SSD) | 61.25 | 61.25 |
| Water (effective) | Appr 9 | Appr 9 |
| Total | 110 + water | 110 + water |

*Note:* The quantities for test 1 are based upon the use of an approximate method of design, the results of which are shown in Table 1.10.

It is important to treat the estimated water content from a 'paper' mix design only as a guide to the amount needed for the required consistence. When the slump has been measured and, if necessary, adjustment made to the water content and the slump retested, the total amount of water added should be recorded, the plastic density measured and cylinders or cubes made for strength tests.

*Note:* If aggregates are used in the tests in conditions other than SSD it will be necessary to correct the weights of aggregate and water. This is necessary to allow for absorption of water when dry aggregates are used or to allow for the moisture contents in excess of the SSD condition when using saturated aggregates.

When initial tests are made it is essential to judge whether the cohesion is adequate for the intended purpose. If the concrete has a tendency to segregate, the mix should be redesigned with a higher proportion of fine aggregate or, if this does not solve the problem, use of a finer sand, a plasticizer, air entrainment, an increased cement content or a change of cementitious material should be considered.

Slight over-cohesion is essential to take account of variations that may occur in full-scale production but if the concrete is excessively cohesive, the proportion of fine aggregate should be reduced to avoid the risk of a mortar layer.

## 1.2.1 Calculations of quantities per cubic metre from initial tests

Quantities of materials per cubic metre can be calculated from the mix proportions used in initial tests and the plastic density determined as described in BS EN 12350–1, Part 6, as follows:

$$\text{Cement content, } (C) = (c)\ (\text{kg}) \times \text{scale factor}$$

where the scale factor is

$$\frac{(PD)}{\text{Total test batch weight (kg)}}$$

$(c)$ is the test batch weight of cement and $(PD)$ is the measured plastic density. Similarly, calculations may be made for the quantities of the other materials.

The process can be tabulated, as, for example, in Table 1.4, where the results are shown of making Test 1 from Table 1.3, the measured plastic density $(PD)$ being 2360 kg/m$^3$. It will be observed from Table 1.4 that, in this case, because the measured values for water demand of 8.85 kg and for plastic density of 2360 kg/m$^3$ were slightly different from the values assumed in the paper mix design in Table 1.3, the materials contents also differ slightly, e.g. 372 compared with 375 kg/m$^3$ for cement content.

**Table 1.4** Results from the initial test and scaled up values for production

| Materials | Quantities (kg) | |
|---|---|---|
| | Test 1 | Scaled-up values (per cubic metre) |
| Cement | 18.75 | 372 |
| Fine aggregate (SSD) | 30 | 596 |
| Coarse aggregate (SSD) | 61.25 | 1216 |
| Water (free) | 8.85 | 176 |
| Total | 118.85 | 2360 (= *PD*) |

Scale factor = 2360/118.85 = 19.86

## 1.2.2 Adjustments to allow for moisture content of aggregates

For production purposes it is necessary to increase the batch weights of the fine and coarse aggregates by their respective free moisture contents, to ensure the correct SSD weights. The amount of water to be added is reduced correspondingly. Assuming 6 per cent free moisture in the fine aggregate and 2 per cent in the coarse aggregate the adjustments to the batch figures would be as shown in Table 1.5. It will usually be necessary to round the final figures to 5 or 10 kg, depending on the weigh scale divisions of the batching equipment and the batch size.

**Table 1.5** Adjustments to batch data to allow for moisture contents of aggregates

| Material | Batch data ($kg/m^3$) | Moisture (kg) | Adjusted batch data ($kg/m^3$) |
|---|---|---|---|
| Cement | 372 | | 372 |
| Fine aggregate | 596 (SSD) | 6% × 596 = 36 | 632 |
| Coarse aggregate | 1216 (SSD) | 2% × 1216 = 24 | 1240 |
| Water | 176 (effective) | 60 | 116 |
| Total | 2360 | | 2360 |

## 1.2.3 Estimation of yield and volume delivered

The compacted volume to be expected from a given total weight of materials batched or concrete delivered, can be calculated simply as

$$\text{Volume}\,(\text{m}^3) = \frac{\text{Total weight of materials or concrete}}{(PD)}$$

where $(PD)$ is the plastic density determined on a fully representative sample to BS EN 12350–1, Part 6. For example, if the difference between the gross and tare weights of a truck mixer is 11 824 kg and the plastic density is measured as 2360 $kg/m^3$ then the delivered volume = 11 824/2360 = 5.01 $m^3$ which can be compared with the volume declared on the delivery docket.

In the plastic density test, about 0.5–1.5 per cent of air may remain entrapped in the concrete after full compaction. In practice, a higher value, say 1–2 per cent, may remain entrapped in the construction, so that a very slight over-yield could be observed but this will usually be masked by the wastage which can occur. With air-entrainment effects are more complex and may lead to a reduction or an increase in total residual air and the yield.

## 1.2.4 Estimation of volume of concrete for ordering purposes

In assessing overall quantities of materials required on-site, an allowance should be made for loss of concrete at the mixer and during transport and placing. These losses may be considerable, and due allowance should be made depending on site conditions, and the effectiveness of any measures taken to minimize losses. Similarly, when ordering ready-mixed concrete allowances should be made for losses subsequent to discharge.

### 1.2.5 Design of an initial test programme

The design of an initial test programme will depend on the specific requirements, complexities and novelties of the specification, the number of different concretes to be designed, the availability of materials and knowledge of their properties, as well as the experience of the designer and whether a data bank of designs for similar projects is available. The two extreme approaches can be considered to be

1 A comprehensive programme intended to establish successively narrowing boundaries within which the requisite solution or solutions can be found.
2 Intuitive initial designs followed by progressive modifications based on expectancy of the effects of such changes.

Method 1 is likely to take much longer but provides useful evidence for justifying the final solution. Method 2 is faster but may be more difficult to explain to other parties and to justify the final choice. Both methods and intermediate versions may be applicable to different situations.

### 1.2.6 Full-scale tests

The principles applicable to laboratory tests apply equally to full-scale tests with the addition of conditions simulating as near as possible those expected to apply to production, transport and to construction, particularly if the tests are to cover performance of the fresh concrete during handling, placing and finishing or to cover the subsequent properties of the hardened concrete in the structure.

### 1.2.7 Data bank

The availability of data from previous laboratory or field trials and from production may save considerable time and effort in enabling instant designs to be available for production or in reducing the number of trials. Naturally, it is necessary to take into account any probable differences in specifications, materials or conditions applying to the new work compared with the data bank.

## 1.3 Comprehensive mix design of ready-mixed concrete based on laboratory trials

The BRMCA method of concrete mix design described by Dewar (1986) and Murdock *et al.* (1991) has been developed over many years by members of the British Ready Mixed Concrete Association. It has been adopted throughout the industry and used in the production of many millions of cubic metres of concrete. The major stages in the method are shown in Figure 1.2.

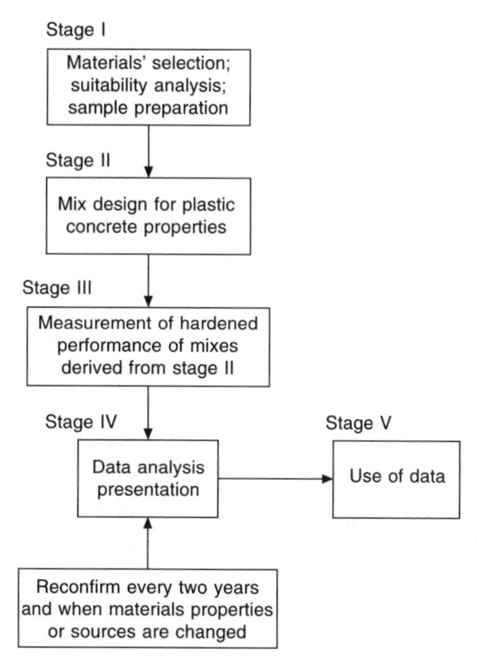


**Figure 1.2** Major stages of the BRMCA method of mix design.

A key feature of the method is that in stage II (Figure 1.3) mixes are designed for optimum performance in the plastic state to ensure that the concrete is generally suitable for transport, handling, compacting and finishing. Only when the mixes have been designed to achieve these properties are the hardened properties of the mixes determined in stage III of the laboratory method (Figure 1.4) or from tests in production.

*Note:* In the alternative production batch method, concrete is sampled from at least 100 production batches over a wide range of cement contents during a period of 12 months.

The analysis and recording of data in stage IV (Figure 1.5) are based on cement content as the primary parameter to which all other constituents and concrete properties are directly related. The key relationships derived in Figure 1.5, and illustrated for typical materials in Figure 1.6, are those between:

- cement content and the major specified concrete properties (Figures 1.6(a)–(d) and (f));
- cement content and the weight of each constituent (Figure 1.6(e)).

## 1.3.1 Use of base data from the BRMCA mix design method

For use in concrete production the base data shown in Figure 1.6(e) are stored in tables or graphs and in batch books or computer memory, in increments of cement content of either 5 kg/m$^3$ or 10 kg/m$^3$ for the typical production range of mixes of 100–450 or 500 kg/m$^3$. The base data provide the ideal proportioning of all mix constituents to achieve specified concrete properties, enable preparation of the correct batching instruction for

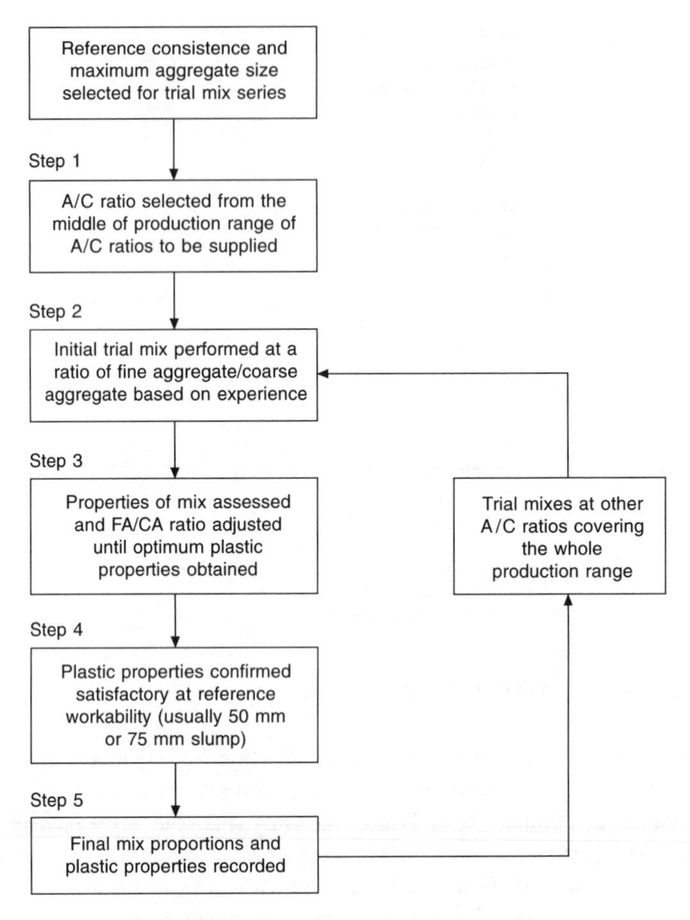


**Figure 1.3** BRMCA mix design method stage II: design for plastic properties.

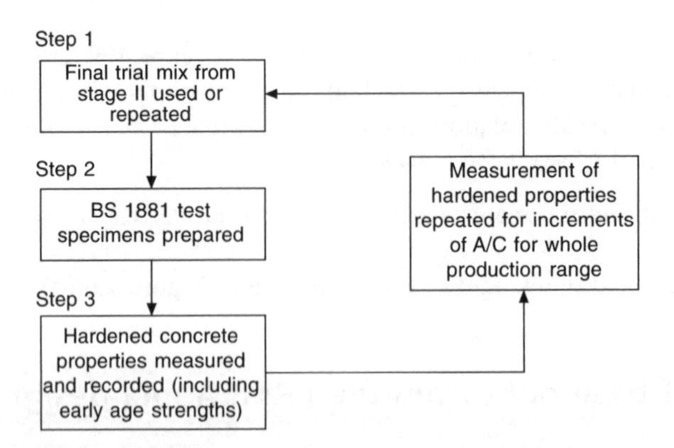


**Figure 1.4** BRMCA mix design method stage III: performance of hardened concrete.

the production of all concretes, and provide the basis for altering mix proportions to achieve different values of consistence and special concrete properties. They enable the key parameters for every combination of materials to be identified, e.g. water/cement ratio; plastic density.

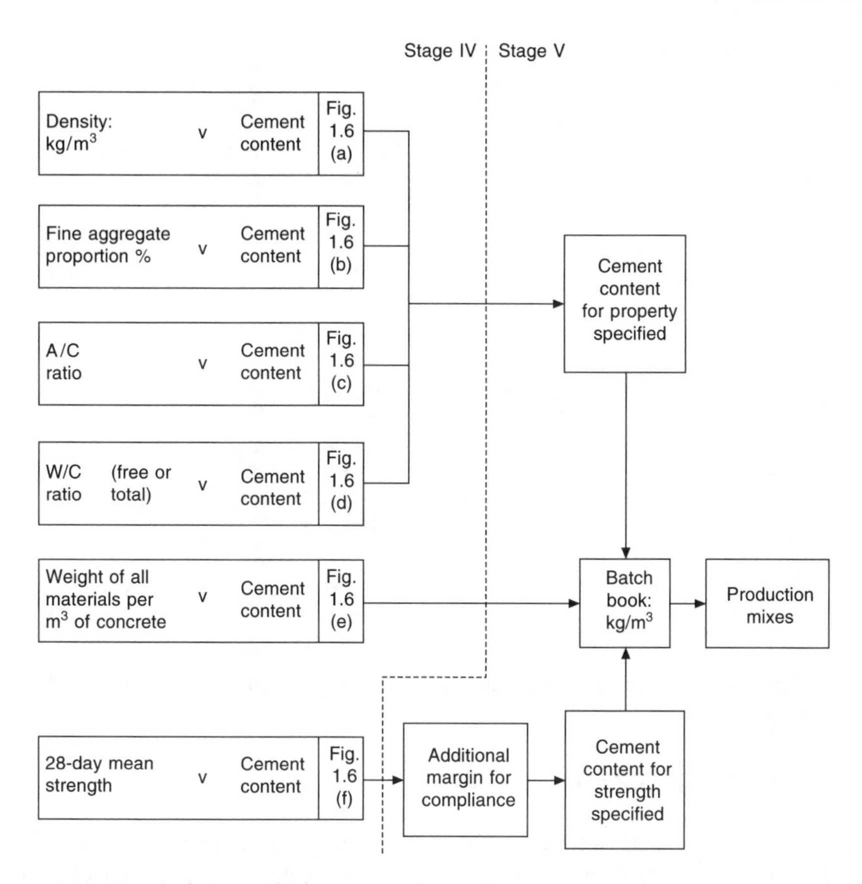


**Figure 1.5** BRMCA mix design method stage IV and V: analysis, presentation and use of mix design data.

The relationship between cement content and strength allows the selection of a mix to satisfy any specified strength requirement when account is taken of the design margin as described earlier.

In preparing batch books for production purposes, adjustments are made to allow for the batching of wet aggregates.

Where designed concretes are supplied the producer has the responsibility of controlling the quality of the concrete so that the conformity tests are satisfied. To achieve this objective the producer operates an approved control system, e.g. QSRMC (2001), which compares the results of tests on samples of materials and concrete with the performance of the materials and concrete on which the mix design was based. Thus, for example, Figure 1.6(f) is updated by the producer as necessary on the basis of current control data. The concrete control system also provides a current value of standard deviation for use in adjusting the design margin.

The following is an example of the use of base data from the BRMCA method of mix design for determining the batch weights per cubic metre of concrete for a designed concrete specification:

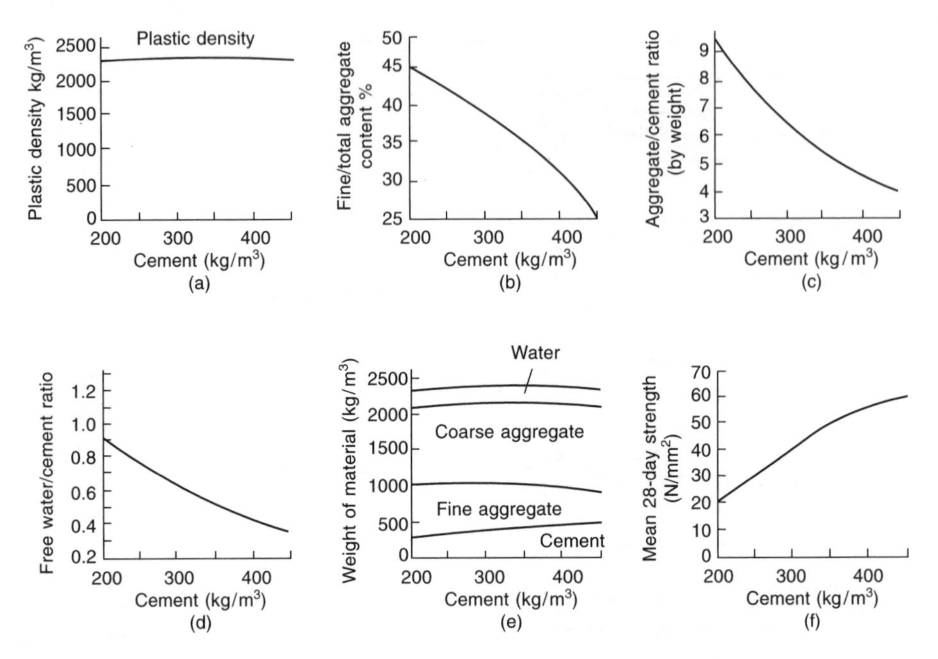


**Figure 1.6** Typical relationships for one material combination: 20 mm maximum aggregate size, 75 mm reference slump.

## 1.3.2 Mix design example

The following steps are taken to select batch quantities for a designed concrete to BS EN 206–1/BS 8500 from typical relationships derived using the BRMCA method, with specified compressive strength class C25/30, target slump 75 mm and maximum effective (free) water/cement ratio 0.50 (see Figure 1.7).

*Step 1.* Select target mean cube strength (*TMS*)

$$(TMS) = 30 + (SD \times k)$$

$k = 2$ (See Table 1.2). Use $SD = 5.0$ N/mm$^2$ from quality control information. Then

$$(TMS) = 30 + (5.0 \times 2) = 40 \text{ N/mm}^2$$

*Step 2.* Read off cement content for (*TMS*) of 40 N/mm$^2$; 300 kg/m$^3$.
*Step 3.* Read off cement content for free water/cement ratio of 0.50: 360 kg/m$^3$.
*Step 4.* Use higher cement content from step 2 or step 3: 360 kg/m$^3$.
*Step 5.* Read off mix quantities per cubic metre for cement content of 360 kg/m$^3$

| | |
|---|---|
| cement: | 360 kg/m$^3$ |
| fine aggregate (saturated and surface dry): | 620 kg/m$^3$ |
| coarse aggregate (saturated and surface dry): | 1200 kg/m$^3$ |
| effective water: | 180 kg/m$^3$ |

*Step 6.* Read off the expected mean cube strength 50 N/mm$^2$

The values for aggregate and water contents need adjustment to allow for free moisture in wet aggregates.

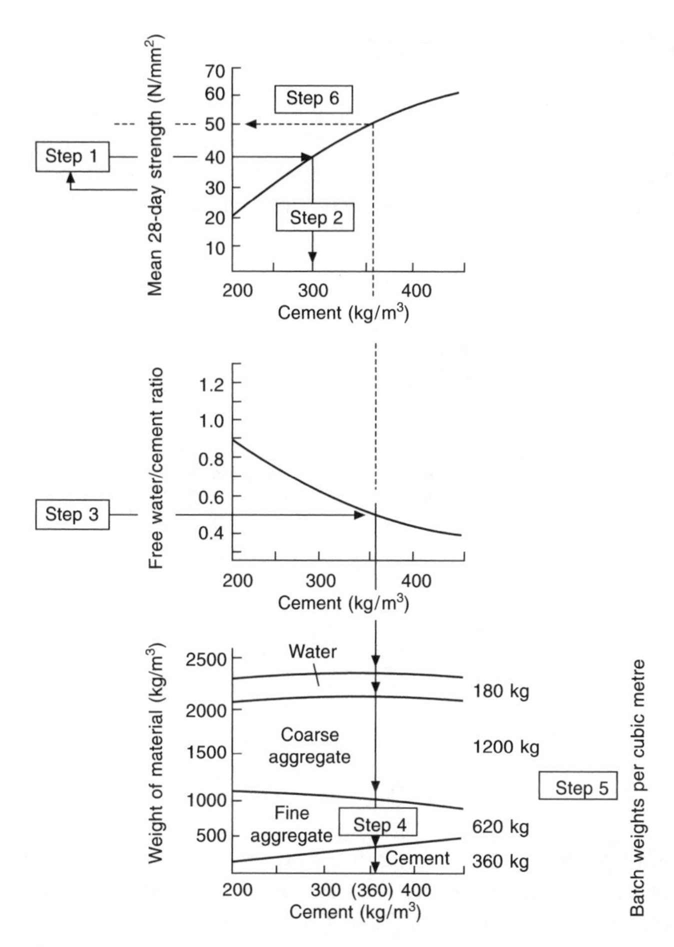


**Figure 1.7** Example of selection of concrete quantities for a particular material combination: 20 mm maximum aggregate size, 75 mm slump.

## 1.4 Comprehensive mix design of concrete based on materials properties

This section describes one comprehensive method of mix design (Dewar, 1999), based on a theory of particle mixtures, computer modelling and measured properties of materials. The method simulates in the computer, the BRMCA method, and has been validated against it. The aim is the elimination of, or at least minimizing the necessity for, any initial tests of concrete except possibly for initial validation and periodic spot checking. When materials properties are subject to routine testing and review and the concrete properties are assessed continuously in a control system, even periodic spot checks of concrete in the laboratory should not be necessary.

*Note:* Other comprehensive methods include: de Larrard (1999) and Day (1999), which enable initial designs to be prepared for a wide range of situations and may also enable some reduction in initial tests of concrete.

## 1.4.1 Principles and test methods

The basic concepts of the author's Theory of Particle Mixtures are that, as shown in Figure 1.8, when two particulate materials of different sizes are mixed together, the volume of voids between the larger particles will be reduced, but the structure of both coarse and fine materials will be disrupted by particle interference creating some additional voids partially offsetting the reduction.

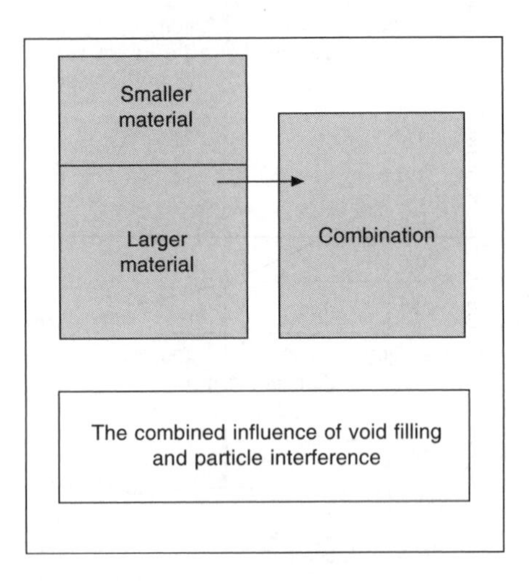


**Figure 1.8** The basic principle of the theory of particle mixtures.

For modelling purposes, the main relevant properties of each component are Mean Size, Voids Ratio and Relative Density, the latter being required only to convert data from mass to volume. The test methods in Table 1.6 were selected on the basis of perceived relevance to these properties, their common use in industrial practice and the availability of data. The details of the main methods are defined in BS EN 933 and BS EN 1097 for testing aggregates, and BS EN 196 for cements.

**Table 1.6** Test methods for the author's method of design

| Property | Test method | |
|---|---|---|
| | For powders | For aggregates |
| Mean size | Particle size distribution (or fineness) | Grading |
| Voids ratio* | Vicat test for water demand | Bulk density (loose ssd)** |
| Relative density | Relative density | Relative density (ssd)** |

* Ratio of sum of air and water voids to volume of solids
** Saturated and Surface Dry

The mean size is estimated from the particle size distribution on a logarithmic basis and approximates to the size at which 50% of material passes. For powders, when the particle size distribution is not available, estimates of mean size may be made from fineness measurements.

*Note:* For some very fine materials, e.g. silica fume, the particle size distribution may be uncertain, because of agglomeration, and the effective mean particle size may be much greater than for materials which are not agglomerated.

The size ratio of two materials to be combined in a mixture is an important factor determining the extent of particle interference. The size ratio $r$ is defined as

$$r = \frac{\text{mean size of the smaller material}}{\text{mean size of the larger material}}$$

Mixtures of materials having a low size ratio will have better filling of voids with reduced particle interference. Examples of mean sizes and size ratios for materials for concrete are shown in Table 1.7.

**Table 1.7** Typical mean sizes and size ratios for materials for concrete

| Material | Mean size (mm) | Size ratio $r$ |
|---|---|---|
| Cementitious | 0.015 | |
| | | 1/40 = 0.025 |
| Fine aggregate | 0.6 | |
| | | 1/20 = 0.05 |
| Coarse aggregate | 12 | |

The voids ratio of a particulate material is defined as the ratio of voids volume to solids volume under a stated method and energy level of compaction. For composite wet materials such as mortar and concrete, the voids volume is the sum of the volumes of free water, any liquid admixtures and air external to the aggregate. Air or water inside the aggregate pore structure is considered part of the solid volume. Factors affecting packing of particles, such as the range in particle size about the mean size and the shape and texture of particles, are accounted for in the assessment of the voids ratio of a material.

For aggregates, voids ratio may be estimated from bulk density and relative density tests. For powders, the interparticle forces in air are substantially higher than in a saturated environment such as fresh concrete. Thus, it is more relevant for powders to be tested in a water medium at a consistence corresponding to typical concrete, as in the Vicat test used for testing cement. Typical values for the properties of common concreting materials are shown in Table 1.8.

**Table 1.8** Typical properties of materials for concrete

| Material | Mean size (mm) | Voids ratio | Relative density |
|---|---|---|---|
| Cement | 0.013 | 0.83 | 3.20 |
| Fine aggregate | 0.50 | 0.70 | 2.60 |
| Coarse aggregate | 11 | 0.80 | 2.55 |

To obtain mixtures having low voids ratios requires low voids ratios of the component materials and also a low size ratio in order to minimize particle interference.

Theoretical relationships and experimental results for voids ratios are compared in Figure 1.9 for mixtures of two single-sized aggregates. The existence of sharp changes of slope and the validity of assuming straight lines between the change points B–D are demonstrated in the overall plot from A–F. Point E is usually, as in this case, the least distinct of the change points.

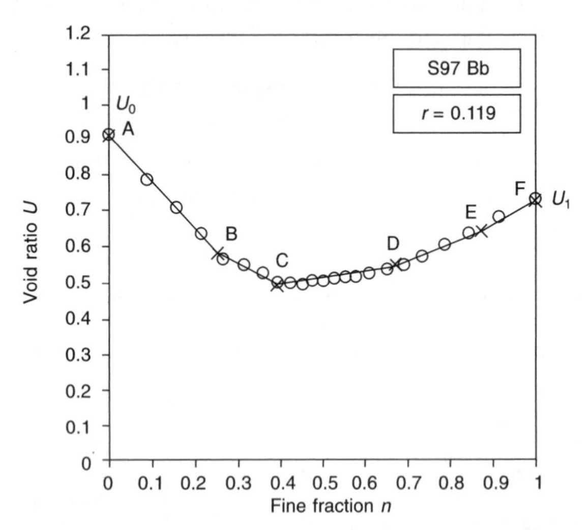


**Figure 1.9** Voids ratio diagram demonstrating straight-line relationships and change points for a mixture of two single-sized aggregates.

Similar agreement is found when the theory is extended to cover graded aggregates, mortars and concretes. Figure 1.10 compares theoretical and actual water demands of mortar at 50 mm slump estimated from voids ratios of the components and their size ratio.

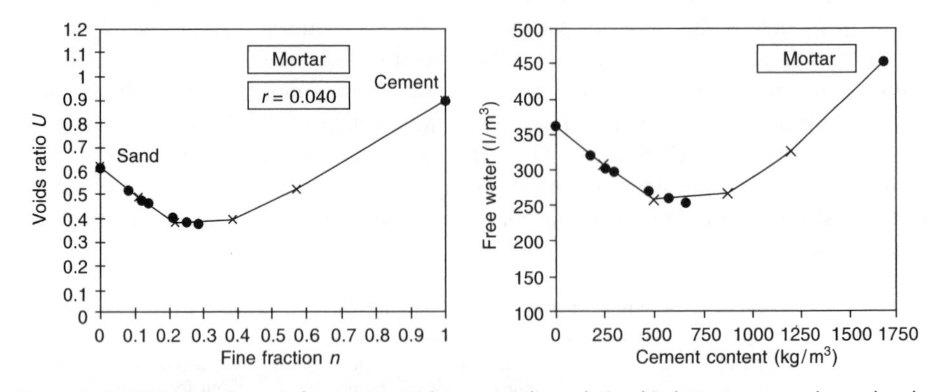


**Figure 1.10** Voids ratio diagram for mortars and corresponding relationship between water demand and cement content showing comparison between theory (lines) and laboratory data (dots).

To produce similar diagrams for concrete it is necessary first to calculate the diagram for mortars as in Figure 1.11 and then to take the mortars at each change point A–F and simulate the addition of coarse aggregate to create concrete. Formulae and computer software have been developed to make this process automatic.

As for mortar voids ratio diagrams, concrete displays a series of change points depending on the voids ratios and mean sizes of the coarse aggregate and the particular mortar solids.

In Figure 1.11 it will be observed that at one of the change points, 'd' in this example, the water demand is a minimum. Concretes to the left of 'd' will be progressively more uncohesive and those to the right will be progressively over-cohesive. Experienced concrete technologists, using the BRMCA trial mix method, will normally judge concretes at point 'x', just to the right of 'd', to have adequate cohesion for most ordinary concreting purposes. For more onerous placing conditions it may be necessary to move 'x' further to the right.

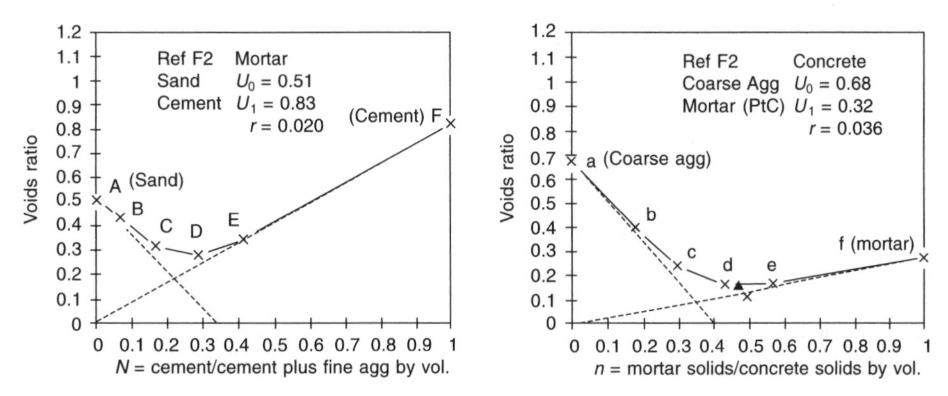


**Figure 1.11** Void ratio diagrams for mortar, and for concrete made with the mortar at point C and a particular coarse aggregate.

The process is repeated for combinations of coarse aggregates with the other mortars, A, B, D, E and F, to complete the development of the concrete series 'a–f' at 50 mm slump.

It is then necessary to convert the data for voids ratios, cement solids/mortar solids and mortar solids/concrete solids into the more familiar cement, fine aggregate, coarse aggregate and water contents per cubic metre together with other essential properties such as plastic density and water/cement ratio.

The modelling process has been further developed to take account of entrapped and entrained air, slumps other than 50 mm, influences of admixtures, mineral fillers, cementitious additions and the effects of all these factors and others on strength.

Some indication of the accuracy attainable by such modelling may be seen in Figures 1.12 and 1.13 which compare laboratory tests data and computerized simulations based on materials properties.

Figure 1.14 demonstrates the ability of the simulation process to enable discrimination between different sets of aggregates with regard to their effects on water demand of concrete.

Figure 1.15 shows that the simulation of the effects on water demand of using fly ash (pulverized-fuel ash) follows the effects obtained in the laboratory.

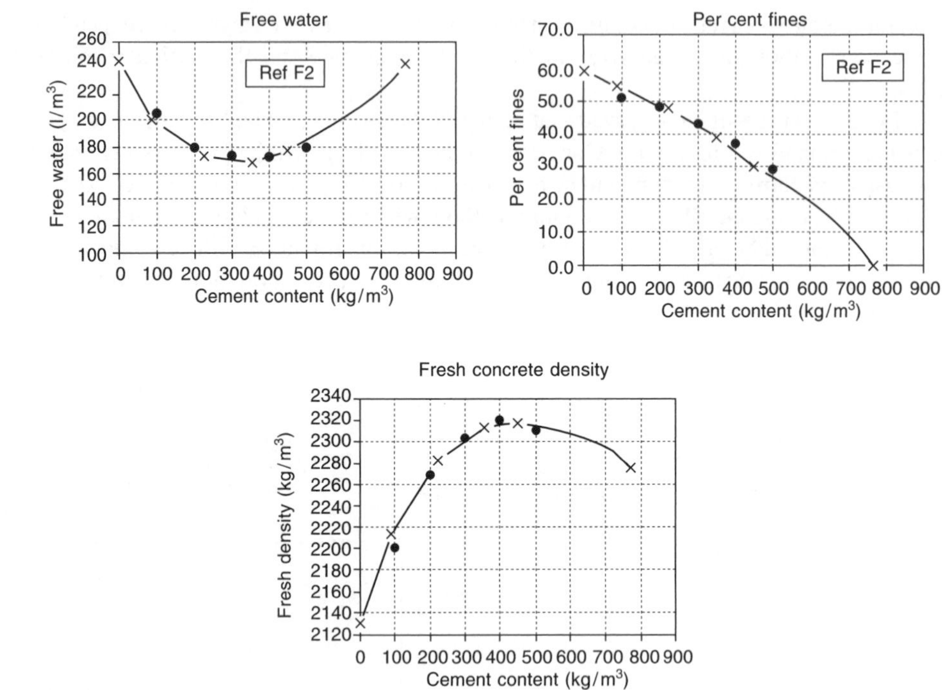


**Figure 1.12** Example of comparison between laboratory test data and simulated relationships for one set of materials.

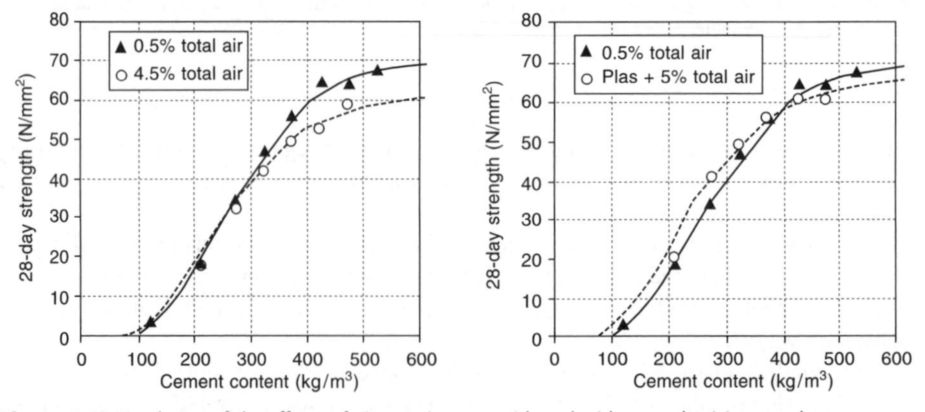


**Figure 1.13** Simulation of the effects of air entrainment, with and without a plasticizer, on the relationships between strength and cement content of concrete, in comparison with laboratory test data (Concrete Series P1).

## 1.5 MixSim – a computerized comprehensive method of mix design

MixSim is commercially available software, developed and marketed by SP Computing and by Questjay Ltd on the basis of the research and experience described in Dewar

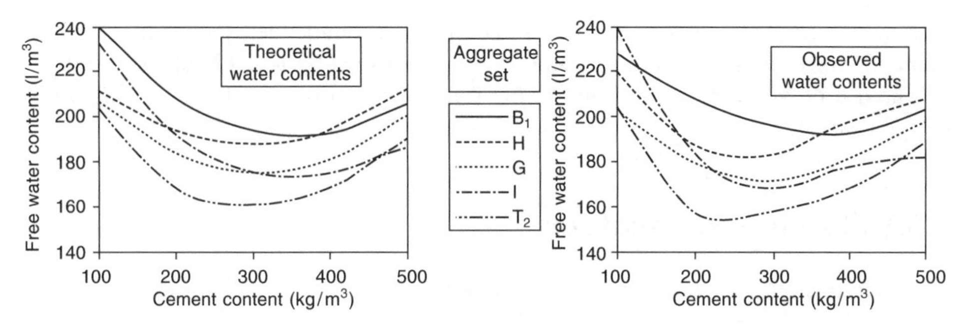


**Figure 1.14** Example of the ability of the author's simulation process to discriminate between 5 sets of materials with respect to water demands of concrete.

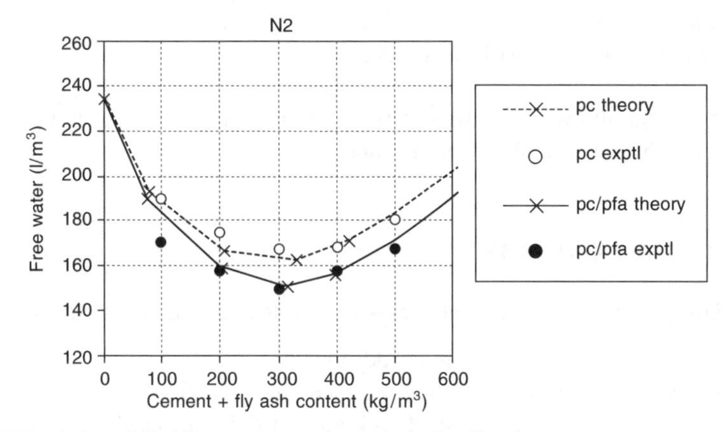


**Figure 1.15** Simulation of the effects on water demand of using fly ash.

(1999), as outlined in the previous section. The aim of MixSim is to simulate in the computer the laboratory concrete mix design process, enabling designs for properties of fresh and hardened concrete to be made more comprehensively, faster and more accurately on the basis of materials test data, and only the minimum of concrete data, as may be essential for determination of strength. The system has been extensively validated against data obtained using a wide range of materials in the BRMCA method of making a series of initial tests, as described earlier in Section 1.3.

MixSim meets the UK requirements of the Quality Scheme for Ready Mixed Concrete (QSRMC), Regulations, Part 4, July 2001, regarding computer modelling for mix design purposes. Using MixSim, concrete designs may be prepared by combining as many as three cementitious components, three fine aggregates or fillers and three coarse aggregates together with three liquid admixtures. Account is taken of any required air-entrainment, strength producing benefits of additions and water reducing benefits of plasticizers. Concretes are adjusted automatically for adequate cohesion. Provision is made for input of any concrete test data for comparison purposes.

Output data from MixSim include tables and graphs relating to water demand, per cent fines, plastic density, compressive strength (cubes or cylinders) and batch quantities taking account of batch size, moisture contents of aggregates and specified slump.

*Note:* Users wishing for more information are recommended to obtain *Computer Modelling of Concrete Mixtures* by J.D. Dewar, published by E & FN Spon, (1999).

The use of MixSim is illustrated in the following sections 1.5.1–1.5.7 using a mix design example.

## 1.5.1 MixSim – key features

The main operating sections of the software are

Materials database
Mix details screen:
  Materials; Conditions; Specifications; Selected concrete; Selected materials quantities
Trials data
Tables; Graphs; Mix Comparison; Batchbook.

The Utilities Menu allows the setting of the program for the use of either cylinders or cubes for strength. The default setting is cubes.

## 1.5.2 Materials database

The Materials Database (not illustrated here) separates the materials into four types

P Powders – cements and cementitious additions
F Fine aggregates – sands, crushed fine material, and mineral fillers
C Coarse aggregates – gravels, crushed rock and artificial materials
Liquid admixtures – including plasticizers and air-entraining agents

The properties, including cost data, are recorded in the database for each material. Drop-down calculation sheets enable the required properties to be evaluated from basic test data.

## 1.5.3 Mix details

Figure 1.16 shows the complete Mix Details screen for designing concretes. The materials and their details (in the upper left-hand section of the screen) are selected from the database by choosing the appropriate materials code from a drop-down menu. Up to three materials can be selected in each subsection. In the example, three powders, three fine aggregates and three coarse aggregates have been selected from the database for possible inclusion.

It is then necessary to enter the desired percentages of each material within each subsection. Under each of the three subsections, the properties of the three composite materials (i.e. Powder, Fine Aggregate and Coarse Aggregate) are then displayed automatically. In the example, two each of the powders, fine aggregates and coarse aggregates have been utilized for the particular design shown.

Admixtures are selected similarly (in the upper right-hand section of the screen)

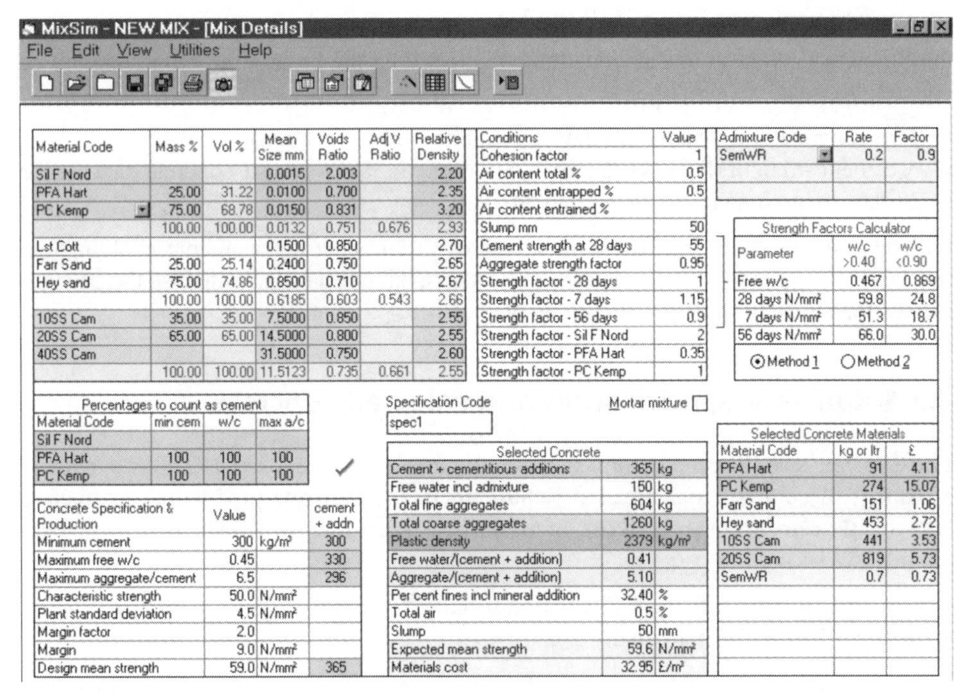


**Figure 1.16** Screen display – mix details.

together with dosage rates (litres per 100 kg of cement per m$^3$ of concrete) appropriate for the intended water reduction or air entrainment or other desired benefit. In the example, one of the three available admixtures, a water-reducing plasticizer, has been selected from the database.

It is also necessary to input a value corresponding to the water reduction expected for the Admixture dosage rate selected, e.g. 0.90 if a 10 per cent water reduction is expected. A corresponding adjustment is made automatically to the voids ratios of the composite particulate materials as shown in the upper left-hand section of the screen.

The upper central Concrete Conditions section allows the input of a number of factors including cohesion factor, air content, slump and strength required for generation of tables of concrete quantities and properties. Of course, some of the factors (e.g. slump and total air) are usually specified by the user of the concrete.

MixSim optimizes the proportioning of the three composite materials to minimize water demand and to ensure that the cohesion is optimum. For optimization, the Cohesion Factor is selected normally as 1. Otherwise, the Cohesion Factor should be selected between 1 and 2 or exceptionally between 0.5 and 3. High values may be appropriate for concrete to be pumped, when a client requires greater cohesion for particular placing conditions or when the local market traditionally requires more cohesive concrete.

For the example, the normal cohesion factor of 1 has been selected, air-entrainment is not specified and an allowance is made for 0.5 per cent entrapped air. The target slump is 50 mm.

Various strength factors are required, e.g. Cement Strength, which is the BS EN196 mortar prism strength at 28 days, Aggregate Strength Factor, Age Factors and also Addition Factors to take account of the contribution of Additions to strength.

On the right, beneath the admixture section is a small section which enables some of the Strength Factors to be calculated automatically on the basis of a limited amount of concrete test data, using either

- Two concrete data sets for 28-day strength and free water/cement
- A cement strength (BS EN196 Part 1, Strength) at 28 days and 1 concrete data set for 28-day strength and free water/cement.

For the example, two sets of recent laboratory trials data are available enabling the bracketed strength factors to be determined. Other available data have enabled the strength factors for the additional powders to be determined.

## 1.5.4 Concrete specification and selected concrete

The Concrete specification section shown in Figure 1.16 at the lower-left allows input of a number of common specification requirements and concrete production parameters affecting the cement content and cost of the particular concrete. The production parameters are

- Concrete strength – Standard deviation usually 2–5 N/mm$^2$
- Strength Margin Factor usually 2–2.5

For the example, values have been specified for three durability parameters together with a 28-day characteristic strength of 50 N/mm$^2$, the current plant standard deviation is 4.5 N/mm$^2$ and the margin factor is 2.

The values for Target Total Air Content and Target Slump are entered in the Conditions section. In the example air-entrainment is not specified and the target slump is required to be 50 mm.

The primary specification aspect determining the final choice of concrete can be identified in the last column of the Concrete Specification section which shows the (Cement + additions) contents needed to meet the various specified requirements. The highest (Cement + additions) content is adopted automatically in the Selected Concrete section (see later) and is the key factor determining the quantities, properties and cost of the selected concrete. In the example, the design mean strength determines the necessary cement plus additions content.

Above the Specification section a table is provided, allowing input of the percentage of each powder permitted by the specification to count towards cement content, aggregate/cement and water/cement for durability. In the example, both of the selected cementitious materials are permitted by the specification to count fully towards the cement content for durability.

The Selected Concrete section to the right of the Concrete Specification section displays automatically the quantities of materials and properties of the concrete designed to satisfy the specified requirements.

Quantities of each individual material are shown in a separate section at the far right which includes the contribution of each material to the total materials cost shown at the foot of the section.

Comparison between the Concrete Specification and Selected Concrete sections will show that for the example all the specified requirements are expected to be met by the selected concrete.

Investigations of varying the choice of materials or their proportions may be made by returning to the Materials Section of the screen and observing the effects on cost or any other property. This is considered again in Section 1.5.6.

## 1.5.5 Trials data

Up to 17 sets of concrete test data may be input in the Trials Data screen (not shown here).

## 1.5.6 View

The View screens are divided into four sections, and each section into a number of separate aspects. The sections are

Simulation Stages; Results Tables; Graphs; Mixes

Figure 1.17 shows the sections of the Results Tables dealing with quantities of materials, and the properties of the Concretes generated by the program for 17 simulated concretes. Graphs are provided as follows:

Water demand<br>
Per cent fines<br>
Plastic density<br>
Free water/cement<br>
Aggregate/cement<br>
} versus cement content

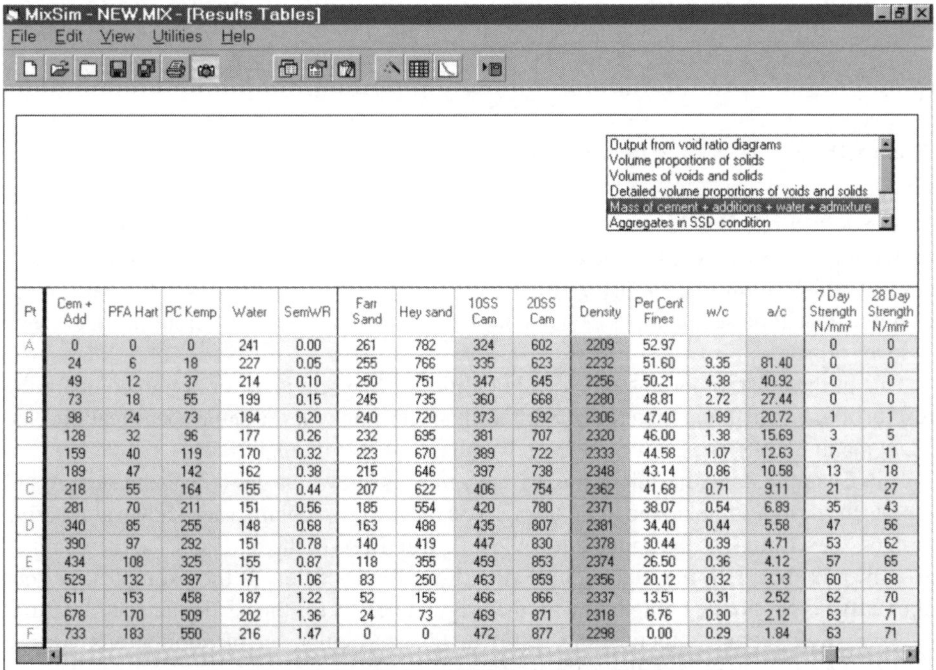

MixSim - NEW.MIX - [Results Tables]

File Edit View Utilities Help

Output from void ratio diagrams
Volume proportions of solids
Volumes of voids and solids
Detailed volume proportions of voids and solids
Mass of cement + additions + water + admixture
Aggregates in SSD condition

| Pt | Cem + Add | PFA Hart | PC Kemp | Water | SemWR | Farr Sand | Hey sand | 10SS Cam | 20SS Cam | Density | Per Cent Fines | w/c | a/c | 7 Day Strength N/mm² | 28 Day Strength N/mm² |
|---|---|---|---|---|---|---|---|---|---|---|---|---|---|---|---|
| A | 0 | 0 | 0 | 241 | 0.00 | 261 | 782 | 324 | 602 | 2209 | 52.97 | | | 0 | 0 |
| | 24 | 6 | 18 | 227 | 0.05 | 255 | 766 | 335 | 623 | 2232 | 51.60 | 9.35 | 81.40 | 0 | 0 |
| | 49 | 12 | 37 | 214 | 0.10 | 250 | 751 | 347 | 645 | 2256 | 50.21 | 4.38 | 40.92 | 0 | 0 |
| | 73 | 18 | 55 | 199 | 0.15 | 245 | 735 | 360 | 668 | 2280 | 48.81 | 2.72 | 27.44 | 0 | 0 |
| B | 98 | 24 | 73 | 184 | 0.20 | 240 | 720 | 373 | 692 | 2306 | 47.40 | 1.89 | 20.72 | 1 | 1 |
| | 128 | 32 | 96 | 177 | 0.26 | 232 | 695 | 381 | 707 | 2320 | 46.00 | 1.38 | 15.69 | 3 | 5 |
| | 159 | 40 | 119 | 170 | 0.32 | 223 | 670 | 389 | 722 | 2333 | 44.58 | 1.07 | 12.63 | 7 | 11 |
| | 189 | 47 | 142 | 162 | 0.38 | 215 | 646 | 397 | 738 | 2348 | 43.14 | 0.86 | 10.58 | 13 | 18 |
| C | 218 | 55 | 164 | 155 | 0.44 | 207 | 622 | 406 | 754 | 2362 | 41.68 | 0.71 | 9.11 | 21 | 27 |
| | 281 | 70 | 211 | 151 | 0.56 | 185 | 554 | 420 | 780 | 2371 | 38.07 | 0.54 | 6.89 | 35 | 43 |
| D | 340 | 85 | 255 | 148 | 0.68 | 163 | 488 | 435 | 807 | 2381 | 34.40 | 0.44 | 5.58 | 47 | 56 |
| | 390 | 97 | 292 | 151 | 0.78 | 140 | 419 | 447 | 830 | 2378 | 30.44 | 0.39 | 4.71 | 53 | 62 |
| E | 434 | 108 | 325 | 155 | 0.87 | 118 | 355 | 459 | 853 | 2374 | 26.50 | 0.36 | 4.12 | 57 | 65 |
| | 529 | 132 | 397 | 171 | 1.06 | 83 | 250 | 463 | 859 | 2356 | 20.12 | 0.32 | 3.13 | 60 | 68 |
| | 611 | 153 | 458 | 187 | 1.22 | 52 | 156 | 466 | 866 | 2337 | 13.51 | 0.31 | 2.52 | 62 | 70 |
| | 678 | 170 | 509 | 202 | 1.36 | 24 | 73 | 469 | 871 | 2318 | 6.76 | 0.30 | 2.12 | 63 | 71 |
| F | 733 | 183 | 550 | 216 | 1.47 | 0 | 0 | 472 | 877 | 2298 | 0.00 | 0.29 | 1.84 | 63 | 71 |

**Figure 1.17** Screen display – results tables – materials quantities per $m^3$ of concrete.

28-day strength versus free water/cement
28-day strength versus cement content
28-day strength versus 7-day strength
Materials cost versus cement content

The water demand screen, as in Figure 1.18, demonstrates that the water demand of concrete for the specified slump is not constant over the entire range of cement content, as is often assumed for convenience in simplified mix design methods.

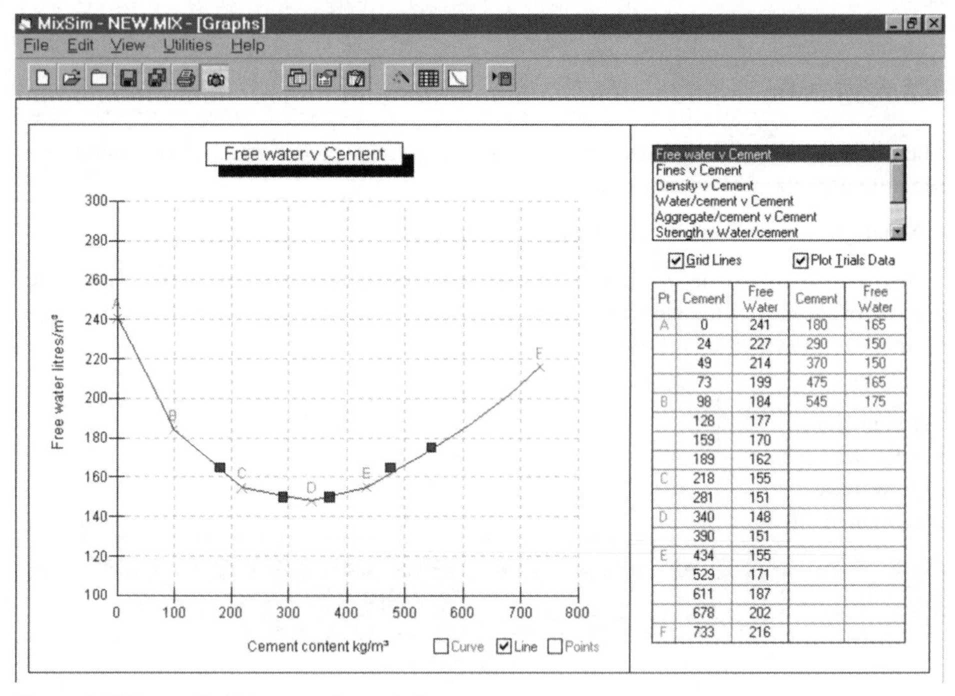


| Pt | Cement | Free Water | Cement | Free Water |
|---|---|---|---|---|
| A | 0 | 241 | 180 | 165 |
| | 24 | 227 | 290 | 150 |
| | 49 | 214 | 370 | 150 |
| | 73 | 199 | 475 | 165 |
| B | 98 | 184 | 545 | 175 |
| | 128 | 177 | | |
| | 159 | 170 | | |
| | 189 | 162 | | |
| C | 218 | 155 | | |
| | 281 | 151 | | |
| D | 340 | 148 | | |
| | 390 | 151 | | |
| E | 434 | 155 | | |
| | 529 | 171 | | |
| | 611 | 187 | | |
| | 678 | 202 | | |
| F | 733 | 216 | | |

**Figure 1.18** Screen display – water demand of concrete.

The screen relating 28-day strength and cement content as in Figure 1.19, is usually important because strength is often a key aspect of specification determining cement content which in turn has a major effect on the cost of the concrete, as is the case in the example. It will be seen that, as usual, at high cement contents the concrete strength tends to level off.

One of the most important and useful facilities of MixSim is the option to compare the results of using different materials or different proportions for the same specification. The View menu is used to select the Mixes Database, as in Figure 1.20.

The Example concrete, identified as 'new.mix' is shown at the left of the screen and repeats the design summarized in Figure 1.16. In the next two columns are shown alternative designs, 'new1.mix', as 'new.mix' but without fly ash (PFA) or the fine sand and 'new2.mix', as 'new.mix' but with the inclusion of silica fume and a filler. In the examples, neither alternative is economic compared with the original design for the particular specification.

*Note:* Key factors affecting the result in this particular set of simulations are the specified requirements for durability and the costs of the materials.

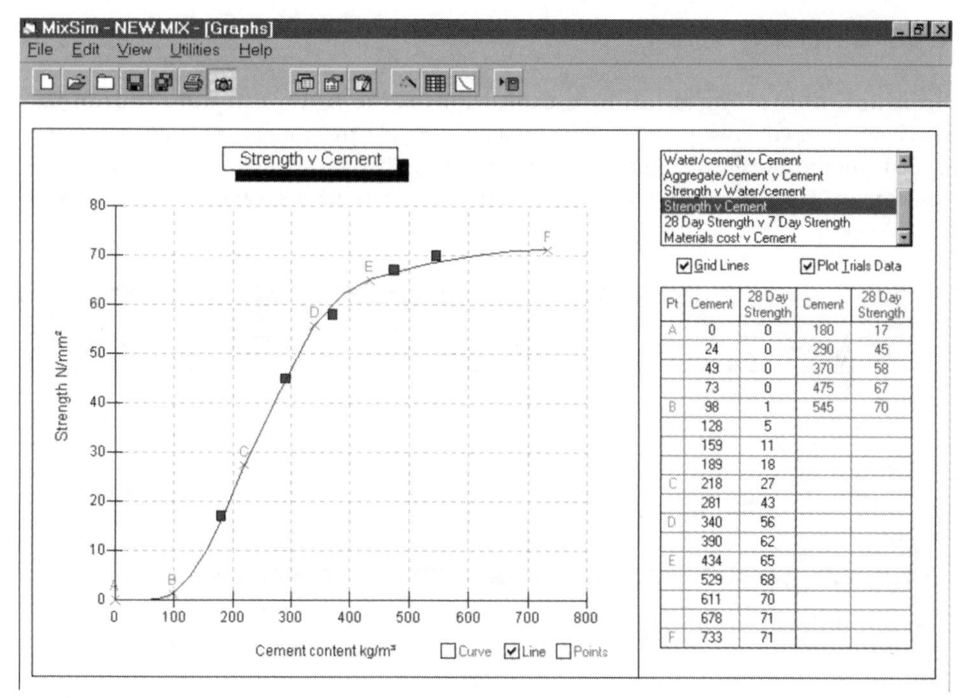


| Pt | Cement | 28 Day Strength | Cement | 28 Day Strength |
|---|---|---|---|---|
| A | 0 | 0 | 180 | 17 |
| | 24 | 0 | 290 | 45 |
| | 49 | 0 | 370 | 58 |
| | 73 | 0 | 475 | 67 |
| B | 98 | 1 | 545 | 70 |
| | 128 | 5 | | |
| | 159 | 11 | | |
| | 189 | 18 | | |
| C | 218 | 27 | | |
| | 281 | 43 | | |
| D | 340 | 56 | | |
| | 390 | 62 | | |
| E | 434 | 65 | | |
| | 529 | 68 | | |
| | 611 | 70 | | |
| | 678 | 71 | | |
| F | 733 | 71 | | |

**Figure 1.19** Graphs – 28-day strength versus cement content of concrete.

MixSim - NEW.MIX - [Mix Comparison]

File Edit View Utilities Help

| File name : | new.mix | new1.mix | new2.mix |
|---|---|---|---|
| Specification code | spec1 | spec1 | spec1 |
| Cement | 365 | 433 | 327 |
| Cement 1 | | | Sil F Nord 26 |
| Cement 2 | PFA Hart 91 | | PFA Hart 82 |
| Cement 3 | PC Kemp 274 | PC Kemp 433 | PC Kemp 219 |
| Free water | 150 | 195 | 135 |
| Admixture 1 | SemWR 0.7 | | SemWR 0.7 |
| Admixture 2 | | | |
| Admixture 3 | | | |
| Fine agg 1 | | | Lst Cott 68 |
| Fine agg 2 | Farr Sand 151 | | Farr Sand 102 |
| Fine agg 3 | Hey sand 453 | Hey sand 556 | Hey sand 509 |
| Coarse agg 1 | 10SS Cam 441 | 10SS Cam 407 | 10SS Cam 437 |
| Coarse agg 2 | 20SS Cam 819 | 20SS Cam 756 | 20SS Cam 812 |
| Coarse agg 3 | | | |
| Plastic density | 2379 | 2348 | 2391 |
| Water/(cem+addn) | 0.41 | 0.45 | 0.41 |
| Agg/(cem+addn) | 5.10 | 3.97 | 5.89 |
| Per cent fines | 32 | 32 | 35 |
| Slump | 50 | 50 | 50 |
| Total air | 0.50 | 0.50 | 0.50 |
| Exp mean strength | 59.6 | 64.9 | 64.8 |
| Materials cost | 32.95 | 35.73 | 37.88 |

Clear grid

Files: new.mix, new1.mix, new2.mix

Folders: c:\, mixsim~2, mixdata

Drives: c:

**Figure 1.20** Screen display – mixes database for comparing concretes.

## 1.5.7 Batching data

Batching Data as in Figure 1.21 for each material may be viewed and printed out for full-size batches of concrete in selected increments of total cementitious material. Allowance is made for the moisture contents of the aggregates. The first column displays the Powder content per cubic metre. All other columns display the Materials contents for the selected batch size.

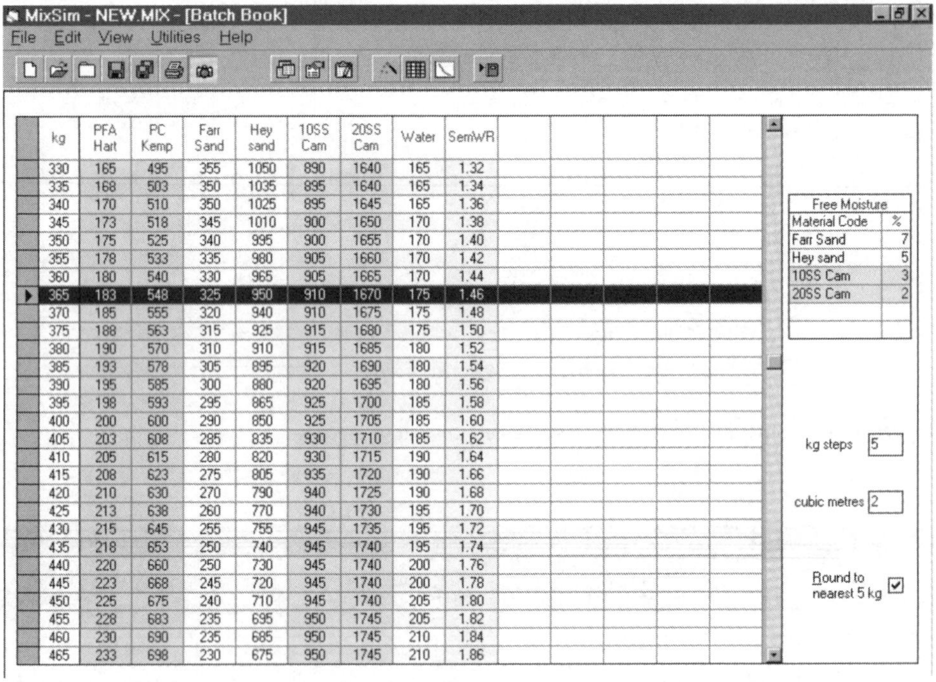

MixSim - NEW.MIX - [Batch Book]

File Edit View Utilities Help

| kg | PFA Hart | PC Kemp | Farr Sand | Hey sand | 10SS Cam | 20SS Cam | Water | SemWR |
|---|---|---|---|---|---|---|---|---|
| 330 | 165 | 495 | 355 | 1050 | 890 | 1640 | 165 | 1.32 |
| 335 | 168 | 503 | 350 | 1035 | 895 | 1640 | 165 | 1.34 |
| 340 | 170 | 510 | 350 | 1025 | 895 | 1645 | 165 | 1.36 |
| 345 | 173 | 518 | 345 | 1010 | 900 | 1650 | 170 | 1.38 |
| 350 | 175 | 525 | 340 | 995 | 900 | 1655 | 170 | 1.40 |
| 355 | 178 | 533 | 335 | 980 | 905 | 1660 | 170 | 1.42 |
| 360 | 180 | 540 | 330 | 965 | 905 | 1665 | 170 | 1.44 |
| 365 | 183 | 548 | 325 | 950 | 910 | 1670 | 175 | 1.46 |
| 370 | 185 | 555 | 320 | 940 | 910 | 1675 | 175 | 1.48 |
| 375 | 188 | 563 | 315 | 925 | 915 | 1680 | 175 | 1.50 |
| 380 | 190 | 570 | 310 | 910 | 915 | 1685 | 180 | 1.52 |
| 385 | 193 | 578 | 305 | 895 | 920 | 1690 | 180 | 1.54 |
| 390 | 195 | 585 | 300 | 880 | 920 | 1695 | 180 | 1.56 |
| 395 | 198 | 593 | 295 | 865 | 925 | 1700 | 185 | 1.58 |
| 400 | 200 | 600 | 290 | 850 | 925 | 1705 | 185 | 1.60 |
| 405 | 203 | 608 | 285 | 835 | 930 | 1710 | 185 | 1.62 |
| 410 | 205 | 615 | 280 | 820 | 930 | 1715 | 190 | 1.64 |
| 415 | 208 | 623 | 275 | 805 | 935 | 1720 | 190 | 1.66 |
| 420 | 210 | 630 | 270 | 790 | 940 | 1725 | 190 | 1.68 |
| 425 | 213 | 638 | 260 | 770 | 940 | 1730 | 195 | 1.70 |
| 430 | 215 | 645 | 255 | 755 | 945 | 1735 | 195 | 1.72 |
| 435 | 218 | 653 | 250 | 740 | 945 | 1740 | 195 | 1.74 |
| 440 | 220 | 660 | 250 | 730 | 945 | 1740 | 200 | 1.76 |
| 445 | 223 | 668 | 245 | 720 | 945 | 1740 | 200 | 1.78 |
| 450 | 225 | 675 | 240 | 710 | 945 | 1740 | 205 | 1.80 |
| 455 | 228 | 683 | 235 | 695 | 950 | 1745 | 205 | 1.82 |
| 460 | 230 | 690 | 235 | 685 | 950 | 1745 | 210 | 1.84 |
| 465 | 233 | 698 | 230 | 675 | 950 | 1745 | 210 | 1.86 |

| Free Moisture | |
|---|---|
| Material Code | % |
| Farr Sand | 7 |
| Hey sand | 5 |
| 10SS Cam | 3 |
| 20SS Cam | 2 |

kg steps 5

cubic metres 2

Round to nearest 5 kg ☑

**Figure 1.21** Screen display – batchbook with the selected concrete highlighted.

The expected water content to be added, excluding free moisture in the aggregates and that in any admixtures, is indicated for the intended slump as input on the Conditions screen.

The highlighted batching data are for the original example considered throughout this section for which the total cement content is 365 kg/m$^3$.

# 1.6 Special concretes

There is no obvious universal definition for special concretes other than those for which the designer of the mix is without either experience, data or confidence. Thus, the definition will vary from individual to individual and from organization to organization. This said, there are some concretes for which the term special may be considered to apply more or less generally, such as

Airfield and road paving
Underwater/tremied
Dry lean
Fibre reinforced
Floor screeds
Floors
Heavyweight
*Note:* It may be necessary to give special attention to cohesion as well as to take account of the effects on volume of the higher density of aggregate compared with normal density aggregate. The density of the concrete may affect the maximum quantitities of aggregates or concrete that may be handled. See also ACI (1991).
High performance
*Note:* See Neville and Aitcin (1998).
High strength
*Note:* See Price (1999).
Lightweight
*Note:* Depending on the use to be made of the concrete, it may be necessary to incorporate both fine and coarse lightweight aggregates or only a coarse lightweight aggregate with a dense fine aggregate. The water absorption value for the lightweight aggregate may be difficult to determine with precision; the moisture condition of the aggregate is a critical property affecting workability, rate of loss of workability, water demand, w/c and strength. As a consequence, it may be necessary to ensure aggregates are fully saturated before use in trials and in full-scale production.
No-fines
*Note:* See Murdock *et al.* (1991, pp. 386–8) and Neville (1995, pp. 711–13).
Precast, e.g. blocks; kerbs; reconstructed stone
Pumping
*Note:* See Masterton and Wilson (1997).
Roller compaction
*Note:* See Hansen and Reinhardt (1991, pp. 39–54 , 129–134 and 141–5) and Marchand *et al.* (1997)
Self-compacting/self levelling/flowing
*Note:* For self-compacting concrete see Gaimster and Gibbs (2001). For flowing concrete see Neville (1995, pp. 757–8).

All of these concretes can be designed by the methods described in this chapter, provided allowance is made for the special features involved, for which advice may be available from construction experts, specialist publications and materials suppliers, e.g. admixture manufacturers.

## 1.7 Simplified mix design methods

Two methods are considered under the heading of simplified methods

- DoE (1988)
- ACI (1991).

These two methods are outlined to show the main principles adopted.

A third method, for medium-strength concretes only, and possibly still in wide use, is the Basic Mix Method of Owens (1973) to which readers are referred, by which a single mix is first designed using a data bank provided. This mix is then tested in the laboratory and the results are adjusted and then extrapolated using simplifying assumptions to provide a range of designs on either side of the one tested. From the range, the mix satisfying all specified requirements is selected.

## 1.7.1 Simplified mix design – the DOE (1988) method

The following is an introduction to the 1988 edition of *Design of Normal Concrete Mixes* which is the result of combined effort by the Building Research Station, Transport and Road Research Laboratory and the British Cement Association. It is published by the Department of the Environment and is often referred to as the DoE method.

The 1988 version provides for designs using Portland cement to BS 12 i.e. CEM I and also includes modifications, to which the readers are referred for detailed information, allowing for the use of air entrainment, pfa and ggbs. The following assumptions are made for Portland cement concretes:

1. A single relationship for all CEM I Portland cements to BS EN 197–1 between 28-day strength and effective (free) water/cement ratio when using a particular coarse aggregate (Figure 1.22).
2. A single relationship as in (1) for all uncrushed coarse aggregates and a different relationship for all crushed coarse aggregates (Figure 1.22).
3. Values of free water content (see Table 1.9) are assumed constant over the full range of cement contents for a given maximum size and type of aggregate (crushed or uncrushed) and level of consistence.

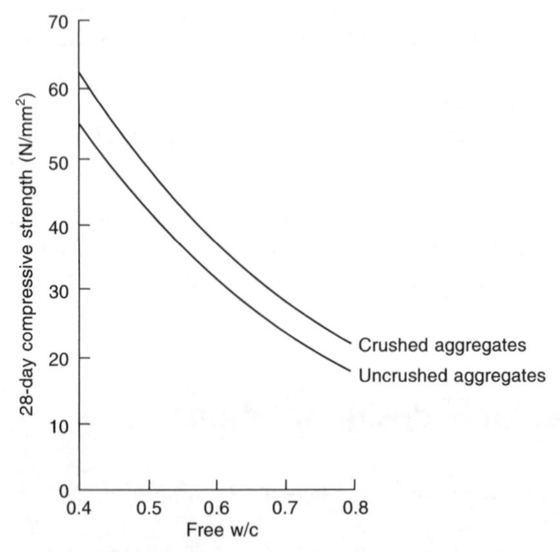


**Figure 1.22** Assumed relationships between strength at 28 days and free w/c for concretes made with CEM I Portland cement to BS EN 197–1 (adapted from DoE, 1988).

*Note:* The implications of such simplifying assumptions are considered in Section 1.1.4.

**Table 1.9** Assumed effective water contents of concrete made with uncrushed aggregates

| Maximum aggregate size (mm) | Effective water ($kg/m^3$) | | | |
|---|---|---|---|---|
| | Target Slump (mm) | | | |
| | 25 | 50 | 75 | 100 |
| 10 | 185 | 205 | 215 | 220 |
| 20 | 165 | 180 | 190 | 195 |
| 40 | 145 | 160 | 170 | 175 |

*Note:* Add 25–40 $kg/m^3$ for crushed coarse and fine aggregates.

A comprehensive series of graphs is provided enabling the proportion of fine to total aggregate to be estimated, dependent on maximum aggregate size, effective w/c, consistence and fine aggregate grading (per cent passing 600 μm). These have been simplified for this introduction in Figure 1.24 to a single set of graphs of per cent fine aggregate plotted against effective w/c, for different sizes of fine aggregate. A set of multiplying factors enables the use of a maximum aggregate size other than 20 mm and a target slump other than 50 mm.

*Note:* The tabulated multiplying factors in Figure 1.24 are used to adjust the per cent fine aggregate estimated from the graphs. If the maximum aggregate size is 20 mm and the target slump is 50 mm the multiplying factor is 1 and no adjustment is necessary.

Examination of Figure 1.24 will show that *less* fine aggregate is required when free w/c is low, the fine aggregate is finer (i.e. higher per cent passing 600 μm), maximum aggregate size is higher and target slump is lower, all of which are in keeping with practical experience. The net result is that

(a) (*W*/*C*) can be selected easily from Figure 1.22 to meet any strength requirement, using assumptions (1) and (2) (see section 1.1.4) and taking account of the design margin.
(b) Table 1.9 for consistence and aggregates determines the water content (*W*)
(c) Using assumption (3) (see section 1.1.4) and (a) and (b) above

$$(C) = \frac{(W)}{(W/C)}$$

(d) Plastic density (*PD*) is estimated, using Figure 1.23, from the relative densities of the aggregates and the water content, assuming no effect of cement content.
(e) Using the valid rule (5) (see section 1.1.4) (*C*) + (*A*) + (*W*) = (*PD*), then (*A*) = (*PD*) – (*C*) – (*W*)
(f) (*A*) is then divided into fine aggregate (*FA*) and coarse aggregate (*CA*) using Figure 1.24.

Trial mixes to confirm or modify the design are identified as essential parts of the process (Table 1.10).

## Example

Compressive Strength Class C28/35 concrete; CEM I cement to BS EN 197–1; 20 mm maximum size uncrushed gravel and sand;

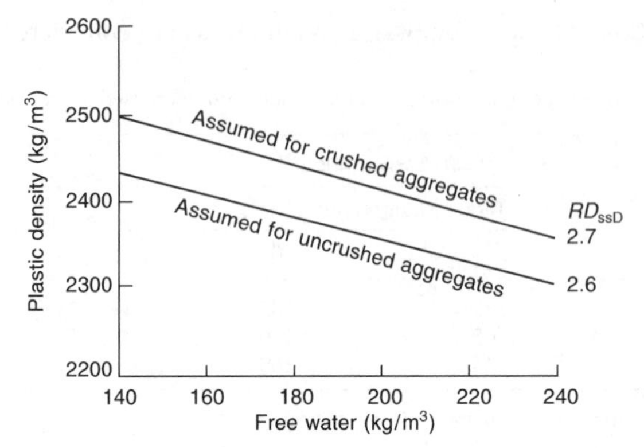


**Figure 1.23** Assumed relationships between plastic density and free water content (adapted from DoE, 1988).

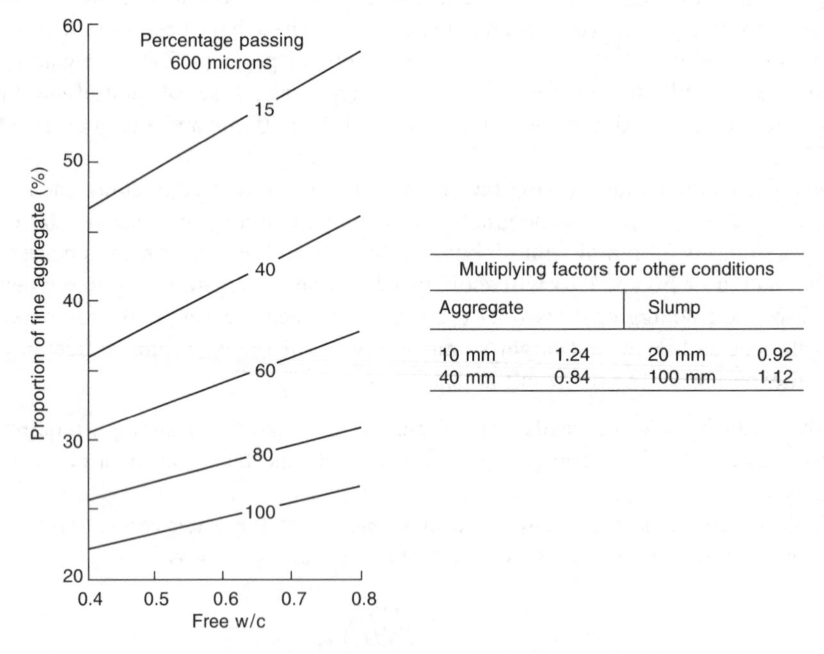


| Multiplying factors for other conditions | | | |
|---|---|---|---|
| Aggregate | | Slump | |
| 10 mm | 1.24 | 20 mm | 0.92 |
| 40 mm | 0.84 | 100 mm | 1.12 |

**Figure 1.24** Example of relationships between percentage of fine aggregate and free w/c for concrete of 50 mm slump made with 20 mm aggregate. Multiplying factors appropriate for other conditions are also shown. (adapted from DoE, 1988).

50 per cent passing 600 μm in sand; maximum free w/c 0.60; minimum cement 300 kg/m$^3$; 50 mm slump; assume a strength design margin of 10 N/mm$^2$.

1 Required average strength = 35 + 10 = 45 N/mm$^2$
2 Free w/c (from Figure 1.22) = 0.48, (which is less than maximum value specified)
3 Free water content (from Table 1.9) = 180 kg/m$^3$
4 Cement content = 180/0.48 = 375 kg/m$^3$ (which is greater than minimum value specified)
5 Plastic density (from Figure 1.23) = 2380 kg/m$^3$
6 Aggregate (SSD) content = 2380 – 375 – 180 = 1825 kg/m$^3$

**Table 1.10** Summary of a DoE (1988) mix design

| Materials | Quantities (kg) | |
|---|---|---|
| | Per $m^3$ | Per 0.05 $m^3$ trial mix |
| Cement | 375 | 18.75 |
| Sand (SSD) | 600 | 30 |
| Gravel (SSD) | 1225 | 61.25 |
| Water (free) | 180 | Appr 9* |
| Total | 2380 | 110 + water* |

*Water content of trial mix should be adjusted to produce the required 50 mm slump (see Table 1.3).

7 Per cent fine aggregate (from Figure 1.24) = 33
8 Fine aggregate (SSD) content = 33/100 × 1825 = 600 kg
9 Coarse aggregate (SSD) content = 1825 – 600 = 1225 kg
10 Summary of mix design (Table 1.10).

For more comprehensive explanation and the full range of options readers are referred to DoE (1988).

## 1.7.2 Simplified mix design – the ACI (1991) method

ACI (1991) recommends the relationship in Table 1.11 between cylinder strength at 28 days and water–cement ratio. The corresponding relation for cube strength is very similar to that shown in Figure 1.22 for uncrushed aggregates with CEM cement to BS EN 197–1.

**Table 1.11** ACI relationship between strength and w/c

| Mean compressive strength N/mm$^2$ at 28 days | | |
|---|---|---|
| Cylinders | (Cubes)* | Water/cement |
| 40 | (50) | 0.42 |
| 35 | (44) | 0.47 |
| 30 | (38) | 0.54 |
| 25 | (31) | 0.61 |
| 20 | (25) | 0.69 |
| 15 | (19) | 0.79 |

*Assuming cube strength = 1.25 × cylinder strength

Water demands are summarized in Table 1.12; they are 5–10 kg/m$^3$ higher than the DoE (1988) values for uncrushed aggregates in Table 1.9.

As for the DoE (1988) method, cement content ($C$) may be estimated from the water cement ratio (w/c) and water content:

$$(C) = \frac{(W)}{(W/C)} \text{ kg}$$

Again as for the DoE (1988) method, the aggregate content ($A$) is estimated from an assumed plastic density ($PD$) as in Table 1.13 by subtracting ($C$) and ($W$):

**Table 1.12** ACI recommendations for free water content

| Maximum aggregate size (mm) | Free water ($kg/m^3$) Slump 50 | 75 | 100 | 125 |
|---|---|---|---|---|
| 10 | 210 | 220 | 230 | 235 |
| 20 | 190 | 195 | 200 | 205 |
| 40 | 165 | 175 | 180 | 185 |

**Table 1.13** ACI first estimate of plastic density

| Maximum aggregate size (mm) | Plastic density ($kg/m^3$) |
|---|---|
| 10 | 2285 |
| 20 | 2355 |
| 40 | 2420 |

$$(A) = (PD) - (C) - (W) \text{ kg}$$

The ACI method of estimating the split between fine and coarse aggregate differs from the DoE method. The bulk volume of dry-rodded coarse aggregate per cubic metre of concrete is assessed from Table 1.14 dependent on the maximum size of coarse aggregate and the fineness modulus of the sand.

The coarse aggregate (SSD) Content, (*CA*), is then estimated as the product of the volume (*V*) from Table 1.14 multiplied by the rodded bulk density (SSD) of the coarse aggregate in $kg/m^3$

$$(CA) = (V) \times (CA \text{ rodded bulk density SSD}) \text{ kg}$$

**Table 1.14** ACI recommendations for bulk volume of dry-rodded coarse aggregate

| Maximum aggregate size (mm) | Bulk volume (*V*) of coarse aggregate $m^3$ per cubic metre of concrete Fineness modulus of sand 2.4 | 2.6 | 2.8 | 3.0 |
|---|---|---|---|---|
| 10 | 0.50 | 0.48 | 0.46 | 0.44 |
| 20 | 0.66 | 0.64 | 0.62 | 0.60 |
| 40 | 0.76 | 0.74 | 0.72 | 0.70 |

*Notes:* The values for (*V*) may be increased by up to 10 per cent for lower consistence workability and similarly reduced for very high consistence.

The fineness modulus is the sum of the cumulative percentages retained on each sieve, from the 150 μm sieve up to the largest sieve used, divided by 100. Coarser materials have higher fineness moduli.

### *Example*

Using the same example as for the DoE method, but using the ACI method and corresponding tables. Compressive strength class C28/35 (equivalent to 28 $N/mm^2$ cylinder strength); CEM I cement; 20 mm maximum aggregate size; 50 per cent passing 600 μm equivalent to 2.6 fineness modulus; assume a rodded bulk density (SSD) of the coarse aggregate of 1600 $kg/m^3$; maximum free w/c = 0.60: minimum cement 300 $kg/m^3$; 50 mm slump; assume a strength design margin of 8 $N/mm^2$ for cylinders (10 $N/mm^2$ for cubes).

1 Required average (cube) strength 45 $N/mm^2$ (36 cyl str)
2 Free w/c from Table 1.11 = 0.46
3 Free water content from Table 1.12 = 190 $kg/m^3$
4 Cement content = 190/0.46 = 415 $kg/m^3$
5 Plastic density from Table 1.13 = 2355 $kg/m^3$
6 Aggregate (SSD) content = 2355 – 415 – 190 = 1750 $kg/m^3$
7 Coarse aggregate content from Table 1.14

$$0.64 \times 1600 \times \frac{110^*}{100} = 1125\ \text{kg/m}^3$$

*10% increase to allow for lower slump (50 mm)
8 Fine aggregate content = 1750 – 1125 = 625 $kg/m^3$
9 Summary of mix design (see Table 1.15).

No significance is attached to the differences between the designs by the two methods, either design providing a good basis for one or more initial tests, as described previously.

*Note:* The ACI, in a range of publications, also provides guidance on mix design for: no slump concrete; air-entrained concrete; light- and heavyweight concrete, and for the use of large maximum size aggregate (up to 150 mm).

**Table 1.15** Summary of ACI mix design

| Material | Quantities (per $m^3$) | | DoE method (for comparison) See Table 1.10 | |
|---|---|---|---|---|
| Cement | 415 | | 375 | |
| Sand (SSD) | 625 | 1750 | 600 | 1825 |
| Gravel (SSD) | 1125 | | 1225 | |
| Water (free) | 190 | | 180 | |
| Total | 2355 | | 2380 | |

## 1.8 Ready-to-use mix designs

BS 8500 caters for safe designs for instant use, and of particular value for site mixing, by the provision of standardized prescribed concretes as shown in Table 1.16. BS 8500 should be consulted for full details and options.

For structural design purposes the compressive strength classes in Table 1.17 may be assumed. Some uses to which these designs may be put are listed in Table 1.18. For full details see BS 8500.

**Table 1.16** Mix proportions for standardized prescribed concretes from BS 8500

| Standard mix | Nominal maximum size of aggregate (mm) | | 40 | | 20 | |
|---|---|---|---|---|---|---|
| | Target Slump (mm) | | 75 | 125 | 75 | 125 |
| | Constituents | | | | | |
| ST1 | Cement or combination | (kg) | 200 | 220 | 230 | 255 |
| | Total aggregate | (kg) | 1990 | 1930 | 1925 | 1860 |
| ST2 | Cement or combination | (kg) | 230 | 255 | 265 | 285 |
| | Total aggregate | (kg) | 1960 | 1905 | 1895 | 1840 |
| ST3 | Cement or combination | (kg) | 265 | 285 | 295 | 330 |
| | Total aggregate | (kg) | 1930 | 1875 | 1865 | 1800 |
| ST4 | Cement or combination | (kg) | 310 | 330 | 330 | 365 |
| | Total aggregate | (kg) | 1900 | 1840 | 1835 | 1775 |
| ST5 | Cement or combination | (kg) | 350 | 375 | 375 | 395 |
| | Total aggregate | (kg) | 1870 | 1805 | 1800 | 1740 |
| ST1<br>ST2<br>ST3 | Fine aggregate (per cent by weight of total aggregate) | | 30–45 | | 35–50 | |
| ST4<br>ST5 | Fine aggregate (per cent by weight of total aggregate) | | | | | |
| | Grading limits CP | | 30–40 | | 35–45 | |
| | Grading limits MP | | 25–35 | | 30–40 | |
| | Grading limits FP | | 25–30 | | 25–35 | |

*Note:* The values given for aggregate content are for the saturated and surface dried conditions and may be adjusted to allow for the characteristics of the aggregates.

**Table 1.17** BS 8500 standardized prescribed concretes and indicative strengths

| Standard mix | Compressive strength class for structural design N/mm$^2$ (= Mpa) |
|---|---|
| ST1 | C6/8 |
| ST2 | C8/10 |
| ST3 | C12/15 |
| ST4 | C16/20 |
| ST5 | C20/25 |

## 1.9 Summary

This chapter summarizes various approaches to concrete mix design in use today throughout the world and covering a wide range of needs. The methods described include ready to use designs for specific purposes, simplified methods requiring minimal input data, comprehensive methods involving laboratory or field tests and comprehensive computerised methods enabling fast and accurate designs. Numerical examples are provided throughout the chapter.

**Table 1.18** Some examples of standardized prescribed concretes advised by BS 8500 as suitable for particular purposes

| Use | Standard mix | Slump class (mm) |
|---|---|---|
| Kerb bedding and backing | ST1 | 51 |
| Blinding | ST2 | 53 |
| Strip footings | ST2 | 53 |
| Mass concrete foundations | ST2 | 53 |
| Trench fill foundations | ST2 | 54 |
| Over-site concrete below suspended slabs | ST2 | 53 |
| Garage floors with no embedded metal | ST4 | 52 |

*Note:* See BS 8500 for further examples.

## References


ACI (1991) *Standard Practice for Selecting Proportions for Normal, Heavyweight and Mass Concrete.* American Concrete Institute (ACI 211.1-91).

British Standards Institution

BS EN 196 *Methods of testing cement. Part 1. Strength.* 1995.

*Methods of testing cement. Part 3. Determination of setting time and soundness.* 1995.

*Methods of testing cement. Part 6. Fineness.* 1992.

BS EN 197–1 Cement – Part 1: composition, specifications and conformity criteria for common cements, 2000.

BS EN 206–1 Concrete – Part 1: specification, performance, production and conformity, 2000.

BS EN 933–1 Determination of particle size distribution, 1997.

BS EN 1097–3 Determination of loose bulk density, 1998.

BS EN 12350 Testing fresh concrete, 2000.

BS 8500 Concrete – complementary standard to BS EN 206–1 Part 1: Method of specifying and guidance 2002 Part 2 specification for constituent materials and for concrete 2002.

Day, K.W. (1999) *Concrete Mix Design, Quality Control and Specification* (2nd edn). E&FN Spon, London.

Dewar, J.D. (1986) Ready-mixed concrete mix design. *Municipal Engineer*, **3**, Feb, 35–43.

Dewar, J.D. (1992) *Manual of Ready Mixed Concrete*, (2nd edn). Chapman and Hall, London.

Dewar, J.D. (1999) *Computer Modelling of Concrete Mixtures.* E&FN Spon, London.

DoE (1988) *Design of Normal Concrete Mixes.* Department of the Environment, HMSO, London.

Gaimster, R. and Gibbs, J. (2001) *Self-compacting concrete. Part 1. The material and its properties.* Current Practice Sheet No 123. Concrete Vol. 35 No 7, July/August, pp 32–34.

Hansen, K.D. and Reinhardt, W.G. (1991) *Roller-compacted Concrete Dams.* McGraw-Hill New York.

Harrison T.A. (2003) The new concrete standards – getting started. The Concrete Society CS 149.

Larrard, F. de. (1999) *Concrete Mixture Proportioning: a scientific approach*, E&FN Spon, London.

Marchand, J., Gagne, R., Ouellet, E. and Lepage, S. (1997) Mixture proportioning of roller compacted concrete – a review. *Advances in Concrete Technology. Proceedings of third CANMET/ACI international symposium.* Auckland, New Zealand. ACI SP-171, pp 457–486.

Masterton, G.G.T. and Wilson, R.A. (1997) *The planning and design of concrete mixes for transporting, placing and finishing.* Report 165, CIRIA.

Murdock, L.J., Brook, K.M. and Dewar, J.D. (1991) *Concrete Materials and Practice* (6th edn). Edward Arnold, London.

Neville, A.M. (1995) *Properties of Concrete* (4th edn) Longman Group Limited, Harlow.

Neville, A.M. and Aitcin, C.-P. (1998) High performance concrete – An overview. *Materials and Structures*, **31**, March, 111–7.
Owens, P.L. (1973) *Basic Mix Method. Selection of proportions for medium strength concretes.* Cement and Concrete Association.
Price, W.F. (1999) *High-strength concrete.* Current Practice Sheet No. 118. Concrete, June, pp 9–10.
QSRMC (2001) *Quality Scheme for Ready Mixed Concrete*, Quality and Product Conformity Regulations, Hampton, Middlesex, UK. February 1995. Amendments: various dates from 1995 to July 2001.

## Further reading

ACI (1991) *Standard Practice for Selecting Proportions for Normal, Heavyweight and Mass Concrete.* American Concrete Institute, (ACI 211.1-91). Detailed guidance on a simplified mix design procedure used widely in the USA and elsewhere.
Day, K.W. (1999) *Concrete Mix Design, Quality Control and Specification* (2nd edn). E &FN Spon, London. Comprehensive computerised mix design and quality control approach based on the author's UK, Australian and Far Eastern experience.
Dewar, J.D. (1992) *Manual of Ready Mixed Concrete* (2nd edn). Chapman and Hall, London. Includes UK readymixed concrete industry approach to comprehensive mix design based on laboratory trial mixes.
Dewar, J.D. (1999) *Computer Modelling of Concrete Mixtures.* E&FN Spon, London. Describes the theory, proving trials and method for the author's comprehensive computerized approach to concrete mix design.
DoE (1988) *Design of Normal Concrete Mixes.* Department of the Environment, HMSO, London. Detailed guidance on a simplified mix design procedure used widely in the UK and elsewhere.
Harrison, T.A. (2003) The new concrete standards – getting started. The Concrete Society. CS 149. Essential changes to definitions and terms.

# Part 2

# Special concretes

# 2

# Properties of lightweight concrete

*John Newman and Phil Owens*

## 2.1 Introduction

Lightweight concretes can be produced with an over-dry density range of approximately 300 to a maximum of 2000 kg/m$^3$, with corresponding cube strengths from approximately 1 to over 60 MPa and thermal conductivities of 0.2 to 1.0 W/mK. These values can be compared with those for normal weight concrete of approximately 2100–2500 kg/m$^3$, 15 to greater than 100 MPa and 1.6–1.9 W/mK. The principal techniques used for producing lightweight concrete can be summarized as follows:

- Omitting the finer fraction of normal weight aggregate to create air-filled voids using a process pioneered by Wimpey in the UK in 1924 (no-fines concrete).
- Including bubbles of gas in a cement paste or mortar matrix to form a cellular structure containing approximately 30–50 per cent voids (aerated or foamed concrete).
- Replacing, either wholly or partially, natural aggegates in a concrete mix with aggregates containing a large proportion of voids (lightweight aggregate concretes).

These are shown diagrammatically in Figure 2.1.

The properties of lightweight concrete can be exploited in a number of ways from its use as a primarily structural material to its incorporation into structures for the enhancement of thermal insulation. This chapter is concerned mainly with lightweight aggregate concretes and, in particular, those made with lightweight aggregates within a Portland cement-based matrix (i.e. closed structure lightweight aggregate concretes).

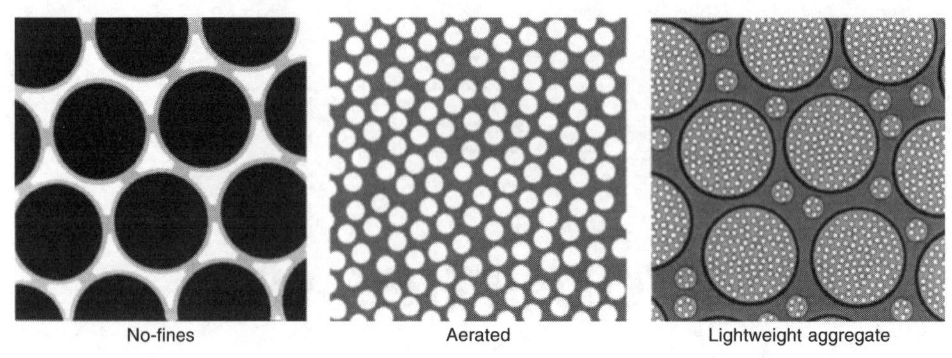


**Figure 2.1** Basic forms of lightweight concrete.

## 2.2 No-fines concrete (NFC)

### 2.2.1 Production

The mix consists of cement, water and course aggregate with fines (sand) omitted. After prior moistening with water the aggregates are mixed with the cement and mix water. This results in each particle of coarse aggregate being coated with a layer (up to about 1.3 mm) of cement paste which bonds it to adjacent particles in point-to-pint contact to leave interstitial voids (Figure 2.2).

**Figure 2.2** No-fines concrete.

The voids are interconnected to produce a porous open-textured concrete with reduced density, strength and shrinkage. Density depends mainly on the type and grading of the aggregate. The lowest densities are achieved with single-sized coarse aggregate. Maximum aggregate size can range from 7 to 75 mm[1] but is usually from 10 to 20 mm. The aggregate should contain no more than 10 per cent of undersized material, no particle less than 5 mm, no flaky or elongated particles and be clean to allow good cohesion with the cement paste. The aggregate should be gravel or hard and crushed aggregate without sharp edges which increase the likelihood of local crushing under load. Lightweight aggregate can be used to further decrease concrete density. For normal weight aggregates, aggregate/cement ratios from 6 to 10 produce densities of between 1200 and 1900 kg/m$^3$

while lightweight aggregate with aggregate/cement ratios of between 3 and 8 give densities of 800 to 1400 kg/$m^3$.

Workability can be checked only visually (organoliptically) and no compaction, other than localized rodding, should be used during placing. Since NFC does not segregate it can be dropped from considerable heights and can be placed in lifts of up to 3 storeys[2]. Although the fresh mix exerts little pressure on formwork (about one-third of that for normal concrete[1] and even less for lightweight aggregate), the striking time must be sufficient to allow the material to cohere. Attention should be paid to curing due to the relatively thin layer of cement paste (Malhotra, 1976) and particular care must be taken with blended cements containing PFA or GGBS since no-fines concrete dries rapidly[1].

## 2.2.2 Properties

### *Density*

Table 2.1 indicates typical properties of NFC made with normal weight and light-weight aggregates.

**Table 2.1** Typical properties of NFC made with normal weight and lightweight aggregates

| Item | Normal weight aggregate | Lightweight aggregate |
|---|---|---|
| Aggregate/cement ratio (by mass) | 6–10 | 3–8 |
| Water/cement ratio (by mass) | 0.38–0.45 | 0.38–0.45 |
| Air dry density (kg/$m^3$) | 1200–1900 | 800–1400 |
| Cube strength (MPa) | 3–7 | 3–7 |

Density depends on the type and size of aggregate with lower maximum aggregate sizes giving higher strengths probably due to the larger number of contact points in a given volume[1].

### *Compressive strength*

In the UK the cube strength for NFC is determined using the method described in BS 1881:Part 113[3]. For a given type of aggregate, compressive strength depends mainly on density which, in turn, is governed mainly by cement content[4]. Figure 2.3 shows the relationship between cube strength and density for concretes made with normal weight aggregates. Strength increases with age in a similar manner to normal concrete.

### *Tensile strength*

The flexural strength is approximately 30 per cent of the compressive strength[5] which is a higher proportion than for normal concrete. Data[6] shows the tensile strength, flexural strength and bond strength to be 12 per cent, 23 per cent and 19 per cent respectively of the cube strength at 28 days.

### *Modulus of elasticity (E value)*

As for normal concrete, $E$ value increases as the strength increases and has been found[5] to be 10 GPa for a cylinder strength of 5 MPa.

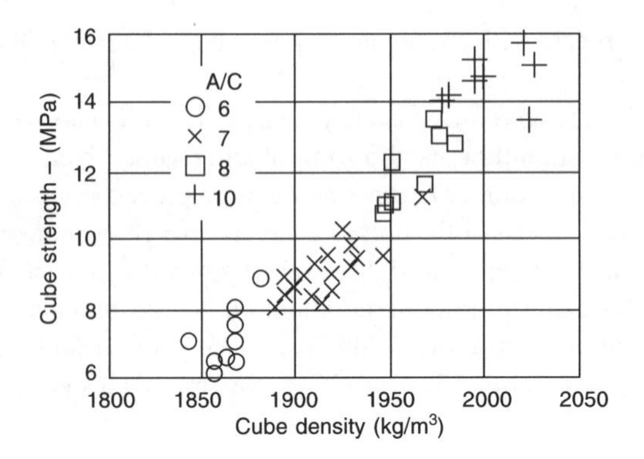


**Figure 2.3** Relationship between 28-day cube strength and density for NFC with various A/C ratios by volume.

## *Drying shrinkage*

Drying shrinkage is significantly lower than for normal concrete since the cement paste layer is thin and the aggregate provides considerable restriant. A typical value is 120 microstrain increasing to 200 microstrain at low humidities[7]. Due to the thin paste layer and the open-textured nature of the concrete the shrinkage rate is high.

## *Thermal properties*

### Thermal expansion

Due to the structure of NFC the thermal expansion is approximately equal to that of the aggregate alone. Typical values are about 0.6–0.8× those of normal concrete[7] and 9 microstrain/°C can be assumed for estimation pruposes[1].

### Thermal conductivity

The coefficient of thermal conductivity of NFC is between 0.69 and 0.94 $J/m^2s$ for normal weight aggregate and about 0.22 $J/m^2s$ for lightweight aggregate[7]. As for normal concrete the coefficient increases with moisture content.

## *Penetrability*

NFC made with dense aggregate can absorb up to 12 per cent of water by mass but in normal conditions absorption is about one fifth of this value[7]. NFC is not subject to significant capillary action and maximum capillary rises of up to 2.5× the nominal size of aggregate have been reported. The large pores in NFC allow it to drain water.

## *Freeze/thaw resistance*

Freeze/thaw resistance is high due to the low capillary action but will be low if the pores in the concrete are saturated.

# 2.3 Aerated and foamed concrete

## 2.3.1 Introduction

Concretes containing intentionally entrained voids in the hardened cement paste or mortar matrix to form a cellular structure of low density are known as aerated, cellular, gas, foamed or foam concretes. In this section they will be divided into two principal types, namely, aerated and foamed.

## 2.3.2 Production

### *Aerated concrete*

Designed for factory production, gas (hydrogen) bubbles are introduced into cement paste or mortar usually made with Portland cement of suitable consistence by aluminium powder (0.2 per cent by mass of cement) which reacts both with $Ca(OH)_2$ and alkalis released into solution. The gas bubbles expand the mixture to the required density after which the concrete is cured either in steam at atmospheric pressure or in steam at 180°C under high pressure in an autoclave. Pre-treated reinforcement can be included in units which are cut to the required size soon after curing. Autoclaved aerated concrete has better strength, volume stability and durability than non-autoclaved. Relevant Standards include BS EN 678, 679, 680, and 1351[8–11]. Figure 2.4 shows an example of autoclaved aerated concrete.

**Figure 2.4** Typical autoclaved aerated concrete.

### *Foamed concrete*

Foamed concrete is classified as having an air content of more than 25 per cent. The air can be introduced into a mortar or concrete mix using two principal methods. First, a preformed foam from a foam generator can be mixed with other constituents in a normal mixer or a readymixed concrete truck. Second, a synthetic- or protein-based foam-producing admixture can be mixed with the other mix constituents in a high-shear mixer. In both methods the foam must be stable during mixing, transporting and placing. The resulting bubbles in the hardened concrete should be discrete and usual bubble size is between 0.1 and 1 mm. Typical mixes are given in Table 2.2[12].

**Table 2.2** Typical foamed concrete mixes

| | | | | | |
|---|---|---|---|---|---|
| Wet density (kg/m$^3$) | 500 | 525 | 600 | 1200 | 1200 |
| Cement content (kg/m$^3$) | 160 | 340 | 340 | 340 | 340 |
| Foam volume (%) | 72 | 73 | 69 | 44 | 39 |
| Filler type | PFA | – | Sand | Sand | PFA |
| Filler content (kg/m$^3$) | 160 | 0 | 66 | 635 | 486 |
| Cube strength at 28 days (MPa) | 1.0 | 2.0 | 2.0 | 6.0 | 7.5 |
| Cube strength at 91 days (MPa) | 1.4 | 2.2 | 2.2 | 7.0 | 10.0 |

## 2.3.3 Properties

### *Autoclaved aerated concrete*

Methods for testing autoclaved aerated concrete are given in a RILEM[13] publication. The determination of dry density, compressive strength, flexural strength, shrinkage and *E*-value is described in BS EN 678, 679, 680, 1351 and 1352[8–11, 14], respectively. Typical properties are shown in Table 2.3. The average drying shrinkage of typical autoclaved aerated concrete is approximately 0.02 per cent.

**Table 2.3** Typical properties of autoclaved aerated concrete

| Dry density (kg/m$^3$) | Compressive strength (wet) (MPa) | Flexural strength (MPa) | *E*-value (GPa) | Thermal conductivity (3% moisture) (W/mK) |
|---|---|---|---|---|
| 450 | 3.2 | 0.65 | 1.6 | 0.12 |
| 525 | 4.0 | 0.75 | 2.0 | 0.14 |
| 600 | 4.5 | 0.85 | 2.4 | 0.16 |
| 675 | 6.3 | 1.00 | 2.5 | 0.18 |
| 750 | 7.5 | 1.25 | 2.7 | 0.20 |

Other relevant standards are BS EN 989 (bond with reinforcement)[16], 990 (corrosion protection to reinforcement)[17], 1353[18] (moisture content) and 1355 (creep under compression).[19]

### *Foamed concrete*

Typical properties are shown in Table 2.4[20]

**Table 2.4** Typical properties of formed concrete

| Dry density (kg/m$^3$) | Compressive strength (wet) (MPa) | *E*-value (GPa) | Thermal conductivity (3% moisture) (W/mK) | Drying shrinkage (%) |
|---|---|---|---|---|
| 400 | 0.5–1.0 | 0.8–1.0 | 0.10 | 0.30–0.35 |
| 600 | 1.0–1.5 | 1.0–1.5 | 0.11 | 0.22–0.25 |
| 800 | 1.5–2.0 | 2.0–2.5 | 0.17–0.23 | 0.20–0.22 |
| 1000 | 2.5–3.0 | 2.5–3.0 | 0.23–0.30 | 0.15–0.18 |
| 1200 | 4.5–5.5 | 3.5–4.0 | 0.38–0.42 | 0.09–0.11 |
| 1400 | 6.0–8.0 | 5.0–6.0 | 0.50–0.55 | 0.07–0.09 |
| 1600 | 7.5–10.0 | 10.0–12.0 | 0.62–0.66 | 0.06–0.07 |

## 2.4 Lightweight aggregate concrete

### 2.4.1 Introduction

The variety of purpose of lightweight aggregate concrete is recognized by RILEM/CEB who have proposed the classification given in Table 2.5 (RILEM, 1978)[21]. This chapter discusses only those concretes within Class I (i.e. structural lightweight concrete).

**Table 2.5** Classification of lightweight concretes

| Property | Class and type | | |
|---|---|---|---|
| | I<br>Structural | II<br>Structural/ insulating | III<br>Insulating |
| Compressive strength (MPa) | >15.0 | >3.5 | >0.5 |
| Coefficient of thermal conductivity (W/mK) | – | <0.75 | <0.30 |
| Approximate density range (kg/m$^3$) | 1600–2000 | <1600 | <<1450 |

Although structural lightweight concrete is usually defined as a concrete with an oven-dry density of no greater than 2000 kg/m$^3$ [22,23,24,25] there are variations in certain parts of the world. For example, in Australia[26] structural lightweight concrete is considered to be a concrete made with lightweight coarse aggregate and normal weight fines resulting in a saturated surface-dry density of not less than 1800 kg/m$^3$. In Norway[27] any combination of any types of aggregate can be used for structural concrete provided the resulting concrete (a) has an oven-dry density of 1200–2200 kg/m$^3$ and (b) a strength grade of no greater than 85 MPa if the mix contains lightweight aggregate. In the USA[28] structural lightweight aggregate concrete is considered to be concrete with an air-dry density of less than 1810 kg/m$^3$. Two main classes of lightweight concrete are considered, namely, concrete made with lightweight coarse and fine materials and that made with lightweight coarse aggregate and natural fines. Interpolation between these classes is permitted. Perhaps the most radical approach is that adopted by the Japanese[29,30] who refer to lightweight concrete but do not specify any density values and properties are only provided for concrete made with lightweight coarse and fine aggregates. In Russia and the CIS lightweight concrete is defined in terms of its compressive strength. An addendum to SNiP 2.03.01–84[31] covers a strength range from 2.5 to 40 MPa with densities from 800 to 2000 kg/m$^3$ depending on the type of aggregate used. In the relevant European standard[32] lightweight aggregate concrete is classified according to oven-dry density (Table 6).

**Table 2.6** Density classes for lightweight aggregate concrete

| Density class | 1.0 | 1.2 | 1.4 | 1.6 | 1.8 | 2.0 |
|---|---|---|---|---|---|---|
| Oven dry density (kg/m$^3$) | 901–1000 | 1001–1200 | 1201–1400 | 1401–1600 | 1601–1800 | 1801–2000 |

## 2.4.2 Properties of lightweight aggregate for structural concrete

Such diverse materials as clay, shale, slate, slag, PFA, etc., after processing using differing techniques produce aggregates with remarkably similar chemical compositions and structures[33]. Typically they have a ceramic-like dense matrix with included air voids and most have a relatively dense exterior shell with a more voided interior (Figure 2.5).

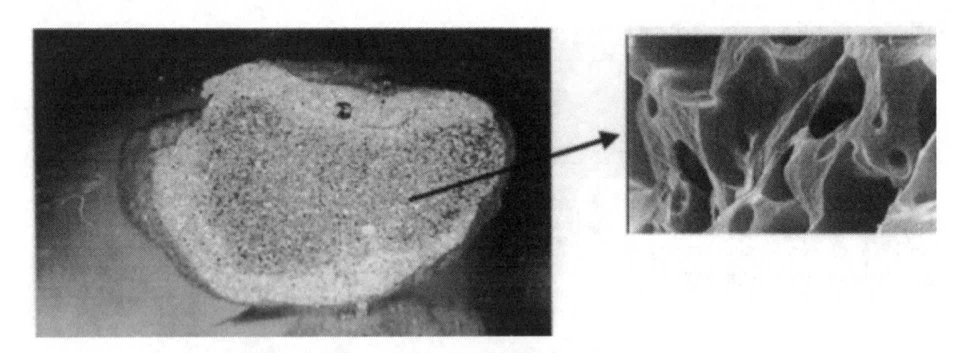

**Figure 2.5** Cross-section and photomicrograph of sintered PFA aggregate.

In the UK all lightweight aggregates (LWAs) except foamed slag (BS 877: Part 2)[34], air-cooled blastfurnace slag (BS 1047)[35] and clinker (BS 1165)[36] are covered by BS 3797: Part 2[37] which will be replaced by BS EN 13055-1[38]. Sampling and testing methods are described in BS 3681: Part 2[39]. Some typical properties of LWAs used for structural concrete are shown in Table 7[40].

**Table 2.7** Typical properties of lightweight aggregates

| Type | Shape | Water Absorption (%) | Oven-dry loose bulk density ($kg/m^3$) |
|---|---|---|---|
| Expanded clay | Rounded | 12–14 | 350–500 |
| Expanded slate | Angular/irregular | 10–15 | 560–720 |
| Expanded shale | Rounded | 12–14 | 500–800 |
| Pumice | Angular/irregular | 30–40 | 500–800 |
| Sintered PFA | Rounded | 9–15 | 800–850 |

These should be compared with dense aggregates which have a dry loose bulk density in the range 1200–1900 $kg/m^3$ (extra dense up to 2600 $kg/m^3$) and a water absorption of approximately 0.5–2%.

Manufactured lightweight aggregates are usually free from deleterious chemicals and do not induce harmful reactions. However, they should be checked for carbon content (loss on ignition ~4%) and sulphates ($SO_3$ ~ 1%).

## 2.4.3 Properties of structural lightweight aggregate concrete

### *Fresh concrete*

The increased absorption, decreased density and range of available lightweight aggregates should be considered when designing a concrete mix. Of particular importance is water absorption. All aggregates, whether natural or manufactured, absorb water at a rate which diminishes with time. Such absorption is important in that for unsaturated or partially saturated aggregate it will influence such properties of fresh concrete as workability (including pumpability) and density and also affect such hardened properties as density, thermal insulation, fire resistance and freeze/thaw resistance.

For an individual aggregate particle the amount of water absorbed and the rate of absorption depend primarily on (a) the pore volume, (b) the distribution of pores within the particle and (c) the structure of the pores (i.e. whether connected or disconnected). For lightweight aggregate particles, which have a relatively large pore volume, the rate of water absorption is likely to be much greater than for natural dense aggregates. However, the characteristics of the surface zone of aggregate particles have a large influence on absorption such that the disparity between natural and lightweight aggregates may not be as large as expected from the differences in density. For example, the sintered 'shell' around some particles (such as those of expanded or sintered aggregates) containing small, relatively disconnected pores, impedes the absorption process.

The water absorption of aggregates is usually expressed as the proportion of the oven-dry mass absorbed after 30 min and 24 h. For lightweight aggregates the 24 h value generally lies within the range 5–15% of the dry mass compared with about 0.5–2% for most natural aggregates. An approximate estimation of the correction to be made for water absorbed by lightweight aggregates during and immediately after mixing can be made on the basis of the 30 min absorption value which typically lies within the range 3–12%[41]. It should be noted that, as for natural aggregates, water absorptions for the finer grades of a material will generally be higher than those for the coarser.

Suitable mix proportioning procedures are described in detail elsewhere[41,42,43,44]. For a given grade of concrete the resulting mixes generally contain a higher cement content than for normal weight concrete, and the maximum attainable strength is governed by the type of aggregate used. It should be noted that 'workability' is underestimated by the slump test[45] and, as for normal weight concrete, cohesion is improved by the use of air entrainment. For pumped concrete workability should be assessed by the flow test[46].

In view of the effect of water absorption of lightweight aggregate, allowance should be made for a possible loss in workability between mixing and placing and during transportation. Admixtures are essential for pumping since water can be forced into aggregate particles by the pressure in the pipeline.

The processes of compaction and curing are no different from normal weight concrete but lightweight aggregate concrete is more tolerant of poor curing owing to the reserve of water held within the aggregate particles. When finishing the surfce of lightweight aggregate concrete due consideration should be given to the possibility of flotation of the lightweight particles.

### *Density*

The oven-dry density of structural lightweight aggregate concretes can range from approximately 1200 to 2000 kg/m$^3$ compared with 2300 to 2500 kg/m$^3$ for normal weight concretes. As the behaviour of lightweight aggregate concrete is closely related to its density, and as the density as well as the strength and durability are important to the designer, it is essential to define what is meant by the following terms[43].

| | |
|---|---|
| Fresh concrete density | The bulk density of concrete when compacted to a practicable minimum air void volume. |
| Oven-dry density | The bulk density after drying for 24 h in air at 105°C. |
| Air-dry density | The density in equilibrium with a dry environment (moisture content of approx. 5–10% by vol.). Approximately equal to fresh density less 100–200 kg/m$^3$ for LW coarse and fine aggregates and fresh density less 50–100 kg/m$^3$ LW coarse and dense fine aggregate. |
| Saturated density | Approximately equal to fresh density plus 100–120 kg/m$^3$. |

It is suggested that the fresh concrete density is used as a basis for comparison.

The principal factors influencing density are[43].

| | |
|---|---|
| Cement content | A 100 kg/m$^3$ increase in cement gives approximately a 50 kg/m$^3$ increase in density. |
| Relative density of aggregates | A substitution of lightweight for dense fines increases the density by approximately 150–200 kg/m$^3$. |
| Entrained air | The resulting changes to the mix proportions decrease density by approximately 90 kg/m$^3$. |
| Moisture content of aggregates | A concrete made with water-saturated or partially saturated aggregates will have a higher fresh density. |
| Environmental conditions | Density changes in response to wetting or drying. |

## *Strength*

### Compression strength (uniaxial)

As for normal weight concrete, a wide range of aggregates produces a corresponding range of cube strengths. When comparing lightweight aggregate concrete with normal weight concrete it is important to consider the types of constituent materials in both cases (Figure 2.6[47]).

Factors affecting strength include:

- *Strength and stiffness of aggregate particles*
  Weaker particles require stronger mortars and thus higher cement contents. The 'ceiling' strength of concrete depends upon the type of aggregate. Excellent particle-matrix bond and similarity of particle and matrix moduli ensure that the matrix is used efficiently[40] (Figure 2.7).
- *Water/cement ratio*
  This has the same effect on strength as for normal weight concrete and the same range of water/cement ratio is used[43]. However, the reduction of free water/cement ratio due to the water absorption of lightweight aggregate is difficult to predict and thus the specification of effective water/cement ratio for mixes is not practicable since it is difficult to measure and verify. Free water contents are the same as for normal weight concrete (say 180–200 1/m$^3$) but aggregate absorption requires higher total water

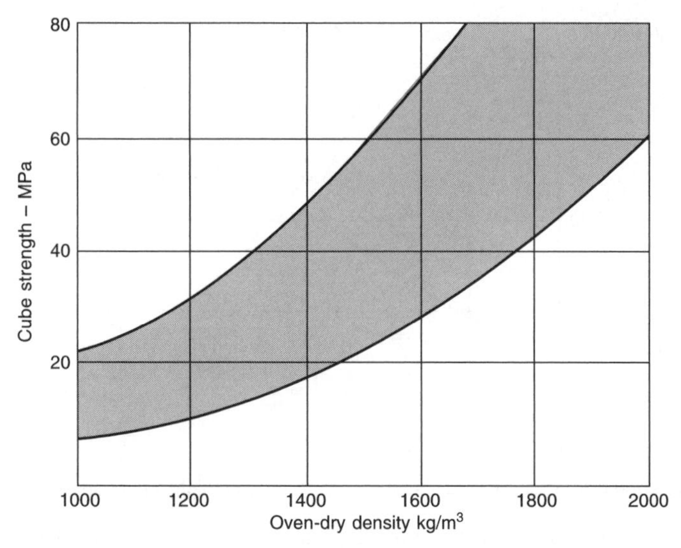


**Figure 2.6** Cube strength versus oven-dry density for lightweight aggregate concretes.

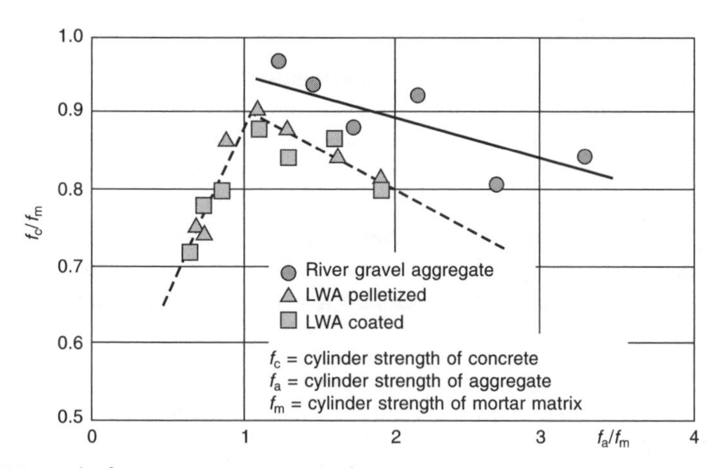


**Figure 2.7** Strength of concrete versus strength of aggregate.

contents (say 250–300 $1/m^3$). For lightweight aggregate concrete it is more relevant for mix design purposes to relate strength to cement content[41].

- *Cement content*

  For a given workability, strength increases with cement content, the increase depending on the type of aggregates used (Figure 2.8[43]). Generally, greater cement contents are required than for normal weight concrete (approximately 10% greater below about 35–40 MPa and 20–25% above). Although the increase in strength for a given increase in cement content depends on the type of aggregate used and the cement content itself, on average for lightweight aggregate a 10% higher cement content will give approximately a 5% higher strength.
- *Age*

  Strength–age relationships are similar to those for normal weight concrete. If concrete dries then hydration will cease but the situation is better for lightweight aggregate

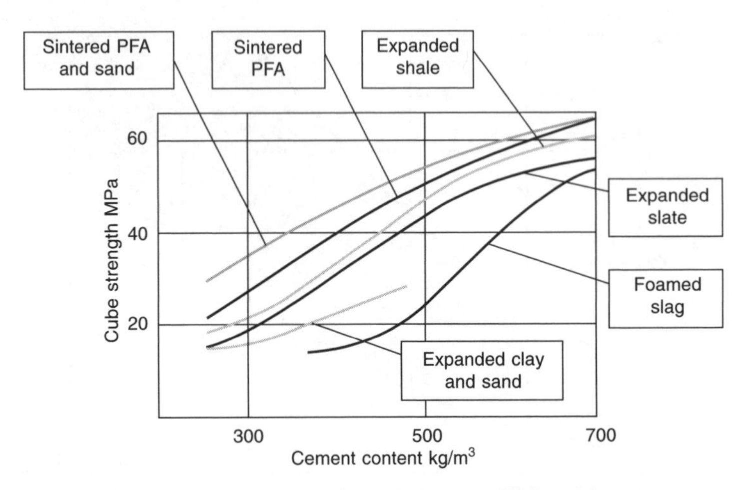


**Figure 2.8** Cube strength versus cement content for various types of lightweight aggregate.

concrete than for normal weight concrete owing to the reserve of water available in aggregate pores. Thus lightweight aggregate concrete is more tolerant of poor curing than normal weight concrete.

- *Density*
  The density of compacted concrete is affected mainly by the aggregate particle density which is related to particle porosity and hence particle strength. Thus, aggregates of different density will result in different concrete strengths as well as densities.

## Tensile strength

Tensile strength is important when considering cracking in concrete elements. The factors influencing compressive strength also influence tensile strength. The principal differences between lightweight aggregate concrete and normal weight concrete are due to:

- *Fracture path*
  This normally travels through, rather than around, lightweight aggregate particles. The behaviour is similar to normal weight concrete made with crushed aggregates in that the flexural/compressive strength ratio is higher (Figure 2.9).

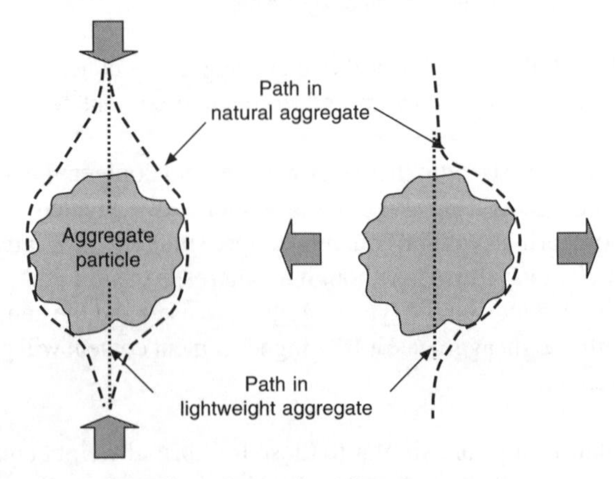


**Figure 2.9** Fracture paths in lightweight and normal weight concretes.

- *Total water content*
  This is higher for lightweight aggregate concrete due to the absorption of the lightweight aggregate. Thus, in drying situations greater moisture gradients can cause a significant reduction in tensile strength although this effect is somewhat alleviated by the effects of increased hydration (Figure 2.10).

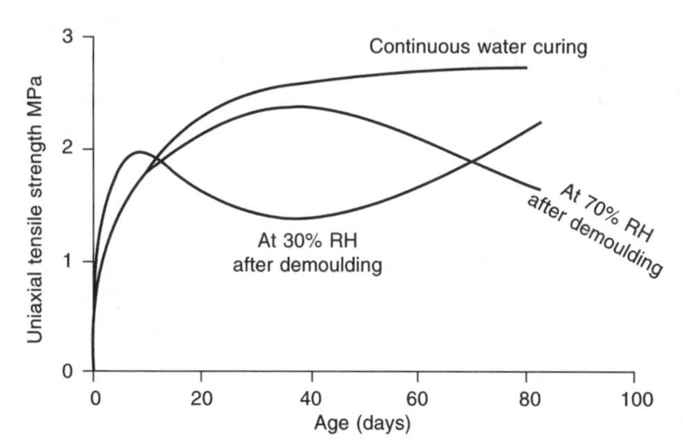


**Figure 2.10** Effect of drying on tensile strength.

These effects should be considered when testing. The flexural strength is more affected than cylinder splitting strength.

For continuously moist-cured concretes the measured tensile strength as a proportion of the compressive strength is similar for both lightweight and normal weight concrete. However, for dried concrete the tensile strength of lightweight aggregate concrete is reduced below that for normal weight concrete of the same compressive strength. This reduction also influences other properties such as shear, bond, anchorage, etc.

BS 8110: Part 2[48] requires that for Grade 25 concrete (i.e. concrete with not more than 5% of cube strength results less than 25 MPa) and above, the design shear strength should not exceed 80% of the value for normal weight concrete (a Table is given for Grade 20 concrete). In no case must the shear stress exceed the lesser of $0.63 f_{cu}$ or 4 MPa. However, a comprehensive investigation covering a wide range of lightweight aggregates has shown conclusively that this reduction is not justified for either shear[49] or punching shear[50].

## Compression strength (multiaxial)

All structural concretes show an increase in the axial stress required to cause failure if a lateral stress is applied simultaneously. Lightweight aggregate concrete exhibits the same trend but the strength enhancement is somewhat lower[51]. In view of this the following relationships have been proposed for lightweight aggregate concrete[52].

$$\sigma_{11} \leq 0.67 f_{cu} + 2\sigma_3$$

where $\sigma_1 \geq \sigma_2 \geq \sigma_3$ and,
$\sigma_{11}$ = maximum principal compressive stress that can be sustained
$f_{cu}$ = characteristic cube strength
$\sigma_3$ = minor principal compressive stress.

The above relationship should be compared with that for normal weight concrete of:

$$\sigma_{11} \leq 0.67 f_{cu} + 3\sigma_3$$

Expressions are also given in the above reference for the limit states of serviceability and ultimate.

## Impact strength

The lower modulus of elasticity and higher tensile strain capacity of lightweight aggregate concrete provides better impact resistance than normal weight concrete. This has been used to advantage in various situations, including the protection of underwater structures[53,54] and stair treads[55]. More research is required if the full capabilities of the material are to be exploited.

## Strength/density ratio

By the very nature of structural lightweight aggregate concrete its strength/density ratio assumes a greater importance than for normal weight concrete. The following examples clearly demonstrate the advantages to be gained by using structural lightweight aggregate concrete for two types of structure.

1. A bridge in Germany constructed using both lightweight and normal weight concrete each containing approximately the same proportions of retarding and plasticizing admixtures by mass of cement[56] (Table 2.8).

**Table 2.8** Comparison between lightweight and normal weight concretes for a bridge in Germany

| Item | LWC | NWC |
|---|---|---|
| Specification requirements | | |
| Strength class | LB45 | B55 |
| Density (kg/m$^3$) | <1800 | 2400 |
| Mix proportions (kg/m$^3$) | | |
| Cement | 400 | 400 |
| Sand | 497 | 494 |
| Gravel | 83 | 706 |
| Lightweight aggregate | 623 | – |
| PFA | 50 | 50 |
| Water | 170 | 172 |
| Fresh properties | | |
| Density (kg/m$^3$) | 1880–1930 | 2380–2420 |
| Flow (mm) | 290–350 | 280–350 |
| Cube strength at 28 days (MPa) | | |
| Mean | 73.3 | 69.3 |
| Standard deviation | 4.7 | 7.5 |
| Characteristic | 65.6 | 57.0 |
| Strength/density ratio | | |
| Based on mean strength | 38 | 29 |
| Based on characteristic strength | 34 | 24 |

2. A comparative study of the use of lightweight aggregate concrete for offshore structures[57]:

The above examples clearly demonstrate the improved strength/density ratios for the lightweight concrete despite the similar fresh and hardened properties for both types of concrete. Of particular interest is the improved control of the lightweight concrete as

**Table 2.9** Comparison between lightweight and normal weight concretes for an offshore structure

| Item | LWC | NWC |
|---|---|---|
| Mix characteristics | | |
| Cement content (kg/m$^3$) | 410 | 400 |
| Free-water/cement ratio | 0.40–0.42 | 0.40–0.42 |
| Fresh properties | | |
| Density (kg/m$^3$) | 1875 | 2420 |
| Slump (mm) | 150–200 | 150–230 |
| Cube strength (MPa) | | |
| Range | 60–70 | 60–70 |
| Standard deviation | 4 | 4–5 |
| Other hardened properties | | |
| Indirect tensile strength (MPa) | 4.0 | 4.0 |
| Flexural strength (MPa) | 6.0 | 6.0 |
| $E$-value (kN/mm$^2$) | 23.5 | 29.0 |
| Ultimate creep at 100% RH & 25MPa (%) | 0.8 | 0.8 |
| Ultimate shrinkage at 65% RH & 20°C (%) | 0.05 | 0.05 |
| Thermal expansion/°C | 8–10 | 9–11 |
| Permeability (m/s) | $10–100 \times 10^{-13}$ | $10–100 \times 10^{-13}$ |
| Strength/density ratio | | |
| Based on mean strength | 35 | 27 |

exemplified by the lower standard deviations. Such observations confirm the view that structural lightweight aggregate concrete should be treated no differently from normal weight concrete.

## *Deformation*

### General

Lower stiffnesses of lightweight aggregate particles and higher cement contents result in larger deformations. However, these effects are alleviated by lower concrete density so small-scale tests in the laboratory can provide pessimistic data compared with behaviour on site. The stress–strain relationships for lightweight aggregate concrete are more linear and 'brittle' than for normal weight concrete (Figure 11[47]).

Such a response is probably attributable to the greater compatibility between the lightweight aggregate particles and the surrounding cementitious matrix. In the case of normal weight concrete the formation and propagation of small microscopic cracks, or microcracks (2–5 μm), have long been recognized as the causes of fracture and failure of concrete and the marked non-linearity of the stress–strain curve, particularly near the ultimate stress level. Although some of these discontinuities exist as a result of the compaction process of fresh concrete, the formation of small fissures or microcracks in concrete is due primarily to the strain and stress concentrations resulting from the incompatibility of the stiffnesses of the aggregate and cement paste/mortar components. The fracture process in normal weight concrete starts with stable fracture initiation (up to approximately 50% of ultimate stress) followed by stable and then unstable fracture propagation (from approximately 75 to 85% of ultimate stress)[58]. Although fundamentally the same process occurs in lightweight aggregate concrete the stable fracture initiation stage is extended and the unstable fracture propagation stage is reduced, so complete disruption occurs abruptly at ultimate. It should be noted that this behaviour is not accounted for in most codes of practice.

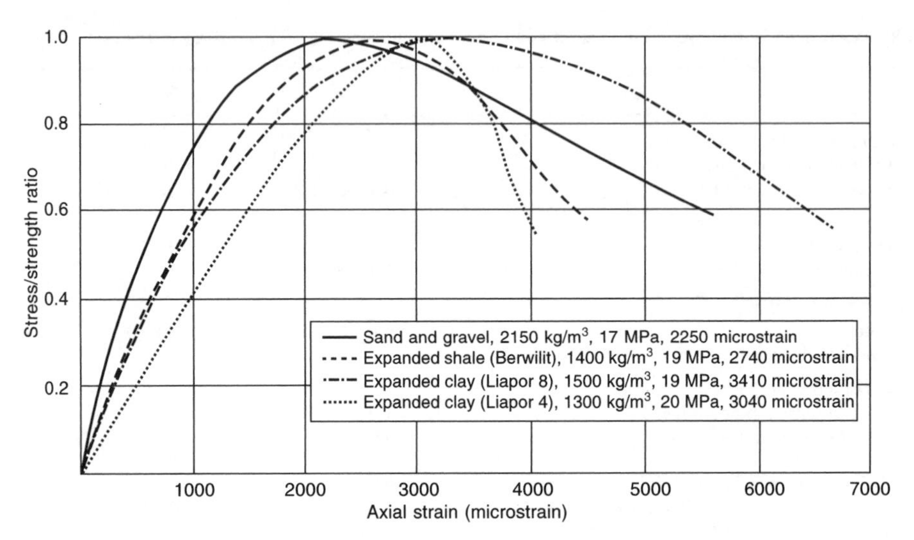


**Figure 2.11** Stress–strain relationships for lightweight and normal weight concretes.

BS 8110: Part 2[48] allows deflections to be calculated as for normal weight concrete with appropriate values for modulus of elasticity, free shrinkage and creep for lightweight concrete made with the given aggregate. Alternatively, heavily loaded members may be checked against the limiting span/effective depth ratios to 85% of those for normal weight concrete.

## Modulus of elasticity (*E*-value)

For any concrete its stiffness depends on the stiffnesses of the various constituents and their relative volumetric proportions in the mix. Simplifying concrete as a two-phase material consisting of coarse aggregate particles embedded in a mortar matrix, the *E*-value of the composite will decrease with (a) a decrease in the stiffness of the mortar which, in turn, depends on the volume proportions of cement, water and fine aggregate and the type of fine aggregate and (b) a decrease in the stiffness of the coarse aggregate. Since the moduli of lightweight aggregate particles are generally lower than those of natural dense aggregates and the fact that most lightweight aggregate concretes contain higher cement contents it follows that the overall moduli of lightweight aggregate concretes will be lower than normal weight concretes. It also follows that concretes made with lightweight coarse and lightweight fine aggregate will be lower than those made with lightweight coarse aggregate and natural dense fines.

It should be noted that although the *E*-value of concrete is not directly related to strength and density, a useful empirical relationship which provides an approximate *E*-value for most purposes is[43].

$$E = D^2\sqrt{f_{cu}} \times 10^{-6} \text{ kN/mm}^2$$

where $E$, $D$ and $f_{cu}$ are modulus, nominal density and cube strength respectively.

For calculating elastic deformation using BS 8110: Part 2[48], the *E*-values are derived from the values for normal weight concrete ($E_{c.28} = K_o + 0.2 \times f_{cu.28}$ where $K_o$ varies from 14 to 26 N/mm$^2$) by multiplying by $(D/2400)^2$, where $D$ is the density of lightweight concrete. Where more accurate values are required tests should be carried out on the given concrete.

The lower $E$-values for lightweight concretes will give rise to increased deformations for structural elements under a given load, although the effect will be reduced by the lower dead loads of lightweight concrete elements themselves. However, under dynamic conditions such as impact or load fluctuation; reduced stiffness can be beneficial.

### Creep

Generally higher creep strains are produced in lightweight aggregate concrete than in normal weight concrete due to the lower $E$-value of aggregate and higher proportion of matrix. The basic creep of lightweight aggregate concrete is approximately 1.00–1.15× that of normal weight concrete, but drying creep is often significantly higher. Thus, in full-scale structures, creep is seldom as large as predicted from laboratory tests due to differences in exposure conditions, curing, reinforcement, restraint, stress level, size, shape, etc.

### Shrinkage

As for creep, the laboratory data for shrinkage are pessimistic and do not relate to full-scale situations. Although shrinkage is generally greater for lightweight aggregate concrete made with lightweight fines than for normal weight concrete (×1.0–1.5) it is similar for lightweight aggregate concrete made with dense fines. Creep and shrinkage occur simultaneously and perhaps due to this effect shrinkage cracking is rare in lightweight aggregate concrete due to relief of restraint by creep, continuous supply of water from aggregate pores and better tensile strain capacity. This is recognized in BS 8110: Part 2, Table 3.2[48]. Realistic values of drying shrinkage for full-scale structures subject to external exposure are 200–300 microstrain. For internal exposure these values may increase to 300–500 microstrain. A value of 350 microstrain could be assumed if no other information is available.

### Bond and anchorage

The bond strength of lightweight aggregate concrete is similar to that of normal weight concrete but, as for normal weight concrete, excessive water contents should be avoided. Anchorage in lightweight aggregate concrete is lower than in normal weight concrete due to the lower bearing strength caused by the relative weakness of lightweight aggregates which requires careful detailing. BS 8110[48] requires that the bond stresses used to determine lap and anchorage lengths for lightweight aggregate concrete be taken as 80% of those for the same grade of normal weight concrete. However, research shows that this factor is conservative[59]. For prestressed concrete transmission lengths are approximately 20–25% longer.

## *Fatigue*

A comprehensive survey of test results under repeated compressive load showed that lightweight concrete above a density of 1500 kg/m$^3$ had the same susceptibility to fatigue as normal weight concrete when expressed as a Wohler or S–N diagram[60]. For both types of material the diagram could be described by the following equation:

$$f_{max}/f_{min} = 1 - 0.0685 \times (1 - R) \times \log^{10} N$$

where, $f_{max}$ and $f_{min}$ are highest and lowest compressive (cylinder) stresses
$R$ is the ratio of lowest to highest compressive stress
$N$ is the number of load cycles to failure.

The above conclusion has been confirmed by other work[61] which showed lightweight aggregate concrete to perform at least as well as, and in some cases slightly better than, normal weight concrete of the same grade. However, other work has indicated that lightweight concrete made with sintered PFA aggregate performed less well than gravel concrete[62].

Information compiled by the CEB[63] shows that for repeated tensile stress states the fatigue strength of lightweight concrete is the same as, or slightly higher than, that of normal weight concrete[64,65] while the fatigue strength under tensile-compressive states of stress is slightly lower[65]. Fatigue tests carried out on prestressed concrete beams[66] showed the onset of, and resistance to, cracking was not less than that of normal weight concrete. As in the case of impact behaviour, more data are required on this subject.

## *Durability*

### Freeze/thaw behaviour

As for normal weight concrete the performance of lightweight aggregate concrete under freeze/thaw conditions depends mainly on the mix proportions, the type of aggregate and its moisture content and the level of air entrainment. Laboratory tests have shown that for the majority of aggregate types both in the pre-soaked and air-dry condition non-air-entrained lightweight aggregate concrete is potentially more durable under freeze/thaw conditions than equivalent strength non-air-entrained normal weight concrete[67,68], particularly when natural fines are used[69]. For air-entrained concrete made with pre-soaked aggregates the performance of lightweight aggregate concrete is not significantly different from that of normal weight concrete, while air-entrained concrete made with lightweight aggregate in the air-dry condition shows a significant improvement over similar normal weight concrete[67]. High-strength lightweight aggregate concretes (54–73 MPa) were found to exhibit 'outstanding performance' under standard freeze/thaw testing[70]. Prolonged exposure to simulated arctic offshore conditions was needed to cause significant damage, and behaviour was dependent mainly on moisture content and moisture condition of the aggregates.

### Chemical resistance

Most lightweight aggregates are stable since they are fired at approximately 1200°C and do not deleteriously react with alkalis. However, most structural grades of lightweight aggregate concrete are made with dense fines and these should be checked for potential reactivity. In addition, the matrix has a lower free water/cement ratio and higher cement content which should reduce penetration.

### Abrasion resistance

As with normal weight concrete, the resistance to abrasion of lightweight aggregate concrete increases with compressive strength[42]. However, if the matrix of lightweight aggregate concrete is abraded to expose the aggregate particles, it will deteriorate relatively rapidly. Resistance can be improved by combining low density coarse aggregate with a natural fine aggregate, improving the quality of the matrix and the use of surface treatments.

## Water absorption

Most lightweight aggregates exhibit significantly higher water absorption than normal weight aggregates. This results in lightweight aggregate concretes having higher absorption than typical normal weight concretes on a mass basis[42] although the difference is not as large as expected since the aggregate particles in lightweight aggregate concrete are surrounded by a high-quality matrix. It has been shown that water absorption of lightweight concrete does not relate directly to durability[71] and this may be due to the inappropriateness, when comparing concretes, of using absorption values based on mass rather than volume[43].

## Permeability

Permeability or, more correctly, penetrability is the principal factor influencing durability. However, porosity and permeability are not synonymous since size of pores and their continuity must be taken into consideration.

Lightweight aggregate concrete is not necessarily more permeable than normal weight concrete[72] since porous lightweight aggregates are surrounded by a matrix which is less cracked[73] owing to (a) a lack of restraint from aggregate particles, (b) excellent aggregate-matrix bond[74,75] due to the surface characteristics of the particles and pozzolanic action, (c) increased hydration of cement due to improved curing and (d) fewer heat of hydration effects[76]. It has also been suggested that improved permeability could be the result of a coating of dense cement paste around lightweight aggregate particles[77]. Researchers have confirmed this behaviour in measuring lower water permeabilities for air-cured lightweight aggregate concretes compared with normal weight concretes[78,79] and much lower gas permeabilities[79,80]. Tests on concretes cured under a variety of conditions from oven drying to sealed show structural lightweight aggregate concrete to have similar or lower oxygen permeability when compared with normal weight concrete of the same volumetric mix proportions[81]. This behaviour has been confirmed in additional tests[82] (Table 2.10).

**Table 2.10** Durability-related properties of concretes made with lightweight and normal weight aggregates

| Aggregate type | Resistivity (ohm-m) | Water permeability ($10^{-12}$ m$^2$) | Oxygen permeability ($10^{-16}$ m$^2$) |
|---|---|---|---|
| Expanded clay (UK) | 650 | 5 | 0.5 |
| Sintered PFA | 350 | 5 | 0.4 |
| Expanded clay (German) | 600 | 15 | 0.4 |
| Granite | 500 | 85 | 1.0 |

Water and chloride penetration tests carried out on lightweight aggregate concretes with cube strengths from 88 to 104 N/mm$^2$ showed (a) penetrabilities which were low and independent of the porosity of the lightweight aggregate used[83] and (b) an optimum cement content above which penetrability was increased. However, some test results on lightweight aggregate concretes with strengths between 50 and 90 MPa show water and gas permeability to be slightly lower for lightweight aggregate concrete than for normal weight concrete[75].

## Carbonation

Carbonation is the reaction between carbon dioxide in the atmosphere, moisture and the minerals present in the cement paste. This reaction reduces the alkalinity of the concrete and can lead to shrinkage; but, more importantly, if it reaches any embedded metal such as reinforcement, it can promote the processes of corrosion.

Most lightweight aggregate particles are more porous and penetrable than normal weight aggregate particles and this, in common with normal weight aggregates of higher porosities[84], will allow more diffusion of gases such as carbon dioxide. However, if lightweight aggregate particles are well distributed in a good-quality matrix the rate of carbonation should be similar to that for normal weight concrete. Thus, it is essential to ensure that no continuous paths through particles exist between the surface and the reinforcement (the difference between the maximum and mean carbonation depth is approximately $0.5 \times$ maximum size of aggregate[85,86]. For this reason, and from considerations of corrosion, BS 8110[48] requires greater cover to lightweight aggregate concrete. However, tests show the resistance to carbonation of lightweight aggregate concrete to be slightly better to slightly worse than for normal weight concrete of the same cement content or strength[68,80]. Carbonation depth decreases with an increase in cement content and with the use of dense fines[85]. Exposure tests in a polluted atmosphere carried out on a range of concretes[87] indicated that for cement contents above 350 kg/m$^3$ the carbonation depth was small for both lightweight aggregate concrete and normal weight concrete.

### Corrosion of steel

The higher cement content in lightweight aggregate concrete provides a highly alkaline environment to inhibit corrosion. This, together with the increased likelihood of achieving good compaction, reduces the corrosion risk[43]. However, low cement contents (less than approximately 300 kg/m$^3$) may lead to early corrosion. Although BS 8110[48] requires greater cover for lightweight aggregate concrete, performance in practice suggests that this may not be required in most cases.

## *Thermal behaviour*

### Thermal expansion

Coefficients of thermal expansion/contraction for lightweight aggregate concrete are less than for normal weight concrete made with the majority of aggregate types. As shown in Table 2.11, the range of coefficients for lightweight aggregate concrete is similar to that for normal weight concrete made with limestone aggregate[43].

**Table 2.11** Thermal expansion coefficients for lightweight and normal weight concretes

| Type of concrete | Coefficient of thermal expansion ($10^{-6}$/°C) |
|---|---|
| LWAC | 7 to 9 |
| NWC | 10 to 13 |
| NWC with limestone aggregate | 8 to 9 |

It has been suggested and confirmed by testing[79] that the combination of low thermal expansion and high tensile strain capacity results in a lower likelihood of lightweight aggregate concrete cracking under thermal stress, the critical temperature changes being approximately twice those for normal weight concrete. This phenomenon is taken into account by BS 8110[48] where the temperature reduction allowed for lightweight concrete is 1.23 and 2.68 times the value for limestone and gravel concretes respectively.

## Thermal conductivity

Air in the cellular structure of structural lightweight aggregates reduces the rate of heat transfer compared with natural aggregates. Thus the inclusion of lightweight aggregate within a cementitious matrix improves thermal conductivity. For this reason the good thermal properties of lightweight concretes have been used widely to improve thermal insulation in buildings and structures. The thermal conductivities of lightweight aggregate concretes vary primarily due to concrete density, aggregate type and moisture content[88]. Although this variation is known to exist, in the UK it is assumed that there are empirical relationships between conductivity and density for a given moisture content and class of material[89].

For concretes, whether lightweight or normal weight, conductivity values are given for densities from 400 to 2400 kg/m$^3$ and moisture contents of 3% (protected environments) and 5% (exposed environments). These relationships are shown in Figure 2.12 and they appear to conform to best fit curves of the form $k = c^\lambda$, where $k$ is the thermal conductivity, $c$ is a constant and $\lambda$ is the bulk dry density.

The values of $c$ derived from the CIBSE data are 0.085 and 0.093 for moisture contents of 3% and 5% respectively.

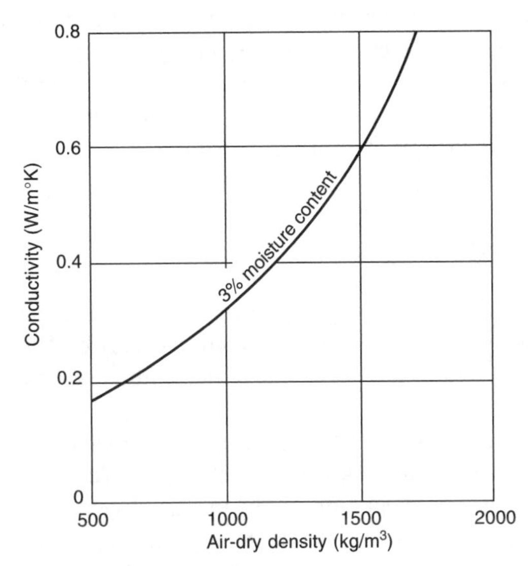


**Figure 2.12** Relationship between thermal conductivity of concrete at 3 per cent moisture content and density.

## Temperature extremes

Lightweight aggregate concretes exhibit excellent behaviour under fire and cryogenic conditions. This is significantly better than normal weight concrete and their high-temperature characteristics are acknowledged in BS 8110[48] which requires lower covers to steel.

Structural lightweight aggregate concretes are suitable for high fire ratings and in this respect are superior to normal weight concretes since they (a) exhibit a lower strength reduction at high temperature, (b) provide more insulation and (c) have lower thermal expansions which result in less spalling.

Benefits can be achieved using lightweight aggregate concrete under cryogenic conditions such as for the storage of liquid gases[90,91] due to its lower penetrability, higher strain capacity and, consequently, greater crack resistance and the enhancement of these properties at low temperatures.

### Acoustic behaviour

The insulation of airborne sound transmission in solid homogeneous units such as walls and floors is improved as their mass increases. Thus, to achieve the same insulation the thickness of lightweight concrete units would need to be greater than normal weight concrete. However, tests have shown[92,93,94] that units made with lightweight concrete containing aggregates having a closely textured surface can perform better than predicted by their mass. For example, the sound insulation of a 200 mm thick wall made with expanded clay aggregate was found to be the same (52 dB) as that for a normal weight concrete wall of the same thickness. Also a 175 mm thick wall made with concrete (density 1600 kg/m$^3$) containing sintered pulverized-fuel ash gave the same insulation as a normal weight concrete wall (density 2400 kg/m$^3$) of the same thickness containing gravel aggregates.

### *Fire resistance*

Structural lightweight aggregate concretes are more resistant to fire than normal weight concrete[42,43] since (a) they experience a lower reduction in strength at high temperature due to the inherent stability of lightweight aggregate, (b) they provide more insulation due to their improved thermal conductivity (enabling cover to be reduced) and (c) they have reduced thermal expansions, causing less spalling.

## 2.4.4 Experience in use

Although the properties of lightweight aggregate concrete can be indicated by data obtained from laboratory testing, the performance of the material can only be demonstrated adequately by its performance in the field and, preferably, using full-scale elements or structures under service conditions.

A survey carried out on the use of lightweight aggregate concrete for marine applications[95] showed its successful use in shipbuilding since the First World War and its long-term durability for bridges, jetties and pontoons in the marine environment. A similar survey[96] also catalogued the successful use of lightweight aggregate concrete in a variety of fixed and floating structures and discussed its advantages and disadvantages. It was shown that lightweight aggregate concrete could withstand the rigorous conditions in marine and offshore situations and gave details of novel projects in which it could be used successfully.

Lightweight aggregate concrete has been used satisfactorily in bridges[97] and samples extracted from various bridge structures in North America showed the excellent performance of the material, due in part to the enhanced bond at the aggregate/paste interface and reduced microcracking as a result of the compatibility between aggregate and mortar[73]. It was concluded from this study that lightweight aggregate concrete had been able to resist satisfactorily several decades of severe exposure conditions. These views were confirmed by another report[98] which concluded that the use of lightweight aggregate concrete in highway structures was likely to result in improved durability and, consequently, reduced maintenance costs.

Long-term observations of a wide variety of lightweight aggregate concrete structures in Japan[99] showed that after 13–20 years of service no reduction was found in strength and modulus of elasticity and no increase in salt penetration and cracking. It was concluded that there was no inferiority compared with normal weight concrete.

Exposure tests carried out to compare lightweight aggregate concrete with normal weight concrete showed that reinforced lightweight aggregate concrete can be durable with regard to carbonation and corrosion for at least 28 years in an aggressive environment[100]. It was concluded that aggregate/cement ratio was more significant than the choice of aggregate and that lightweight aggregate concrete performed better than normal weight concrete at the same aggregate/cement ratio. Different conclusions may have been drawn had the comparisons been made on the basis of concrete grade and if the mixes tested had been made with modern materials.

Experimental work was carried out in the USA on air-entrained concrete slabs made with various types of normal weight aggregate and an expanded shale lightweight aggregate[101]. These were subjected to external exposure conditions including the repeated application of de-icing salt solution. After 202 freeze/thaw cycles and 123 applications of de-icer only 1 out 17 lightweight aggregate concrete slabs was affected (with a slight blemish) while only 4 out of the 27 normal weight concrete slabs survived, these 4 having been surface treated.

A comprehensive examination of 40 structures in the UK[102] indicated that lightweight aggregate concrete is no less durable than normal weight concrete but might be more sensitive to poor workmanship. Problems identified in older structures, such as rate of carbonation and early age microcracking, would be overcome by modern practices of using cement contents of 400 kg/m$^3$ or more, natural fine aggregate and admixtures that allow water reduction.

## References


[1] Crosswell, S.F. (1994) No-fines concrete. *Fulton's Concrete Technology,* 7th edn, Ed. Addis, B.J., Portland Cement Institute, Midrand, South Africa, 291–296.

[2] Brook, K.M. No-fines concrete, *Concrete*, **16**, No. 8, 1982, 27–28.

[3] BS 1881: Part 113: Method for making and curing no-fines test cubes.

[4] McIntosh, J.D., Botton, R.H. and Muir, C.H.D. (1956) No-fines concrete as a structural material. *Proc. ICE*, Part 1, No. 6, 677–694.

[5] Malhotra, V.M. (1976) No-fines concrete – its properties and applications. *JACI*, **73**, No. 11, 628–644.

[6] Jain, O.P. (1966) Proportioning no-fines concrete. *Indian Concrete Journal*, **40**, No. 5, May, 182–189.

[7] Neville, A.M. (1995) *Properties of concrete*, 4th edn, Longmans, p. 713.

[8] BS EN 678: Determination of the dry density of autoclaved aerated concrete.

[9] BS EN 679: Determination of the compressive strength of autoclaved aerated concrete.

[10] BS EN 680: Determination of the drying shrinkage of autoclaved aerated concrete.

[11] BS EN 1351: Determination of flexural strength of autoclaved aerated concrete.

[12] Cox, L. and van Dijk, S. (2002) Foam concrete: a different kind of mix. *Concrete*, **36**, No. 2, Feb. 54–55.

[13] RILEM (1993) *Autoclaved aerated concrete*: *Testing and design*, E & FW Spon, London.

[14] BS EN 1352: Determination of static modulus of elasticity under compression of autoclaved aerated concrete or lightweight aggregate concrete with open structure.

[15] Building Research Establishment (1989) Autoclaved aerated concrete, *Digest 342*, Garston, UK.

[16] BS EN 989: Determination of the bond behaviour between reinforcing bars and autoclaved aerated concrete.

[17] BS EN 990: Test methods for verification of corrosion protection of reinforcement in autoclaved aerated concrete and lightweight aggregate concrete of open structure.

[18] BS EN 1353: Determination of moisture content of autoclaved aerated concrete.

[19] BS EN 1355: Determination of creep strains under compression of autoclaved aerated concrete or lightweight aggregate concrete of open structure.

[20] Aldridge, D. (2002) Foamed concrete. *Concrete*. **36**, No. 2, 20–22.

[21] RILEM (1978) *Functional classification of lightweight concretes: Recommendation LC2*. 2nd edition.

[22] RIEM (1975) Terminology and definitions of lightweight concrete, Recommendation LCI, 1st edition.

[23] CEB (1978) *CEB-FIP Model Code for Concrete Structures*, Vol. II, 3rd edn. CEB, Paris.

[24] Eurocode No. 2 Design of concrete structures, Part I. General rules and rules for buildings, Final text, October, (1991), EEC.

[25] Eurocode No. 2 Part 1–4 (1992) The use of lightweight aggregate concrete with closed structure, Draft, June, EEC.

[26] AS 3600: Concrete structures, Standards Association of Australia, Sydney, 1988.

[27] NS 3473: Concrete structures, Norwegian Council for Building Standardisation, 1989.

[28] ACI 318-89, Requirements for reinforced concrete, ACI, Detroit, 1989.

[29] Standard specification for design and construction of concrete structures, Part 1 (Design), Japan Society of Civil Engineers, Tokyo, 1986.

[30] JASS 5 (revised 1979): Japanese architectural standard for reinforced concrete, Architectural Institute of Japan, March 1982.

[31] SNiP 2.03.01-84: Manual for the design of concrete and reinforced concrete structures from normal weight concrete and lightweight concrete without prestressing the reinforcement, USSR.

[32] ENV 206. Concrete-Performance, production, placing and compliance criteria, Final draft, February 1989 (BSI Document 89/11639).

[33] Bremner, T.W. and Newman, J.B. (1992) Microstructure of low density concrete aggregate, *Proc 9th Congress of FIP*, Vol. 3, Stockholm, June, FIP, Slough.

[34] BS 877 (1973) Specification for foamed or expanded blastfurnace slag lightweight aggregate for concrete, Part 2. BSI, London.

[35] BS 1047 (1983) Specification for air-cooled blastfurnace slag aggregate for use in construction. BSI, London.

[36] BS 1165 (1985) Specification for clinker and furnace bottom ash aggregates for concrete. BSI London.

[37] BS 3797 (1976) Specification for lightweight aggregates for concrete, Part 2. BSI, London 18 BS 3681: Methods for sampling and testing of lightweight aggregates for concrete, Part 2: 1976, BSI, London.

[38] BS EN 13055–1. Lightweight aggregates, Part 1. Lightweight aggregates for concrete, mortar or grout.

[39] BS 3681 (1976) Methods for sampling and testing of lightweight aggregates for concrete: Part 2. BSI, London.

[40] Newman, J.B. (1993) Properties of structural lightweight aggregate concrete. In *Structural lightweight aggregate concrete*, 1st Edition, Ed. Clarke, J.L., Blackie Academic & Professional, 1993, pp 19–44.

[41] Lydon, F.W., *Concrete Mix Design*, 2nd edition, Applied Science Publishers, 1982.

[42] ACI Committee 213, Guide for structural lightweight aggregate concrete, *Journal ACI*, 64, No. 8, 1967, 433–69.

[43] I. Struct. E./Concrete Society, *Guide to the Structural Use of Lightweight Aggregate Concrete*, I. Struct. E; London, October 1987.

[44] ACI Committee 211-2-91. Recommended practice for selecting proportions for structural lightweight aggregate concrete, 1992.

[45] BS 1881 (slump test): Method for determination of slump, Part 102: 1983, BSI, London.
[46] BS 1881 (flow test). Method of determination of flow, Part 108: 1984, BSI, London.
[47] FIP Manual of lightweight aggregate concrete, 2nd Edition, Surrey University Press, FIP, 1983.
[48] BS 8110: Structural use of concrete: Part 2. Code of practice for special circumstances, BSI, London, 1985.
[49] Clarke, J.L. and Birjandi, F.K. (1990) Shear resistance of lightweight aggregate concrete slabs, *Magazine of Concrete Research*, **42** No. 152, 171–6.
[50] Clarke, J.L. and Birjandi, F.K., Shear resistance of lightweight aggregate concrete slabs, *Magazine of Concrete Research*, **42** No. 152, 171–6.
[51] Hanson, J.A. Strength of structural lightweight aggregate concrete under combined stresses. Portland Cement Association, *Development Bulletin D61*, 1963.
[52] Hobbs, D.W., Newman, J.B. and Pomeroy, C.D. Design stresses for concrete subjected to multiaxial stress, *The Structural Engineer*, **55** (No. 4; 1977), 151–64.
[53] Jensen, J.J. Impact of falling loads on submerged structures, *Proc of Int. Symp. on Offshore Structures*, Rio de Janeiro, Brazil, October 1979, pp. 1215–32.
[54] Jensen, J.J. and Hoiseth, K., *Impact of dropped objects on lightweight concrete*, Nordic Concrete Research, Publication No. 2, Nordic Concrete Federation, Oslo, December 1983, pp. 102–13.
[55] Bailey, J.H., Bentley S., Mayfield, B. and Pell, P.S., Impact testing of fibre-reinforced stair treads, *Magazine of Concrete Research*, **27** (No. 92; September 1975).
[56] Grube and Knop, personal communication.
[57] Norwegian contractors, personal communication.
[58] Newman, J.B. Concrete under complex states of stress, Chapter 5, *Developments in Concrete Technology-1*. (ed. Lydon, F.D.), Applied Science Publishers, 1979.
[59] Clarke, J.L. and Birjandi, F.K. *The Bond of Reinforcement in Lightweight Aggregate Concrete*, C&CA Services Report 1.039.00.2, BCA.
[60] Tepfers, R. and Kutti, T. Fatigue strength of plain, ordinary and lightweight concrete, *Journal of* ACI. Proc. 76 (No. 5; May 1979), 635–52.
[61] Waagaard, K. and Kepp, B. Fatigue of high strength lightweight aggregate concrete, *Proc. of Symp. on Utilisation of High Strength Concrete*, Stavanger, Norway, June 1987, pp. 291–306.
[62] Sparks, P.R. The influence of rate of loading and material variability on the fatigue characteristics of concrete, A.C.I. Special Publication SP-75, *Fatigue of Concrete Structures*, Detroit, 1982, pp. 331–41.
[63] CEB, *Fatigue of Concrete Structures*, State of the Art Report, CEB Bulletin d'Information, No. 108, Lausanne, 1988, pp. 300.
[64] Saito, M. Tensile fatigue strength of lightweight concrete, *Journal of Cement Composites and Lightweight Concrete*, **6** (No. 3; 1984), 143–9.
[65] Cornellissen, H.A.W. *Fatigue of Concrete, Part 4, Tensile and Tensile-Compressive Stresses*, IRO-MATS/CUR Report, 1987.
[66] Howells, H. and Raithby, K.D. Static and repeated loading tests on lightweight prestressed concrete bridge beams, TRRL Laboratory Report 804, 1977, Transport and Road Research Laboratory, Crowthorne, UK.
[67] Klieger, P. and Hanson, J.A. Freezing and thawing tests of lightweight aggregate concrete, *Journal of ACI*, **57** (No. 7; January 1961), 779–96.
[68] Dhir, R.K. Durability potential of lightweight aggregate concrete, *Concrete* (April 1980), 10.
[69] Lydon, F.D. Some freeze–thaw test results from structural lightweight aggregate concrete, *Precast Concrete*, **12** (1981).
[70] Whiting, D. and Burg, R. Freezing and thawing durability of high strength lightweight concretes, SP 126–4, pp. 83–100.
[71] Hanson, J.A. Replacement of lightweight aggregate fines with natural sand in structural concrete, *Journal of ACI, Proc.* **61** (No. 7; July 1964), 779–94.
[72] Soroka, I. and Jaegermann, C.H. Permeability of lightweight aggregate concrete, Paper No. 2.3.4, 2nd CEB/RILEM Symp. on Moisture Problems in Buildings, Rotterdam, September 1974.

[73] Hornain, H. and Regourd, M. Microcracking of concrete, *Proc. 8th Int. Congress on the Chemistry of Cement*, Rio de Janeiro 1986, Vol. 5, pp. 53–9.

[74] Holm, T.A., Bremner, T.W. and Newman, J.B. Lightweight aggregate concrete subject to severe weathering, *Concrete International-Design and Construction*, 6 (No. 6; June 1984), 49–54.

[75] Bamforth, P.B. The properties of high strength lightweight concrete, *Concrete*, 21 (No. 4; April 1987), 8–9.

[76] Lydon, F.D. and Al-Momen, M.H. The effects of moderate heat of hydration on some early and later age properties of concrete, *RILEM Int. Conf. on Concrete of Early Ages*, Paris, April 1982, p. 97.

[77] Nishi *et al.* Watertightness of concrete against sea water, *Journal of Central Res. Lab.*, Onoda Cement Co., Tokyo, 32 (No. 104; 1980), 40–53.

[78] Dhir, R.K., Munday, J.G.L. and Cheng, H.T. Lightweight concrete durability, *Construction Weekly* (25 August 1989), 11–13.

[79] Bamforth, P.B. The Performance of High Strength Lightweight Concrete using Lytag Aggregate, Taywood Engineering Research Report No. 014H/85/282, July 1985.

[80] Lydon, F.D. and Mahawish, A.H. Strength and permeability results for a range of concretes, *Cement and Concrete Research*, 19 (1989), 366–76.

[81] Ben-Othman, B. and Buenfeld, N.R. Oxygen permeability of structural lightweight aggregate concrete, *Protection of Concrete* (eds Dhir, R.K. and Green, J.W.), E & FN Spon, 1990, pp. 725–36.

[82] Denno, M.G. *The durability of high strength lightweight aggregate concrete*, PhD Thesis, University of London, 1996.

[83] Zhang, Min-Hong and Gjorv, O.E. Permeability of high-strength lightweight concrete, *ACI Materials Journal*, 88 (No. 5; September/October 1991), 463–9.

[84] Collins, R.J. Carbonation-comparison of results for concretes containing PFA, cementitious slag, or alternative aggregates, *Materials Science and Technology*, 3 (December 1987), 986–92.

[85] Bandyopadhyay, A.K. and Swamy, R.N. Durability of steel embedded in structural lightweight concrete, RILEM, *Materiaux et Constructions*, 8 (No. 45, 1975).

[86] Schulze, W. and Gunzler, J. Corrosion protection of the reinforcement in lightweight concrete, *Proc. Ist Int. Congress on Lightweight Concrete*, London, May (1968), C&CA, 1970.

[87] Grimer, F.J. Durability of steel embedded in lightweight concrete, *Concrete*, 1 (No. 4; April 1967), 125–31.

[88] FIP State of Art Report, Principles of Thermal Insulation with Respect to Lightweight Concrete, FIP, Report FIP/8/1, C&CA, Slough, England, 1978.

[89] CIBSE Guide A3, Thermal Properties of Building Materials, CIBSE, London, 1986.

[90] Bamforth, P.B., Murray, W.T. and Browne, R.D. The application of concrete property data at cryogenic temperature to LNG tank design, *2nd Int. Conf. on Cryogenic Concretes*, Amsterdam, October 1983 (organised jointly by UK and Dutch Concrete Societies).

[91] Berner, D., Gerwick, B.C. and Polivka, M. Prestressed lightweight concrete in the transport of cryogenic liquids, *Oceans '83 Conference*, MTS and IEEE, San Francisco, August/September 1983.

[92] Schule, W. Functional properties of concrete and concrete structures, *Zement-Taschenbuch, 1968/68*, Verein Deutscher Zementwerke EV, Wiesbaden, Bauverlag GmbH, 1968, pp. 331–56.

[93] Veronnaud, L. Physical properties of lightweight concrete, *Revue des Materiaux de Construction et des Travaux Publics* (No. 62; November 1970), 338–42.

[94] Forder, C. Lightweight concrete's place in the insulation spectrum, *Concrete* (January 1975).

[95] FIP, State of Art Report, Lightweight Aggregate Concrete for Marine Structures, FIP, 1982.

[96] Concrete Society, Technical Report No. 16, Structural Lightweight Aggregate Concrete for Marine and Offshore Applications, The Concrete Society, May 1978, 29 pp.

[97] Raithby, K. and Lydon, F.D. Lightweight concrete in highway bridges, *Int. Journal of Cement Composites and Lightweight Concrete*, 2 (No. 3; May 1981), 133–46.

[98] Concrete Society, Design and Cost Studies of Lightweight Concrete Highway Bridges, Report of an investigation by the Lightweight Concrete Committee, April 1985.

[99] Ohuchi, T. *et al.* Some long term observation results of artificial lightweight aggregate concrete for structural use in Japan, *RILEM/ACI Int. Symp. on Long Term Observation of Concrete Structures*, Budapest, September 1984, pp. 273–82.

[100] Nicholls, I.C. and Longland, J.T. The Durability of Reinforced Lightweight Aggregate Concrete, Dept of Environment, Building Research Establishment, Nore No. 80187, August 1987, 32 pp.

[101] Walsh, R.V. Restoring salt-damaged highway bridges, *ASCE, Civil Engineering*, 37 (No. 5. May 1967), 57–9.

[102] Mays, G.C. and Barnes, R.A. (1991) The performance of lightweight aggregate concrete structures in service. *The Structural Engineer*, **67** No. 20, October, 351–61.

# 3

# High strength concrete

*Bill Price*

## 3.1 Aims and objectives

High Strength Concrete is a relatively recent development in concrete technology made possible by the introduction of efficient water-reducing admixtures and high strength cementitious materials. This chapter will discuss the materials technology underlying the development of high strength concrete, examine the selection of optimum constituent materials and consider the concrete mix design. The properties of both fresh and hardened high strength concrete will be discussed, highlighting any particular differences in behaviour compared to more conventional concretes. Finally, the production and use of high strength concrete on-site, illustrated by examples worldwide, will be examined.

It is intended that this will lead to an understanding of the potential benefits and limitations of high strength concrete, together with the expertise required to produce and use the material in a practical and effective manner.

## 3.2 Introduction

When considering high strength concrete (HSC) we must first define what we mean by 'high strength'. The perception of what level of compressive strength constitutes 'high strength' has been continually revised upwards over the past 20 years or so (FIP-CEB, 1990) and may well continue to rise in the near future. A simple definition would be 'concrete with a compressive strength greater than that covered by current codes and standards'. In the UK this would include concrete with a characteristic compressive strength of 60 MPa or more, but in Norway the design code already includes concrete with characteristic cube strengths up to 105 MPa (Helland, 1996) as does the forthcoming Eurocode (CEN, 2002), so even this simple definition is not really adequate.

Therefore for the purposes of this chapter, concrete with a characteristic concrete (cube) strength greater than 80 MPa will be considered as 'high strength'. Furthermore we are concerned here with essentially conventional type concretes that can be used with currently accepted construction techniques and not more specialized cement-based materials such as reactive powder concretes (RPC) (O'Neil *et al.*, 1996) or heavily fibre reinforced concretes (Aarup, 1994; Naaman, 2000).

It should be noted that there is an increasing tendency to consider the two terms 'high strength' and 'high performance' as being synonymous. In the author's view, however, high strength concrete is only **one** of the extensive possible range of high performance concretes (which could include high workability concretes, lightweight concretes or even concretes with enhanced durability) and the two terms should be kept separate. The remainder of this chapter considers high strength concrete alone.

## 3.3 Materials technology of HSC

In order to achieve high compressive strength, it is important to understand the factors that govern the strength of concrete, i.e.:

- The properties of the cement paste
- The properties of the transition zone between the paste and the aggregate
- The properties of the aggregate
- The relative proportions of the constituent materials.

All these factors must be optimized in order to make significant increases in compressive strength. Throughout this chapter the term 'cement' will be used to describe all the cementitious materials in the concrete and not just Portland cement alone.

### 3.3.1 Paste properties

In conventional concrete technology, the strength of the paste is a function of its water/cement ratio. This is true also for high strength concrete but it is also the effect of the porosity within the paste, the particle size distribution of the crystalline phases and the presence of inhomogeneities within the hydrated paste that must be considered in detail.

A reduction in water/cement ratio will produce a paste in which the cementitious particles are initially closer together in the freshly mixed concrete. This results in less capillary porosity in the hardened paste and hence a greater strength. This reduced capillary porosity also favours the formation of fine-textured hydration products that have a higher strength than the coarser equivalents. The capillary porosity can also be reduced by optimizing the particle size distribution of the cementitious materials in order to increase the potential packing density. Special high strength cements are available and the inclusion of finely divided reactive materials such as silica fume will also contribute to an increase in packing density and reduced capillary porosity.

It should be noted that even commercially available high strength concretes have free water/cement ratios as low as 0.22 (Burg and Ost, 1992). This is well below the theoretical minimum for full cement hydration. However, the hydration of the cementitious particles within the paste is sufficient to 'glue' together the unhydrated cores of the particles and to reduce the interstitial porosity between these hydrated particles.

The role of superplasticizers in enabling workable concretes to be produced at very low water/cement ratios (and without the need for excessively high cement contents) is critical. Furthermore the effect of superplasticizers in preventing the flocculation of Portland cement particles and distributing material such as silica fume homogeneously through the freshly mixed concrete leads to a reduction in inhomogeneities within the paste and hence improved paste strength. The strength of the paste will be limited by the flaws that form the weakest link, be they inhomogeneities or capillary pores. In order to improve the strength of the paste as a whole, all such flaws must be minimized.

## 3.3.2 Transition zone properties

When fracture surfaces of failed conventional concretes are examined, it is often observed that the failure has occurred, either with the paste itself or, more often, at the interface between the paste and the coarse aggregate particles. Whilst it is possible to increase the strength of the paste significantly as described above, if the transition zone to the aggregate is weak, the strength of the concrete will not increase commensurately. In conventional (say, 40 MPa) concretes, this transition zone is quite large and is characterized by a high porosity and large crystalline hydration products (such as Portlandite $Ca(OH)_2$). Reducing the water/paste ratio and the incorporation of silica fume into the concrete both contribute to reducing the width and improving the strength of the transition zone (Mindess *et al.*, 1994). The rapid conversion of $Ca(OH)_2$ to CSH by silica fume is thought to be of particular importance. Reduced bleeding within the paste also reduces the potential for accumulation of water around aggregate particles.

## 3.3.3 Aggregate properties

When the transition zone between the paste and the aggregate is improved the transfer of stresses from the paste to the aggregate particles becomes more effective. Consequently the mechanical properties of the aggregate particles themselves may be the 'weakest link' leading to limitation of achievable concrete strength. Fracture surfaces in HSC often pass through aggregate particles rather than around them.

Crushed rock aggregates are generally preferred to smooth gravels as there is some evidence that the strength of the transition zone is weakened by smooth aggregates (Aitcin and Mehta, 1990). The aggregate should have a high intrinsic strength and granites, basalts and limestones have been used successfully, as have crushed glacial gravels.

During the crushing process, aggregate particles may be severely microcracked. The number of microcracks will be greater in larger particles, consequently it is common practice to use smaller particles (10–14 mm nominal size) for high strength concrete (Mehta and Aitcin, 1990a).

It is assumed that small aggregate particles will contain less internal flaws and hence produce a higher concrete strength. It must be stressed that the selection of appropriate sources of aggregate is much more critical for high strength concrete than for conventional concretes.

## 3.4 Materials selection and mix design

It should be recognized that there is no single or unique composition for high strength concrete. HSC can be made with a range of materials and mix designs which will produce slightly differing properties.

### 3.4.1 Cements

HSC can be produced with most available Portland cements, but those cements that are particularly coarsely ground are usually unsuitable (Mehta and Aitcin, 1990a). Special cements have been developed for HSC in Norway which are more finely ground and with lower tricalcium aluminate ($C_3A$) content (Helland, 1996) but elsewhere normal commercial products are generally employed.

Silica fume is almost ubiquitous in HSC as it has approximately three times the cementing efficiency (on a weight for weight basis) as Portland cement. This facilitates the achievement of high strength without excessive cement contents. Silica fume is available in Europe (and elsewhere) both in the form of a water-based slurry and as a densified powder (Concrete Society, 1993). To be effective it should always be used in conjunction with a superplasticizer. It is usually incorporated into concrete at 5–15 per cent by weight of total binder.

PFA and GGBS have also been used successfully together with Portland cement and silica fume (Burg and Ost, 1992), albeit at lower levels than in conventional structural concrete. The reasons for use include improvements in pumping performance and reducing in heat evolution. The use of metakaolin (a highly reactive pozzolan) has also been proposed for HSC (Calderone *et al.*, 1994; Ryle, 2000) although not yet used very extensively. As different sources of cementitious materials may interact with different efficiency, trials to establish the optimum combination and sources of materials may often be required.

All cements should comply with appropriate national or international standards.

### 3.4.2 Admixtures

The role of admixtures is much more significant in HSC than for more conventional concretes. To produce workable concretes at very low levels of water/cement ratios (typically below 0.30), without needing unacceptably high cement content, requires the use of superplasticiszers. Melamine-based, naphthalene-based and polycarboxylate ether-based superplasticizers have been used successfully, either individually or in combination. The dosage rates of the superplasticizers can be very high (up to 3 per cent by weight of cement) in order to achieve the required workability. It should be noted that there is generally a saturation dosage of superplasticizers above which no further increase in workability will occur. This can easily be determined by the use of a flow cone (Aitcin *et al.*, 1994). The efflux time is measured at the same free water/cement ratio for a series of admixture dose rates. This will enable the maximum effective level of admixture addition (i.e. minimum efflux time) to be identified (see Figure 3.1).

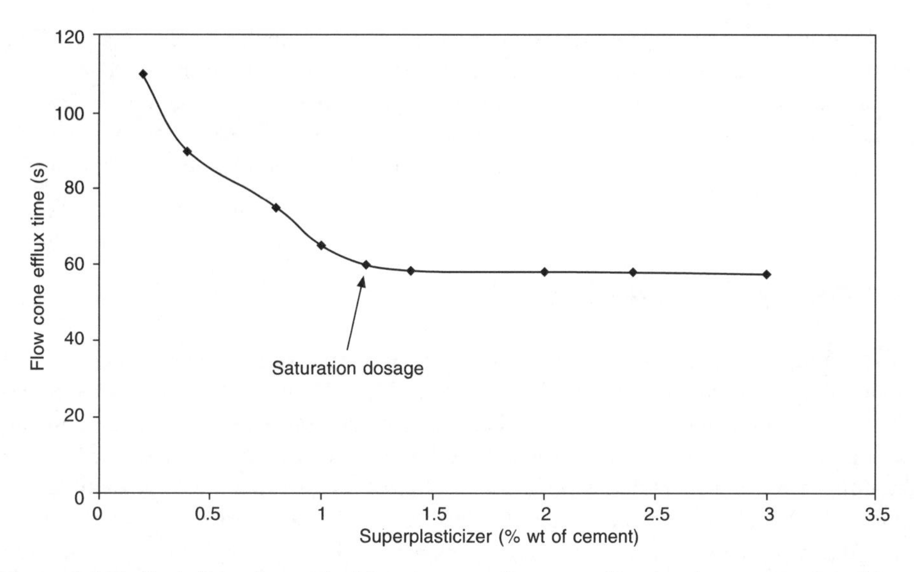


**Figure 3.1** Idealized effect of superplasticizer dosage on flow cone efflux time (at constant w/c ratio).

Lignosulphonate-based plasticizers may also be combined with melamine superplasticizers in order to extend their workability retention. Compatibility between different admixtures used in combination as well as compatibility between admixtures and different cement types, must be considered when materials are selected (again flow cone tests may be useful).

## 3.4.3 Aggregates

Fine aggregates for HSC should be selected to reduce the water demand. Rounded particles are thus preferred to crushed rock fines where possible. The silt, clay and dust content of both fine and coarse aggregates should be kept as low as possible.

As most HSC concrete mixes contain a large amount of fine material in the cement (often greater than 500 kg/m$^3$), it is accepted practice to utilize slightly coarser gradings of fine aggregate than is normal for conventional structural concrete (Aitcin, 1998). The finest fractions of the fine aggregate are no longer essential to increase workability or prevent segregation; a coarser grading (fineness modulus 2.7 to 3.0 or BS 882 Class C) (British Standards Institution, 1992) is therefore appropriate. The gradings curve of the fine aggregate should, however, generally be smooth and free of gap grading to optimize the water demand.

The requirements for coarse aggregates have been examined earlier. However, the particle shape should ideally be equidimensional (i.e. not elongated or flaky) and the grading should once again be smooth with no gaps in the grading between fine and coarse fractions. A maximum aggregate size of 10–14 mm is usually selected (Mehta and Aitcin, 1990a) although aggregates up to 20 mm may be used if they are strong and free of internal flaws or fractures. This can, however, only be evaluated from trial mixes.

As the influence of aggregates on the performance of high strength concrete is of particular significance it may not be possible to achieve the required strength on a project using local aggregate supplies alone. Importation of aggregate supplies or blending materials from a number of sources may be required in order to optimize performance. During the construction of HSC offshore platforms in Norway, a consistent fine aggregate grading was maintained by blending up to eight separate size fractions (Ronneberg and Sandvik, 1990). However, in most cases single sources or blends of two or three sources will be satisfactory.

## 3.4.4 Concrete mix design

Whilst a number of studies have considered the development of a rational or standardized method of concrete mix design for HSC (de Larrard, 1999; Mehta and Aitcin, 1990b), no widely accepted method is currently available. The main requirements for successful and practical HSC are a low water/cement ratio combined with high workability and good workability retention characteristics. In the absence of a standard mix design method, the importance of trial mixes in achieving the desired concrete performance is increased.

The following factors should, however, be considered when designing a high strength concrete mix (see Table 3.1):

- The appropriate free water/cement ratio should be selected either from experience or by reference to published data. This will typically be in the range 0.25–0.30.
- The cement composition should be selected to maximize strength and other performance requirements. At its simplest this will be Portland cement blended with 5–10 per cent silica fume.
- Proportion coarse and fine aggregates to give a smooth overall grading curve in order to keep the water demand low. The proportion of fine aggregate is generally around 5 per cent lower (as a proportion of total aggregate) than for normal strength concrete. Care must be taken, however, not to make the mix too deficient in fine aggregate, particularly where the concrete is to be pumped.
- Use the saturation dosage of admixture (or admixtures), determined with a flow cone, to produce workability. It should be noted (see section 3.5) that most HSC is also high workability concrete, of, say, 600 mm flow table spread.

Trial mixes should be made and strength, workability and workability retention measured. Modifications can then be made to the mix to optimize the concrete's performance.

**Table 3.1** Commercial HSC mix designs from North America (data from Burg and Ost, 1992)

| | 1 | 2 | 3 | 4 | 5 |
|---|---|---|---|---|---|
| Cement (kg/m$^3$) | 564 | 475 | 487 | 564 | 475 |
| Fly ash (kg/m$^3$) | – | 59 | – | – | 104 |
| Microsilica (kg/m$^3$) | – | 24 | 47 | 89 | 74 |
| Coarse agg. (kg/m$^3$) | 1068 | 1068 | 1068 | 1068 | 1068 |
| Fine agg. (kg/m$^3$) | 647 | 659 | 676 | 593 | 593 |
| Water (L/m$^3$) | 158 | 160 | 155 | 144 | 151 |
| Superplasticizer (L/m$^3$) | 11.61 | 11.61 | 11.22 | 20.12 | 16.45 |
| Retarder (L/m$^3$) | 1.12 | 1.04 | 0.97 | 1.47 | 1.51 |
| Free water/cement ratio | 0.281 | 0.287 | 0.291 | 0.220 | 0.231 |
| 90-day cylinder strength (MPa) | 86.5 | 100.4 | 96.0 | 131.8 | 119.3 |

# 3.5 Properties of HSC

## 3.5.1 Fresh concrete

Normal practice (particularly in North America, but also in Europe), is to produce high workability HSC. Slumps in excess of 200 mm are common, particularly where HSC is used in areas of congested reinforcement. In most cases, however, the flow table is a more appropriate way of assessing the workability of the concrete on site than is the slump test.

It has been found that HSCs often appears to require more effort to compact than a more conventional concrete of a similar slump (often termed 'sticky'). This is probably due to a combination of a high cement content and high levels of admixture. HSC is essentially thixotropic in that whilst it flows easily under the influence of vibration, flow ceases once the vibration is removed.

HSC is also characterized by significantly lower bleeding than more conventional concretes. If the concrete contains a high silica fume content (> 10 per cent of total cement), bleeding may be eliminated altogether. The absence of bleeding can lead to difficulties with finishing and also increase the importance of effective early curing in order to prevent plastic cracking.

As the total content of cementitious materials in HSC is typically high (often in excess of 500 kg/m$^3$), the heat of hydration of the concrete would also be expected to be high. In fact, whilst the heat generation is higher than for lower strength concrete, it does not rise in proportion to cement content. The low water content of a typical HSC may not enable all the cementitious material to hydrate fully. Consequently the inhibition of continued hydration also acts to limit the generation of heat. However, if HSC is used in massive sections, the normal precautions will still be required to minimize thermal cracking (Bamforth and Price, 1995).

## 3.5.2 Hardened concrete

### *Strength*

HSC is obviously characterized by high ultimate compressive strength. When measured on standard water cured cubes, however, the rate of early strength gain is similar to that of lower strength concrete. If metakaolin is used, the strength gain may be slightly more rapid than for microsilica-based HSC (Calderone *et al.*, 1994). In some cases when retarding admixtures or very high superplasticizer levels are used the early strength gain may even be lower than normal.

Another characteristic of HSC (particularly when containing silica fume) is that continued strength gain beyond 28 days is often very small, and this is even more so when *in-situ* strength is considered. However, long-term strength gain is dependent on the type and combination of cementitious materials in the concrete.

The build-up of heat within structural elements accelerates the hydration of the cement and hence the development of strength. Using temperature-matched curing techniques to monitor the development of *in-situ* strength has indicated that *in-situ* strength can rise rapidly from about 8 hours after casting (see Figure 3.2). Greater than 100 MPa has been recorded at an age of 24 hours (Price, 1996).

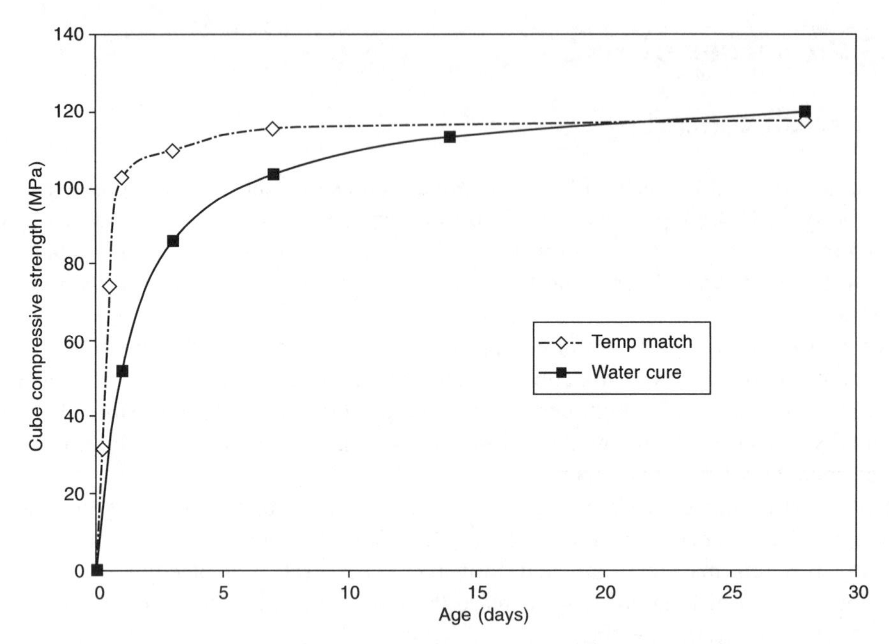


**Figure 3.2** Strength development of a high strength concrete slab (after Price, 1996).

As with conventional normal strength concrete, the tensile strength of HSC increases as compressive strength rises. However, care should be taken in extrapolating existing relationships between compressive and tensile strengths, as the tensile strength does not increase pro-rata with compressive strength (Ahmad and Shah, 1985). Factors such as aggregate shape and composition will also have an influence.

## *Elastic modulus and stress–strain behaviour*

The elastic modulus of HSC is generally higher than that of normal strength concrete. The increase of elastic modulus is not pro-rata to increases in compressive strength and some existing relationships between these properties are thought to overestimate the elastic modulus at compressive strengths over 100 MPa (American Concrete Institute, 1992). The effects of the shape and mineralogy of the coarse aggregate also has a significant influence over elastic modulus (Aitcin and Mehta, 1990). Stiffer aggregates such as siliceous flints etc. will achieve a much higher modulus than softer granites and limestones at a given level of compressive strength (Nilsen and Aitcin, 1992). It is recommended that if elastic modulus is an important factor in the design, the modulus of the concrete is actually measured on the concrete proposed for use in the project (Concrete Society, 1998).

It is generally recognized that the stress–strain behaviour of HSC differs from that of normal strength concrete. In HSC, the ascending part of the stress–strain curve becomes steeper and more linear, remaining linear to a higher proportion of the ultimate stress (see Figure 3.3). The increased compatibility in elastic modulus between a high strength binder and aggregate particles reduces the degree of microcracking around the aggregate during loading. This in turn results in increased linearity of the ascending limb. The strain at maximum stress is slightly higher than for normal strength concrete but the descending

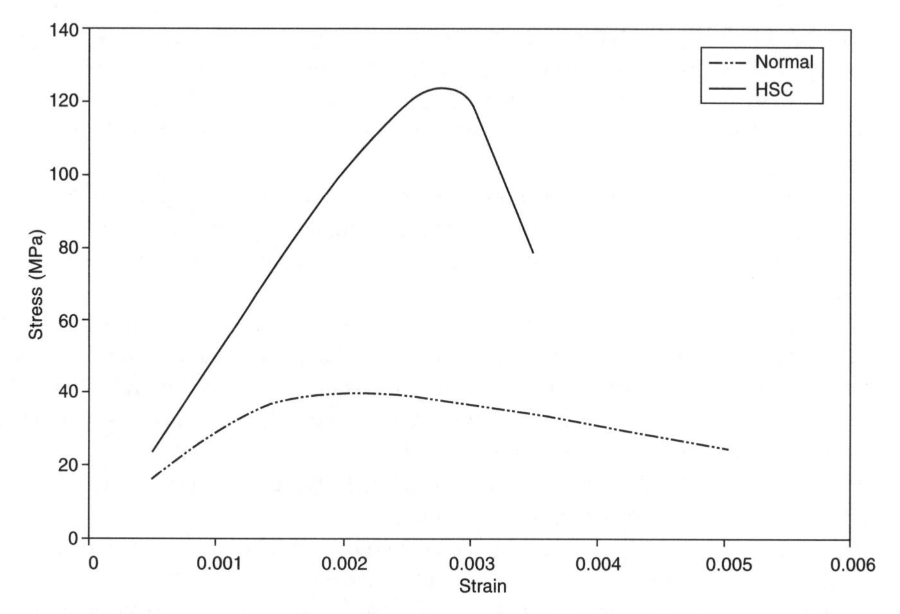


**Figure 3.3** Idealized comparison of stress–strain curves of high strength and normal strength concrete.

portion of the curve is, however, significantly steeper. The brittle behaviour of HSC has implications in terms of secondary reinforcement details, for ensuring ductile behaviour of structures (Concrete Society, 1998).

## *Creep and shrinkage*

The creep of HSC (expressed as either specific creep or creep coefficient) appears to be significantly lower than that of normal strength concrete. However, the available information on creep of HSC is relatively limited and further research is required.

Relatively little information is available on the drying shrinkage characteristics of HSC. However, in general due to the low initial water content of the concrete and its low intrinsic vapour permeability, drying shrinkage is thought to be lower than for normal strength concrete.

On the other hand, autogenous shrinkage of HSC can be significant. Autogenous shrinkage is a reduction in volume occurring without loss of water to the atmosphere (Aitcin, 1999). The combination of low initial water content and silica fume leads to self-desiccation when sufficient water for continued hydration is not available and hence shrinkage. Autogenous shrinkage in HSC is also more rapid than in normal strength concrete. In one study, the autogenous shrinkage of a 100 MPa concrete was 110 microstrain, compared with only 40 microstrain for a 40 MPa concrete (de Larrard and Le Roy, 1992). In certain circumstances, the high autogenous shrinkage may be a significant influence on the proposed design.

### Durability

It is not possible to give a detailed description here of the durability aspects of HSC. However, this can be summarized as follows:

- The low free water/cement ratio required to produce HSC also generally confers enhanced durability on the concrete. When combined with the use of silica fume (present in most, if not all, HSC), significant reductions in water permeability and chloride ingress have been observed (Helland, 1996). HSC is often used in parking structures in North America in order to prevent deterioration resulting from the extensive use of de-icing salts.

- Other areas, in which HSC has found applications on durability grounds, are as a consequence of its improved abrasion resistance compared to normal strength concrete (Gjorv *et al.*, 1990) and its increased resistance to attack by aggressive chemicals (Nischer, 1995).
- When used in severe freezing and thawing conditions, some air entrainment should still be used even though adequate protections will be achieved at a lower air content than that required by normal strength concrete (Hammer and Sellevold, 1990). Air entrainment is more difficult in low water content, superplasticized pastes and higher than normal amounts of air-entraining admixtures may be needed (Mailvaganam, 1999) to establish a satisfactory air void system. In typical UK conditions, non-air-entrained HSC will be resistant to frost damage.
- Although HSCs generally contain high cement contents, the presence of silica fume is thought to prevent ASR (Gudmundsson and Olafsson, 1996). The very low internal moisture content of the concrete will also prevent the swelling and expansion of any gel formed if potentially reactive aggregates are used.

## 3.6 Production and use of HSC

### 3.6.1 Production

The key to successful production of HSC is maintaining a consistent and low water/cement ratio together with effective mixing. HSC has been produced successfully in both wet batch and dry batch plants but in all cases, stringent control of all sources of water in the mix is critical. These include:

- Added mix water
- Water in liquid admixtures or silica fume slurry
- Free moisture on fine and coarse aggregates. It should be noted that small changes in the moisture content of the fine aggregate have a proportionately greater effect on water/cement ratio and hence strength of HSC, than it does for normal strength concrete.
- Other sources of water such as washout or cleaning water in mixers and transport vehicles.

In order to maximize the workability retention of HSC, it was formerly common practice in the UK and elsewhere to add only a part of the total admixture dose to the concrete at the batching plant. When the concrete arrived on-site, the workability was assessed (usually visually) and further superplasticizer (**not** water) added, to increase the workability to the required level and keep the workability retention time as long as possible. The introduction of new-generation superplasticizers with significantly improved workability retention, particularly those based on polycarboxylate ethers (Concrete Society, 2002), has significantly reduced the need for this procedure.

When transporting high workability HSC it is common practice to carry reduced loads, i.e. 4 $m^3$ of concrete in a 6 $m^3$ capacity drum. This reduces the wear and tear on the drum and ring gear (caused by the increased density of HSC) and also reduces the risk of spillage during transport.

Production of HSC requires particular attention to detail as factors causing only second-order effects in normal strength concrete can have major implications at very high strength levels. HSC is relatively a high-cost material, but it can produce overall savings when used in appropriate situations. Producers should be selected on the basis of proven experience with HSC specifically, rather than general experience or cost alone.

## 3.6.2 Use on-site

Properly proportioned HSC can be easily placed by skip and has been successfully pumped over large distances at only slightly higher pump pressures than normal (Page, 1990). The possibility of limited workability retention time must, however, also be kept in mind.

The behaviour of HSC is different in certain respects from conventional strength concrete. As mentioned earlier, it tends to be very cohesive and requires more effort to compact at a given level of slump. The concrete also tends to stop moving as soon as the vibrator is removed (i.e. it is thixotropic). Consequently poker vibrators must be inserted at closer spacings and immersed longer than normal. Particular attention must be given to the interface between two batches of concrete (these will not just flow together) and in instances where small amounts of concrete are used to fill low areas etc., vibration will be required in these locations in order to prevent cold joints.

The low bleed of typical HSC can cause problems with finishing, as the concrete tends to stick to trowels etc. If finishing is delayed too long, the concrete surface often dries and tears. To avoid early age (plastic) cracking, prompt and effective curing, for as long as the concrete is plastic, is essential. Covering exposed surfaces in wet hessian and polythene sheeting is preferred to curing membranes application and early application of water curing is also suggested as a means of reducing autogeneous shrinkage (Aitcin, 1999). Finishing must follow compaction and be itself followed by curing in a continuous sequence with no delays. HSC is not very tolerant of delays and plastic cracking is almost inevitable, unless finishing and curing are promptly carried out.

## 3.6.3 Testing

### *Fresh concrete*

Test regimes for acceptance of HSC on-site do not differ significantly from normal strength concrete. The flow table test (British Standards Institution, 2000a) is much more appropriate (and sensitive) for high workability HSC than the slump test (British Standards Institution, 2000b). A high rate of sampling is prudent at the start of any project but this may be relaxed as construction procedures.

### *Hardened concrete*

Concrete test specimens should be made in rigid moulds and compacted by vibration. The use of 100 mm cubes (or 100 mm diameter × 200 mm cylinders) is common with HSC as a means of reducing the required load capacity of testing machines. Test specimens should be kept moist and at the correct temperature until transferred to the laboratory curing tank. Test specimens should not be moved until they have properly hardened, as high levels of admixture addition may delay setting. ACI 363.2R (American Concrete Institute, 1998) recommends specimens should not be moved until at least 16 hours and not more than 48 hours. If temperature-matched curing is being considered, sealing the test specimens from contact with external water during curing will provide a more accurate estimate of *in-situ* strength (Price and Hynes, 1996).

Cores and cylinders (American Concrete Institute, 1998) should have the ends ground rather than capped as the evidence is that the use of low strength capping compound increases variability and reduces the measured strength (American Concrete Institute, 1998). High strength capping materials applied as thin (2 mm) caps have been used successfully (American Concrete Institute, 1998). In general testing HSC requires more attention to detail, as small changes in procedure can cause proportionally very large changes in the measured strength.

Compression test machines must be of sufficient load capacity such that cube failure occurs within the optimum working range. The lateral stiffness of the machine should also be high enough to allow the load to be maintained right up until failure (American Concrete Institute, 1998). As explosive failures are common with HSC, testing machines should incorporate safety guards to protect personnel from flying concrete fragments.

## 3.7 Examples of use of HSC

High strength concrete has been utilized in many structures around the world (CEB, 1994), but perhaps the most common use of the material has been in the columns of high-rise buildings, particularly in North America and Australia. HSC columns often with reduced reinforcement are an economical solution to providing heavily loaded elements in high-use buildings. The dimensions of columns can also often be reduced by using HSC although possible buckling of very slender elements needs consideration (Concrete Society, 1998). This enables the amount of rentable floor space to be maximized and also minimizes interruption of parking spaces in basement garages.

In Chicago, a number of tall buildings have utilized HSC in columns included 311 S. Wacker Drive (see Figure 3.4) where high strength columns (Design strength 83 MPa (cylinder)) were used at the base of the building, with lower strength concrete (69 and 62 MPa) being used in the more lightly loaded upper floors (Russell, 1994). The HSC was used in conventionally reinforced columns and nearby all the concrete was placed using pumps.

In Seattle, a different form of construction with HSC has been employed. Large-diameter (3 m) steel tubes form the core of the building with smaller steel tubes around the perimeter. These tubes contain shear studs on the internal face but not reinforcement. High strength concrete (Design strength 97 MPa (cylinder)) is pumped into the tubes from the bottom of each storey and without any vibration. This forms a very economic and stiff structure. During the construction of 2 Union Square in Seattle, a 58-storey structure (Russell, 1994), the designer also wished to achieve an elastic modulus of at least 50 GPa. Consequently the actual strength of the concrete was much higher than the design strength in order to produce the desired modulus. Long-term compressive cylinder

**Figure 3.4** 311 South Wacker Drive, Chicago – high strength concrete used in columns.

strengths in excess of 130 MPa (approximately equivalent to a cube strength of 145 MPa) were measured during construction.

In addition to well documented examples in North America, HSC has been used in all tall buildings in Australia, Germany and South-east Asia (CEB, 1994).

Whilst the most widespread application of HSC technology has been in high-rise buildings, offshore structures have been utilizing concrete with ever increasing strength for many years. Bridges and in particular prestressed concrete bridges have also used the benefits of HSC. One experimental footbridge near Tokyo (Price, 1999) incorporates 100 MPa concrete to achieve a span/depth ratio of 40:1. HSC can enable the span of bridge beams to be increased or to reduce the number of beams required for a given span. This can, in turn, lead to a lower unit cost for a given length of bridge.

In the UK, HSC has not been used to construct any major structures (high-rise structures being comparatively rare), but has found applications in providing chemical resistance, or resistance to abrasion in industrial facilities and in the remediation of structures. Combining the use of HSC with temperature matched curing to monitor *in-situ* strength development is particularly effective for reducing the time before roads and bridges can be reopened to traffic.

The use of HSC will probably always remain a very small proportion of overall concrete consumption worldwide. However, the properties of HSC offer additional options to designers and contractors to utilize concrete in overcoming engineering challenges.

## 3.8 Summary

High strength concrete (HSC) is made possible by the development of high strength cementing materials and superplasticizing admixtures. The selection of constituent materials is critical in achieving a high compressive strength, but compressive strength in excess of 100 MPa is now routinely achievable in the UK and elsewhere. Maintaining a consistent low water/cement ratio is the most important factor for successful production.

HSC can be used with most conventional construction techniques, but special attention must be given to avoiding delays during placing, finishing and curing. HSC is also characterized by an increased elastic modulus and tensile strength as well as lower drying shrinkage. Autogenous shrinkage, however, can be very high. The main applications of HSC have been in the columns of high-rise buildings, offshore structures and long-span bridges.

## References

Aarup, B. (1994) Ministruct – Minimal Structures using ultra high strength concrete. *Proceedings of International Conference 'Concrete across borders'*, Odense, June.

Ahmad, S.H. and Shah, S.P. (1985) Structural properties of high strength concrete and its implications for precast prestressed concrete. *PCI Journal*, Nov.–Dec., 91–119.

Aitcin, P.C. (1998) *High Performance Concrete*. E&F N Spon, London, p. 199.

Aitcin, P.C. (1999) De-mystifying autogenous shrinkage. *Concrete International*, **21**, No. 11, November, 54–56.

Aitcin, P.C. and Mehta, P.K. (1990) Effect of coarse aggregate characteristics on mechanical properties of high strength concrete. *ACI Materials Journal*, **87**, No. 2, 103–107.

Aitcin, P.C., Jolicoeur, C. and MacGregor, J.G. (1994) Superplasticisers: how they work and why they occasionally don't. *Concrete International*, **16**, No. 5, May, 45–52.

American Concrete Institute (1992) *State of the art report on high strength concrete.* Farmington Hills, ACI 363 R-92 (Reapproved 1997).

American Concrete Institute (1998) *Guide to quality control and testing of high strength concrete.* Farmington Hills, ACI 363.2R-98.

Bamforth, P.B. and Price, W.F. (1995) *Concreting deep lifts and large volume pours*. CIRIA Report 135, London.

British Standards Institution (1992) Specification for aggregates from natural sources for concrete. BS 882.

British Standards Institution (2000a) Testing fresh concrete. Slump test. BS EN 12350-2.
British Standards Institution (2000b) Testing fresh concrete. Flow table test. BS EN 12350-5.
Burg, R.G. and Ost, B.W. (1992) *Engineering properties of commercially available high strength concretes.* Skokie, PCA Report RD 104.
Calderone, M.A., Gruber, K.A. and Burg, R.G. (1994) High reactivity metakaolin: A new generation admixture. *Concrete International*, **16**, No. 11, Nov. 37–41.
CEB (1994) *Application of high strength concrete.* CEB Bulletin d'information No. 222.
CEN (2002) *Eurocode 2: Design of concrete structures – Part 1: General rules and rules for buildings* (Revised final draft). pr EN 1992-1-1.
Concrete Society (1993) *Microsilica in Concrete*. Conc. Soc. Tech. Rept No. 41, Slough.
Concrete Society (1998) *Design guidance for high strength concrete.* Conc. Soc. Tech. Rept No. 49, Slough.
Concrete Society (2002) *A guide to the selection of admixtures*. Conc. Soc. Tech. Rept. No. 18 (2nd end), Slough.
de Larrard, F. (1999) *Concrete Mixture Proportioning: A scientific approach.* London: E&FN Spon, London.
de Larrard F. and Le Roy, R. (1992) The influence of mix composition on mechanical properties of high performance concrete. In Malhotra, V.M. (ed.), *Fly ash, silica fume, slag and natural pozzolans in concrete.* Istanbul, May, 965–986.
FIB-CEB (1990) *High strength concrete. State of the art report.* CEB Bulletin d'information No. 197.
Gjorv, O.E., Baerland, T. and Ronning, H.R. (1990) Abrasion resistance of high strength concrete pavements. *Concrete International*, **12**, No. 1, Jan. 45–48.
Gudmundvson, G. and Olafsson, H. (1996) Silica fume in concrete – 16 years of experience in Iceland. In Shayan, A. (ed.), *Proceedings of the 10th International Conference on Alkali-silica in Concrete*. Melbourne, 562–569.
Hammer, T.A. and Sellevold, E.J. (1990) Frost resistance of high strength concrete. In Hester, W.T. (ed.), *Proceedings of the 2nd International Symposium on High Strength Concrete.* Berkeley pp. 457–487.
Helland, S. (1996) Application of high strength concrete in Norway. *ACI SP167*, 27–54.
Mailvaganam, N.P. (1999) Admixture compatibility in special concretes. *ACI SP186*. 615–634.
Mehta, P.K. and Aitcin, P.C. (1990a) Microstructural basis of selection of materials and mix proportions for high strength concrete. In Hester W.T. (ed.), *Proceedings of the 2nd International Symposium on High Strength Concrete.* Berkeley, pp. 265–286.
Mehta, P.K. and Aitcin, P.C. (1990b) Principles underlying production of high performance concrete. *Cement, Concrete and Aggregates*, **12**, No. 2, Winter, 70–78.
Mindess, S., Qu, L. and Alexander, M.G. (1994) The influences of silica fume on the fracture properties of paste and micro-concrete. *Advances in Cement Research,* **6**, No. 23, 103–107.
Naaman, A.E. (2000) HPFRCCs: Properties and applications in repair and rehabilitation. *ACI SP182*, 1–16.
Nilsen, A.V. and Aitcin, P.C. (1992) Properties of high strength concrete containing light normal and heavyweight aggregate. *Cement, Concrete and Aggregates*, **12**, No. 1, Summer, 8–12.
Nischer, P. (1995) High performance concrete – Resistance against dissolution attack. In Sommer, H. (ed.), *Proceedings of RILEM Workshop Durability of High Performance Concrete,* Vienna, pp. 177–181.
O'Neil, E.F., Dauriac, C.E. and Gilliland, S.K. (1996) Development of reactive powder concrete (RPC) products in the United States Construction Market. *ACI SP176*, 249–262.
Page, K.M. (1990) Pumping high strength concrete on world's tallest buildings. *Concrete International*, **12**, No. 7, Jan., 26–28.
Price, W.F. (1996) Stronger, bigger, better: Construction and testing of a large high strength concrete slab. *Concrete*, Jan./Feb., 28–29.
Price, W.F. (1999) High strength concrete. *Concrete*, June, 9–10.

Price, W.F. and Hynes, J.P. (1996) In-situ strength testing of high strength concrete. *Magazine of Concrete Research*, **48**, No. 176, 189–198.

Ronneberg, H. and Sandvik, M. (1990) High strength concrete for North Sea platforms. *Concrete International*, **12**, No. 7, Jan. 29–34.

Russell, H.G. (1994) Structural design considerations and applications. In Shah S.P. and Ahmad S.H. (eds), *High Performance Concretes and Applications*. Edward Arnold, London, pp. 313–340.

Ryle, R. (2000) Metakaolin: a highly reactive pozzolan for concrete. *ICT Yearbook 1999–2000*, pp. 19–26.

## Further Reading

Aitcin, P.C. (1998) *High Performance Concrete.* E & FN Spon, London.

Nawy, E.G. (1996) *Fundamentals of High Strength, High Performance Concrete.* Longman, London.

Shah, S.P. and Ahmad, S.H. (eds) (1994) *High Performance Concretes and Applications*. Edward Arnold, London.

In addition to these books, the proceedings of the regular international conferences/symposia on 'Utilization of High Strength/High Performance Concrete' are valuable reference sources:

*Proceedings of the International Symposium 'Utilization of High Strength Concrete'*, Stavanger, June 1987 (published by Tapir).

*Proceedings of the 2nd International Symposium 'Utilization of High Strength Concrete'*, Berkeley, May 1990 (published as ACI SP121, 1990).

*Proceedings of the 3rd International Symposium 'Utilization of High Strength Concrete'*, Lillehammer, June 1993 (published by the Norwegian Concrete Association).

*Proceedings of the 4th International Symposium 'Utilization of High Strength/High Performance Concrete'*, Paris, May 1996 (published by RILEM).

*Proceedings of the 5th International Symposium 'High Strength/High Performance Concrete'*, Sandefjord, June 1999 (published by the Norwegian Concrete Association).

*Proceedings of the 6th International Symposium 'Utilization of High Strength/High Performance Concrete'* Leipzig, June 2002 (published by IMB-Leipzig).

# 4

# Heat-resisting and refractory concretes

*Ron Montgomery*

## 4.1 Introduction

The definition of refractory concrete will vary depending on the specific reference. However, a good general definition is:

Concrete which is suitable for use at high temperatures composed of hydraulic cement (calcium aluminate cement) as the binding agent. Combined with heat resistant, refractory aggregates and or fillers.

The boundary between heat-resistant and refractory concrete is somewhat arbitrary, but is probably about 1000°C although some definitions of refractory concrete start at 1500°C. However, there is, in fact, a more or less continuous spectrum of high-temperature resistant concretes, extending from about 300–400°C (the limit of concretes bound with Portland cements) to 2000°C or more, using high range calcium aluminate cements (CAC) containing 80 per cent alumina.

## 4.2 Calcium aluminate cement (CAC) versus Portland cement (PC)

Although they were not developed specifically for high-temperature use, the heat-resistant and refractory properties of CACs quickly became apparent. Early applications began from about the 1930s and significant growth of these applications took place from the 1950s onward. Ordinary Portland cement (OPC) has limited use at high temperatures for a number of reasons. At temperatures of ~500°C the hydrated lime or Portlandite ($Ca(OH)_2$), which forms a significant proportion of hydrated Portland cement, will dehydrate to form quicklime (CaO) (Robson). This is a reversible reaction:

$$Ca(OH)_2 \Leftrightarrow CaO + H_2O$$

Exposure to moisture (atmospheric moisture is sufficient) leads to rehydration of the quicklime which is an expansive reaction and will disrupt the concrete (Figure 4.1). Thus any application where there is cycling between high temperatures and ambient temperatures is excluded. Furthermore, Portland cement is high in lime and silica, which form low melting point compounds at the service temperature of the refractory concretes.

**Figure 4.1** Portland cement and calcium aluminate cement concrete cylinders after 8 cycles of 6 hours at 500°C, followed by 24 hours in humid conditions.

Hydrated lime is not formed during the hydration of CAC. Thus it is not subject to the disruption caused by the rehydration of quicklime. The hydrates that are formed do dehydrate progressively as the temperature increases to about 300°C and above. However, the compounds formed are themselves stable and at even higher temperatures (>1000°C) begin to react with the refractory aggregates (see section 4.4) to form new stable phases.

## 4.3 Refractory limits of calcium aluminate cements

It is often said that the refractoriness (i.e. the softening point or service temperature limit) of CAC is governed by the alumina content. The higher the alumina, the more refractory the cement. Whilst this is true in the first instance, refractoriness is also dependent on the presence or not of compounds that form low melting point eutectics – CaO, $SiO_2$, $Fe_2O_3$.

Lower grade grey CAC (~39 per cent $Al_2O_3$) contains relatively high amounts of these compounds – about 36 per cent CaO, 5 per cent $SiO_2$ and 16 per cent $Fe_2O_3$. (Table 4.1) (Hewlett, pp. 709–778).

**Table 4.1** Typical analyses of calcium aluminate cements compared to Portland cement

| Colour range | Grey Portland | Grey CAC 'low' range | Buff CAC 'mid' range | White CAC 'high' range | White CAC 'high' range |
|---|---|---|---|---|---|
| $Al_2O_3$ | 4–6 | 36–42 | 48–60 | 65–75 | 80–82 |
| CaO | 63–67 | 36–40 | 36–40 | 25–30 | 15–20 |
| $SiO_2$ | 19–23 | 3–8 | 3–8 | < 0.5 | < 0.2 |
| $Fe_2O_3$ | 2–3.5 | 12–20 | 1–3 | < 0.5 | < 0.2 |
| $TiO_2$ | < 0.5 | ~2 | ~3 | – | – |
| MgO | ~1 | ~1 | ~0.1 | – | – |

As the alumina ($Al_2O_3$) content increases, in the higher grades of CAC, then the relative amounts of compounds that form low melting point eutectics decrease, leading to higher refractoriness (Figure 4.2) (Lafarge Aluminates Ltd). Ultimately these compounds are present in only trace quantities (< 1 per cent of total) and high-range CAC (~70 per cent $Al_2O_3$) is composed of almost pure calcium aluminates.

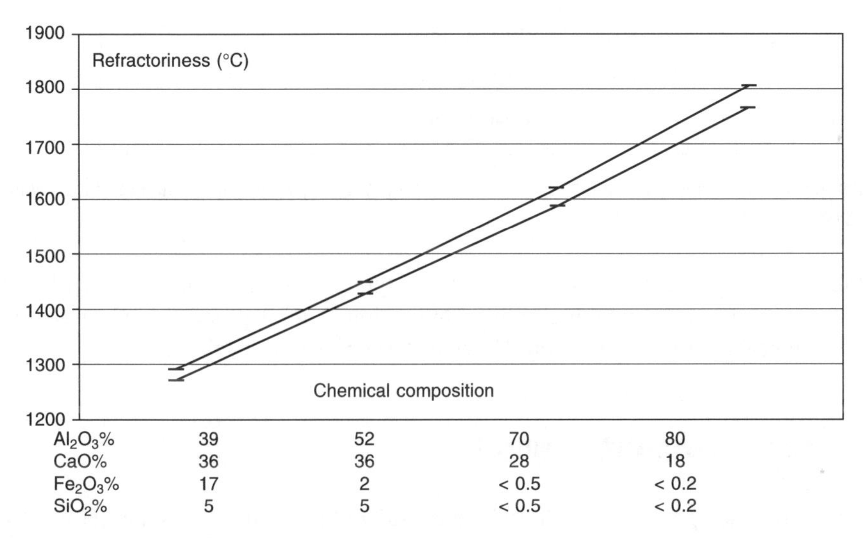


**Figure 4.2** Relationship between refractoriness and chemical composition of calcium aluminate cements.

In order to extend the refractoriness further, even more free alumina is added to 70 per cent $Al_2O_3$ CAC to increase the alumina content to ~80 per cent. This, with very few exceptions, is the current upper limit of the alumina content of modern CACs.

## 4.4 Refractory and heat-resisting aggregates

The aggregates used in heat-resisting and refractory concretes are generally not those used in conventional concrete. There is, if anything, a much wider range of these specialist aggregates available, most of them being synthetic or derived from heat treating naturally occurring materials.

Much as with the cements, the temperature resistance of these aggregates depends on their nature and chemistry (Figure 4.3) and many of the same regulations regarding alumina content etc. hold true. However, the applications of heat-resisting and refractory concretes are extremely diverse and resistance to high temperature is not the only factor that needs to be taken into account. Such properties as thermal insulation, abrasion resistance; resistance to molten metals and slags, expansion/contraction etc. need to be considered. The final properties of the concrete depend very much on the correct combination of cement and aggregate and, as will be seen later, reactive fillers such as finely ground aluminas and fume silica.

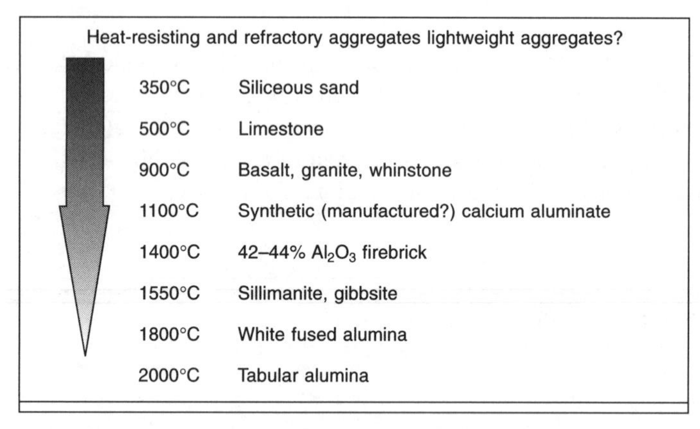


**Figure 4.3** Service temperature limits of heat resisting and refractory aggregates, when combined with the appropriate calcium aluminate cement.

Table 4.2 gives a non-comprehensive list of typical aggregates used in heat-resistant and refractory concretes together with an indication of their properties when combined with the appropriate CAC (Robson; Hewlett; Lafarge Aluminates Ltd).

## 4.5 Heat-resistant concretes

Concretes for use up to about ~1000°C are considered to be 'heat resistant'. However, as stated earlier, this temperature is not an immutable boundary but a convenient marker. Many heat-resistant concretes are used at temperatures well below 1000°C and sometimes above this temperature. The important thing is that the choice of cement and aggregate are appropriate to the application.

It is likely that concrete technologists or engineers will come across the need to use heat-resisting, rather than true refractory concretes, at some time, as their applications are

**Table 4.2** Properties of heat-resisting and refractory concretes[1]

| Type of aggregate<br>Insulating concretes | Type of CAC<br>Colour | %<br>$Al_2O_3$ | Approximate service temperature limit[2] (°C) | Approximate fired density[3] ($kg/m^3$) | Indicative thermal conductivity[4] (W/m.°K) |
|---|---|---|---|---|---|
| Vermiculite | Grey | 39 | 900 | 550 | 0.17 |
| Perlite | Grey | 39 | 900 | 600 | 0.18 |
| Diatomite | Grey | 39 | 900 | 850 | 0.25 |
| Pumice | Grey | 39 | 900 | 1200 | – |
| Expanded Clay | Grey | 39 | 1100 | 1450 | 0.50 |
| Sintered PFA | Grey | 39 | 1100 | 1500 | 0.35–0.58 |
| Expanded Chamotte | Grey | 39 | 1200 | 1500 | 0.46–0.58 |
| | Buff | 50 | 1350 | 1500 | – |
| Bubble Alumina | White | 70 | 1700 | 1500 | 0.45–0.60 |
| | White | 80 | 1800 | 1300 | 0.65–1.00 |
| *Heat-resistant concretes* | | | | | |
| Siliceous Sand | Grey | 39 | 350 | 2200 | 1.4 |
| Limestone | Grey | 39 | 500 | 2200 | – |
| Crushed House Brick | Grey | 39 | 800 | 1920 | 1.1 |
| Basalt, Granite | Grey | 39 | 900 | 2400 | 1.4 |
| Emery | Grey | 39 | 1000 | 2500 | – |
| Alag™ | Grey | 39 | 1100 | 2600 | 1.2–1.5 |
| Firebrick (35% $Al_2O_3$) | Grey | 39 | 1200 | 1920 | 1.1 |
| Olivine | Grey | 39 | 1200 | 2650 | – |
| *Dense refractory concretes* | | | | | |
| Firebrick (44% $Al_2O_3$) | Grey | 39 | 1300 | 1900 | 0.9–1.1 |
| Chamotte (42–44% $Al_2O_3$) | Grey | 39 | 1300 | 2050 | 0.9–1.1 |
| | Buff | 50 | 1450 | 2030 | 0.7–1.0 |
| | White | 70 | 1500 | 2000 | 0.5–0.7 |
| | White | 80 | 1520 | 2100 | 0.8 |
| Silliminite, Gibbsite | Grey | 39 | 1350 | 2160 | |
| | Buff | 50 | 1450 | 2160 | 1.5 |
| | White | 70 | 1600 | 2160 | |
| Calcined Bauxite | Grey | 39 | 1400 | 2560 | |
| | Buff | 50 | 1500 | 2560 | 1.7 |
| | White | 70 | 1700 | 2560 | |
| Brown Fused Alumina | Grey | 39 | 1400 | 3000 | 1.5–1.8 |
| | Buff | 50 | 1550 | 3000 | 1.5–1.8 |
| | White | 70 | 1650 | 3000 | 1.2–1.7 |
| | White | 80 | 1750 | 3050 | – |
| White Fused Alumina | White | 70 | 1800 | 3000 | 2.30 |
| | White | 80 | 1900 | 3050 | – |
| Tabular Alumina | White | 70 | 1900 | 2850 | 1.4–1.8 |
| | White | 80 | 2000 | 2900 | 1.9–2.3 |

[1] Robson (pp. 196, 197, 216); Lafarge Aluminates Ltd.
[2] Maximum hot face temperatures, assuming a thermal gradient through the mass of the concrete.
[3] Density after firing to 1000°C unless service temperature limit is lower.
[4] Coefficients of thermal conductivity vary considerably depending on the temperature at which they are measured.
The figures given are indicative values for 500–1000°C.

often more closely linked to conventional concrete applications. These specialist concretes are often still mixed and placed on-site, as readymixed concretes containing CACs are not generally available. It is important to remember, however, that most of the factors governing conventional concretes will apply to heat-resistant concrete. Questions such as the relationships between shrinkage, strength, abrasion resistance, durability etc. and the water/cement ratio, compaction and, to a lesser extent, cement content will all hold true.

However, there are other important parameters that must also be taken into account if the concrete is to perform correctly. Normally grey CAC (39 per cent alumina) will have sufficient temperature resistance for most heat-resisting applications up to 1000°C. Indeed in the case of some aggregates, it will be the properties of the aggregate, rather than the cement, that will limit the service temperature of the concrete, as these may be lower than the CAC itself. e.g. vermiculite, perlite, basalt and some heat-treated aggregates are themselves less refractory than grey CAC. In refractory concretes (see section 4.11) the reverse is often the case, and the higher refractoriness of the aggregate will extend the temperature range of a given CAC.

## 4.6 Insulating concretes

The thermal insulating properties of these concretes are primarily associated with their density and the major factor affecting density would be the density of the aggregate. Lightweight aggregates vary in density from the extremely light, (e.g. Perlite; bulk density 100–110 kg/m$^3$) to moderately light (e.g. Sintered PFA or expanded clay; bulk density 600–800 kg/m$^3$). The insulating properties (i.e. the thermal conductivity) of concretes made with these aggregates will be in the range 0.15–0.5 W/m.°K (Figure 4.4) (Robson).

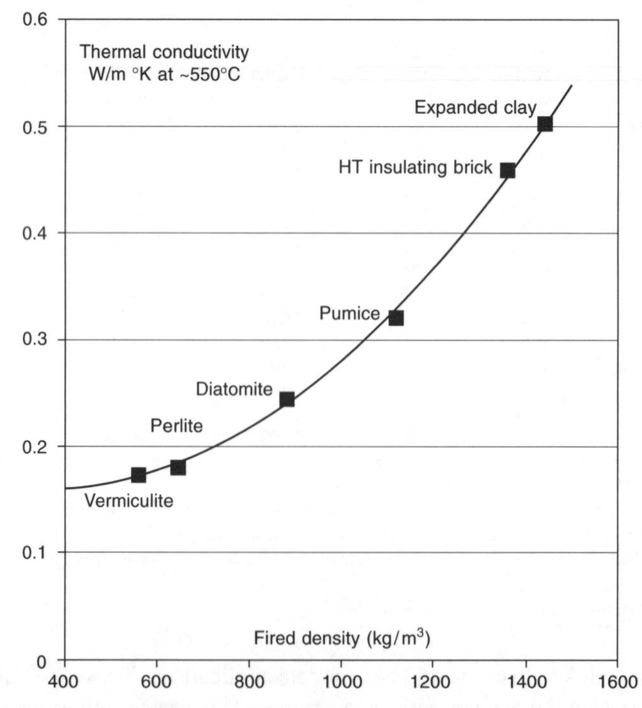


**Figure 4.4** Relationship between fired density and thermal conductivity of heat-resisting insulating concretes (Robson, p. 217).

This type of concrete would often be used in conjunction with a dense refractory concrete in heat-retaining vessels. The dense or 'hot face' concrete would then be on the inside of the vessel, with the insulating concrete behind. In this way the thermal gradients and heat losses can be closely controlled. Since CAC would be the cement binder for both of these concretes, they can be cast monolithically. Even when the hot face concrete requires a higher range CAC (e.g. 50–70 per cent alumina) due to the higher temperature inside the vessel in question, this does not pose a problem, as high- and low-range CACs are compatible with each other.

As with heat-resisting concrete, grey CAC (39 per cent $Al_2O_3$) would normally be sufficiently heat resisting for insulating concrete, since the service temperatures of these materials rarely exceed 1000°C. There are, however, exceptions and very high temperature insulating concretes can be made with specialist aggregates such as bubble alumina or micro-porous calcium hexa-aluminate ($CaO.6Al_2O_3$). In this case the high-purity 70 per cent or 80 per cent $Al_2O_3$ CAC would be necessary, as service temperatures would be higher than the range of grey CAC.

## 4.7 Abrasion and heat-resisting concretes

The service conditions of heat-resistant and refractory concretes often also require a degree of abrasion resistance, due to direct physical contact (e.g. wheeled or tracked vehicles, billets of hot metal etc.) or erosion due to gas-borne particulates. Many dense, high-temperature concretes, although they may not be designed specifically for abrasion, have a reasonable degree of abrasion resistance.

There are, however, some notable applications, where the severe conditions require that the concrete must withstand heat, thermal shock and heavy abrasion (e.g. floors in foundries and hot metal treatment or handling areas, fire training areas and constructions). For these applications there is a particularly suitable type of synthetic aggregate, which is made by crushing and grading calcium aluminate clinker (Lafarge Aluminates Ltd). This aggregate has similar chemistry and mineralogy to grey CAC.

In concrete, this aggregate bonds hydraulically with the CAC matrix, increasing both the strength and resistance to abrasion and erosion. Furthermore, the hardness of the aggregate is high, and so wear rates are very low. Such concrete is often used where there is a multiplicity of constraints – heat, abrasion, thermal shock (high and low temperatures) and chemical attack.

## 4.8 High-temperature refractory concrete

True refractory concretes, such as are used in heat containment, metallurgy, ceramics and cement industries, are usually highly specialist and are prepared pre-packed ready for site mixing and installation. The technology involved with these 'castable' refractories is more akin to refractory technology than concrete technology and these are usually manufactured and installed by specialist companies.

There are today a number of categories of refractory castable, and these tend to be classified by their cement content or their method of installation. These are Conventional Castables (CC), Low Cement Castables (LCC), Ultra Low Cement Castables (ULCC),

Self-Flowing Castables (SFC) (the last can be low or medium cement content) and Gunning Castables.

Conventional castables have cement contents of about 15–25 per cent by weight and are therefore similar in this respect to normal concretes. The particular combination of CAC and refractory aggregate more or less governs the upper service temperature limit of the concrete (see Table 4.2). This type of refractory concrete was the only type of hydraulically bound castable in use from the 1930s until about the 1980s and still represents a significant percentage of the market.

The high-temperature performance of conventional castables is, to some extent, limited by their chemistry and in particular the lime (CaO) content (see section 4.3 above). Thus a conventional castable containing 15 per cent of a high-range CAC (70–80 per cent $Al_2O_3$) with an aggregate composed entirely of alumina (e.g. tabular alumina, a refractory aggregate made from pure sintered alumina and very widely used in high-temperature refractory concretes) would still contain 2.5–4.5 per cent CaO, dependent on the cement used. This limits the upper service temperature at which such concretes can be used.

With the availability of sub-micron-sized fillers such as Fume Silica (FS), the development of LCCs and ULCCs having lower cement contents, and therefore lower lime contents, became possible (see section 4.9).

## 4.9 Low- and ultra-low-cement castables

The lime content of CAC was one of the factors governing the high-temperature performance of refractory concretes. An 80 per cent alumina CAC contains about 18 per cent lime (CaO) in the form of calcium aluminates. Cements having higher alumina contents (i.e. 90 per cent) and therefore lower lime contents were tried, but never really gained commercial success. Thus, the way to reduce the lime content of the refractory concrete (and hence improve its high-temperature performance) was to reduce the cement content.

Low-cement castables were developed during the 1970s (Clavaud *et al.*, 1985) using particle-packing theories that had been developed in the 1920s and 1930s. However, these theoretical Particle Size Distributions (PSD) were not possible in practice without aggregates (or more correctly fillers) having particle sizes of 0.1 to 1 microns. Fume silica began to be available at around this time, initially as a by-product of the ferro-silicon industry, and then as a specifically manufactured product in its own right (Clavaud *et al.*, 1985; Elkem website; Hewlett, pp. 676–708).

This material, together with micronized aluminas, allowed the PSD of the concretes to follow the theoretical curves down to ~0.1 micron, which allowed cement contents to be decreased to 5–8 per cent. These concretes became known as Low-Cement Castables (LCC). Further development in this technology, improvement of dispersion by additives etc. allowed further reductions in cement contents down to ~2 per cent. Hence the term Ultra-Low-Cement Castables (ULCC).

The great advantage of LCCs and ULCCs is that their mechanical performance at high temperature was greatly superior to that of conventional castables (Clavaud *et al.*, 1985). This allowed refractory concretes to be used in applications which were hitherto the domain of pre-fired refractory bricks.

## 4.10 Self-flow castables

Self-flow Castables (SFC) are the refractory equivalent of self-compacting concrete. The installation of LC and ULC castables was sometimes difficult, particularly if the cast was large and complex in shape and often not in ideal site conditions. The continued development of flow-enhancing additives, together with close attention to the PSD of the whole concrete, not just the submillimetre sizes, has allowed the development of concretes that flow into place, without vibration being necessary to compact or de-air the concrete. The development of SFCs did ease these problems, but these have not become as widespread in use as the LC and ULC castables.

## 4.11 Installation of heat-resisting and refractory concretes

### 4.11.1 Placing and compaction

As with conventional concretes, the placing and compaction of heat-resisting and refractory concretes is extremely important. The methods used are identical to those used with conventional concrete, thus no specialist equipment or skills are necessary. Similarly materials for moulds and shuttering are standard, but it may be necessary to pay more attention to dimensions if interlocking precast sections are involved.

With careful attention to the mix design of the mortar or concrete, application by gunning is commonly undertaken. This work is normally conducted by specialist contractors for areas where access is difficult for regular casting and may also be carried out at elevated temperatures.

### 4.11.2 Curing

When using any CAC-based concretes proper curing is of the utmost importance. Poorly cured concrete will lead to dusty and friable surfaces and possible failures in service. The objective of curing is to maintain the moisture in the concrete so that proper hydration takes place. Again the methods used are similar to conventional concrete, but due to the rapid hardening and high heat evolution of CAC concretes, it is important to start curing 3–4 hours after placing and continue until at least 24 hours.

### 4.11.3 Drying and firing

After curing there is still a considerable amount of water left in the ‘green’ (unfired) concrete. This water must be allowed to escape as the concrete is heated for the first time if spalling is to be prevented. Drying-out time is extremely important before the concrete goes into service for the first time. Natural drying out or forced drying at up to ~100°C, to eliminate as much of the free water as possible, is ideally advised before exposure to higher temperatures.

After drying, as the concrete is heated from 100°C to about 350°C, the water of hydration is driven from the concrete. This is the water combined within the hydrated cement and will not be eliminated by drying out at 100°C.

Heating schedules during first firing are very important. The exact details of the schedules often vary with application, thickness, type of concrete etc. However, a good rule of thumb for conventional castables is no more than 50°C per hour up to 500°C, with a hold at this temperature for about 6 hours. Then continue, perhaps at a slightly higher rate, up to the service temperature (private communications). For particularly thick sections (>100 mm) a hold at different temperatures may be advisable.

If reasonable drying-out and first firing schedules are not possible, or indeed if the sections are very large (>500 mm) it is prudent to incorporate some artificial means of escape for the water. An addition of a small amount of more porous aggregate or the use of organic fibres (e.g. polypropylene) will form 'natural' paths for the water vapour to escape.

After the first firing to the full service temperature, no further special heating schedules should be necessary, unless the concrete is allowed to become saturated again, due to prolonged exposure to water (e.g. external storage over a winter period). Intermittent wetting should not be problematical e.g. fire training areas, where heat-resisting CAC-based concretes are known to give satisfactory service.

## 4.11.4 Reinforcement

Steel reinforcement in concrete intended for use at high temperatures must be carefully considered. At temperatures that are very moderate with respect to heat-resisting and refractory concretes, ~300°C, the difference of the thermal expansion of the steel and the concrete, becomes such that the normal bond between them is reduced or lost. At higher temperatures, particularly with heavy-gauge steel, this may also lead to cracking or spalling. Above 400°C, the tensile properties of ordinary steels begin to decrease rapidly as the temperature increases and the advantages gained from its presence are progressively lost (Robson).

Thus any substantial reinforcement that may be necessary, in floor slabs for example, should be as far away as possible from the hot face and the temperature at the level of the steel should not exceed 300°C. Lighter gauge meshes may be employed, but these should never be above the mid-point in the cast.

If necessary, steel fibres may be used in heavy industrial areas. These may be of normal mild steel but stainless steel fibres have more tolerance to high temperatures.

## 4.11.5 Shrinkage and thermal expansion

It is not unusual or abnormal for refractory concretes to exhibit cracks after the first firing. These are due to dehydration shrinkage and possibly ceramic reactions between the cement and aggregates at high temperatures. In normal service these cracks will close when the concrete is reheated to its service temperature. After the first firing, the thermal expansion is reversible on cooling and reheating (Figure 4.5) (Robson). These cracks should not cause a problem unless detritus is allowed to accumulate in the cracks between firings. Subsequent firings could then increase the width of the crack.

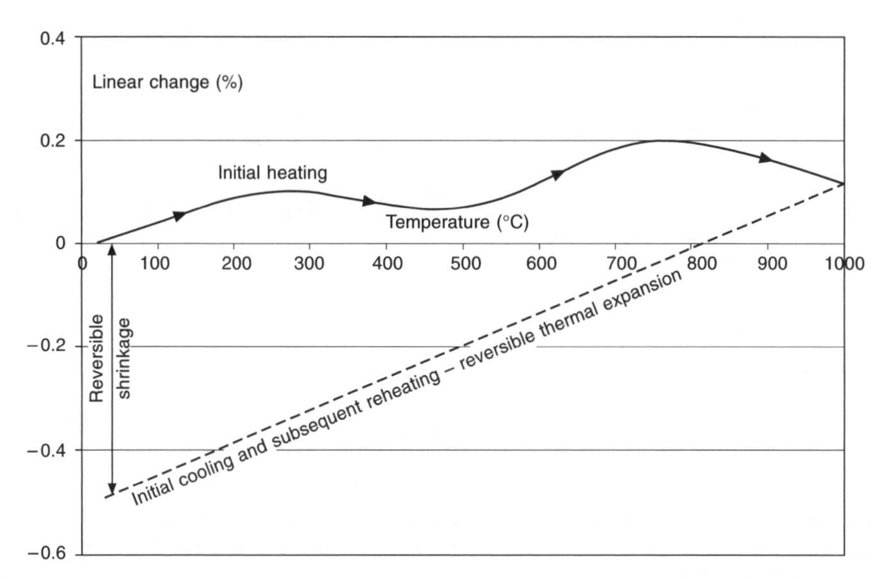


**Figure 4.5** Linear movement during first firing and subsequent heating and cooling cycles (Robson, p. 214).

## 4.11.6 Strength after firing

The strength development of conventional castables before firing is similar to that of normal CAC concrete. Hardening begins soon after setting (3–4 hours) and about 90 per cent of the full green strength is developed in 24 hours.

During the initial drying and firing cycle, strength changes take place that are associated with the loss of free and combined water and at higher temperatures, the reaction of the CAC with the aggregates used. Figure 4.6 (Lafarge Aluminates Ltd; American Concrete Institute) shows the typical strength development of both conventional and low cement castables during the initial firing.

Typically a conventional castable loses some strength up to about 500°C, due to the loss of the hydraulic bond in the cement. On further heating there is little change until the point is reached when a ceramic bond begins to form between the cement and aggregate phases. This occurs at 900–1200°C depending on the type of aggregate and cement used (see sections 4.3 and 4.4). In effect the formation of the ceramic bond occurs as the concrete reaches its softening point, thus the strength when tested at the working temperature decreases progressively as the temperature is raised. However, if the concrete is cooled and tested cold, the strength is increased due to the formation of the ceramic bond (see Figure 4.6) (Lafarge Aluminates Ltd; American Concrete Institute).

A Low-Cement Castable exhibits higher strengths when tested both hot and cold and shows higher ‘refractoriness’ due to the lower content of liquid forming phases (see section 4.3). LCCs have particularly good performance at working temperatures, which is one of the reasons for their widespread use in applications such as the iron and steel industries.

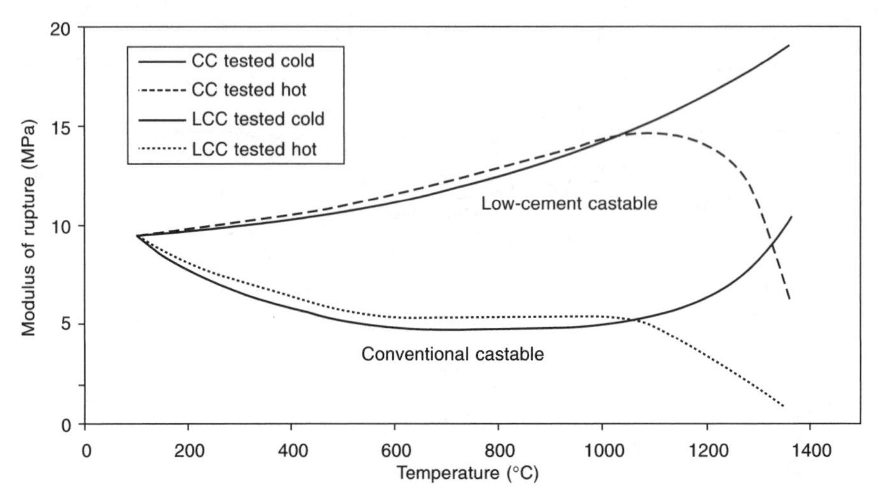


**Figure 4.6** Changes in the modulus of rupture (flexural strength) of typical conventional and low-cement castables when tested hot and cold, after drying and firing to temperatures from 300°C to 1400°C (American Concrete Institute; Mathieu).

## 4.12 Applications

The applications of true refractory concretes such as used in the iron and steel and non-ferrous metals industries, where process temperatures reach 1800°C or more, are beyond the scope of this text. The examples given below are more of an indication of where heat-resisting concretes may be encountered by the concrete technologist or engineer.

### 4.12.1 Domestic flues, fireplaces and chimneys

The linings of domestic flues are usually composed of precast, heat-resistant concrete, made with grey CAC and a kiln burnt aggregate, e.g. sintered PFA. Such concretes will resist temperatures up to 1000–1100°C although, in normal use, the flue temperature may only reach a few hundred degrees Centigrade. In a chimney fire, however, the temperature could reach 900°C or more.

The elements that comprise the fire hearth itself may be subject to temperatures of up to 1200°C and these are often cast with a low range refractory concrete e.g. grey CAC with firebrick or chamotte. See Table 4.2.

The chimneys leading from many industrial processes or incinerator chimneys are often cast with or lined with concretes based on CAC. This can be for heat-resistant purposes, as much as for the risk of chemical attack, due to acidic or other aggressive agents in the flue gases.

### 4.12.2 Foundry floors

Any industrial area that may be subjected to molten metal spillage requires a concrete that will resist thermal shocks as well as sustained high temperatures. This sort of area is also required to resist impact and abrasion. Often, heavy-grade industrial use such as this requires not only the use of CAC but also hard, heat- and abrasion-resistant aggregates. Some basalts and certain granites can withstand this level of treatment, but the exceptional qualities of CAC combined with a synthetic calcium aluminate aggregate (Lafarge Aluminates Ltd) have proven to be particularly suitable in these types of application.

### 4.12.3 Fire training areas

This is a particularly demanding application for any concrete. Fire training areas are exposed to regular thermal cycling, both when fires are set and when they are extinguished. The material being burnt may be of many different types including hydrocarbons, wood and paper, furniture, tyres etc. These burning materials, besides creating intense heat, may produce chemicals which will attack the surrounding concrete.

The two most common forms of fire training area are either large flat surfaces (usually outside, but occasionally inside buildings) and full-scale rooms or even two-storey buildings, including staircases, where firefighting crews can train or try new firefighting equipment.

If built with conventional concrete, these structures would quickly become unserviceable. Again the particular properties of concretes based on CAC are required, possibly with synthetic calcium aluminate aggregate (Lafarge Aluminates Ltd).

## References

American Concrete Institute Report 547R-79-Refractory Concrete.

Clavaud, B., Kiehl, J.P. and Radal, J.P. (1985) A new generation of low-cement castables. Lafarge Refractories, Venissieux, *Fr. Adv. Ceram.*, 13 (New Dev. Monolithic Refract.), 274–284.

Elkem website: www.refractories.elkem com – click on 'Elkem Microsilica'.

Hewlett, P. (ed.), *Lea's Chemistry of Cement and Concrete* (4th edn). Arnold, London, pp. 709–778.

Lafarge Aluminates Ltd. Technical literature.

Mathieu, A. Aluminous cement with high alumina content and chemical binders.

Robson, T.D. *High Alumina Cements and Concretes*. John Wiley, New York.

# 5

# High-density and radiation-shielding concrete and grout

*Eric Miller*

## 5.1 Objectives

- Principal applications and uses of high-density concrete and grout including use as biological shielding in the nuclear industry
- Currently available specifications
- Description of the main sources and properties of aggregates used in high-density concrete
- Summary of mix design procedures for high-density concrete and grout
- Description of the fresh and hardened properties of high-density concrete and grout
- Comparisons and differences in procedures for batching, transporting, placing and curing of high-density concrete with those for standard-density concrete
- Outline recommendations for specifying high-density concrete and grout

## 5.2 Introduction

High-density, or heavy density, concrete accounts for a relatively small section of concrete production within the construction industry with specialist applications traditionally within the nuclear industry. Consequently the development of these materials has been limited to two periods, the early nuclear power programme, in the USA, during the 1950s and 1960s, and in the UK to a considerable extent during the 1980s and 1990s. There has, however, been increased use within the UK in non-nuclear applications.

This restricted use has meant that both the availability and properties of aggregates can be fundamentally different from conventional materials. This offers the concrete technologist additional challenges when using these for producing concrete or grout. In addition, the properties of both fresh and hardened products can be different from those of normal concrete and therefore require careful evaluation before use.

This chapter discusses the selection of aggregates, concrete mix design and production requirements together with a summary of the properties of high-density concrete and grouts. Guidance is also given for specifying these materials. It will demonstrate that these materials can be successfully produced to exacting standards by the application of standard concrete technology principles.

## 5.3 Uses and applications

### 5.3.1 Nuclear industry

The principal conventional application of high-density concrete is in the nuclear industry where it is used as dense biological shielding against protection against various types of ionizing radiation in both power generation plants and in reprocessing facilities. It should, however, be appreciated that the majority of shielding concrete is high-quality, standard density concrete and high-density grades with an increased density are generally used as a compromise between available space and cost parameters. For example, the substantial increased cost of high-density concrete in a relatively thin-walled processing cell may more than offset the cost of larger mechanical plant items, particularly if these are fabricated in stainless steel. Such cost increases can also offset potentially larger volume buildings or instances when rebuild work is planned where space is at a premium. Examples of other industries where radiation is encountered include university and industrial research, and numerous medical uses such as linear accelerators as used in hospital oncology departments.

In general terms, the effectiveness of a biological shield is proportional to the density per unit thickness, though the chemistry of the concrete in terms of elemental analysis can be equally important. This is, however, complicated due to the different types and intensities of radiation such as beta and gamma rays and neutrons. In some instances, for example, the nature of the radiation may generate significant quantities of neutrons that require a more specialized approach to control effective shielding. In such cases it is necessary to capture these neutrons to prevent further secondary effects within the concrete which as a result could generate heat internally and possibly give rise to potential structural and, in turn, shielding problems. The science of nuclear physics is outside the scope of this chapter and is in the specialist realm of the nuclear physicist, who must always be consulted in such matters. However, the concrete technologist can make a positive contribution through working with both designers and physicists to resolve some of the practical applications encountered in the construction of facilities where shielding concrete is needed.

One aspect of the solution involves the ability to guarantee within the concrete the long-term presence of a minimum quantity of chemically bound water or, more strictly speaking, hydrogen, together, if necessary, with quantities of chemically light elements such as boron. These are discussed further in section 5.4.5. It is therefore of vital interest to the specialist-shielding physicist to know the complete chemical or elemental analysis of shielding concrete, whether this is standard or high-density. As an example, Table 5.1 illustrates and compares typical values for the elemental analysis of a standard concrete and an iron ore, magnetite, concrete. It should be noted that there are significant differences in elemental analysis of the cement type, whether OPC or PFA or GGBS and consequently, a concrete supplier should not change cementitious type once a mix design has been approved for a particular project without reference to the designer.

**Table 5.1** Examples of mix designs and elemental composition

| Concrete type | Standard density | Magnetite |
|---|---|---|
| **Mix composition (weight ratio)** | | |
| Cement | 1 | 1 |
| Fine aggregate | 2.0 | 3.0 |
| Coarse aggregate | 3.3 | 6.9 |
| Water | 0.47 | 0.49 |
| Typical density ($kg/m^3$) | 2420 | 4000 |
| **Elemental composition (%wt)** | | |
| Iron | 0.9 | 54.8 |
| Calcium | 24.5 | 5.3 |
| Silicon | 14.5 | 2.8 |
| Aluminium | 1.4 | 0.9 |
| Sulphur | 0.1 | 0.1 |
| Carbon | 5.7 | 0.1 |
| Oxygen | 49.9 | 33.0 |
| Hydrogen | 0.8 | 0.5 |
| Other minor constituents | 2.2 | 2.5 |

*Notes*:
1 Standard concrete using 50–50 OPC + GGBS, limestone aggregate and natural sand
2 Magnetite concrete using 50–50 OPC + PFA
3 Elemental composition calculated from typical oxide analyses of constituent materials
4 Hydrogen content of superplasticizer ignored.

## 5.3.2 Other uses

Counterweights form a significant use for high-density concretes. These vary from those found in some heavy construction plant such as excavators weighing in excess of 5 tonnes to small balance weights found on domestic items such as washing machines. The offshore oil industry also finds uses for these materials. Undersea mats for covering pipelines are a common use as are pipelines coated with magnetite concrete. At least one North Sea gravity platform constructed in the UK has considerable quantities located in its legs for structural stabilization. Occasional uses are found in bridge counterweights, in foundations where a restricted volume but high mass is required.

Considerable volumes have also been used in the construction of submerged structures; the Conway Tunnel in North Wales is a typical example.

## 5.4 Definitions and standards

For practical purposes densities exceeding about 3000 kg/m$^3$ can be regarded as high-density concrete and thus use aggregates not normally utilized for standard structural concrete.

There is some inconsistency regarding terms. BS 6100 'Building and Civil Engineering Terms' defines high-density aggregate as material whose bulk density exceeds 4000 kg/m$^3$; this value is considered to be excessive by the author as many high-density aggregates have values significantly below this value. BS 8110 defines high-density aggregates as material with a particle density exceeding 3000 kg/m$^3$. EN 206–1 'Concrete-Performance, Production, Placing and Compliance Criteria' defines these materials as concrete with an oven-dry density greater than 2800 kg/m$^3$ and aggregate as having a particle density appreciably greater than 3000 kg/m$^3$.

There are no British Standards for either high-density aggregates or concrete. There are, however, internal standards for high-density aggregates, concrete and grout for shielding purposes published internally by British Nuclear Fuels plc, a major designer and user of these materials. These define aggregate and mix requirements together with advice for production and testing a range of concrete and grouts having densities between 2.8 and 8.6. These may be structural or non-structural.

There are two American Standards for aggregates, ASTM C637 and ASTM C638. Both were published in 1984 and additionally refer to other complimentary ASTM standards for relevant testing methods. Two American Concrete Institute publications refer to the production aspects of these materials (American Concrete Institute Publication 304R-85 and American Concrete Institute Special Publication 34).

There are French specifications written by the Commissariat à L'Energie Atomique, (CEA).

## 5.5 Aggregates

Table 5.2 describes various ores and manufactured materials, which are used as aggregates together with their principal properties and individual features. Kaplan (1989) presents a more detailed study of these materials including discussion on the nuclear properties. They are divided into two arbitrary categories, 'naturally occurring' and 'man-made'.

### 5.5.1 Naturally occurring aggregates

*Iron ores* of various types have been used for many years abroad but have only recently been used in the UK using imported ore, principally magnetite with smaller quantities of haematite as coarse and fine materials. Other examples are the minerals limonite and goethite though these are rarely used. Their densities are not as high as magnetite or haematite, but these ores do contain varying quantities of chemically bound water in their structure. This property can have advantages in certain nuclear applications where high-density neutron shielding is required.

**Table 5.2** Summary of *typical* properties of high-density aggregates

| | Type | Description | Sizes | Density Relative/bulk (kg/m$^3$) | Notes |
|---|---|---|---|---|---|
| Naturally occurring ores | Barytes | Barium sulphate ore $BaSO_4$ min 94% purity | 20 & 10 mm coarse Zones C & M fines | 4.2 2300 | Be wary of chemical impurities. High-fines fines may cause set retardation. Avoid excess handling of coarse – may breakdown in size. |
| | Magnetite | Iron oxide $Fe_3O_4$, 72.4% Fe | 20 & 10 mm coarse Fines 6 mm down | 4.9 3000 | Good particle shape, high-density fines but clean. |
| | Haematite | Iron oxide $Fe_2O_3$, 70% Fe | 20 mm coarse fines may be non-standard | 4.9 2900 | Can have an effluent problem due to intense red coloration |
| | Limonite | Iron oxide (hydrated) $2Fe_2O_3.3H_2O$, 60% Fe | Source dependent | 3.7 2200 | Inconsistent gradings. Variable relative density, high-density silt contents and water absorption possible. Not normally produced for concrete aggregates. |
| | Geothite | Iron oxide (hydrated) FeO(OH), 63% Fe | Source dependent | 3.5 2100 | |
| | Ilmenite | Iron titanium oxide $FeOTiO_2$, variable Fe | Source dependent | 4.6 approx. 2600 | |
| | Serpentine | Hydrous Mg-silicate | Source dependent | 2.6 1500 | Retains $H_2O$ at high-density temperatures, good neutron attenuator. |
| Manufactured | Iron Shot | Chilled iron or steel shot/grit | 8 to 0.1 mm single size or graded | 7.6 4800 | Grit normally added to increase cohesiveness of mix. |
| | Lead Shot | Pure lead processed in shot tower | 4 to 0.1 mm single sizes | 11.3 7000 | Careful mix design and cement selection – lead is soluble in alkalies and can form a gel. |
| | Fergran | Steelworks Scrap Materials | 20–5 mm coarse | 6.9 3900 | Poor particle shape, some oxide layer. |
| | Feriron | | 2 mm fine | 7.1 4100 | Good shape tending to spherical. |
| | Chilcon | Metallic ferroslag, slag material 80% iron 20% silicon | 20–5 mm coarse C-M fines | 6.3 3900 | Good particle shape. Extremely high-density 10% fines of 450 KN. Can exhibit good visual impact for exposed applications. |

*Barytes* is a naturally occurring ore, barium sulphate ($BaSO_4$), found in veins, is often associated with lead and zinc mineral ore deposits. A purity greater than 90 per cent is generally necessary to achieve a minimum concrete density of 3400 kg/m$^3$. This can vary with different aggregate sources having varying relative densities between 4.0 and 4.3. The mineral can often be tainted with brown coloration due to iron staining. In the UK barytes deposits are found in Northern England, principally the Pennines and to a lesser extent in Derbyshire, also in Scotland. There are also large deposits in Morocco, Germany, the USA and China.

## 5.5.2 Synthetic or man-made aggregates

*Iron shot* can be chilled iron or steel and is available from a limited number of suppliers in a variety of standard sizes as both shot and grit, normally with a maximum size of 8 mm. Suppliers can offer pre-blending options to produce either a maximum density or specific values required by their customers.

*Chilcon*, a ferroslag by-product of the silica fume and fused alumina industries is an inert material and available in limited supply within the UK, but is imported in significant quantities from China and Brazil. It has the appearance of metallic lustre limestone, but is much heavier. It has a very high 10 per cent fines value, in excess of 400 KN and is also used in floor toppings imparting very high wear properties. It is very important to use the correct grade of this material, one with a minimum iron content of 80 per cent. This is necessary to achieve maximum density and also to ensure chemical purity and stability as grades with higher silicon content may chemically react with alkali to liberate small volumes of hydrogen. Potential users should therefore undertake detailed risk assessments of specific materials prior to use.

One specialist aggregate supplier can provide a range of ferrous materials, which are essentially waste or scrap iron products. 'Fergran' is a nominal coarse 15–1 mm roughly shaped waste iron product while 'Feriron' is a nominal 2 mm rounded iron fine aggregate. These have been evaluated and successfully used to manufacture concrete, both structural and non-structural grades, and the supplier can offer advice on mix designs. Clean, large iron 'plums'can also be used in appropriate locations. It is essential that all these materials are free from organic contamination if used for aggregates.

*Lead shot* is available in a variety of sizes between 4 and 1 mm and normally supplied in single sizes in 12.5 kg bags. A full technical assessment for their use is necessary which must include the relevant safety regulations for the use of lead materials.

## 5.5.3 General considerations when assessing high-density aggregates

Aggregate gradings may not be ideal nor always comply with normal-density aggregate standards such as those specified in BS 882. This should not be considered detrimental to the production of high-density quality concrete. Potential users, including specifiers and contractors, must accept that these materials are, in general, rarely manufactured for concrete production. More important is grading consistency and this fact should be emphasized when discussing requirements with suppliers who should be able to agree and offer satisfactory, workable limits on which mix designs can be based.

The chemical properties of all high-density aggregates must be thoroughly evaluated before use with due consideration given to chemical reactivity, particularly in highly alkaline environments as found in cement pastes. Also aspects of long-term durability such as alkali–aggregate reactivity, sulphate and chloride attack together with other impurities should be evaluated. Barytes, for example, does not have sulphate issues due to its insolubility in water though it may contain small quantities of heavy metals such as lead and zinc, both of which can retard set due to chemical activity with cement paste. Its mineralogical properties indicate that it can be mechanically weak and friable and, depending on the source, it can reduce in size with prolonged mixing, and this aspect should also be fully evaluated prior to use. Further, lead shot is reactive in alkaline conditions and requires special mix designs and cement systems such as very high GGBS content cement or high alumina cement to satisfactorily counteract this reactivity.

It is recommended that users insist on receipt of full chemical analysis of all potential materials including similar minerals from different sources.

A specific aggregate type produces a concrete density commensurate with its inherent density. With suitable technical evaluation and planning it is, however, possible to combine different aggregate types, preferably at the supplier's works, to produce an aggregate, and in turn concrete with a predicted, specified density. For example, blending different proportions of magnetite and chilcon can produce 'tailor-made' concrete densities between about 3900 and 5300 kg/m$^3$ as illustrated in Table 5.3. Standard density sand can be used with a high-density coarse aggregate to give densities between about 2.5 and up to 4.7. Conversely, fine high-density aggregates can be blended with standard density coarse aggregates; for example, 20–5 mm limestone blended with magnetite fines can produce a concrete density of about 3.0. If this practice is to be adopted then the technologist must ensure that both the chemical and physical properties of the two aggregate types are fully compatible.

It must be appreciated that in some cases many of these materials may only be available in relatively short supply, magnetite being the exception in the UK. Long delivery times may consequently be encountered. Hence good resource planning is crucial. Also cost factors are important since all materials are *considerably* more expensive than conventional aggregates. In general, cost increases with density, but not necessarily in direct proportion.

Aggregates for grouts include magnetite, chilcon, iron shot and even lead shot. These are even more specialized though there is at least one supplier who offers a range of pre-

**Table 5.3** Example illustrating a range of concrete densities possible by blending two aggregate types

| Concrete density (kg/m$^3$) materials | 3800 | 4000 | 4200 | 4400 | 4600 | 4800 | 5000 | 5200 |
|---|---|---|---|---|---|---|---|---|
| % wt/wt magnetite | 100 | 95 | 80 | 63 | 48 | 32 | 16 | 0 |
| % wt/wt ferroslag | 0 | 5 | 20 | 37 | 52 | 68 | 84 | 100 |

*Notes*:
1 Percentages will depend on individual material grading
2 Blend of each aggregate type would normally consist of ideal coarse and fines combination
3 Trial mixes are necessary to confirm water demand, superplasticizer dose and workability
4 Minor adjustments may be necessary to achieve optimum density and yield
5 Ensure aggregate types are fully compatible.

blended grouts ready for use with densities of around 2.8, 3.5 and 4.3. Again off-the-shelf availability should not be assumed.

Other specialized aggregates find use in the nuclear industry. Serpentine, a naturally occurring magnesium–silicate ore, has a density of only about 2.6. It has the property of having stable water of crystallization within the structure, even at relatively high temperatures of about 450°C. This is similar to limonite and goethite iron ores, but stable to higher temperatures and all have been used in radiation shielding concrete because of this chemical stability of the combined water under the higher temperature environments likely to be encounted in these applications. Boron ores also can be used when high-energy neutrons are present. This material is generally in the form of the ore colemanite, a hydrated boron oxide. Care must be taken with boron materials, as they are likely to cause severe retardation in all concrete mixes even when used as a coarse aggregate.

## 5.6 Mix design

Mix design requirements will frequently include minimum density values as well as strength. Methods available for standard density concrete can be used with care for naturally occurring materials including magnetite and barytes, also for chilcon by making suitable adjustments to the SSD relative densities. Mixes can generally be successfully designed to comply with the various requirements detailed in EN 206–1/BS 8500. However, the somewhat non-standard gradings of these aggregates, together with some of the irregular gradings of the synthetic, or 'man-made' materials will often require a number of trial attempts in order to produce a cohesive mix with minimal segregation and bleeding. Good concrete technology experience coupled with occasional trial and error may be necessary at times to achieve satisfactory results particularly when blending materials that have significantly differing relative densities. For other aggregate types such as shot mixes the principle of maximizing aggregate bulk density is recommended. However, a balance between optimum density, time and cost needs to be considered, as these mixes are often used in only small volumes and do not therefore justify extensive mix development.

It should be noted that it could appear to be misleading when considering the aggregate/cement ratio (a/c) of a mix design unless this value is fully appreciated. For example, iron shot concrete has an a/c of about 12 but the cement content is approximately 400 kg/m$^3$; (a standard concrete with an a/c of 12 has a cement content in the order of 170 kg/m$^3$). Similarly, and often because of the unusual gradings that may be encountered, the fines percentages may appear unusual. Magnetite concrete may have a fines content of only around 25 per cent but can be pumped; this is due to the high fines content of the fine grade together with the relatively high fines content sometimes found in the coarse aggregate.

To minimize particle segregation and bleeding, both which can be problematic, a low free water/cement ratio (w/c) is advised, particularly in structural grades. This naturally improves durability. The use of superplasticizers is therefore recommended which can in addition marginally increase concrete density.

It is important that producers and users fully evaluate laboratory trial mix designs particularly when the aggregates are new. Whenever possible, workability and placing trials should be carefully assessed prior to use in permanent works. Also, despite the seemingly high costs, full-scale trials are recommended using the actual production plant and should be witnessed by all involved personnel including the engineer, concrete supplier and contractor. The concrete produced in this full-scale trial can be used by the concrete placing gang in a typical small section or as temporary works to assess the handling characteristics of the mix.

The cement can be OPC or a blend of this with either GGBS or PFA; the latter will not reduce density but may enhance otherwise poor workability. HAC can also be used under appropriate circumstances, particularly for lead shot to reduce the chemical reactivity of lead in high alkaline conditions. Some materials, barytes for example, tend to have a very coarse fine grading and in order to improve cohesiveness of such mixes it may be advantageous to incorporate higher cementitious contents, or a higher percentage PFA/GGBS content than required for normal strength or w/c criteria. This may seem an expensive way of modifying the fines content but it should be appreciated that the cement fraction is the cheapest ingredient in these mixes and easier to include than an additional small quantity of another fine, inert filler. Lower cement contents and high free w/c ratios can be used for non-structural applications such as counterweights enclosed in a steel box. In such applications an overall weight will often be an overriding criterion and the presence of an upper, less dense mortar fraction will normally not be an issue.

The use of the normal range of additives is recommended with due consideration given to the cement type. It will be found that use of superplasticizers is beneficial in reducing water to minimize bleeding and maintain a cohesive mix that has minimum segregation.

It has been found with some of the metallic aggregate that concrete cube strengths can be significantly higher compared with those found for a standard aggregate mix using similar cement contents and w/c ratios. For example, ferroslag aggregate concrete having a cement content of about 350 kg/m$^3$ and free w/c 0.45 can attain strengths of 80 + MPa.

Mix designs and specifications should consider the relationship between fresh density and hardened density. It is suggested that oven-dry values at 100°C are unrealistic under normal circumstances unless the concrete is to be used under these conditions. An appropriate density air-dried at say 50°C for 24 hours may represent a realistic condition though for thick sections, fully cured, this is likely to apply mainly at the surface rather than throughout the full section. In this context the importance of correct curing is emphasized. In some locations surfaces may eventually receive a decorative coating which will minimize long-term moisture loss.

## 5.7 Production, transporting and placing

### 5.7.1 Production

Standard batching procedures can be used for naturally occurring aggregates such as barytes and magnetite with volume modifications to allow for the increase in aggregate relative density. These, and occasionally other types, are sometimes required in significant volumes to justify large-scale batching through conventional plant. It is, however, often necessary to use smaller drum mixers, e.g. 21/14 (approximately 0.5 m$^3$) reverse-drum type, or even a $1^1/_2$ cubic foot capacity in the case of lead shot. To aid production, aggregates can be purchased from some suppliers pre-weighed in 'dump sacks' or other containers to suit the size of the batching plant, the additional cost being more than offset by an almost nil wastage. This type of storage provides additional protection from external contamination. Iron shot and other metallic aggregates can also be purchased in 25 kg bags whilst lead shot is normally delivered in 12.5 kg bags. This makes batching *very* tedious! Consideration must be given to storage; under cover is recommended to prevent contamination. Ferrous materials should be stored under cover in dry conditions.

It is recommended that the density and volume ratio of mixes compared with standard concrete be taken into account in storage bays and batching equipment. For example, a 1 $m^3$ capacity batching plant should not batch more than around 0.5 $m^3$ magnetite concrete based on relative weights and possible differences in weight distribution. The plant manufacturer should be consulted over this matter if necessary. The use of dump sacks, previously mentioned, for supplying relatively small amounts of aggregate has obvious advantages. It is noted that one supplier has the flexibility to blend precise quantities of coarse and fine aggregates of different types to offer tailor-made pre-mixed 'all-in blends as well as provide 'tailor-made' densities as previously noted in section 5.5.3.

Batching times will be longer per cubic metre, due to smaller unit batch volumes, but individual batch mixing times should be similar to those for standard aggregate concretes. Care must be taken with barytes to assess any potential for particle breakdown, since this material can be weak and friable. Conversely mixing times may need to be extended for materials such as fergran, which has a very rough and irregular shape. It is advisable to pre-grout mixers when small quantities are produced.

Workability is difficult to assess and some experience is necessary. Barytes and iron ore concrete can be measured using the slump test, though the differing rheological properties of these mixes require careful objective assessment. Magnetite mixes can be designed to have flows up to 600 mm whereas metallic aggregate mixes, particularly the iron aggregates such as fergran, may appear to have very low slump values estimated at about 20 mm otherwise serious segregation is likely during compaction. These mixes do, however, vibrate better than would be expected from their visual assessment. The workability of these and similar mixes should be judged on experience only, slump tests are not considered appropriate.

Water control is important, as this can affect density and segregation as well as strength. Shot mixes have consistent aggregate shape and grading and can accordingly be produced to very close tolerances where water control is even more important.

## 5.7.2 Transporting, placing and curing

Transporting can be by dumper or conventional truck mixers on a reduced volume basis and commensurate with density. This affects costs and pour times and small volumes may necessitate mixer drums to be pre-grouted. Place using skips, funnels or tremie tubes depending on access and allow for reduced volumes in skips. Smaller volumes, for example lead shot, will need to be transported in very small volumes (consider that a standard 10 litre bucket will weigh nearly 90 kg).

Barytes and iron oxide mixes can be designed for pumping, even over some considerable horizontal and vertical distances. It is recommended that pumping contractors are made aware of any requirement to pump these materials. Chilcon with natural sand fine aggregate has been pumped with difficulty. Mixes containing iron aggregates should not be pumped as damage to pumps is likely to occur.

Placing can be by conventional poker vibrators, generally a large size such as 3 inches, though a significantly reduced 'radius of action' is found. This results in longer compaction times as the poker will need to be inserted at closer intervals. Similarly, concrete layers should be reduced in thickness and under no circumstances should pokers be used to move concrete horizontally.

It is sometimes found that a surface laitence layer may be formed particularly when placing the concrete in high lifts. Contrary to normal concrete practice it is recommended this layer should either (a) be removed and replaced with concrete or (b) topped up with aggregate blend to maintain the density. The end use of the structure will decide if it is very important to maintain a full height density profile or whether only a total weight is necessary.

All high-density concrete should be cured in a similar way to standard concrete. Metallic aggregate concretes, if not contained, should be protected from surface corrosion using appropriate durable coatings.

### 5.7.3 Summary

With careful planning, minimal problems should occur in the production and placing of high-density concrete but the significant cost increases incurred for these, often small-volume mixes, must be allowed and planned for. In addition small-volume production, particularly shot mixes tends, to be labour intensive. For example, 1 $m^3$ is considered a large pour for lead shot concrete and may take several hours. The additional costs for all trial mixes together with those for technical supervision should not be underestimated.

## 5.8 Concrete properties

### 5.8.1 Fresh concrete

The density of high-density concrete is important, if not more so at times, than strength and a suitable testing regime should be adopted to continually monitor this property using fresh density tests on-site before the concrete is placed. Cube or cylinder tests are similar to normal concrete. Generally use 100 mm cubes as 150 mm can be very heavy and are of significant cost. Hence additional technical staff will be required to undertake the increased amount of quality control necessary both at the mixing plant and on-site. Fresh density tests can be evaluated using a fine aggregate density pot; the normal 0.01$m^3$ pot will be very heavy (approaching 90 kg for lead shot concrete!). Test methods are similar to those for standard density products though additional tamping may be required to fully compact mixes containing poorly shaped aggregates.

High-density concrete can be prone to segregation and bleeding. These properties can generally be managed with correct mix design though users and producers should be aware of this particularly during compaction.

### 5.8.2 Hardened concrete

High-density concrete can be used for both structural and non-structural purposes, as well as for infill. The exception is lead shot concrete, which must be placed in appropriately designed boxes and only used in non-structural locations. Structural designers should establish relevant structural properties, particularly regarding compressive and tensile strengths, shear characteristics, elasticity behaviour and shrinkage values. For example, shear strengths for barytes generally has a lower value compared to standard aggregates whereas this property can be reversed for the poorly shaped scrap iron aggregates. Table 5.4 summarizes some typical properties of a variety of mix types.

Very high cube strengths can be achieved with some metallic and ferroslag (chilcon) concrete. Values of 80 MPa at 28 days have been achieved for mixes with 350 kg/m$^3$ cement and 0.45 w/c.

The high coefficient of thermal expansion and relatively low thermal conductivity properties of barytes concrete are examples that illustrate the necessity to fully investigate all additional properties of these concretes, and the relevance of these will depend on the intended end use. Thermal conductivity and other related properties can be of specific interest in the design of some nuclear facilities.

Kaplan (1989) gives comprehensive details of structural, nuclear and other properties for a wide variety of concrete types.

## 5.9 High-density grouts

These are mentioned briefly, as they are even more specialized than high-density concrete. They can be developed 'in-house' with care and a good knowledge of grout technology but can sometimes be commercially available from specialist grout firms who can offer grouts with a range of densities from 2900 kg/m$^3$ through to 4300 kg/m$^3$. A cohesive, pourable lead shot grout with a density of 6600 kg/m$^3$ has been successfully produced. All these products may require reasonable flow properties, be stable with minimal bleeding and segregation and of course be shrinkage compensated. It is necessary to utilize special additives such as thickening agents to stabilize these products. Required volumes are generally small and production is best achieved by mixing in dustbins or similar containers with powerful drill and whisk-type paddle.

## 5.10 Quality management

High-density concrete and grouts require careful planning at the design, tender assessment and pre-production stages of a contract as well as during actual production on-site. Costs can be very high compared with standard concrete and grouts as volumes are often small to be successfully produced in conventional ready- or site-mixed concrete plants. As previously mentioned, workability can be difficult to assess, hence the need for trial mixes which should be attended by site personnel including those actually placing the material. A higher degree of quality control is needed, with fresh densities being frequently measured prior to mixes being placed. This requires appropriate equipment that is not usually available on site. Full method statements and quality plans for all stages, including materials procurement, planning, production, placing and testing, should be prepared and approved prior to starting work. It is strongly recommended that all parties involved are clearly aware of their own responsibilities at each stage of the work. The use of these checks will often prevent what could potentially be very costly mistakes for a small volume of concrete or grout.

**Table 5.4** Summary of *typical* properties of high-density concretes

| Property \ Concrete type | Barytes | Magnetite | Chilcon coarse natural sand | Chilcon coarse & fines | Fergran coarse feriron fines | Iron shot | Lead shot | Typical standard concrete comparison |
|---|---|---|---|---|---|---|---|---|
| Typical cement content ($kg/m^3$) | 330+ | 200–400 | 360 | 350 | 300 | 400 | 500 | 330 |
| Typical free w/c (+ s/plasticizer) | 0.50 | 0.7–0.45 | 0.45 | 0.45 | 0.45 | 0.43 | 0.34 | 0.50 |
| Typical density ($kg/m^3$) | 3500 | 4000 | 4400 | 5200 | 5300 | 5900 | 8600 | 2300–2400 |
| 28-day compressive strength (100 mm cube) (MPa) | 20–35 | 20–40 | 70.0+ | 80.0+ | 45–55 | 30–45 | 20.0 (NB: Non-structural) | 30–40 |
| Indirect tensile strength (200 × 100 mm cylinders) (MPa) | 2.7 | 4.0 | 5.0 | 5.2 | 6.0 (est.) | 4.5 | N/A | 3.5 |
| Shear characteristics (compared with standard aggs) | Lower | Similar | Better | Better | Better | Slightly lower | N/A | |
| Young's modulus (GPa) | 20 | 20–35 | 40 | 50 | 50–60 (est.) | 70 (est.) | N/A | 20–30 |
| Flexural strength (MPa) | 3.7 | 6.0 | 6.0+ | 8.5 | N/A | 6.5 | N/A | 5.5 |
| Dry shrinkage (%) | 0.04 | 0.04 | Similar | Similar | 0.03 (est.) | 0.02 | N/A | 0.04 |
| Thermal conductivity (W/mk) | 1.1 | Similar | 2.5 (est.) | 2.5 (est.) | 25–30 (est.) | 33 | N/A | 1.9 |
| Coefficient of linear expansion ($\times 10^{-6}$µm/°C 30–90°C) | 22 (high) | 10.0 | N/A | N/A | 12 (est.) | 12.5 | N/A | 8–15 |

*Notes*:
1 Results are regarded as typical and will vary with mix design and/or material sources changes. They should be verified for specific use.
2 Mechanical properties are typical for structural mixes and will differ if lower cement contents are used for non-structural mixes.

## 5.11 Specifications

When high-density concrete or grout is required it is advisable that the design engineer issues addendum specifications to the standard concrete section, as some aspects of their use are fundamentally different from standard concrete production. Points for consideration are summarized in the list below:

*Aggregates* Source limitations and delivery
- Cost
- Appropriate physical, chemical and petrological properties
- (Note that alternative sources of the same generic aggregate type may have fundamental differences in some properties)

*Concrete* Strength
- Other appropriate mechanical and physical properties
- Low free w/c ratio
- Mandatory use of superplasticizers
- Laboratory and full-scale trial mixes to confirm fresh and hardened properties including fresh density

Designers should consider the implications of non-compliant mixes, with particular respect to marginal densities. This may be important and some statistical safety margin may need to be included on a basis similar to those used for compressive strength. A final point to consider is the potential expense of remedial work. Some grades of high-density concrete, particularly those containing iron aggregate, will be extremely difficult to remove from a structure, even in very small volumes. This fact also has relevance at design stage, for example if bolts or other fastenings are required as drilling can be slow and expensive.

## 5.12 Summary

A range of high-density concrete and grouts can be satisfactorily manufactured and used for construction purposes. They can be used as mass fill or in structural locations. These are normally regarded as specialist products for nuclear shielding purposes but have significant other uses, for example counterweights, bridge foundations and under-water pipeline coverings. They are normally required to fulfil the dual compliance for strength and density and as such their quality must be closely monitored and verified during production. Further, high-density concrete and grouts are significantly more expensive than normal products and as such justify the application of a high degree of quality management to all aspects of their use from design and specification through to all aspects of production and testing.

## References

Many references are from the USA and were written specifically for the nuclear industry during the 1950–60 era. They generally refer to aggregate properties and mix designs for specific sources of the USA. These references have not been included.

American Concrete Institute Publication 304R-85, *Guide for measuring, mixing, transporting and placing concrete.*

American Concrete Institute Special Publication 34, *Preliminary recommendation for design, making and control of radiation-shielding concrete structures.*

ASTM C637, *Standard Specification for Aggregates for Radiation-Shielding Concrete.*

ASTM C638, *Descriptive Nomenclature of Constituents of Aggregates for Radiation-Shielding Concrete.*

Kaplan, M.F. (1989) *Concrete Radiation Shielding, Nuclear Physics, Concrete Properties, Design and Construction.* Longman Scientific and Technical, Harlow. (*Note*: this is a comprehensive treatise on the subject and contains many additional references.)

6

# Fibre-reinforced concrete

*D.J. Hannant*

## 6.1 Introduction

The use of fibres in brittle matrix materials has a long history going back at least 3500 years when sun-baked bricks reinforced with straw were used to build the 57 m high hill of Aqar Quf near Baghdad. In more recent times, asbestos fibres have been used to reinforce cement products for about 100 years, cellulose fibres for at least 50 years, and steel, polypropylene and glass fibres have been used for the same purpose for the past 30 years. The main objectives of the modern engineer in attempting to modify the properties of concrete by the inclusion of fibres are as follows:

(a) To improve the rheology or plastic cracking characteristics of the material in the fresh state or up to about 6 hours after casting.
(b) To improve the tensile or flexural strength.
(c) To improve the impact strength and toughness.
(d) To control cracking and the mode of failure by means of post-cracking ductility.
(e) To improve durability.

This chapter concerns the use of fibres in concrete as opposed to cement because the latter is a more specialist field which is commonly controlled by suppliers of relatively fine-grained, thin sheet materials such as glass-reinforced cement or alternatives to asbestos cement containing polyvinyl alcohol or cellulose fibres.

It is generally accepted that the inclusion of any type of short fibre in a three-dimensional random fibre distribution at practical fibre volumes will not significantly alter the load at which cracking occurs in hardened concrete. Therefore the main benefits of the inclusion of fibres in hardened concrete relate to the post-cracking state. In this context, it is worth considering an understanding of the word 'reinforcement'. If it is assumed that any load-bearing capacity greater than zero is described as reinforcement, then all types of fibres at any volume addition will reinforce hardened concrete. However, if we consider 'reinforcement' to mean carrying a force in excess of the force required to crack the concrete, then less than about 0.4 per cent by volume of short three-dimensional random fibres will not generally provide load capacity in excess of the cracking load in beams and slabs, and two or three times this fibre volume is required to increase the load capacity in uniaxial tension.

## 6.2 Properties of fibres and matrices

The performance of the composite is controlled mainly by the volume of the fibres, the physical properties of the fibres and the matrix, and the bond between the two. Values for bond strength for straight round steel fibres rarely exceed 4 MPa but certain types of mechanical deformation can prevent bond slip altogether resulting in fibre failure. In the range between fibre slip and fibre failure there is a variety of fibre deformation and slip mechanisms depending on the proprietary fibre shape and concrete strength, so that it is not possible to give a generalized figure which could be used in numerical calculations as a 'bond strength'. Bond strength is generally less than 2 MPa for polymer fibres where mechanical deformations are rarely possible due to the small diameter or thickness of fibres or films. The bond strengths will also change with time and with storage conditions which may permit densification of the interface region due to continuing hydration. Typical ranges for other physical properties of fibres and matrices are shown in Table 6.1.

It is apparent from this table that the elongations at break of all the fibres are two or three orders of magnitude greater than the strain at failure of the matrix and hence the matrix will usually crack long before the fibre strength is approached. This fact is the reason for the emphasis on post-cracking performance in the theoretical treatment in section 6.3.

On the other hand, the modulus of elasticity of the fibre is generally less than five times that of the matrix and this, combined with the low fibre volume fraction, means that the modulus of the composite is not greatly different from that of the matrix.

## 6.3 Post-cracking composite theory

The main benefits of the inclusion of fibres in hardened concrete relate to the post-cracking state, where the fibres bridging the cracks contribute to the increase in strength, failure strain and toughness of the composite.

## 6.4 Theoretical stress–strain curves in uniaxial tension

### 6.4.1 Characteristic shapes of stress–strain curves

Fibre-reinforced concretes are generally considered to be most useful when carrying bending or impact loads but, unfortunately, a theoretical analysis of the mechanics of

**Table 6.1** Typical properties of cement-based matrices and fibres

| Material or fibre | Relative density | Diameter or thickness (microns) | Length (mm) | Elastic modulus (GPa) | Tensile strength (MPa) | Failure strain (%) | Volume in composite (%) |
|---|---|---|---|---|---|---|---|
| Mortar matrix | 1.8–2.0 | 300–5000 | – | 10–30 | 1–10 | 0.01–0.05 | 85–97 |
| Concrete matrix | 1.8–2.4 | 10 000–20 000 | – | 20–40 | 1–4 | 0.01–0.02 | 97–99.9 |
| Asbestos | 2.55 | 0.02–30 | 5–40 | 164 | 200–1800 | 2–3 | 5–15 |
| Carbon | 1.16–1.95 | 7–18 | 3-continuous | 30–390 | 600–2700 | 0.5–2.4 | 3–5 |
| Cellulose | 1.5 | 20–120 | 0.5–5.0 | 10–50 | 300–1000 | 20 | 5–15 |
| Glass | 2.7 | 12.5 | 10–50 | 70 | 600–2500 | 3.6 | 3–7 |
| Polypropylene | | | | | | | |
| monoflament | 0.91 | 20–100 | 5–20 | 4 | – | – | 0.1–0.2 |
| chopped film | 0.91 | 20–100 | 5–50 | 5 | 300–500 | 10 | 0.1–1.0 |
| Polyvinyl alcohol | | | | | | | |
| (PVA, PVOH) | 1–3 | 3–8 | 2–6 | 12–40 | 700–1500 | – | 2–3 |
| Steel | 7.86 | 100–600 | 10–60 | 200 | 700–2000 | 3–5 | 0.3–2.0 |

reinforcement in these systems is very complex. A more fundamental and a more easily understood stress system is that of direct tension and a knowledge of the behaviour of fibres in such a system provides a good background by which an engineer can judge the potential merits of a fibre cement composite for a given end use.

Two basic types of tensile stress-strain curve available to the engineer are shown in Figure 6.1. The portion OX defines the elastic modulus of the uncracked composite ($E_c$) which cracks at its normal cracking stress. Curve A is where the fibre volume is very low or else where the fibres are so well bonded that most of them break at cracking of the concrete. Curve B is for the higher volumes of steel fibres which pull out rather than break. The area under these curves defines the toughness of the fibre concrete which is usually well in excess of the area under OX.

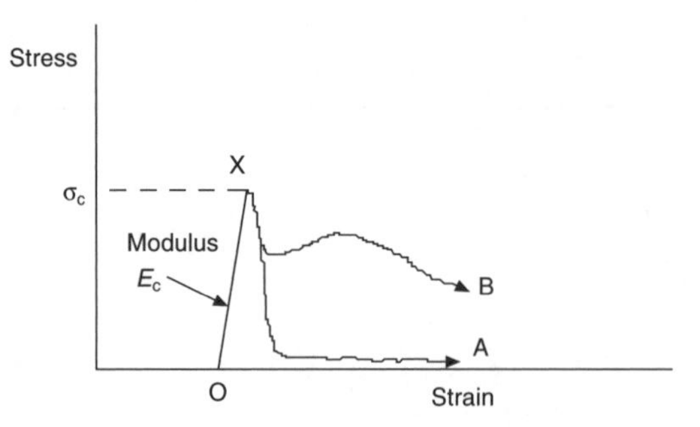


**Figure 6.1** Typical stress–strain curves for short random fibre concrete.

## 6.4.2 Critical volume fraction ($V_{fcrit}$) in uniaxial tension

The critical fibre volume is the volume of fibres which after matrix cracking will carry the load which the composite maintained before cracking. This definition needs to be used with a little care because material which has less than the critical volume of fibre in tension (curve OXB) may have greater than the critical fibre volume required for flexural strengthening. However, it is common practice to assume that the above definition of critical fibre volume refers only to uniaxial tensile stresses.

For the simplest case of fibres all aligned parallel to the applied stress:

$\sigma_c$ = composite cracking stress
$V_{fcrit}$ = critical volume of fibres
$\sigma_f$ = fibre strength or average pull-out stress of fibre depending on whether fibres break or pull-out at a crack.

After cracking, the whole stress is carried by the fibres. Assume that there are just sufficient fibres to support this stress, i.e. fibre volume = $V_{fcrit}$:

$$\sigma_c = \sigma_f V_{fcrit} \tag{6.1}$$

$$V_{fcrit} = \sigma_c / \sigma_f \tag{6.2}$$

Thus, if $\sigma_c = 3$ MPa, $\sigma_f = 300$ MPa, then $V_{fcrit} = 0.01$ or 1 per cent.

There are important points to note about equation (6.2):

1 $V_{fcrit}$ can be decreased by decreasing $\sigma_c$.
2 Poor bond may reduce $\sigma_f$ by allowing fibre pull-out at a fraction of the fibre strength.

## 6.4.3 Short random fibres which pull out rather than break

Factors affecting a realistic estimate of $V_{fcrit}$ and post-cracking strength are:

1 Number of fibres across a crack
2 Bond strength and fibre pull-out or fracture load

### *Number of fibres across a crack*

The situation in a cracked composite may be represented by Figure 6.2. Some fibres may fracture and others pull out depending on anchorage length and proprietary shape. If composite failure is by fibre pull-out, the average pull-out length is l/4. For small crack openings a realistic estimate of the load carried after cracking can therefore be obtained by multiplying the number of fibres crossing a unit area of crack by the average force per fibre.

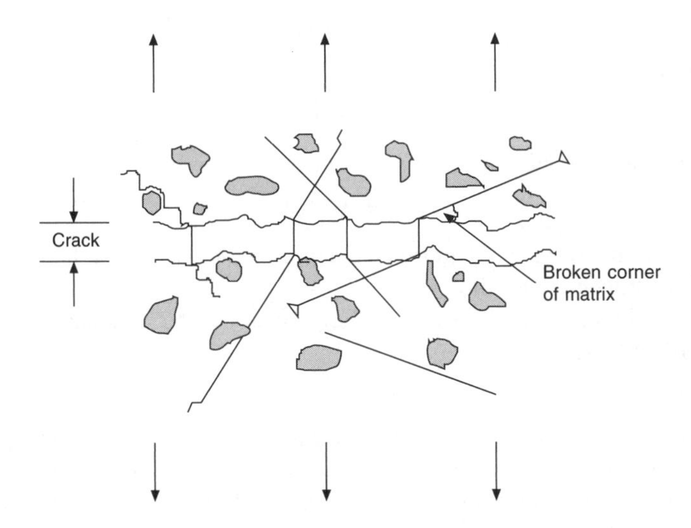


**Figure 6.2** Cracked short fibre composite containing *N* fibres per unit area and showing change in fibre orientation at a crack.

The appropriate number of fibres, *N*, per unit area can be calculated as follows:

$V_f$ = total fibre volume in a unit volume of composite
$A_f$ = cross-sectional area of a single fibre

For fibres aligned in one direction

$$N = \frac{V_f}{A_f} \tag{6.3}$$

For fibres random in two dimensions

$$N = \frac{2}{\pi}\frac{V_f}{A_f} \tag{6.4}$$

For fibres random in three dimensions

$$N = \frac{1}{2}\frac{V_f}{A_f} \tag{6.5}$$

### *Post-cracking composite strength*

The average pull-out force per fibre ($F$) is given by:

$$F = \sigma_f A_f \tag{6.6}$$

where $\sigma_f$ is the stress in the fibre at pull-out or fracture. The ultimate stress ($\sigma_{cu}$) sustained by a unit area of composite after cracking is therefore given by $N$ times $F$, i.e.

$$\sigma_{cu} = N \cdot \sigma_f A_f \tag{6.7}$$

Substituting for $N$ from equations (6.3)–(6.5) in equation (6.7) gives

Aligned fibres in one direction

$$\sigma_{cu} = \sigma_f V_f \tag{6.8}$$

Random two-dimensional

$$\sigma_{cu} = 2\sigma_f V_f/\pi \tag{6.9}$$

Random three-dimensional

$$\sigma_{cu} = \sigma_f V_f/2 \tag{6.10}$$

Thus, the three-dimensional random fibre concrete should have about half the post-crack strength of the aligned composite.

The critical fibre volume for short random fibres can be obtained by re-arranging equation (6.10) to give:

$$V_{fcrit} = 2\sigma_c/\sigma_f \tag{6.11}$$

The average value of $\sigma_f$ at the maximum value of post-crack stress will vary widely depending on the concrete strength, fibre dimensions and anchorage mechanisms and can only be determined experimentally. If the anchorage detail causes the fibres to fail, then $\sigma_f$ is equal to the fibre strength, but some fibres will always pull out due to a short embedded length or local concrete fracture. In these circumstances, an average $\sigma_f$ for good steel fibres can be 300–500 N/mm$^2$ and for polypropylene fibres can be about 200–300 N/mm$^2$.

## 6.5 Principles of fibre reinforcement in flexure

### 6.5.1 Necessity for theory

In many of their major applications, cement-bound fibre composites are likely to be subjected to flexural stresses in addition to direct stresses, and hence an understanding of the mechanism of strengthening in flexure may be of equal importance to the analysis of the direct stress.

The need for a special theoretical treatment for flexure arises because of the large differences which are observed experimentally between the post-cracking flexural strength and the post-cracking uniaxial tensile strengths, both in glass-reinforced cement and in steel-fibre concrete. In both of these materials the so-called flexural strength can be much greater than the uniaxial tensile strength even though, according to elastic theory, they are nominally a measure of the same value. The same situation occurs to a lesser degree with plain concrete.

The main reason for the discrepancy in fibre concrete is that the post-cracking stress–strain curve X–A and XB in Figure 6.1 on the tensile side of a fibre concrete or fibre cement beam are very different from those in compression and, as a result, conventional beam theory is inadequate. The flexural strengthening mechanism is mainly due to this quasi-plastic behaviour of fibre composites in tension as a result of fibre pull-out or elastic extension of the fibres after matrix cracking.

Figures 6.3(a) and (b) show a cracked fibre-reinforced beam with a linear strain distribution and the neutral axis moved towards the compression surface. At a crack the fibres effectively provide point forces holding the section together (Figure 6.3(c)). However, the exact stresses in the fibres are generally unknown and are ignored in flexural calculations and an equivalent composite stress block such as in Figure 6.3(d) is assumed. The shape of this stress block depends on crack width and fibre volume, bond strength, orientation and length efficiency factors. An accurate analysis of such a system presents formidable problems but a simplified treatment which is satisfactory for many practical situations is given below.

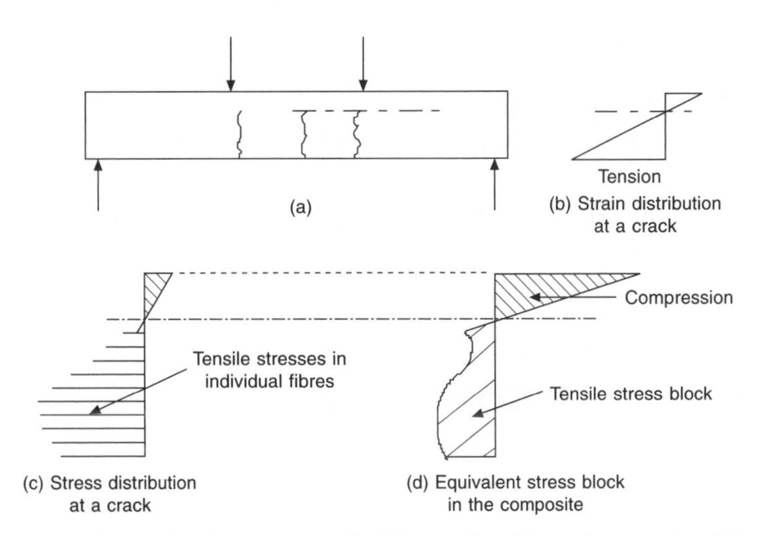


**Figure 6.3** Strain and stress distributions in a cracked fibre-reinforced beam. (*Note:* Scales of (c) and (d) are different.)

## 6.5.2 Analysis using a rectangular stress block in the tensile zone of a beam

The analysis which follows is based on a simplified assumption regarding the shape of the stress block in the tensile zone immediately after cracking. The stress block for an elastic material in bending is shown in Figure 6.4(a) and this is usually used to calculate the flexural strength ($\sigma_{fl}$) even although it is known to be grossly inaccurate for quasi-ductile fibre composites.

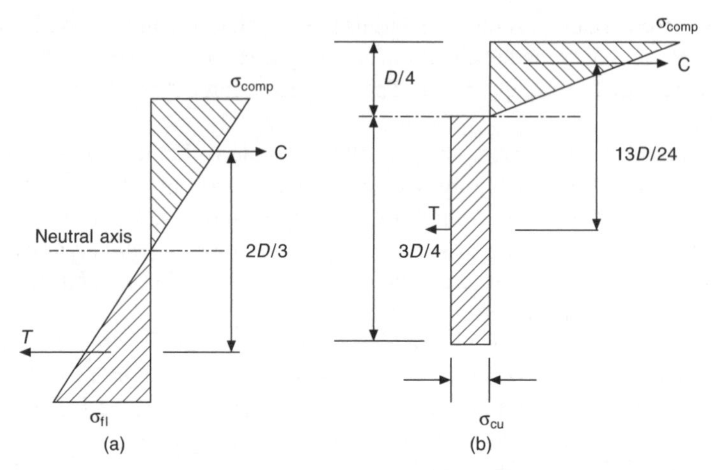


**Figure 6.4** Stress blocks in flexure. (a) Elastic in tension and compression. (b) Elastic in compression, plastic in tension.

Figure 6.4(b) shows a simplifiication of the stress block in Figure 6.3(d) where the fibres are extending or are pulling out at approximately constant load across small crack widths (~0.5 mm). The post-cracking tensile strength of the composite is $\sigma_{cu}$ and $\sigma_{comp}$ is the compressive stress on the outer face of the beam. A conservative estimate for the distance of the neutral axis from the compressive surface is $\frac{1}{4}D$ and using this assumption the moments of resistance of the two stress blocks can be compared:

$$\text{moment of resistance} = \frac{1}{6}\sigma_{fl}D^2 \quad \text{for Figure 6.4(a)} \tag{6.12}$$

$$\text{moment of resistance} = \frac{13}{32}\sigma_{cu}D^2 \quad \text{for Figure 6.4(b)} \tag{6.13}$$

In order that the two beams represented in Figure 6.4 can carry the same load, their moments of resistance should be equal, i.e.

$$\frac{1}{6}\sigma_{fl}D^2 = \frac{13}{32}\sigma_{cu}D^2 \tag{6.14}$$

Therefore

$$\sigma_{fl} = 2.44\sigma_{cu} \tag{6.15}$$

Equation (6.15) implies that a fibre concrete with less than half the critical fibre volume in tension, as shown in Figure 6.5, will not exhibit a decrease in flexural load capacity immediately after cracking, implying that the critical fibre volume in flexure has been achieved. This is the case for the higher fibre volumes of the most efficient fibres in steel fibre concrete.

This type of simplified analysis explains why the post-cracking flexural strength for fibre concretes may be quoted to be about twice the tensile strength. For this reason it is preferable to avoid flexural tests on fibre-reinforced concrete wherever possible when the post-crack tensile strengths are required for design purposes.

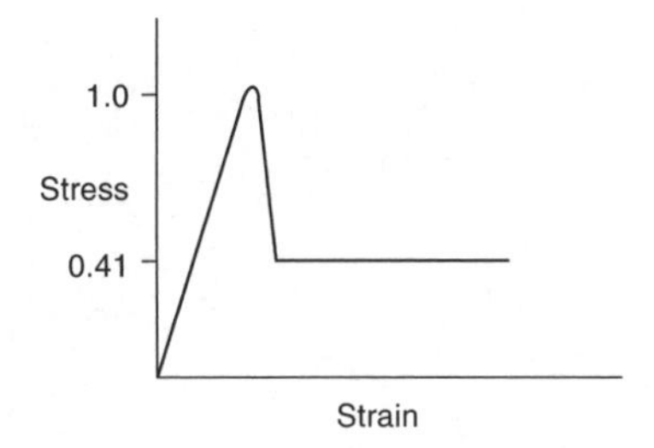


**Figure 6.5** Stress–strain curve in uniaxial tension. This would give no decrease in flexural load after cracking.

## 6.6 Steel fibre concrete

Concrete reinforced with chopped steel fibres in volumes generally less than 1 per cent has a tensile stress–strain curve of the type shown as OXB in Figure 6.1. The reason for this is that it is physically very difficult to include sufficient fibres in the mix to exceed the critical fibre volume which, for short random three-dimensionally oriented fibres is given by equation (6.11).

This critical fibre volume for short steel fibres generally exceeds 1 per cent. Apart from the economics, it is physically difficult to include these fibre volumes at an aspect ratio of more than 50 because concrete contains about 70 per cent by volume of aggregate particles which obviously cannot be penetrated by fibres. Also, the fibres tend to end up in a three-dimensional random distribution when mixed in a rotary mixer which, together with their short length, makes them very inefficient at reinforcement in any given direction of tensile stress. Nevertheless, useful properties in the composite have been achieved by many practical systems. A great variety of fibre shapes and lengths are available depending on the manufacturing process, a few of the more common types being shown in Figure 6.6. Cross-sectional shapes include: circular (from drawn fibres); rectangular (from slit sheet or milled from ingots); sickle shaped (from the melt extract process); and mechanically deformed in various ways to improve the bond strength. Fibre lengths range from 10 to 60 mm with equivalent diameters between 0.5 and 1.2 mm. Mild steel, high tensile steel and stainless steel fibres are available.

It should be realized that the average fibre pull-out length is $l/4$ which for the longest 60 mm fibres, is only 15 mm. This length is insufficient to allow efficient use to be made of the high tensile strength of drawn wire unless devices such as bends, crimps or flattened ends are used to improve anchorage efficiency.

## 6.7 Mix design and composite manufacture

Mix designs for rotary mixers generally recommend the use of fibres with an equivalent length/diameter ratio of 70 or less and typical weights of fibre are between 20 kg/m$^3$ and 60 kg/m$^3$. These weights are equivalent to fibre volumes between 0.25 per cent and 0.76 per cent.

It is important to use a relatively high proportion of fines, for example a typical mix may contain 800 kg/m$^3$ of river sand and 300–350 kg/m$^3$ of cement often with the further

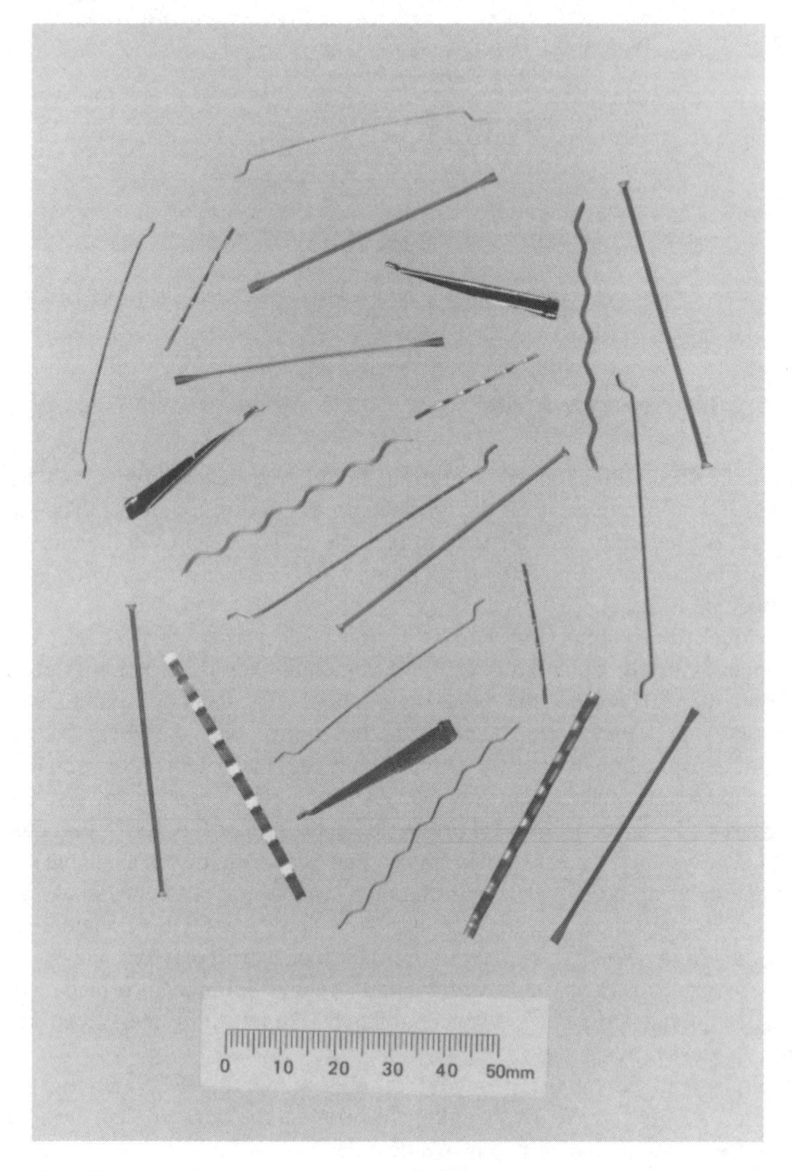


**Figure 6.6** Steel fibres showing a variety of types of mechanical deformation.

addition of pulverized-fuel ash. Aggregates larger than 20 mm should be avoided and it may be preferable to limit the aggregate fraction larger than 14 mm to 15–20 per cent. Free water cement ratios of less than 0.55 are preferable and workability is commonly improved by the addition of plasticizers or superplasticizers to give slumps of more than 100 mm.

Longer fibres give better reinforcement but reduce the workability so that a compromise must be reached usually at equivalent $l/d$ ratios between 40 and 80 with fibre lengths between 20 mm and 60 mm.

Generally the fibres are added last to the fresh concrete, care being taken to ensure that no clumps are added and the fibres are rapidly removed from the entry point to the mixer. Alternatively they may be added onto the aggregate on the conveyor belt. Proprietary systems exist for introducing fibres into a readymix truck by blowing the individual fibres at high velocity into the back of the truck. Collated fibres which are glued together with a water-soluble glue also considerably assist the batching and mixing actions with bags of fibres being added to the mixer drum, the paper bags themselves also being water-soluble.

Guniting or sprayed steel fibre concrete is another manufacturing process which is widely used for tunnel linings and rock slope stabilization. Fibre lengths typically range from 25 mm to 40 mm. As with normal concrete, the greater the fibre aspect ratio and fibre volume, the better is the performance of the sprayed concrete but the more difficult it is to mix, convey and spray. Sprayed mixes generally have less than 10 mm maximum sized aggregate with cementitious contents in excess of 400 kg/m$^3$ and fibre contents between 30 kg/m$^3$ and 80 kg/m$^3$. Cement replacements and admixtures are widely used.

In a third manufacturing process, fibre volumes up to 20 per cent are pre-placed into a mould before mixing and are then infiltrated with a fine-grained cement-based slurry. This gives very high strength and toughness in localized regions such as beam/column intersections. Flexural strengths up to 60 MPa and tensile strengths up to 16 MPa are said to be possible with this system.

## 6.8 Properties

Steel fibres provide virtually no increase in the compressive or uniaxial tensile strength of concrete. The main benefits in uniaxial tension result from the control of crack widths due to shrinkage or thermal effects in slabs and tunnel linings and this is not an easily quantifiable parameter but relates to post-cracking fibre pull-out or fracture forces. Post-cracking uniaxial tensile strengths of 0.5–1.5 MPa are possible at commonly used fibre volumes, at crack openings up to about 2 mm. For the lower fibre volumes, the post-cracking flexural strength calculated from elastic theory is generally less than the matrix cracking strength but nevertheless exceeds the post-cracking tensile strength as a result of the increased area of the tensile part of the stress block as shown in Figures 6.3 and 6.4. The post-crack flexural performance is a most important part of the commercial uses of steel fibre concrete enabling reductions of thickness to be made in sections subject to flexure or point loads. A typical stress/deflection curve for beams with a high fibre volume is shown in Figure 6.7 and demonstrates the post-cracking characteristics in bending which result from the ductile characteristics of the tensile stress block even although the fibre volume is less than the critical fibre volume in tension. Impact strength and toughness, defined as energy absorbed to failure, are greatly increased. The increased toughness results from the increased area under the load deflection curve in tension and flexure. A variety of toughness indices have been proposed in the literature depending on the flexural deflection which is chosen to represent a typical serviceability limit. Improved fatigue resistance is often claimed but this is a complex parameter which depends so heavily on fibre type and volume that no generalized improvements can be stated.

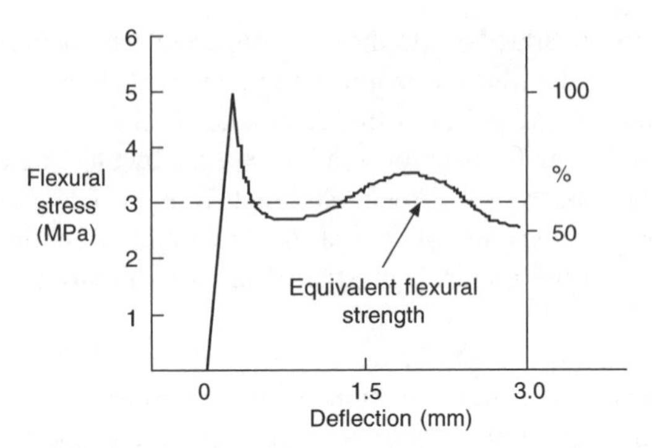


**Figure 6.7** Typical stress–deflection curve for steel fibre concrete used to determine equivalent flexural strength in the post-crack state.

### 6.8.1 Durability

Steel fibres are generally well protected in uncracked concrete where the high alkalinity provides a passive layer on the fibre surface. Even when the fibres are near the surface in a carbonated zone, serious corrosion takes many years to occur and surface spalling is rare. The main durability problem is likely to occur where load-bearing carbon steel fibres are exposed across cracked sections in the presence of chlorides, where they will readily corrode and it would be wise in such conditions to use stainless steel fibres.

## 6.9 Testing

### 6.9.1 Fresh concrete

Fibre distribution and dosage can be checked by wash out tests on random samples and some manufacturers have automatic plant to perform these measurements. Standard tests for determining the fibre distribution in both fresh and hardened states are available (EFNARC, 1996).

### 6.9.2 Hardened concrete

From the point of view of specification and control testing, there are additional complexities compared with conventional concrete. Neither the cube strength nor the flexural strength is greatly changed by the addition of fibres at normal fibre dosages and therefore other methods have been devised for design purposes and to assess whether the product, as placed, fufils the specification requirements.

A method which is gaining acceptance is to assess the equivalent flexural strength by measuring the load–deflection curve up to 3 mm deflection for 150 mm by 150 mm by 600 mm long beams tested in 1/3rd point bending (JCI, 1984; ASTM, 1997). A jig is necessary to eliminate support deflections and an approximately constant deflection rate must be achieved after the maximum load. This is not normally possible with load-controlled hydraulic machines and therefore a closed-loop servo control from the deflection-measuring devices is used. Otherwise, errors due to immediate post-cracking instability caused by energy release from the test system may give results which are not repeatable by other test machines. By this means, the post-cracking characteristics and toughness index can be determined from the equivalent flexural strength up to 3 mm deflection (see Figure 6.7).

## 6.10 Applications

A major use of steel fibres is to use them as a replacement for conventional steel mesh in industrial ground-floor slabs. Fibre dosages of between 15 kg/m$^3$ and 60 kg/m$^3$ are commonly used in floors with slab thicknesses between 120 mm and 200 mm. The fibres are particularly beneficial in the laser screed process for large-area pours (> 1000 m$^2$/day) because they avoid interruptions to the construction process caused by placing formed joints and mesh reinforcement. Joints are normally still necessary and induced joints should be cut to about 1/4 of the slab depth with a diamond saw within the first 24 hours after casting at between 5 m and 10 m centres.

Alternatively, where joints have to be formed, steel-edged permanent shutters with lateral movement capability are beneficial in preventing edge breakdown. When sawn joints have formed due to restrained contraction, the fibres restrict joint opening and maintain aggregate interlock at the joint surface. Also when subjected to flexural stresses, the load at which cracks become visible on the top surface can be increased compared to plain concrete slabs.

The main advantages claimed by specialist users of the laser screed process are: elimination of mesh which makes faster, more simplified construction, reduction of manpower costs, saving of dowel bars, tie wire etc., better performance of joint arrises induced by diamond blade cutting, and more uniform joint opening than when no mesh or fibres are used.

The typical stress maintained across a joint after cracking may be calculated from equation (6.10) of section 6.4.3, i.e. for 30 kg/m$^3$ of steel fibres (0.38 per cent by volume), 60 mm long by 1 mm in diameter at an average fibre stress of 400 N/mm$^2$ the maximum post-cracking tensile strength, $\sigma_{cu}$, will be

$$\sigma_{cu} = \sigma_f V_f / 2 = 400 \cdot 0.0038/2 = 0.76 \text{ N/mm}^2$$

This apparent composite post-cracking strength is equivalent to a force of about 114 kN/m (11.4 T/m) width of a 200 mm thick ground-floor slab sawn to a depth of 50 mm to induce cracks. This force can be compared with a force of 65 kN/m (6.5 T/m) width supplied by a typical mesh reinforcement consisting of one layer of fabric. Hence the replacement of fabric with steel fibres for the control of induced crack widths appears to be a practical proposition.

There are a few specialist systems with fibre dosages of 60 kg/m$^3$ or more in which the need for sawcut joints is eliminated completely and this prevents any problems with curling or spalling of the slabs. Also, steel fibre-reinforced concrete has been used for pile-supported floor slabs, which are essentially structural load-bearing systems, although the steel fibres may be complemented by traditional reinforcement at local areas of stress concentration such as across the pile heads.

Improved abrasion resistance is sometimes claimed for steel fibre concrete and this is specified for various flooring applications in BS 8204 (1987), the measuring technique being described in Concrete Society Technical Report No. 34 (1994). Concrete grades between C40 and C60 are required to ensure adequate abrasion resistance which is mainly controlled by concrete strength and curing procedures.

In order to reduce the number of fibres lying directly on the surface of the slab, dry shake toppings are recommended (ACIFC, 1999).

Other major applications of steel fibre concrete are *in-situ* tunnel lining and rock-slope stabilization using the gunite technique and the use of steel mesh in these situations is being steadily replaced. Precast tunnel segments have also proved a success.

One of the most successful uses of stainless steel fibres has been in castable refractory concretes for use at temperatures up to 1500°C. In these products, calcium aluminate cement (CAC) is used and initial cost is not the prime consideration provided that product life can be increased, typically by 100 per cent.

## 6.11 Polypropylene fibre-reinforced concrete

Extensive use has been made in the construction industry of small quantities (0.1 per cent by volume) of short (<25 mm long) fibrillated or monofilament polypropylene fibres as shown in Figure 6.8 to alter the properties of the fresh concrete, notably to reduce the extent of plastic shrinkage cracking should it occur. The fibres also have some minor

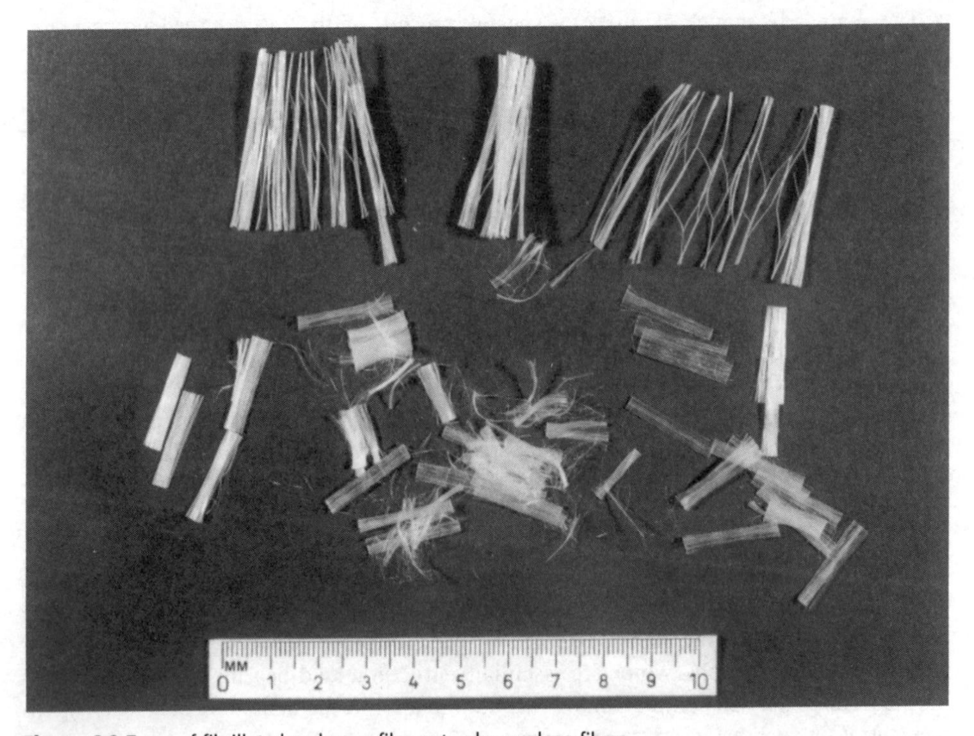


**Figure 6.8** Types of fibrillated and monofilament polypropylene fibres.

effects on hardened concrete properties, but they should never be used as a replacement for structural reinforcing bars.

## 6.12 Mix design and manufacture

In the most used mixes (0.1 per cent fibre volume) the fibre volume is so low that mixing techniques require little or no modification from normal practice. Usually the fibre comes prepackaged in 0.9 kg water-soluble bags which are placed in the mixer for each cubic metre of concrete mixed. The fibres are released and dispersed during the normal mixing cycle. Also on the market are water-soluble bags containing a combination of steel and polypropylene fibres which give some of the benefits of each fibre type.

## 6.13 Properties of fresh concrete

Badly made and cured concrete may suffer from sedimentation of the aggregate particles with consequential bleeding of water to the surface of the cast layer. With excessive sun or wind, this water may rapidly evaporate with an increase in suction and shrinkage on a horizontal surface which can cause substantial cracks to form known as plastic shrinkage cracks. Small quantities of fibres in the mix may increase the cohesion and prevent sedimentation due to their interlocking network characteristics. The result may be that in some, but not all, cases the quantity of bleed water may be reduced. This will not necessarily have a beneficial effect on plastic shrinkage cracking but may help to limit plastic settlement cracking.

### 6.13.1 Plastic shrinkage cracking

This is a topic which is so complex that there is no agreed mechanism for plain concrete let alone an understanding of the effects of fibres. However, it is an area where the earlier theory can indicate some possibilities.

In order that plastic shrinkage cracks can occur, it is presumed that the tensile stress exceeds the concrete tensile strength, which is very low during the first 4 hours. It has been found that typical concretes may have a tensile strength of about 0.02 N/mm$^2$ at 4 hours. In order that the cracking stress of 0.02 N/mm$^2$ should be sustained by 0.1 per cent of random fibres, the fibre stress can be calculated from equation (6.10) to be

$$\sigma_f = \frac{2 \cdot \sigma_{cu}}{V_f} = \frac{2(0.02)}{0.001} \qquad (6.16)$$

$$= 40 \text{ N/mm}^2$$

Sufficient bond stress with the polypropylene can be developed during the first 4 to 5 hours after placing for fibre stresses of this order to be developed so that the width of plastic shrinkage cracks can be limited by fibres. The fibres also endow the concrete with some post-cracking ductility and increased strain capacity at these very early ages which would have a beneficial effect on plastic shrinkage cracking.

## 6.14 Properties of hardened concrete

For the system containing 0.1 per cent by volume of polypropylene fibres it can easily be shown theoretically that the fibres will have little measurable effect on the tensile or flexural strength of hardened concrete and that they cannot be considered as a primary reinforcement. This is because the most optimistic estimate of critical fibre volume from equation (6.11) is for the unlikely case of all fibres breaking (rather than pulling out) at a stress of 400 N/mm$^2$. Thus for a typical concrete tensile strength of 3 N/mm$^2$

$$V_{\mathrm{fcrit}} = \frac{2(3)}{400} = 1.5 \text{ per cent}$$

More realistically $\sigma_f \sim 200$ N/mm$^2$ and hence $V_{\mathrm{fcrit}} \sim 3$ per cent.

The actual post-cracking stress for 0.1 per cent $V_f$ will depend on the fibre–cement bond strength. Assuming that an average fibre stress of 200 N/mm$^2$ can be sustained then

$$\sigma_{\mathrm{cu}} = \frac{V_f}{2} \cdot \sigma_f$$

or

$$\sigma_{\mathrm{cu}} \sim \frac{0.001}{2} \cdot 200 \sim 0.1 \ \mathrm{N/mm^2}$$

or a maximum of 0.2 N/mm$^2$ if all the fibres break at 400 N/mm$^2$, thus giving minimal post-cracking strength in the hardened state.

This estimate is supported by experimental evidence where the maximum pull-out force across a pre-formed crack is about 0.15 N/mm$^2$. In a ground-floor slab of the same dimensions as described for steel fibres this would result in a restraining force of only 22.5 kN/m (2.25 T/m) width which is much less than for the steel fibre case. Additional benefits to impact strength and frost resistance have also been shown to occur.

### 6.14.1 Durability

Polypropylene is extremely resistant to the alkalis in concrete and the concrete matrix protects the fibres from ultraviolet light which otherwise would cause chain scission and degradation. Little change in fibre strength has been observed up to 18 years in a variety of exposure environments and accelerated tests have predicted a lifetime considerably in excess of 30 years. However, for the early age properties, the durability of the fibres is not critical and it is only second-order effects on hardened properties which are affected by fibre durability.

## 6.15 Applications

Polypropylene fibre is used in concrete in a wide variety of applications in general construction and specifically in ground-floor slabs. Specific uses have included shotcreting, precast products and situations where fire resistance is important. The latter is achieved by the fibres melting and leaving channels in the concrete through which steam can escape, thus improving the spalling resistance. Also, an indication of its growing use in general construction is that nearly 10 per cent of readymixed concrete in the USA is said to contain polypropylene fibres.

Some constructions use combinations of polypropylene and steel fibres where the benfits of both fibre types may be complementary.

## 6.16 Glass fibre-reinforced concrete

There are few uses for glass fibre reinforcement in bulk concrete because the expense of including a sufficient quantity of random glass fibre would rarely be cost effective. Also, significant mixing difficulties would occur for the normal range of concretes at fibre volumes in excess of 1 per cent.

However, fibre quantities as low as 0.03 per cent by volume (0.8 kg/m$^3$) of 12 mm long fibres can offer improved resistance to plastic shrinkage cracking for normal concretes in a similar way to polypropylene fibres. These mixes will inhibit bleeding and plastic shrinkage cracking and will have between 2 million and 200 million fibres per m$^3$ of concrete depending whether the filaments are allowed to disperse from the strands.

## References

ACIFC (1999) *Steel fibre reinforced concrete industrial ground floors*. Association of Concrete Industrial Flooring Contractors. (Obtainable from The Concrete Society.)

ASTM C1018-97 (1997) Standard test method for flexural toughness and first crack strength of fiber-reinforced concrete (using beam with third point loading).

BS 8204: Part 2 (1987) Classification of abrasion resistance and proposed limiting depths of wear for the accelerated abrasion test.

Concrete Society Technical Report No. 34 (1994) Concrete industrial ground floors – A guide to their design and construction.

EFNARC (1996) European specification for sprayed concrete. European Federation of Producers and Applicators of Specialist Products for Structures.

JCI (1984) Method of test for flexural strength and flexural toughness of fibre reinforced concrete. Japan Concrete Institute JCI – SF4.

## Further reading

Aveston, J., Cooper, G.A. and Kelly, A. (1971) Single and multiple fracture. Paper 2 in *The Properties of Fibre Composites, Conference Proceedings of the National Physical Laboratory.* IPC Science and Technology Press.

Aveston, J., Mercer, T.A. and Sillwood, J.M. (1974) Fibre-reinforced cement – scientific foundations for specifications. *Composites Standards Testing and Design. National Physical Laboratory Conference Proceedings.*

Balaguru, P.N. and Shah, S.P. (1992) *Fiber reinforced Cement Composites.* McGraw-Hill, New York.

Bentur, A. and Mindess, S. (1990) *Fibre Reinforced Cementitious Composites.* Elsevier Applied Science, New York.

Hannant, D.J. (2001) *Fibre-reinforced cements and concretes.* In Illston J.M. and Domone, P.L.J. (eds), *Construction Materials – Their nature and behaviour*, 3rd edn, pp. 386–422.

Majumdar, A.J. and Laws, V. (1991) *Glass Fibre Reinforced Cement.* BSP Professional Books.

# 7

# Masonry mortars

*Neil Beningfield*

## 7.1 Aims and objectives

The aim of this chapter is to consider the requirements and specification of building mortars for brickwork, blockwork and stonework (hereinafter termed masonry mortars). The history of the materials is briefly reviewed and their properties, composition, specification and use is then evaluated. The interaction of masonry mortars with their environment is addressed, as is their durability and reaction to aggressive situations, including soluble sulphates and cycles of freezing and thawing. The mechanism and stability of pigmentation is considered. Finally, site practice, common practical problems and their resolution are addressed. These issues are considered in the context of the British and European Standards for specification, design and practice.

## 7.2 Historical background

### 7.2.1 Ancient mortar

Probably the oldest mortar that still survives is that found in Galilee, Israel, near Yiftah'el, which is said to be 10000 years old (Malinowski, 1981). Some other of the oldest examples of our built environment, dating back about 5000 years to the time of Mesopotamia, still survive in the form of masonry ruins. Much of the mortar used therein remains durable even today. The Egyptians are said to have used gypsum mortars and lime mortars in the construction of the pyramids (Neuberger, 1930), and lime mortar was used in the Greek water cistern or megars in 500BC (Malinowski, 1981). Later, the Roman writer Vitruvius (*c*. 27 BC) wrote of formulations for lime mortars. Many examples of Roman mortar survive throughout Europe and, within the UK, Hadrian's Wall still has massive unrestored areas of the original materials. More recently, the Tower of London, some 900 years old, provides further evidence of the potential durability of masonry materials.

### 7.2.2 The development of modern cements

Probably the first to document the manufacture of a calcareous cement compound was Bryan Higgins, who patented a cement for use in external rendering in 1779 and published a paper (Higgins, 1780) on the manufacture of cement. This was followed by the better known work of Smeaton, who calcined calcium carbonate in the form of limestone, with clay impurity to give a material that reacted with and hardened in the presence of water. Smeaton's work formed the basis for his book (1791) and for the rebuilding of the Eddystone Lighthouse in the late 1700s. Against this background must be set the quite recent invention by Aspdin of Portland cement, patented in 1824 (Patent no. 5022). Meanwhile, Vicat in France was working on the production of cements and hydraulic limes and this work culminated in his comprehensive book on the subject (1837).

Have technologists or designers of today anything to learn from these historic examples? It has been suggested that the answer is probably yes, in that structures designed and constructed much more recently have failed to achieve their relatively modest design lives. It would be a bold observer indeed who would extrapolate their potential life to even hundreds of years let alone the millennia achieved by some historic examples.

## 7.3 The requirements of mortar

The requirements of mortars may be considered in two categories, those for the hardened properties and those for the plastic or wet properties.

### 7.3.1 The hardened requirements

These consist primarily of strength, durability, dimensional stability and the ability to exclude water sufficiently and to have adequate thermal and acoustic properties. In addition, adequate fire protection and satisfactory appearance are required. This list of requirements may be shortened and expressed in the form of strength and stability, durability and appearance.

The strength of the mortar is usually taken to be that in compression and flexural strength is sometimes measured, particularly in the case of plastering and rendering mortar where it is of importance. Bond strength may also be considered in tension, flexure, shear or sometimes a combination of two or more of these properties. Other suggested measurements of strength are based on the resistance to penetration of, for example, a shot- (ASTM, 1989) or spring-fired pin, resistance to penetration of a standard nail (Figg), resistance to the pulling out of a standard screw (BRE Digest 421, 1997), and resistance to penetration by a drill bit (RILEM MD3). The most common strength measurement by far, however, is that of compressive strength, using either prisms to produce equivalent cubes or cast cubes (Edgell; BSI 1998).

Sand complying with the soon to be superseded British Standard (BSI, 1986), with the draft new European standards current at the time of writing (BSI, 1999), gauged with Portland cement in the amount prescribed in most European standards and application documents, will generally produce compressive strengths that comply with the requirements of contemporary national standards. The bond strength is a more complex matter, primarily because it is much more difficult to measure, with high variability and relatively high cost a function of most test methods. Nevertheless, so long as extremes of weather and/or unit suction are avoided, and workability is adequate, bond strength will generally prove to be adequate for existing structural design methods.

Durability is an even more complex issue and unlike strength, the development of an internationally accepted test method, or even one accepted in Europe alone, has proved difficult. It is theoretically possible to base a test on cycles of freezing and thawing, either of mortar specimens alone or of specimens of masonry. The problem with the use of mortar specimens alone is that these are usually produced in conditions that are completely different to those experienced in reality when mortar is used to lay bricks or blocks. The mortar is often cast in steel cube moulds, which are impermeable and give rise to vastly differing air void structures than those actually present in masonry.

It has often been shown that the size and form of the air voids has a profound effect on resistance to freeze–thaw damage (Harrison and Bowler, 1990) and the use of specimens cast in impermeable steel cube moulds for an assessment of this property in a real structure is therefore of little merit. To overcome this problem, Harrison and Gaze (1989) suggested casting mortar joints with gauze on either side of the joint, adjacent to the brick. In this way, the joints could be separated and tested on their own, but with their structures having interacted with the brick or unit suction in a way that is relevant to site built masonry. This method was the subject of a draft European test method (BSI 1999, prEN1015-14) which also incorporates a further enhancement due to Harrison. This is to use, additionally, sulphate solution rather than tap water to wet the specimen prior to freezing. With the use of a sulphate solution, the test can also measure the resistance of the mortar to chemical attack by sulphate ions. This is the form in which the test was under appraisal as a potential European test method with the ability to test under both conditions. The other problem with freeze–thaw testing is that the amount of freezing and the presence or absence of an ice front in the specimen is critical and is generally unlikely to mirror site conditions. In practice, most masonry (excepting that in free-standing walls) is wetted from one side only and is dry, or at least generally far drier, on the inside or cavity face. Moreover, in a structure, as opposed to a free standing wall, there is a temperature gradient. In the winter period with a low ambient temperature, there is a temperature of perhaps 15–20°C on the inner face of the inner leaf of masonry and a corresponding temperature gradient. This means that in practice the mechanism of freezing of the outer leaf is associated with the gradual advance of an ice front into the units and the mortar joints, followed by a gradual receding as thawing commences.

Ceram Research (1984) developed a test for bricks, using brickwork masonry to simulate this condition which is also proceeding as a European standard (West, Ford and Peake, 1984). The test consists of wetting the masonry, freezing, and then progressively freezing and thawing from the front face. This results in a frozen front that progressively moves and exerts pressure on an unfrozen section and thus acts to disrupt any potentially non-durable material. The test has now been carried out many thousands of times, for the purpose of appraising the freeze–thaw resistance of ceramic bricks and has also been used, to a lesser extent, with calcium silicate bricks. It has been further appraised for use with aggregate concrete blocks to investigate the durability of the aggregates used therein, in cases where this may have been suspect. In all cases, the results have accorded well with site experience and also with work carried out using exposure panels at a Scottish exposure site (Edgell, 1999). The test has so far been used only sparingly for the assessment of mortar durability Edgell, (1999) but with encouraging results. In this work, the test clearly performed satisfactorily. However, in the majority of the work to test the unit as opposed to the mortar, the latter was selected to be a strong and durable material such that freeze–thaw failure occurred in the bricks and not in the mortar; thus, no data was provided as to the performance of the mortar.

In conclusion, with respect to this test it appears to have great promise, performs well for units, but has not yet been accepted and extensively tested for use as a mortar as opposed to a brick test, its present application.

## 7.3.2 The plastic requirements

The plastic requirements of mortar are also important, and in masonry mortars, sometimes assume perhaps a greater importance in the minds of the operatives than is desirable. This can result in undue influence being exerted over the mortar specification in matters like the choice of sand and perhaps the addition of unauthorized and often unsophisticated and ill-considered admixtures. These may well enhance the working properties, as perceived by the mason – the so-called 'workability' – but will degrade the hardened properties listed earlier. The term 'workability' in masonry mortars is different to that used in the context of concrete. In concrete a test, usually a one point test, commonly for slump, consistence or some related procedure is used and is defined as the workability, at least in the UK and the USA. In mainland Europe it may be more correctly referred to as the consistence because workability should, as in UK parlance, define the sum of those working properties of the mortar that are perceived by the operative. These consist of an amalgam of cohesion, that is internally within the mortar, adhesion to the trowel, brick or block, together with less easily defined terms. The operative is apt to consider the ease of using the material in terms of how 'creamy' or 'buttery' it feels when manipulated with the trowel. He will also sometime tend to grade the material as 'harsh', 'short' or 'dead'. *Harshness* is probably self-explanatory, often a function of relatively large, usually angular particles of sand that are felt quite prominently on the bricklayer's trowel as they individually scrape it, as it is moved through a mass of mortar. Mortar that is said to be *short* may be deficient in part of the sand grading fraction – some dredged material may be claimed to possess this characteristic. Lack of fines is sometimes thought to be the cause of the defect but is often not. Frequently the cause is a preponderance of one sand particle size although it may also occasionally be caused by a lack of or shortage of fines, from which the name presumably derives. *Dead* mortar may contain insufficient plasticizer, either as added admixture as with air entrainment or as a function of the binder or part binder as in the case of hydrated lime.

Care must be taken to ensure that the operative does not attempt to maximize those plastic properties which he believes to be beneficial, whilst ignoring or degrading the properties of importance to the designer, in particular strength and durability. He may do this by adding excessive amounts of plasticizing admixtures, by using inappropriate clayey sand, or by a combination of these procedures. The addition of excess plasticizing agents may take several forms. If sand is mixed on site with the addition of cement and plasticizer, then excess amounts of the latter may be added on purpose to entrain extra air. This will produce a highly air entrained mortar that is light and easy to use, although probably deficient in strength and durability. Additional air may also be entrained in the mix by incorrectly using masonry cement, which is already formulated with an integral air entraining agent, together with an additional plasticizer added to the mixer which will entrain further air. This problem can also occur when premixed lime:sand mortar which is already air entrained, is incorrectly subject to the addition of further admixture to the mixer. Another way in which excess air may be introduced to the mix is by the addition of domestic detergents or 'washing up' liquids to the site mixer. This is relatively common and the discarded containers may often be seen adjacent to site mixing areas.

In summary, a 'well graded sand' with the correct amount of cement and/or lime and adequate but not excessive air entrainment will provide a mortar of acceptable working properties. Addition of excessive air is quite unnecessary. Moreover, greatly excessive air tends to be unstable, with substantial amounts lost within a few minutes of mixing. This in itself causes problems with the site operatives because they have to restore the workability by addition of water and subsequent remixing – a labour-intensive procedure.

## 7.4 Properties of mortar

The properties of mortars may be considered in terms of those parameters that meet or are measured against the requirements considered in the previous section. They may therefore be considered in the same manner and the same two categories, the plastic and the hardened properties.

### 7.4.1 The hardened properties

The hardened property of mortar that is most often tested for is the strengh. Although mortar very rarely fails in direct compression, the compressive strength is the most specified of all of the hardened properties. This is probably primarily because the test procedure is convenient to carry out. Additionally, it is one that is familiar in principle to many engineers and technologists who are used to the testing of concrete in compression. In addition, there is clearly a relationship between strength and binder content so the procedure is sometimes used to provide a convenient surrogate for the latter where prescribed mixes are specified.

The compressive strength has also been related empirically to other parameters that are more difficult to test for, particularly masonry strength, and is thus often used to calculate an empirical assessment of that property. Compressive strength is tested for either by using cubes or by making prisms and producing specimens for testing as equivalent cubes. Cubes of many different sizes are used, BS 4551 (1998) covers 70.7 mm and 100 mm sizes, whilst 150 mm cubes are sometimes used. Much smaller cubes have also been used, including specimens as small as 10 mm sq. taken from mortar bed joints, currently specified in a DIN standard (DIN, 1999). More recently, prisms have been loaded in flexure and the two equivalent cubes so produced tested in compression. This method was adopted in BS 4551 (1999), in the CEN test for 40 × 40 × 160 mm specimens (CEN, 1999), although an earlier RILEM recommendation was for prisms of 40 × 40 × 80 mm (RILEM, 1982). The determination of bond strength for mortar is generally accompanied either by high variability, in the case of simple tests, or high costs in the case of tests on larger specimens. Many years' successful experience lie behind the method given in the BS 5628 Part 1 Code of Practice (BSI, 1992). Unfortunately, this test uses a relatively large number of bricks and to carry out a determination using a significant number of replicates is costly. A simpler test has been developed within CEN and involves the use of testing for initial shear strength as a proxy test (BSI, 2002). However, this test also shows a high variability. A simpler test still uses a bond wrench. This is an arm or similar which is clamped onto a brick after the perpendicular joints have been progressively cut through, usually with a diamond saw. It is loaded to produce a bending moment that causes the joint to fail. This method has also been proposed as a CEN test (BSI, 1998).

In summary, the situation with respect to mortar and masonry strength is that bond strength determination is generally relatively complex and costly, and often with high variability. For this reason, compressive strength is usually used as a surrogate but may not relate directly to practical service conditions. Many other tests for hardened mortar properties have been proposed but most are for specialist applications although recently adopted CEN tests (BSI, 1999) include those for water absorption due to capillary action, water vapour permeability (BSI, 1999), adhesive strength (BSI, 1999) and compatibility of one-coat renders. The water absorption test (BSI, 1999) is a simple one based on immersing preweighed mortar prisms in 5–10 mm of water and then weighing again after a period of immersion. It appears to be easy and reliable. The water vapour permeability (BSI, 1999) is a requirement for plastering and rendering mortars and uses a basic permeability cell. There is not a great deal of international experience of this test but again, it is simple and robust. The adhesion of plastering and rendering mortar is determined using a pull-off test as defined in prEN 1015-12 (BSI, 1999). This test uses normal pull-off testing equipment and is easy and reliable although, in common with all pull-off tests, attention to bonding the specimen to the disc is required. The test must be carried out carefully and the failure mode must be precisely reported.

The compatibility of one-coat rendering systems may be assessed by the use of tests developed in France (BSI, 1999). In this test sample panels are rendered and then subjected to cycles of heating to 60°C by infrared radiation, freezing at –15°C and immersing in water for 8 hourse. The adhesive strength and mortar permeability are determined after a number of cycles of heating, freezing and saturating. This test is little known outside France.

## 7.4.2 The plastic properties

The terminology and separate classification of plastic properties is not integrated worldwide, or even within Europe, and the same general tests may even be used to appraise properties that are given different names. This is seen in the case of flow, used in some countries to determine consistence, whereas in other countries, consistence is first determined using a different test and flow then applied to determine a property said to be a quite different rheological parameter, one more akin to cohesion or adhesion. The reason for this dichotomy is lack of precision in considering exactly what properties are being measured by a particular test, coupled with an imprecise knowledge of exactly what the measured value actually represents. It is not always easy to quantify these matters but the individual properties are discussed hereafter in an attempt to present a logical treatment.

### *Consistence*

The consistence may be defined as a measure of the resistance to penetration of the plastic mortar, by a body under the influence of gravity, or its resistance to flow when exposed to the force of gravity or to some other force as, for example, electrically or mechanically induced vibration. Consistence may be measured by dropping a free falling, independent object into the mortar, as is the case with the British Standard dropping ball test (BSI, 1997). It may also be measured by allowing a sliding or tethered object to drop onto and penetrate the mortar, as with the German DIN plunger test, subsequently adopted as a CEN test (BSI, 1999). Alternatively, vibration may be applied as in the French LCL meter (BSI, 1999). Confusingly, in Germany, in some Scandinavian countries and now within the CEN tests consistence is also measured by using a flow table (BSI, 1999).

In all these cases, the property is a function of the sum of different properties of the mortar but also is proportional to the water content. Regardless of the complexity of the basic determined properties, for a given mix the predominant parameter influencing consistence is water content. Unfortunately, this may lead to confusion in the minds of observers, and to the erroneous belief that it is merely a function of the content – a completely incorrect view. Other rheological properties also have an influence.

### *Flow*

As discussed earlier in section 7.4.1, the flow is used in some countries to determine the consistence, which can be said to be valid. However, it may be used differently in cases where the consistence has already been determined by another test. This will occur, for example, where, using BS 4551 (BSI, 1998) mortar is brought to a standard consistence by a penetration method and is then subjected to a flow test. In this situation, two mortars of precisely the same consistence as determined by the dropping ball penetration test (or by any other non-flow test) might give completely different flow values. This shows that the flow test is not measuring all of the same properties as the penetration test, or if it is, then some are being weighed more than others in the result which is influenced by a mixture of different properties. Other properties, as for example yield stress also influence the result. It is not easy to quantify all the factors contributing to a given flow value. Clearly consistence does have an input but other factors also exert a great influence. Plastic adhesion (to the flow table surface), yield, degree of bleeding and segregation, plastic viscosity (closely related to consistence) are probably the key influences. The important issue, however, is that flow is not simply a function of water content, nor is it necessarily another way of measuring consistence.

### *Water retention*

The water retention or water retentivity is a function of the ability of the plastic mortar to resist the suction of an absorptive substrate. It is inversely proportional to the amount of water removed or the ease with which water is removed by substrate suction. It may be tested for by using standard filter papers as the substrate (BSI, 1998), or in conjunction with a sintered glass background (BSI, 1999). Water retentivity is an important mortar property but not one for which minima are prescribed by current standardization, although the property is closely associated with bond strength, particularly in conjunction with units that tend to either low or high extremes of suction. It has been proposed as a surrogate for bond strength, but this approach is not desirable, as there are clearly other influences.

### *Consistence retention*

The property of consistence retention or consistence retentivity is analogous to water retentivity but is a measure of the effect of substrate suction on consistence, as opposed to the effect on water content. Although at first sight this test could be interchangeable with, and measure a property similar to water retentivity, in practice this is not quite the case. Moreover, the test has a high variability and was dropped from the suite of BS 4551 (BSI, 1998) plastic property tests for that reason.

## 7.5 Constituents of mortar

The constituents of mortar may be considered as two key components, binder and aggregate, together with admixture and sometimes additive.

These are covered by existing and recently superseded British and European Standards as shown in Table 7.1

**Table 7.1** The mortar material standards

| Material | Old BS/British Standard | EN/European Norm |
|---|---|---|
| Portland cement | BS 12 | ENV 197 – 1 |
| Lime | BS 890 | ENV 459 – 1 |
| Sand | BS 1200 | prEN 13139 |
| Pigment | BS 1014 | prEN 934 – 3 |
| Admixture | BS 4887<br>BS 5075 | BS EN 934 – 2 |

Whilst the binder and admixtures will almost certainly comply with the standards shown in the table, there is no certainty that the sand will. Investigations both in the UK (Ragsdale and Birt, 1976) and in Germany have shown that a large amount of the sand used for mortar failed to comply with the current material standards. In some mortar types, for instance that used that used for rendering, the majority failed.

Unfortunately, however, it is not possible to say with certainty that just because a sand fails the standard it has caused a particular failure or problem – millions of tonnes of non-complying sand have been used for the production of mortar that has not failed. This is because the aggregate standards have historically been prescriptive in their treatment of grading and have relied on empirical relationships with performance that have been loose and lacking in sufficient data for their accurate derivation. The requirements of the sand are discussed further hereafter.

## 7.5.1 Sand

The old British Standard grading requirement (BSI, 1986) for mortar sand is shown in Table 7.2. It is important to note that there exists two sand types and grading requirements, type S and type G. These arose following the investigation into the properties and gradings of sands actually in use within the UK and a comparison with those of the British Standard requirement then in existence referred to earlier (Ragsdale and Birt, 1976). This revealed that an extremely large number of sands in common use fell outside the grading range given in the British Standard. At the same time, investigation of the method used to test the material showed that it was common practice to dry sieve the sand. This procedure resulted in an inaccurate test as agglomerates either of clay, of sand particles coated with clay or a mixture of both, were erroneously recorded as individual particles rather than agglomerated and mixed pieces of smaller individual particle size than those actually recorded. When the test procedure was modified to require pre-washing, to separate these agglomerated over-size particles, it was found that the resultant numerical grading requirement effectively widened. In order not to exclude material that previously complied with the Standard, it was therefore necessary for it to be printed in revised form with wider grading limits. These were stated to be temporary, with further confirmation or modification to take place pending results of research which would be instigated. However, after a period of about 10 years, the research results were still not available. The European Standard reached the stage in drafting where, because European law requires that further work on the national standard must be stopped if a European Standard is in progress to cover that topic, work was halted.

**Table 7.2** Masonry mortar sand

| BS sieve (mm) | Type S | Type G |
|---|---|---|
| 6.30 | 100 | 100 |
| 5.00 | 98–100 | 98–100 |
| 2.36 | 90–100 | 90–100 |
| 1.18 | 70–100 | 70–100 |
| 0.6 | 40–100 | 40–100 |
| 0.3 | 5–70 | 20–90 |
| 0.15 | 0–15 | 0–25 |
| 0.075 | 0–5* | 0–8† |

*0–10% for crushed stone sands.
†0–12% for crushed stone sands.

A very simple and shallow survey carried out by the author in 1999 raised the question of whether or not respondents to the survey took account of the relatively new requirement concerning the two gradings S and G defined in the standard. This requirement is that for type G sands, the richer in cement of the two mix designs given in the Code of Practice BS 5628 (BSI, 1998) be adopted, the leaner being permissible if the sand is type S. These alternative leaner and richer mixes are shown in Table 7.6 of this chapter. This small survey, which encompassed standardization bodies, regulatory bodies, specifiers and mortar manufacturers found that there was complete unanimity in *not* complying with the requirement, which has clearly not been taken up. However, as there have not been reports of sand induced mortar failures increasing in the recent period, it appears that the

Code requirement (BSI, 1998) was too onerous and that successful mortar can indeed be produced with the leaner mixes and type G sands. Clearly, there are other variables and this conclusion must be regarded as tentative. It may well be that there is now more cognisance of the value of air entrainment in assisting with durability: Additionally, or alternatively, site care and practice may have improved, or detailing become better specified although the author is unaware of any substantiated reports of one or more of these situations having taken place in practice.

In conclusion therefore, with respect to the British Standards, it appears that type G sand is probably suitable for a large proportion of masonry mortar applications. (For rendering, this is almost certainly *not* the case and the appropriate British Standard grading should be complied with.) The European sand standard situation is a little different to the UK one as it is relies on a less rigid and prescribed concept. Instead of a series of sieve sizes, with the amount of each sand size tightly constrained, the principle is a constraint on the oversize and the undersize, together with a 'declared typical grading'. The oversize is handled in a practical manner in that 98–100 per cent is required to pass 1.4 times the nominal maximum, 100 per cent to pass twice the nominal. Thus, for a 2 mm nominal sized sand, 100 per cent is required to pass a 4 mm sieve size. The basic requirements of the European Mortar Sand Standard are shown in Table 7.3 for nominal size 0–2 mm. There exists also a requirement for a 1 mm nominal size and 4 mm nominal size but the 2 mm will cover the vast majority of current sources in use. In addition, there is a requirement to state a declared grading, with permissible deviations as shown in Table 7.4.

**Table 7.3** Masonry mortar sand

| Type | | Oversize | |
|---|---|---|---|
| | 2D | 1.4*D* | *D* |
| 0/2 | 100 | 98–100 | 85–99 |

**Table 7.4** Tolerances on producer's declared typical gradings for general use aggregates

| Sieve size (mm) | Maximum tolerance in percentage passing by mass* | | |
|---|---|---|---|
| | 0/4 | 0/2 | 0/1 |
| 2 | – | ±5[‡] | – |

* Notwithstanding the tolerances listed above the aggregate shall conform to the requirements of Table 7.1 and Table 7.3

[†] For special purposes the supplier and purchaser can agree reduced grading tolerances.

[‡] If the percentage passing *D* is >99% by mass the supplier shall document and declare the typical grading including the sieves identified in Table 7.2 of the Standard.

It has been suggested that the European concept of *D* and declared grading is more realistic and practical in light of the variability of known sources than the alternative historic treatment prescribed in typical national standards. It should be noted, however, that whichever compliance system is used it is still possible for sand to comply with the standard but to be deficient in some respects. Certainly in respect of the perceived workability of the mortar, it is possible for complying sands to nevertheless be excessively single sized with concomitant tendency to bleed and segregation. Thus compliance with the Standard does not in itself mean that the sand will produce an acceptable mortar.

## *Particle shape*

In addition to the sand grading, the particle shape is of importance. As with many materials where packing is critical, spherical particles are desirable in theory and this thesis is confirmed by regular practical feedback from operatives. Excessive angularity is less desirable and may be noticed by the users and reported as producing material that feels harsh, 'on the trowel'. An excess of flaky particles can cause problems, particularly with certain joint finishes where the operative may report that the joint is 'tearing'. Additionally, non-spherical particles tend, of course, to a higher void ratio which, from first principles, produces a higher water cement ratio and less optimized hardened properties.

## *Impurities*

**Silt and clay** – Certain plate-like clay species are associated with much increased water demand and with a resultant reduction in the values of many of the hardened properties. The presence of silt and clay reduces strength in compression, flexure, tension and bond, with clay having by far the greater effect.

The methylene blue test, used for many years in some continental European countries, has been suggested as a screening test for clay. There is an empirical relationship between methylene blue adsorption and clay content but different clay species give different methylene blue adsorptions, which means that the relationship is dependent on clay type as well as amount. As clay in suspension associates chemically with water, clay is perceived by the operative as being beneficial, because it produces the properties that he views as attributes, 'creaminess', 'smoothness', 'butteriness'. These arise due to the attraction by the clay of ions deriving from the solution present, i.e. from the water in the mix and the soluble components therein. Unfortunately, the presence of these working property attributes, following from the inclusion of clay, is accompanied by a reduction in the optimum hardened properties. Additionally, clay and silt may be present as agglomerated lumps, so called 'balls of clay' rather than as dispersed, discrete or semi-discrete particles. This situation can occur when material is unprocessed or when screening equipment allows slivers or particles to pass between adjacent wires or screening members. These large clay agglomerates, which may typically be 2 to 5 mm in diameter often fail to break and disperse in the mixer, thus remaining as aggregated particles within the cement:sand or cement:lime:sand mortar system. Any lumps that are at, or near, the surface may become saturated with water and in subsequent freeze/thaw cycles may well expand and disrupt the surface of the joint.

**Lignite** – Lignite is present in many, probably the majority geographically, of UK sand sources although some are free of the material. Lignite may cause stains on the surface of the mortar. Particles of lignite may also be absorbent, soft and friable, leading to the same potential problems as those associated with clay agglomerates and durability. When lignite begins to decompose, brown to blackish stains may result, which can run vertically down the structure as they are carried by water run off. These can be so pronounced that they prove quite unacceptable. The commercial degree of acceptability of lignite is very much a function of the historic occurrence of that material in the particular geographical location. Thus there are areas in the UK where the presence of any lignite particles whatsoever in the masonry structure would be considered grossly unacceptable, but other parts where the appearance of minor or relatively minor numbers of particles would be accepted as unavoidable, or virtually so. A similar situation exists in continental Europe.

**Iron pyrites** – Iron pyrites is a mineral form of iron sulphide, sometimes known as 'fools gold', due to the gold colour of some forms. The author estimates that it is present in 50 per cent of all quarries south of a line joining the Wash and the Bristol Channel. Fortunately, this is a rather theoretical treatment and in the vast majority of those quarries the material is not actually present in the product although it may well underlie the worked deposits of material. Very small amounts of iron pyrites are sufficient to cause problems with staining as the material oxidizes on exposure to atmospheric oxygen to give the soluble iron sulphate. This may react further and can lead to a pronounced stain being associated with each particle. As well as staining, the iron pyrities particles can cause 'pop outs' or 'blows' as the reaction with oxygen is expansive. Iron pyrites is not always of the reactive form and the incidence of reactive iron pyrites in most building sand offered for sale in the market place is fortunately rare although some quarries have become known as sources where the problem is particularly prevalent.

**Chlorides** – Chlorides are sometimes regarded as an undesirable impurity in building sand and are often erroneously associated with efflorescence. Sources of dredged aggregate available within the UK almost always have an acceptably low level of chloride when issues of chemical reactivity are considered, although of course, the presence of unprotected steel in masonry mortar is rare – wall ties, lintels and other components are either in non-reactive materials or are galvanized or otherwise protected. Efflorescence should not be a permanent consequence of chlorides because those that are likely to be present are very readily soluble so that any deposit arising on the surface would be rapidly washed and weathered away, almost certainly within a period of a few days or weeks at most.

## 7.5.2 Cement

The UK cement standards were some of the first in the masonry construction materials sector to be integrated within European Standardization. Much data and successful site experience exists to validate the use of ordinary Portland cement to BS12 in mortars for masonry, plastering and rendering but there is a paucity of data concerning the performance of some of the cements included within the European cements covered by the EN. Sulphate resisting cement with a constraint placed on the maximum permissible tricalcium aluminate content has a successful history of use in mortars within the UK however, and can be used with confidence.

### *Masonry cement*

Masonry cements are blends of cement with fillers, admixtures and sometimes other binders. Traditionally, the compositional requirement in the British Standard was for not less than 75 per cent Portland cement, which gave rise to formulations of inert filler, admixture and Portland cement, with the amount of the latter being as close as practically possible to the 75 per cent value for economic reasons. The inert filler is generally finely divided calcium carbonate, although silica may also be used. In North America, the added component is often lime rather than inert material and it will thus contribute a little to the long-term strength development by carbonation. Masonry cements based on Portland cement and lime have recently been introduced into Europe. The problem with masonry cement where the air entraining agent is already added at the manufacturing plant, is that it requires to be added in an amount such that air content maxima are not exceeded. Many factors affect the amount of air entrained by a given amount of air entraining agent, the most important ones being listed hereafter.

### The factors affecting the amount of air entrained:

- cement content
- lime content
- sand grading
- amount and type of clay present in the sand
- temperature
- mixing time
- time of addition of admixture into mixing cycle
- water hardness
- aggregate shape
- presence of additives.

This list is not exhaustive but it can clearly be seen that there exists a large number of variables influencing the final air content. Excessive air is not desirable and maximum values are given in virtually every European specification as well as in American, UK and other national standards. This means that the amount of air entraining agent formulated into masonry cement has to be set at a conservatively low value in order to ensure that the maximum values are never exceeded, even if many of the causative factors listed may produce a high value in only a few specific examples in practice. The effect of high air contents in mortar, that is those exceeding a level of about 20 per cent, is to reduce strength but importantly, to reduce flexural, tensile and bond strength. From first principles, an increase in air content will decrease the effective binder per unit volume of material and this effect can be seen in Table 7.5, where two examples of air entrained mortars are considered; one based on a Portland cement and the other on masonry cement, where the cementitious fraction is in compliance with the British Standard at about 75 per cent. It can clearly be seen that the Portland cement fraction per unit volume decreases markedly as the air content increases, particularly in the case of the masonry cement. Nevertheless, masonry cements do provide a convenient way of introducing some air entrainment and some fine fines into a site made mortar. A further category of cement that finds application in masonry mortars is that known as 'improved cement'. These materials have a modest

**Table 7.5** The effect of air content on cement content in kg/m$^3$ for a 1:1:6 mortar and 1:6 mortar

| Air content | Cement content/m$^3$ | |
|---|---|---|
| | Portland cement | Masonry cement |
| 0 | 220 | |
| 10 | 200 | 142 |
| 20 | 180 | 128 |
| 30 | 156 | 110 |

amount of air entrainment, less than the optimum and are made to fulfil a range of general purpose applications on the building site. They clearly do not possess any optimized properties but show increased durability and plasticity when compared to non-modified cement and may thus be useful for some general purpose applications.

**Cement replacements** – Cement replacements as, e.g. slag, pulverized-fuel ash, etc. are used in some mortars. There is a lack of long-term data concerning their performance in the UK although slag has been used in Portland cement mortars in Scotland and Northern Ireland for several decades without any reported problems, and use throughout the UK is now becoming widespread. Whilst the working properties of the mortar will be enhanced, the long-term hydration obviously relies on the presence of free water and in practice, this is much less likely to be available in mortar than in concrete, due to the far greater surface area of the former that is exposed to atmospheric drying and to the suction of the substrate.

## 7.5.3 Lime

There are many different types of lime, hydraulic lime, quicklime and hydrated lime being the most used in mortars with the latter.

### *Hydraulic lime*

Traditional building mortars were composed of rich mixes of hydraulic lime and aggregate. Mix proportions of 1:$2^1/_2$ to 1:3, sometimes those as rich as 1:2, were generally used. These hydraulic limes resulted from the burning of quarried material that was predominantly calcareous, e.g. chalk or limestone, but that contained silicious and/argillaceous or clay-like impurities. Modern lime production relies on winning product that is substantially free from clay contamination but historically, clay present in the calcareous feed stock resulted in the formation of calcium silicates that had the potential to react with water to produce a hydrated material, thus the nomenclature 'hydraulic' limes. The degree of hydraulicity in these limes varied greatly and some of the purer sources developed little hydraulic strength, relying instead on carbonation for minor strength development. Hydraulic limes are now manufactured in many European countries although often on a small scale, sometimes almost as a cottage industry. Hydraulic limes manufactured on the South Coast of England from the blue limestones found in parts of Dorset are termed blue lias limes and have a long record of use in masonry. The European Standard for lime permits the manufacture of materials known as artificial hydraulic limes.

These materials are not hydraulic limes at all but are mixtures of limes, usually pure or relatively pure substantially non-hydraulic materials, with cement and air entraining agent. Their name is misleading and they are often specified or used erroneously in the belief that they have a natural property that is in some way desirable. In reality, however, they only represent mixtures of binder and air entraining agent and do not possess the qualities of longer and slower strength gain associated with naturally occurring hydraulic limes.

### *Hydrated lime*

Hydrated lime is produced by the burning of calcareous material which drives off the combined carbon dioxide to form quicklime which is then hydrated or reacted with water to produce hydrated lime. The process may be represented as a cycle, as shown in Figure 7.1 below. It is seen that the hydrated lime can react further with carbon dioxide to produce calcium carbonate again, the original starting point, and the process is therefore named the lime cycle. It is this production of calcium carbonate that results in the beneficial property within mortar of self-healing or autogenous healing. It should be noted that the process of carbonation, beneficial in mortars due to the production of calcium carbonate with its traditional cementing action, is in contrast to carbonation in concrete which is often regarded as unhelpful when it reduces the alkalinity adjacent to embedded steel, thus removing the passivating effect of the previous environment.

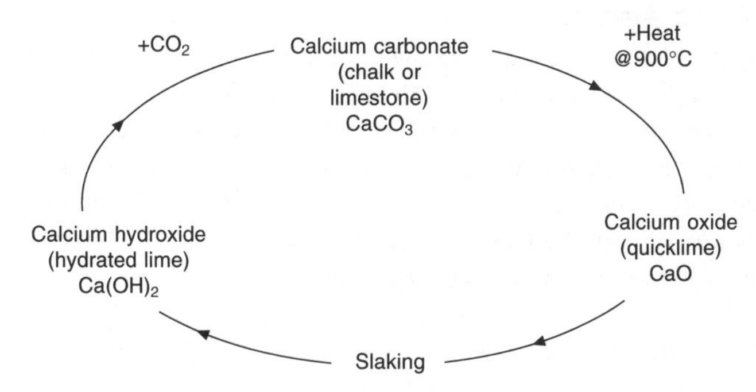


**Figure 7.1** Lime cycle.

## 7.5.4 Admixtures

All, or virtually all, factory produced mortar, and the vast majority of site made mortar, contains admixtures although this fact is often not appreciated. Air entraining agents are by far the most widely used, although they are often erroneously referred to as plasticizers by the user. Other admixtures also find application within mortars. Retarders, water retaining agents, non-air entraining plasticizers, so called 'water proofing agents' and fibres all have a place. Other, specialist admixtures may also be used.

### *Air entraining agents*

The most widely used air entraining agent in mortars has been vinsol resin, a product of the refining of certain timber species' by-products. Used for many years, vinsol resin was an excellent material to use and became almost a commodity product. It produced stable, or relatively stable, air bubbles, was reasonably tolerant of changing chemistry of cements and aggregates and did not readily over-entrain air. However, the availability of this material has now become much less and this has resulted in synthetic surfactants becoming more economic and preferred in the majority of the market place for commercial as opposed to technical reasons. These surfactants have the ability to entrain large amounts of air, are usually more concentrated than the vinsol resin materials but some appear to be much more sensitive to the presence of certain impurities, e.g. clay within the sand. Whatever the source of entrained air, it has been demonstrated conclusively that its presence enhances durability both in respect of freeze–thaw cycles and resistance to sulphate solutions.

### Retarders

Retarders find application in ready-to-use mortars which are usually retarded for a period of between 24 and 48 hours. It should also be noted however, that when laid between absorptive masonry units hydration commences immediately, or relatively so, notwithstanding the overall retardation time period for the batch of material prior to usage. This means that masonry may be laid progressively and continuously as it gains strength, rather than be delayed by issues of retardation and concomitant plastic instability. When hydration does take place, the resulting hydrate is found to be better formed and to result in improved strength as compared to an identical mix without retarder.

### Water retaining agents

Water retaining agents may be employed where mortar is to be used in either hot weather conditions where rapid drying is likely, with high suction substrates or with a combination of both. Thin layer mortars for use with lightweight blocks are a particularly useful application of these materials which are based on a variety of compounds. Amongst the most widely used are cellulose ethers; these are long-chained molecules which reduce the rate at which water is removed from the plastic mortar and sucked into the background. Other materials, for example, polysaccharides are also used.

### Non-air entraining plasticizers

Lignosulphonates and other plasticizers which work by distributing the electrical charges and dispersing agglomerates are used in some mortars but not to a great extent.

### Waterproofing agents

Mortar that is completely waterproof would be difficult to attain and probably unnecessary in practice but so-called 'waterproofing agents', in reality water resisting agents, do find use in some applications. In their most basic form these are based on stearates and oleates and increase the water resisting, or even water repelling, properties of the hardened material. They may be used in the backing coat or undercoat of multi-coat rendering systems to reduce the suction of that layer of rendering so that the subsequent layer has a longer open time and is easier to apply and work.

### Fibres

Polypropylene fibres are used in some rendering mortars and are useful in preventing or reducing the incidence of cracking. They are probably most helpful in preventing early plastic cracking although there are indications that they may also prevent or reduce cracking at later ages. Their use is probably best confined to undercoats as in the finish coat the appearance of hair like fibres on the surface may prove unacceptable. Alkali resisting glass fibres may also find application in this field.

### Pigments

Pigmented mortars have been produced for many years and mortar containing soot and brick dust exists in structures many centuries old. Neither of these materials however, would be considered desirable pigment which are generally best produced using various metal oxides. Iron oxide is used for black pigment as well as for red and yellow. Various shades of brown and orange are made by mixing together these three 'primary' colours. Ultramarine blue is sometimes used but is unstable to alkali and may very quickly lose its colour. Stable blue pigments are available, based on cobalt spinels. Green pigments are available based on chromium dioxide, as are white pigments based on titanium dioxide. All of these pigments are stable. Carbon black has been used in large amounts to pigment both mortar and concrete but its stability is questionable, particularly where it is used in mixtures with yellow or red iron oxide, where the carbon may be preferentially lost, causing a shift in colour with weathering towards that of the iron oxide.

#### *Other admixtures*

A variety of natural substances have been used as mortar admixtures throughout the world and there is historic evidence of such materials having been used for many centuries. Blood, often of oxen or other domestic animals, was frequently used and examination of this material shows that some of the proteins contained therein are similar chemically to the active compounds in some types of the air-entraining agents that are used today. Milk, dung and a number of other materials were found historically to improve the properties of mortar and have been added throughout the years.

## 7.6 Mortar standards and application documents

The British Standard for Mortar, BS 4721, was written for factory-made mortars although it may also be used, and often is, for site-made mortars. It is a prescribed document in that mixes are designated by mix proportion but it also gives minimum strength and other requirements although these are undemanding and generally readily attainable so long as the prescribed mix proportions are present. The Standard relies on BS 4551 the test method standard which was published at the same time, for details of test requirements.

### 7.6.1 BS 4551 – methods of testing mortars

This standard contains test methods for hardened and plastic mortar. For hardened mortar, there are procedures for determining flexural strength and compressive strength with different curing regimes for different mix types. For plastic, i.e. unhardened mortars, there are procedures for consistence and water retentivity, flow and air content and stiffening rate.

### 7.6.2 prEN 998 – The European Standard for mortars

Following the introduction of European standardization, CEN Technical Committee 125 produced 2 mortar standards, pr EN998-1, Plastering and Rendering Mortars, and pr EN998-2, Masonry Mortars. These differ fundamentally from the British Standards in that they are performance standards and define compressive strength minima and other performance characteristics rather than prescribed mix proportions.

### 7.6.3 Application documents

The British Code of Practice is BS 5628, which is the Code of Practice for the Design and Use of Masonry. This has a mixture of designed and prescribed requirements and is in three parts, for 'normal' unreinforced masonry, calculated load bearing masonry and reinforced masonry. The durability of masonry, a key parameter, has historically always been dealt with prescriptively but it is hoped to introduce a design requirement based on objective test criteria in the future. The Building Research Establishment has developed a test which may be used to measure durability in the presence of cycles of freezing and thawing, presence of soluble sulphate solution or both. Following appraisal throughout Europe, this method may become more widely used although it is both lengthy and costly and international comparative work on reproducibility and repeatability is as yet unavailable.

## 7.7 Mortar mix design

The mix design of mortar is based on the concept of filling the voids between the solid particulate aggregate with a binder, a similar concept to that obtaining in many building materials as e.g. concrete, coated stone, etc. In the case of mortar, the volume proportion of binder required to achieve this is such that mixes of between about $1:2^1/_2$ and $1:3^1/_2$ by volume of binder to aggregate were used historically.

These fulfil the requirement of filling the voids in the sand which are between about 25 and 40 per cent by volume. Typically, a binder to sand ratio of 1:3 produces a broadly acceptable mix in terms of the working or plastic properties. As the voids are filled, the mix should be relatively acceptable to the operative and not prone to excessive bleeding or segregation, so long as the sand grading is acceptable.

### 7.7.1 Mix proportions

Table 7.6 below shows the mix proportions by volume prescribed in the British Standard and it is seen that 3 generic binder/mortar types are permitted. These are cement:lime:sand mortars, masonry cement mortars and cement:sand plasticized mortars.

**Table 7.6** Masonry mortars

| Mortar designation | Type of mortar (proportions by volume) | | |
|---|---|---|---|
| | Cement:lime: sand | Masonry cement: sand | Cement:sand plasticizer |
| (i) | $1 : 0$ to $^1/_4 : 3$ | – | – |
| (ii) | $1 : ^1/_2 : 4$ to $4^1/_2$ | $1 : 2^1/_2$ to $3^1/_2$ | 1 : 3 to 4 |
| (iii) | 1 : 1 : 5 to 6 | 1 : 4 to 5 | 1 : 5 to 6 |
| (iv) | 1 : 2 : 8 to 9 | $1 : 5^1/_2$ to $6^1/_2$ | 1 : 7 to 8 |
| (v) | 1 : 3 : 10 to 12 | $1 : 6^1/_2$ to 7 | 1 : 8 |

The common arithmetic function within each of these mixes is that the ratio of binder to sand approximates to 1:3. This is more readily seen by referring to Table 7.7, which shows a simplified form with only the cement:lime:sand mixes.

The strength that these mixes produces varies from about 1 N/mm$^2$ to about 20 N/mm$^2$ and the minima required by British Standard 4721 are shown in Table 7.8, for laboratory

**Table 7.7** Masonry mortars

| | Mix proportions by volume for cement:lime:sand mortar | | |
|---|---|---|---|
| | Cement | Lime | Sand |
| Low in cement | 1 | 2 | 9 |
| ↓ | 1 | 1 | 6 |
| | 1 | $^1/_2$ | $4^1/_2$ |
| Rich in cement | 1 | $^1/_4$ | 3 |

produced mixes. For ready-to-use mortars, there is no 7-day strength requirement and the 28-day strength minima are in the same order as the 7-day values shown in Table 7.8 below.

**Table 7.8** Masonry mortar compressive strength requirement

| Mortar designation | Traditional volume proportions cement:lime:sand | Compressive strength | |
|---|---|---|---|
| | | 7 days | 28 days |
| (i) | 1 : $^1/_4$ : 3 | 10.7 | 16.0 |
| (ii) | 1 : $^1/_2$ : $4^1/_2$ | 4.3 | 6.5 |
| (iii) | 1 : 1 : 6 | 2.4 | 3.6 |
| (iv) | 1 : 2 : 9 | 1.0 | 1.5 |

Although the standard defines compressive strength, other strengths, in particular, flexural, tensile and bond strength, are of great importance for masonry mortars. Although failures in compression are theoretically possible, the vast majority of masonry is adequate in compression in the case of axially/vertically applied loads, with any mechanical failure usually occurring in tension, flexure or bond. Thus, failure in the case of laterally applied forces, e.g. wind loads, impact, etc. is not unknown. Indeed, the compressive strength of mortar is probably not as important as some of these other strengths although they are much less easy to measure. Amongst other concepts, tests using masonry panels in flexure, bond wrenches to apply a turning moment and couplets to pull apart in direct tension, have been suggested and are used for the determination of masonry bond strength but all suffer from high variability. Finally, it should be noted that for some applications mortar can be excessively hard and strong, as in the case of weak, or relatively weak, masonry units where high strength mortars are not generally appropriate.

## 7.8 Basic masonry design for durability

The design of masonry should aim to avoid excessive saturation, particularly where destructive mechanisms such as freeze–thaw cycling or saturation with soluble sulphates may exist. Design codes emphasize the need to consider details such as copings and overhangs, and damp proof courses should be incorporated where appropriate. Exposure is an important consideration and may be considered in terms of the driving rain index or in terms of graded exposure conditions which are defined in the Masonry Code of Practice

(BSI 1973). The mortar joint profile is sometimes said to be of importance but probably only plays a minor role.

## 7.9 Site problems

Although masonry is a relatively easy material in which to design and construct, a variety of site problems can occur, the most common of which are discussed hereafter.

### 7.9.1 Incorrect mix proportions

It is possible to find situations where excessive binder has been used but the vast majority of mortar with incorrect mix proportions has insufficient binder and this is particularly the case where materials are mixed on the building site. Although virtually all codes and application documents refer to the need to measure binder, preferably by weight but if not, then by volume, in the vast majority of cases, that for site-mixed mortar is not measured. The binder is merely added by a shovel, with no attempt at measurement, save perhaps for counting the number of shovels used. Even if this counting method is adopted, it is grossly inaccurate as a shovelful of binder can be much less than a shovelful of sand, as well as being greatly variable. It is strongly recommended that *all* mortars are gauged by measurement, either by volume or preferably by weight.

### 7.9.2 Use of unauthorized admixtures

It is relatively commonplace to find that site-made mortars are adulterated by the use of unauthorized air entraining agents, commonly taking the form of domestic detergent or washing up liquids. Because these entrain air, they produce a mortar of enhanced plastic properties and ease of use but the amount of air entrained is uncertain and may well be excessive, leading to reduced strength, particularly in bond which is most undesirable.

### 7.9.3 Sulphate attack

The tricalcium aluminate present in Portland cement can react with soluble sulphates to form tricalcium sulphoaluminate. Because this reaction is expansive, with the reaction compound ettringite having a much larger volume than the reactants, sulphation can cause spalling, degradation and ultimately failure of the mortar. It is important therefore to avoid excessive saturation where sulphates may be present or to use binder that does not react or that reacts to a minimal degree only. Although in theory sulphate may result from a number of diverse sources, in reality it is probably only present in sufficient amounts to cause failure in either the masonry units or the ground, and the number of masonry units produced with excessive sulphates is small and declining. Assuming that the atmosphere is not contaminated, the problem therefore generally becomes one of ensuring that any sulphates present in the soil are not permitted to come into contact with the mortar, which means the correct detailing of damp-proof courses and related details.

For those relatively rare situations where sulphate may be present in the units or in the atmosphere, protection against saturation in the form of properly designed copings and overhangs will assist and should be afforded at the design stage. There are indications that the presence of entrained air will act as some degree of protection again sulphate attack.

Recently, some examples of the thaumasite form of attack have also been reported in mortars. As this form attacks the calcium silicate hydrates rather than the tricalcium aluminate the use of sulphate resisting cement will not offer any protection.

## 7.9.4 Freeze–thaw cycles

If masonry is saturated then cycles of freezing and thawing may cause degradation and ultimately failure. Again, detailing should aim to avoid this problem with proper provision of damp-proof courses and copings. However, much masonry will inevitably become wet, sometimes saturated, and in these situations the use of air entrainment is extremely beneficial. The standards for most Western nations specify the use of air-entrained mortar where durability is an issue and, as earlier considered within the context of sulfation, this will provide enhanced durability to both sulphates and freeze–thaw cycles.

## 7.9.5 Aesthetic failures

Structural failure and to a lesser extent, degradation due to durability failure, are obvious and unacceptable failure mechanisms but aesthetic issues are also important. Widespread efflorescence or bloom is deemed by some clients to be totally unacceptable and the materials and design should take account of this issue. Workmanship is a key factor and the advice of the old UK Code (BSI 1973) was that 'newly erected brickwork (*for brickwork one could read masonry*) should be covered at end of each working day or when rained off'. This advice is both concise and apposite. Failure to cover newly erected masonry will clearly render it vulnerable to saturation, with the possibility of durability problems but also of disfiguring staining occurring. Hydrated Portland cement remains very sparingly soluble and saturation results in the potential for leaching of a calcareous solution from the material, primarily in the form of dissolved calcium hydroxide. When water from this solution evaporates, solid material may be deposited on the surface of the joint or unit, generally in the form of calcium carbonate as a function of reaction with atmospheric carbon dioxide or with dissolved species in rain water. Failure to properly protect newly erected work often results in a typical horizontal line of white staining, formed by this process of leaching of solution. A related form of white staining may occur where the detailing or poor workmanship results in an area where there is long-term propagation of a localized stain. This may typically occur below a defect in a coping or in a free-standing wall where unfilled or poorly filled vertical joints form reservoirs for water to fill and subsequently leach out soluble compounds. These calcium carbonate stains, in contrast to some other efflorescent stains, may readily be removed by the application of dilute acid treatment, 10 per cent hydrochloric acid being recommended by most authorities.

## 7.10 Summary

Specification and practice should be in accordance with the material standards and with the design and application documents. Design should consider both durability and aesthetics/ weathering issues. Materials should be properly gauged, preferably by weight. Site work should be properly protected during construction.

## References

Aspdin, J. UK Patent No. 5022.
ASTM (1989) C-803 Standard test method for penetration resistance of hardened concrete.
BRE (1997) Digest 421 Measuring the compressive strength of masonry materials: the screw pull out test.
BSI (1998) pr EN1052-5. Methods test for masonry – Part 5: Determination of bond strength by bond wrench method. British Standards Institution, London.
BSI (1998) EN 1052-3. Methods of test for mortar for masonry – Part 3: Determination of initial shear strength. British Standards Institution, London.
BSI (1999) prEN 1018-18. Methods of test for masonry – Part 18: Determination of water absorption coefficient due to capillary action. British Standards Institution, London.
BSI (1986) BS1199, 1200. Building sands from natural sources. British Standards Institution, London.
BSI (1999) BS EN 1015.5. Method of test for mortar for masonry: determination of consistence of fresh mortar by LCL meter, British Standards Institution, London, (deleted work item).
BSI (1999) prEN13139. Aggregates for mortar. British Standards Institution, London.
BSI (1973) CP 121: Part 1. Code of Practice for Walling. Part 1. Brick and block masonry. British Standards Institution, London.
BSI (1992) BS 5628: Part 1:1992. Code of practice for the use of masonry – Part 1: Structural use of unreinforced masonary. British Standards Institution, London.
BSI (1998) BS 4551: Part 1. Methods of testing mortars, screeds and plasters – Part 1: Physical testing. British Standards Institution, London.
BSI (1999) BS EN 1015-3. Methods of test for mortar for masonry – Part 3: Retention of consistence of fresh mortar by flow table. British Standards Institution, London.
BSI (1999) BS EN 1015-4:1999. Method of test for mortar for masonry – Part 4: Determination of consistence of fresh mortar by plunger penetration. British Standards Institution, London.
BSI (1999) BS EN 1015-8. Methods of test for mortar for masonry – Part 8: Determination of water retentivity of fresh mortar. British Standards Institution, London.
BSI (1999) pr EN 1015-11:1999. Methods of test for mortar for masonry – Part 11: Determination of flexural and compressive strength of hardened mortar. British Standards Institution, London.
BSI (1999) pr EN1015-12. Methods of test for mortar for masonry – Part 12: Determination of adhesive strength of hardened rendering and plastering mortar on substrate. British Standards Institution, London.
BSI (1999) prEN1015-14:1999 (Draft) Methods of test for mortar for masonry – Part 14: Determination of durability (sulphate resistance). British Standards Institution, London.
BSI (1999) prEN1015-19. Methods of test for plastering and rendering – Part 19: Determination of water vapour permeability of hardened rendering and plastering mortar. British Standards Institution, London.
BSI (1999) pr EN1015-21. Test methods for one coat rendering mortar – Part 21: Compatibility with background through the assessment of adhesive strength and water permeability after conditioning. British Standards Institution, London.

BSI (2002) BS EN 1052–3: 2002. Methods of test for masonry – Part 3: Determination of initial shear strength. British Standards Institution, London.

DIN (1999) DIN 18555-9. Testing of mortars containing mineral binders – Part 9: Hardened mortars, determination of the mortar compressive strength in the bed joint. Deutsche Institut für Normung, Berlin.

Edgell, G. Personal correspondence.

Edgell, G.J. *et al.* (1999) Effect of air entrainment on freeze/thaw durability of masonry mortars. *Proceedings of the 8th North American Masonry Conference*, Austin, Texas.

Figg, J. Personal correspondence.

Harrison, W.H. and Bowler, G. (1990) Aspects of mortar durability. *Brit Ceram. Trans. J.*, **(89)**, 93–101.

Harrison, W.H. and Gaze, M.E. (1989) Laboratory-scale tests on building mortars for durability and related properties. *Masonry International*, Vol. 3, No. 35. British Masonry Society, Stoke-on-Trent, England.

Higgins, B. (1780) *Experiments and Observations Made with the View of Composing and Applying Calcarious Cements and of Preparing Quick Lime, Theory of These Arts and Specifications of the Authors Cheap and Durable Cement, For Building, Incrustation or Stuccoing and Artificial Stone*. T. Cadell, London.

Malinowski, R. (1981) *Ancient Mortars and Concretes: Durability Aspects*, ICCROM Symposium, Rome.

Neuberger, A. (1930) *The Technical Arts and Sciences of the Ancients*. London.

Ragsdale, C.A. and Birt, J.C. (1976) Building sands: availability, usage and compliance with specification requirements. CIRIA report 59. London.

RILEM (1982) MD3 Determination of mortar strength by the drilling energy method Rilem. Testing methods of mortars and renderings. MR-7 site test for the determination of the compressive strength of mortar cured hydraulically.

Smeaton, J. (1791) *A Narrative of the Building and a Description of the Construction of the Eddystone Lighthouse*. London.

Vicat, L.J. (1837) *Mortars and Cements*, Transl. by J.T., Smith, latest edition 1997. Donhead Publishing, Shaftesbury, England.

Vitruvius *De Architectura* (*c*. 27 BC) 5th edition (1970). Transl. from Harlien manuscript 2767, F., Granger, Heinemann, London.

West, H.W.H., Ford, R.W. and Peake, F. (1984) A panel freezing test for brickwork. *Brit. Ceram. Trans. J.* **(83)**, 112.

# 8

# Recycled concrete

*Rod Collins*

## 8.1 Introduction

Increased concern for environmental protection and for promotion of the principles of sustainable development has led some governments to introduce legislation to encourage the use of recycled aggregates. A favoured method is to lower the selling price of recycled aggregates in relation to natural aggregate, and this is largely achieved by increasing landfill costs. This policy is particularly well developed in the Netherlands and the Copenhagen district of Denmark which now recycle over 80% of demolition waste.

In many countries there is considerable experience in the use of crushed concrete as a hardcore material and as a sub-base for highway construction, but constraints such as contamination with other demolition debris and the cost of transporting the materials means that much potentially usable material is placed in landfill. A worldwide research effort on the use of crushed concrete and masonry waste as an aggregate for concrete has been reviewed by Hansen (1986, 1992) and has been the subject of several international conferences, the latest of which were in Odense, Denmark (Lauritzen 1994) and in London (Dhir *et al.*, 1998). Successful application of crushed concrete as an aggregate in concrete has been achieved in several countries, often in major projects.

## 8.2 BRE Digest 433 and the properties of recycled aggregate

In the UK, demolition waste, and crushed concrete in particular, is being recognized increasingly as a valuable resource. Until the launch of Digest 433 *Recycled Aggregates*

(BRE, 1998), there was little in the way of specification guidance for the use of these materials in the UK. The Digest is intended to bridge the gap between recent UK practice and full integration or recycled aggregates into European Standards for aggregate in 2005. It takes as its basis the Specification by RILEM (1994) and the work of the ad hoc group for recycled aggregates set up by European Standards Technical Committee TC 154 'Aggregates'. The properties of recycled aggregate as laid out in Digest 433 are given below.

## 8.2.1 General description/classification

The relative proportion of concrete to brick masonry is the most critical issue:

- brickwork originated is normally associated with lower density and lower quality use – Class RCA (I)
- concrete originated for relatively high grade uses – Class RCA (II)
- a 50/50 mixture for intermediate uses – Class RCA (III)

Aggregates in Class RCA (II) will be suitable to substitute for Classes RCA (III) and RCA (I), and RCA (III) for RCA (I). The limiting values given in Table 8.1 (taken from Digest 433) reflect this:

**Table 8.1** Classes of recycled aggregate (RCA) (BRE Digest 433)

| Class | Origin (normal circumstances) | Brick content by weight |
|---|---|---|
| RCA (I) | Brickwork | 0–100% |
| RCA (II) | Concrete | 0–10% |
| RCA (III) | Concrete and brick | 0–50% |

## 8.2.2 Blends with natural aggregate

This is normally a consideration for the higher grade applications, e.g. to have negligible effect on the performance of structural grades of concrete, up to 20% of natural coarse aggregate may be replaced with RCA (III). Thus for use in concrete according to Digest 433, RCA (I) is restricted to low grades (up to C20, or C35 if the dry density is above 2000 kg/m$^3$); RCA (III) is restricted to 20/80 blends with natural aggregate, and only RCA (II) has relatively little restriction (up to grade C50). RCA (II) could of course be used in higher grades of concrete if blended 20/80 with natural aggregate, as it would conform with the specification for RCA (III).

Replacing 20% of natural aggregate with RCA may not at first sight seem worthwhile because of the relatively low usage of RCA. However, there is insufficient C&D material to supply anywhere near 20% of the current demand for aggregates, so replacement at this level on a general basis could easily accommodate all available material, making higher percentage blends unnecessary.

## 8.2.3 Impurities

| | Use in concrete as coarse aggregate | Use in road construction – unbound/cement bound | Hardcore fill or granular drainage |
|---|---|---|---|
| Asphalt and tar (as lumps, e.g. road planings, sealants) | Included in limit for other foreign material | 10% in RCA (I)[a] or 5% in RCA (II)[a] or 10% in RCA (III)[a] | 10%[a] |
| Wood (also any material less dense than water) | 1% in RCA (I) or 0.5% in RCA (II) or 2.5% in RCA (III)[b] | Sub-base: 1% Cement-bound: 2% Capping layer: 2% | 2% |
| Glass | Included in limit for other foreign material | Contents above 5% to be documented | Contents above 5% to be documented |
| Other foreign materials (e.g. metals, plastic, clay lumps) | 5% in RCA (I) or 1% in RCA (II) or 5% in RCA (III)[b] | 1% (by volume if ultra-lightweight) | 1% (by volume if ultra-lightweight) |

Sulphates: (acid soluble): 1% for concrete and cement bound; for unbound material see Digest 363.
[a]No limit if physical and mechanical test criteria are satisfied (e.g. CBR). Road recycling will produce aggregates composed of bitumen or mixtures of bitumen and concrete. Suitability for purpose, e.g. production of road sub-base is likely to be more dependent on placement technique, so a performance specification rather than a traditional recipe-type will be advisable.
[b]RCA (III) must not replace more than 20% of natural aggregate. Limits on wood and other foreign matter assume that there will be no contribution from the natural aggregate. Similarly, a limit of 1% acid soluble $SO_3$ should apply to 1:4 mixtures of RCA (III): natural aggregate.

Recipe-type specifications are normally easier to prepare and control, but the supporting test methods may not always be closely related to the required performance. This may unnecessarily restrict the efficient use of resources. There is a long-term aim, expressed by the European Standard Committee for Road Materials (CEN/TC 227) to develop performance-based specifications. With such specifications, there should no longer be a focus on constituent materials (whether recycled aggregate, industrial by-product or natural aggregate) but on suitability of the material as a product for its proposed end use.

The Specification for Highway Works (Department of Transport *et al.,* 1991) gives details for the use of crushed concrete, i.e. Class RCA (II) in granular sub-base material Type 1 or 2. Aggregates for use in pavement concrete or cement bound material (CBM) categories 3, 4 and 5 may be crushed concrete 'complying with the quality and grading requirements of BS 882'. Details are also given for 6F1 and 6F2 capping layers which may contain Classes RCA (I), (II) or (III). Similarly CBM categories 1 and 2 may contain Classes RCA (I), (II) or (III). Mixed concrete/brick materials of Class RCA (III) have been used as sub-base materials in the Netherlands, and there may be opportunities for their use in the UK, subject to the approval of the Resident Engineer for each individual contract.

## 8.2.4 Test methods

BRE Digest 433 recommends a manual sorting method as the simplest method (conceptually), although in practice it tends to be rather tedious and not particularly accurate because of difficulty in categorizing some particles. At least 500 particles should be sorted and the separate heaps weighed (categories from BRE Digest 433):

- concrete and dense or normal weight aggregate
- brick, mortar, lightweight block and lightweight aggregate
- asphalt, bitumen, tar and mixtures of these materials with aggregate
- wood
- glass
- other foreign materials such as metals, clay lumps and plastics

Lightweight block – separate category if more than 1% by volume (approx 5 pieces in 500). Ultra-lightweights – should be noted if more than 0.5% by volume (approx 2 pieces in 500) Gypsum plaster – lumps of gypsum are normally pulverized by the crushing process and end up in the fine aggregate. The gypsum content is limited where required by an overall limit on sulphate (see Table 2).

Chlorides – overall limits in BS 5328 *Concrete* remain unchanged, but an acid-soluble chloride test method (BS 1881 Part 124) should be used rather than the water soluble because chlorides bound by the hardened cement paste in recycled aggregates may subsequently become available in a new matrix.

An alternative way of classifying the composition of recycled aggregates is by density. Materials with an average oven-dry relative density of above 2000 or 2100 kg/m$^3$ may be classified as crushed concrete. RILEM (1991) also proposed the use of dense medium separation (ASTM C123) to exclude an excess of the lighter weight materials: not more than 10% with SSD density less than 2200 kg/m$^3$ (equivalent to OD densities of 2000 or 2100 with water absorption of 10% or 5%) and not more than 1% with SSD density less than 1800 kg/m$^3$ (intended to represent the lower end for brick/ceramic material with ODRD of 1500 and water absorption of 20%).

## 8.2.5 Other properties

Reference has already been made to the absorption value for recycled aggregates which is higher than for most natural aggregates in current use. Values of 5–6% are typical for crushed concrete and would exclude its use in concrete water-retaining structures (BS 8007) and maritime concrete (B6349). Strength as measured by the '10% fines' test (BS 812 Part iii) is typically well over 100 kN for crushed concrete but values down to 70 kN are typical of RCA (I). A lower limit of 50 kN is specified in BS 882. Some natural aggregates, e.g. Jurassic limestones have absorption values and strengths in this range and long-term tests at BRE (Collins, 1986) have indicated that this is no barrier to the production of good concrete.

Concretes containing recycled aggregate will have higher drying shrinkage than is normal with most natural aggregates. This is primarily because of the increased quantity of hardened cement paste (hcp) within the concrete as a whole (hcp within the aggregate + hcp in the new concrete). For most structures this will be of little consequence, but there may be some situations where it may be advisable to consider the structural effects to see if any compensatory factors, such as reinforcement design in long slender elements, should be built in to the design – see Digest 357 (BRE, 1991).

For the control of alkali silica reaction, several differences (positive and negative) need to considered for recycled aggregate.

- Less is usually known about the origin of any reactive components – they could be derived from a wide selection of quarries.
- Alkalis in the demolished concrete/masonry mortar will increase the alkali load in any new concrete.
- The relatively high absorption value of RCA will reduce alkali concentration in the pore water and offer some relief for any alkali silica gel pressures developed. This effect is evident with natural aggregates (Collins and Bareham, 1987).

Before the 1997 revision of BRE Digest 330, a simple route for avoiding problems was to use 50% or more of suitable ground granulated blastfurnace slag (ggbfs) in the cementitious component. This method was used for the BRE Environmental Building and new strong floor for the European Concrete Building Project at the BRE Cardington Laboratory. While the same recommendation could be obtained from the new Digest 330 (BRE, 1999), it is arrived at in a different way. Because of the first two factors listed above, the Digest has classified RCA as highly reactive. This limits the alkali equivalent value ($Na_2O + 0.658K_2O$) of the new concrete to 2.5 kg/m$^3$ rather than 3.5 kg/m$^3$ for normally reactive aggregates. Another approach developed in Digest 433 gives similar result, but in addition caters for the use of RCA (III), for which an alkali equivalent limit in the region of 3 kg/m$^3$ may be more appropriate.

For weathering resistance, sulphate soundness tests are inappropriate because of the reaction between sulphate and hydrated cements. Frost heave testing (BS 812 Part 124) may be used for unbound materials, or for concretes a test on the concrete as a whole (ASTM C666 or method for concrete in BS 5075 Part 2).

## 8.3 Methods of recycling and quality

For a good quality product it is essential to separate out different types of material before it enters the crusher. If time permits, the best place to start any separation is at the demolition site itself. Fixtures such as wood, windows, plumbing and wiring can be removed and recycled separately before the remainder of the structure is demolished. It is particularly important to exclude materials which could be contaminated, e.g. for the demolition of old factories with chimneys, furnaces and concrete contaminated by chemical processing. A fixed recycling plant normally draws material from several demolition sites; each of these sites should have been inspected so that contaminated material is not delivered for recycling. Statutory requirements for the inspection of structures before demolition and the filing of plans to show how each part of the construction is to be recycled or disposed of, as practised in the Netherlands and in Denmark, would greatly assist this process. Input control is a key part of a UK quality control scheme for recycled aggregates (BRE, 2000).

For road recycling contracts, selecting a clean input to the crusher should be easier, although there could be problems with the inclusion of variable amounts of bituminous bound material and contamination with lower quality material from beneath the slabs. Nevertheless, significant quantities of material are derived from demolition and the Highways Agency will expect all recycled aggregate in bound and unbound applications to have been produced under the Protocol (BRE, 2000) which they have helped to develop.

The jaw crusher has almost universally been accepted as the most suitable method of reducing concrete and masonry to aggregate sizes. Other types of crusher are sometimes used, generally for secondary crushing. In England where road sub-base and hardcore are specified to a maximum size of 75 mm a secondary crusher is not normally used, but in the United States where the maximum size is normally in the range 25 mm to 50 mm secondary crushers are more usual. Similarly, secondary crushers are used in Europe where a maximum size of 32 mm is often used for road sub-base or for use in concrete.

In Europe demolition produces similar proportions of masonry (brick) and concrete. In the United States the mineral component is mainly concrete, often accompanied by a heavy concentration of reinforcement. Reinforcement is capable of causing serious problems of clogging and entanglement in the primary crusher. To ensure that a minimum of time is lost through clogging or entanglement, large crushers with large clearances are most effective. This also reduces the need to break up large concrete lumps in the feed material. Small plant may, however, be preferred where mobility is an important factor, e.g. for on-site recycling at relatively small demolition sites, and for road reconstruction projects. In practice most static sites in the UK use plant that is mobile.

In normal circumstances soil is mostly screened off before crushing. After crushing, remaining contamination of wood, reinforcement, plastic etc. is picked off as far as possible by personnel standing by the exit conveyor. Further metal is removed by a magnetic separator. Aggregates for concrete ideally benefit from washing to remove dusty material and small particles of wood. The material is then sieved into the required product sizes. In normal circumstances only the coarse aggregates are supplied for use in concrete. Fine RCA often gives unacceptable increases in water demand and any chemical contamination of the demolition material is likely to be concentrated in this fraction. Nevertheless, there are examples of successful use for fine RCA, some of which are detailed in the last section of this chapter.

## 8.4 Recipe and performance specifications

### 8.4.1 The specification 'environment' in the UK before Digest 433

Despite a large body of laboratory research data showing that recycled aggregates could give excellent performance in concrete, practical use in the concrete industry was restricted by questions of specification, risk, availability and cost. These issues have been highlighted in demonstration projects at BRE (see 8.5 below). In a study (Collins and Sherwood, 1995) on the relationship between specifications and the use of recycled materials and other wastes as aggregates, it was concluded that British Standards in themselves (and in particular BS 5328 for concrete) did not prevent the use of recycled aggregates in concrete. The main reason for exclusion of such material was the wording of contract specifications which in addition demand compliance with BS 882 (*Natural* aggregates for concrete). Only BS 1047 (Blastfurnace slag aggregate) or sometimes BS 3797 (Lightweight aggregate) are allowed as alternatives. To use any other type of aggregate, evidence of suitability for purpose (as required in BS 5328) was not in itself generally sufficient for specifiers who saw no reason to take on any further risks or other costs, however small. The situation is now changing with environmentally aware clients beginning to accept the minimal risks involved and the relatively small additional costs (if any).

### 8.4.2 Product specifications

Another route for the application of recycled aggregates was identified by Collins and Sherwood (1995) and this is potentially simpler to implement. Manufactured products should be guaranteed by the manufacturer as fit for purpose. Thus, in principle, the quality control of any input of recycled material needs to be verified only by the manufacturer of the product. In practice, however, not all products are specified purely on performance and standards often contain both recipes and performance requirements. In these circumstances, proof may be required to the satisfaction of the user that a recipe can be varied without detriment to the product. These specification routes are explored in the first two of the four demonstration projects detailed below.

## 8.5 Applications – demonstration projects

### 8.5.1 Concrete blocks

A partnership project led by BRE has concentrated on the use of demolition waste in precast concrete blocks. The intended use for these blocks was for beam-and-block flooring systems. This was considered to be fairly non-onerous end-use since there are no weathering requirements and the general experience with loading tests on floor is that the margins of safety are very large (at least a factor of 10).

Initial studies in this project were concerned with the requirements for recycled aggregates to be used in conventional blockmaking plants. These plants rely on the free fall of materials in hoppers etc. and thus a sufficient proportion of fines must be removed from the recycled aggregate to ensure that clogging of the plant does not occur. The grading of the recycled aggregates thus needs to be coarser than that needed to produce dense, well-compacted blocks. Some natural sand needed to be added to the recycled aggregate at the mixer, i.e. it was not possible to produce good blocks containing 100% recycled aggregate with conventional plant. Blocks with up to 75% recycled aggregate could be made with no difficulty. (ARC Conbloc using materials supplied from Pinden Plant & Processing.) Floor loading tests were carried out by Kingsway Technology on blocks from three trial runs containing between 50% and 75% recycled aggregates. Beam-and-block floors 2.9 by 3 m were constructed for each trial run and finished with a 50 mm thick 3:1 sand:cement screed which was left to cure for 28 days. Each floor required the use of 6 beams and 70 blocks. Floors were loaded centrally via a 100 mm square plate. Results for ultimate load gave safety factors between 33.7 and 39.0 (in other words between 33.7 and 39.0 times the ‘minimum value to be adopted’ for a fairly non-onerous domestic floor), and deflection was well within limits given in BS 8110 for the structural use of concrete. Normal blocks would give similar results.

### 8.5.2 Precast structural concrete

A further partnership project led by BRE in association with members of the Precast Flooring Federation, Leeds, Nottingham Trent and Sheffield Universities is concerned with the recycling of rejected precast elements within precast works. Wastage of concrete within precast works due to a number of factors such as breakage, poor compaction, malformation and off-cuts can sometimes approach 10%. Although some materials have been crushed and used as hardcore rather than landfilled, there would be further advantages if the material could be fed back into the production line. Use of reclaimed product in precast is already practised in some countries where this is allowed in the specifications. The challenge in the UK where current specifications may preclude its use was to produce an industry code of practice or protocol acceptable within the construction industry as a whole. This was launched along with the quality control scheme in April 2000.

Successful trials have been carried out at five precast works during 1997/98 showing little if any effect from the replacement of 20% of the coarse aggregate and 10% of the fine aggregate by reclaimed product. Reclaimed product is a purer and more consistent material than recycled aggregates from demolition waste and thus requirements for quality control are very much reduced. The incorporation of a small percentage of fine material without detriment also illustrates the purer nature of reclaimed product – most research workers have found the fine material from demolition waste to be too contaminated for very successful use in concrete.

## 8.5.3 The Environmental Building at BRE

This is a new office and seminar facility at the heart of the BRE Garston site designed to act as a model for low energy and environmentally aware office buildings of the next century. J Sisk & Son Ltd constructed the building to the designs of a team led by architects Feilden Cleff in consultation with BRE staff. Recycled aggregates were used under the supervision of structural engineers Buro Happold and staff of the BRE Centre for Concrete Construction. This building incorporates the first-ever use in the UK of recycled aggregates in readymixed concrete. Crushed concrete from Suffolk house, a 12-storey office block being demolished in central London, was used as coarse aggregate in over 1500 $m^3$ of concrete supplied for foundations, floor slabs, structural columns and waffle floors.

### *Concrete in the Environmental Building*

For the foundations, a C25 mix (75 mm slump) for Class 2 ground conditions was specified. According to BRE Digest 363, a minimum OPC-based cement content of 330 $kg/m^3$ and a maximum free water/cement ratio of 0.50 were required. For floor slabs etc., a C35 mix, also with 75 mm slump was specified. RMC trial mixes (all with water-reducing admixture Fosroc Conplast P250) gave the alternatives in Table 8.3. C25 Mix 3 and C35 1 were chosen for use. A high content of ground granulated blastfurnace slag was chosen for the C35 mix for protection against carbonation. All mixes contained 985 $kg/m^3$ of crushed concrete coarse aggregate apart from mixes for pumping in which this was reduced by 50 $kg/m^3$ and the cement content increased by 10 $kg/m^3$.

With the aid of a portable laboratory, RMC Ltd made frequent tests on the concrete delivered to site and showed that the maintenance of quality was adequate (Figure 8.1). The recycled aggregates were also tested frequently for chlorides (~0.015%), sulphates (~0.35%) etc. The absorption value of the recycled aggregates varied about a mean of 5%. Drying shrinkage measured by several test methods was in the region of 0.035–0.045% (i.e. within the requirement of job specification).

**Table 8.3** Trial mixes

| Description | OPC + ggbfs | OPC/ggbfs | Standard 28-day cube strength |
|---|---|---|---|
| C25 Class 2 Mix 1 | 375 kg/m$^3$ | 100/0 | 47.0 Actual for trial mix |
| C25 Class 2 Mix 2 | 375 kg/m$^3$ | 50/50 | 45.0 Actual for trial mix |
| C25 Class 2 Mix 3 | 375 kg/m$^3$ | 30/70 | 36.0 Estimated by interpolation |
| C35 Mix 1 | 385 kg/m$^3$ | 50/50 | 46.0 Actual for trial mix |
| C35 Mix 2 | 400 kg/m$^3$ | 30/70 | 46.0 Estimated by interpolation |

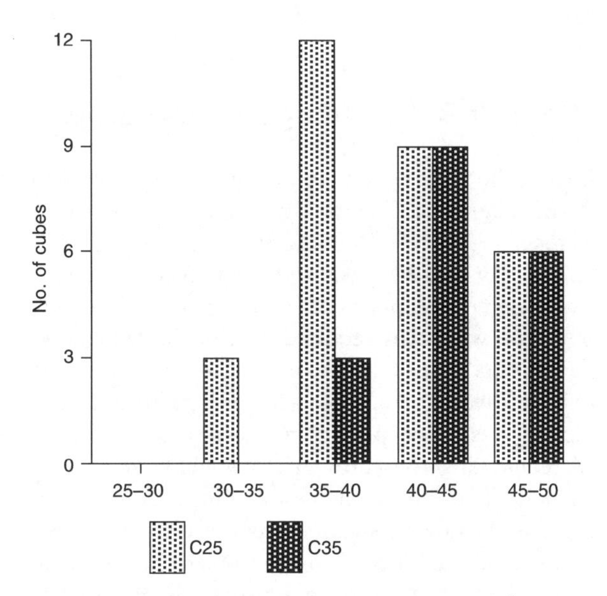


**Figure 8.1** Concrete mix quality control data (compressive strength in N/mm$^2$).

## *Availability of materials*

Because of the problems outlined above with regard to the 'specification environment' at that time, there was no demand for high-quality processing of recycled aggregates in the UK. This meant that under most circumstances, the correct grading and quality of material had to be pre-ordered whereas suitable natural materials were available 'off the shelf' – a 'Catch 22' situation where there were no accepted standards or specifications because no-one used the material, no-one used the material because it was not ready prepared, and no-one was willing to prepare the material because there were no specifications or market.

With regard to the BRE project, material was pre-ordered to a suitable specification and quality. Control on quality was assisted by selecting material from one building. This in itself was an object lesson on availability. Several structures were considered before Suffolk House was chosen. Delays due to legal problems ruled out the first choice, and material from Suffolk House came on stream rather later than originally anticipated. Trial mixes with the material were less than 28 days old before concrete work started on the new building. Results only for 7-day tests were obtained, but fortunately strength development was fairly closely matched to that for trial mixes carried out on samples of other crushed concrete materials previously supplied by the recycling plant.

### *Cost*

The need for trial mixes and quality control has cost implications. Obviously, as a market develops, such costs will be reduced but are likely to remain higher than for natural aggregates. Costs will also be incurred for the higher cement contents often (but not always) required by recycled aggregates. Such costs may eventually be offset by the increasing cost for landfill. An extra cost in the BRE project was for increased transport requirements. This was accepted in order to obtain the most secure supply lines of material and ensure that this 'one-off' demonstration project had the greatest possible chance of success.

## 8.5.4 Strong floor at BRE Cardington laboratory

The new strong floor at the BRE Cardington laboratory was chosen to demonstrate the use of Class RCA (III) aggregate. The heavily reinforced slab is 0.5 m thick and forms the base of the European Concrete Building Project. Although the tender documents suggested that preference would be given to tenders incorporating the use of recycled aggregates, none was offered as supplies of suitable material were not readily available. Since a rapid start to the work was required, one of the tenders which was the most suitable in all other aspects (but of course quoting the use of natural aggregates) was accepted, leaving about 3 weeks to arrange and certify a suitable source of RCA.

500 $m^3$ of concrete was to be placed in one day, 23 February 1996, and about 100 tonnes of RCA (containing up to 50% brick) would be required. The nearest source that could be found in the short time available was King's Cross, London, but the tonnage required in the 20–5 mm size range could not be produced until a few days before casting. Another source of material was located on the other side of London near Dartford. Samples of both materials were taken for trial mixes at RMC (Tables 8.4 and 8.5). Both contained in the region of 40% brick masonry. It was assessed that there would be no problem in achieving a strength of at least 35 $N/mm^2$ at 56 days (as required in the project specification) in any of the concretes with approximately 20% replacement of natural coarse aggregate by RCA. Strength development was of course slower where the ground granulated blastfurnace slag/ordinary Portland cement ratio was 70/30, but the ultimate strength would be similar.

**Table 8.4** Aggregate data for BRE Cardington trial mixes

| Aggregate type | Supplier | Source | Moisture content | Absortion value |
|---|---|---|---|---|
| Natural sand | CAMAS | Little Paxton | 5.0% | 1.6% |
| 20–5 mm natural gravel | St Albans Sand and Gravel | Sandy | 4.5% | 1.7% |
| 20–5 mm mixed RCA | Galliford Roadstone | King's Cross | 5.0% | 9.2% |
| 20–5 mm mixed RCA | Pinden Plant & Processing | Pinden Quarry, near Dartford | 6.5% | 7.2% |

It is also interesting to note that the trial mixes where all the natural coarse aggregate was replaced by the mixed RCA aggregates, the strength measured at 7 days was not much below equivalent mixes with only approximately 20% replacement of the natural coarse aggregate. Also, similar results were obtained from the two substantially different sources of RCA. The King's Cross material almost invariably derives from large demolition sites and is delivered directly, thus the source of materials can be identified directly from the waste transfer notes. This documentation is required by law and must identify the place where the waste was loaded. The majority of material at Pinden arrives via waste transfer stations so that the waste transfer notes identify the waste transfer station and not the multitude of small- and medium-sized inputs to the waste transfer stations. The actual sources of the material would be difficult to trace. However, Pinden Quarry can only accept inert waste, and the fact that a large number of small loads are thoroughly mixed before processing means that the risk of materials deleterious to the perofrmance of concrete being present in sufficient quantity to cause problems must be extremely low.

Mix 1 in Table 8.5 was chosen for use in the strong floor and the production and placement of the concrete went ahead on schedule. As noted for the BRE Environmental Building, the RCA did not affect the pumping and placing of the concrete. There was also no effect on production schedules with up to ten truckloads of concrete being delivered per hour between 7 am and 5 pm. Compliance test data from RMC are given in Figure 8.2. As can be seen there was a considerable 'overkill' on compliance, but in the absence of long-term data for the trial mixes, this could not be avoided. As shown in the next section, such problems could largely be avoided with a quality control system.

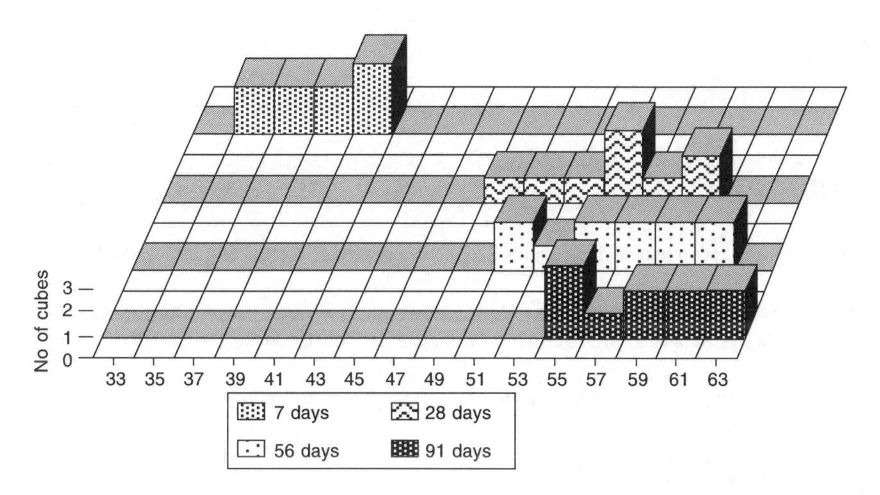


**Figure 8.2** Strength data for concretes used in the strong floor at BRE Cardington (N/mm$^2$).

Although, with hindsight, a much lower cement content could have been chosen, the mix design of the new floor was very similar to the mix design of the original strong floor so that a direct comparison could be made. This shows that the substitution of RCA for up to about 20% of the coarse aggregate does not lead to any significant reduction in the properties of the concrete.

**Table 8.5** Trial mix data for BRE Cardington – approximate proportions and test data

| Mix constituents | Mix 1 | Mix 2 | Mix 3 | Mix 4 | Mix 5 | Mix 6 |
|---|---|---|---|---|---|---|
| Total cementitious content (kg/m$^3$) | 385 | 385 | 385 | 385 | 385 | 385 |
| OPC/ggbfs | 30/70 | 50/50 | 30/70 | 30/70 | 50/50 | 30/70 |
| Natural sand (kg/m$^3$) | 685 | 685 | 730 | 685 | 685 | 650 |
| Natural gravel (kg/m$^3$) | 746 | 746 | – | 746 | 746 | – |
| RCA (Kings Cross) (kg/m$^3$) | 190 | 190 | 825 | – | – | – |
| RCA (Pinden Plant) (kg/m$^3$) | – | – | – | 190 | 190 | 925 |
| Admixture Fosroc P250 (ml/m$^3$) | 150 | 150 | 150 | 150 | 150 | 150 |
| Slump (mm) | 105 | 100 | 110 | 110 | 110 | 100 |
| Free water/cement ratio | 0.47 | 0.47 | 0.45 | 0.5 | 0.49 | 0.5 |
| Total water/cement ratio | 0.57 | 0.58 | 0.68 | 0.6 | 0.59 | 0.7 |
| Average 7 day strength (N/mm$^2$) | 27.5 | 33.4 | 24.6 | 26.2 | 32.1 | 22.8 |

## References


BRE (1991) Shrinkage of natural aggregates in concrete *Digest 357*, Building Research Establishment, Watford.

BRE (1998) Recycled aggregates *Digest 433*. Construction Research Communications, London.

BRE (1999) Alkali-silica reaction in concrete *Digest 330*. Construction Research Communications, London.

BRE (2000) *Quality Control. The production of recycled aggregates.* Construction Research Communications, London.

Collins, R.J. (1986) *Porous aggregates in concrete: Jurassic limestones.* BRE Information Paper IP2/86. BRE, Watford.

Collins, R.J. and Bareham, P.D. (1987) Alkali silica reaction: suppression of expansion using porous aggregate. *Cement and Concrete Research,* **17** (1), 89–96.

Collins, R.J. and Sherwood, P. (1995) *The Use of Waste and Recycled Materials as Aggregates; Standards and Specifications.* HMSO, London.

Department of Transport, Scottish Office Industry Department, Welsh Office and Department of the Environment for Northern Ireland (1991) Specification for Highway Works. HMSO, London.

Dhir, R.K., Henderson, N.A. and Limbachiya, M.C. (ed.) (1998) *Use of Recycled Concrete Aggregate.* Thomas Telford, London.

Hansen, T.C. (1986) Recycled aggregate and recycled aggregate concrete. Second state of the art report. Developments 1978–1984. RILEM Technical Committee 37 DRC. *Materials and Structures* **19** (111), 201–246.

Hansen, T.C. (1992) Recycling of demolished concrete and masonry. RILEM Report No. 6, E & FN Spon, London.
Lauritzen, E.K. (ed.) (1994) Demolition and reuse of concrete and masonry. Proc. 3rd Int. RILEM Symposium. E & F N Spon, London.
RILEM TC 121-DRG (1994) Specifications with recycled aggregates. *Materials and Structures,* **27**, 557–559

## Further reading

Building Research Establishment (1993) Efficient use of aggregates and bulk construction materials – the role of specification. BRE, Watford, Volume 1 – Overview of the effect of specifications on the aggregates industry and the contribution of waste and recycled materials. Volume 2 – Technical data and results of surveys.
British Standards Institution (1985) Guide to the use of industrial by-products and waste materials in building and civil engineering. British Standard BS 6543: 1985. BSI, London.
Limbachiya, M.C., Leelawat, T. and Dhir, R.K. (2000) Use of recycled concrete aggregate in high strength concrete. *Materials and Structures* **33**, 574–580.
Sherwood, P. (1995) *Alternative Materials in Road Construction.* Thomas Telford, London.
Speare, P.R.S. and Ben-Othman, B. (1993) Recycled concrete coarse aggregates and their influence on durability. Concrete 2000 (Proceedings Conference Dunndee), 419–432 E&FN Spon, London.
BRE Information Papers (4–6 pages giving some additional data and references).
IP5/94 Use of recycled aggregate in concrete.
IP1/96 Management of construction and demolition wastes.
IP3/97 Demonstration of reuse and recycling: BRE Energy Efficient Office.
IP14/98 Blocks with recycled aggregate: Beam and Blocks Floors.

# 9

# Self-compacting concrete

*Rob Gaimster and Noel Dixon*

## 9.1 Introduction

The majority of concrete cast relies on compaction to ensure that adequate strength and durability is achieved. Insufficient compaction will lead to the inclusion of voids, which not only leads to a reduction in compressive strength but strongly influences the natural physical and chemical protection of embedded steel reinforcement afforded by concrete. Concrete is normally compacted manually using vibrators, often operated by untrained labour and the supervision of the process is inherently difficult. Although poorly compacted concrete can be repaired, overall durability is more often than not reduced. The consequences of concrete compaction not only affect the material but also have health and safety and environmental risks with operators subjected to 'white finger syndrome' and high levels of noise. Moreover, recent work has shown that the perception of 'full compaction' does not actually produce a homogeneous concrete (Wallevik and Nielsson, 1998).

The concept of self-compacting concrete (SCC) resulted from research into underwater concrete, *in situ* concrete piling and the filling of other inaccessible areas. Before the advent of superplasticizers and other admixtures, the cost of mixes for these purposes were often expensive with high cement contents required to offset associated high water contents. The development of water-reducing superplasticizers meant high-workability and high-strength concrete could be achieved without excessive cement contents, but excessive segregation and bleeding restricted the use of admixtures to flowing concrete having slumps of between 120 mm and 150 mm. These concretes, however, still required a degree of compaction.

Research undertaken into underwater placement technology in the mid-1980s within the UK, North America and Japan led to concrete mixes with a high degree of washout resistance. This, in turn, led to the development of self-compacting concrete with the concept first initiated in Japan in the mid-1980s. It was recognized that the reducing number of skilled workers in the Japanese construction industry was leading to a reduction in the quality of construction work with subsequent knock-on effects on concrete durability (Okamura and Ouchi, 1999).

In the last decade, self-compacting concrete has been developed further, utilizing various materials such as pulverized-fuel ash (PFA), ground granulated blast furnace slag (GGBS) and condensed silica fume (CSF). The development of self-compacting concrete by a number of Japanese contractors has led to increased use of the material throughout Japan, although the overall production is still relatively small compared to conventional concrete. Globally, structures are now incorporating self-compacting concrete where congested areas of reinforcement make use of conventional concrete unfeasible (Khayat, 1999) and self-compacting concrete is now also well established in a number of countries such as Sweden and the USA.

The readymix concrete industry is producing a portfolio of self-compacting concretes for a whole host of applications. This paper, however, concentrates on high-performance SCC, for demanding applications.

Khayat *et al.* (1999) define SCC as:

> a highly flowable, yet stable concrete that can spread readily into place and fill the formwork without any consolidation and without undergoing any significant separation.

Feature/benefit analysis would suggest that the following benefits should result:

- increased productivity levels leading to shortened concrete construction time
- lower concrete construction costs
- improved working environment
- improvement in environmental loadings
- improved *in situ* concrete quality in difficult casting conditions
- improved surface quality

Non-vibrated concrete is already commonplace in the construction industry and is used with acceptable results in, for example, piling and shotcrete applications. Development of SCC has mainly focused on congested civil engineering structures and its acceptance within the market place has primarily grown in solving technically difficult casting conditions. It is a niche product, a problem solver.

Okamura and Ouchi (1999) have commented on the reduction in the number of skilled workers affecting the quality of construction work in Japan. With SCC reducing the dependency of concrete quality on the workforce, further market penetration can be expected.

## 9.2 Materials

Before designing a mix for SCC an understanding is needed of the properties required for self-compaction and how this can be optimized utilizing materials currently available.

The two main requirements are a highly material which has significant resistance to segregation. To achieve a highly mobile concrete, a low yield stress is required and for a high resistance to segregation, a highly viscous material is important. Water can be added to decrease the yield stress but this addition also lowers the viscosity. Addition of a superplasticizer will also lower the yield stress but will only lower the viscosity slightly. The viscosity of a mix can be increased by changes in mix constituents or the addition of a viscosity modifier, but this will increase the yield stress of the paste. Thus being able to find a happy medium between the two parameters is essential. Figure 9.1 illustrates the relationship between shear rate and shear stress.

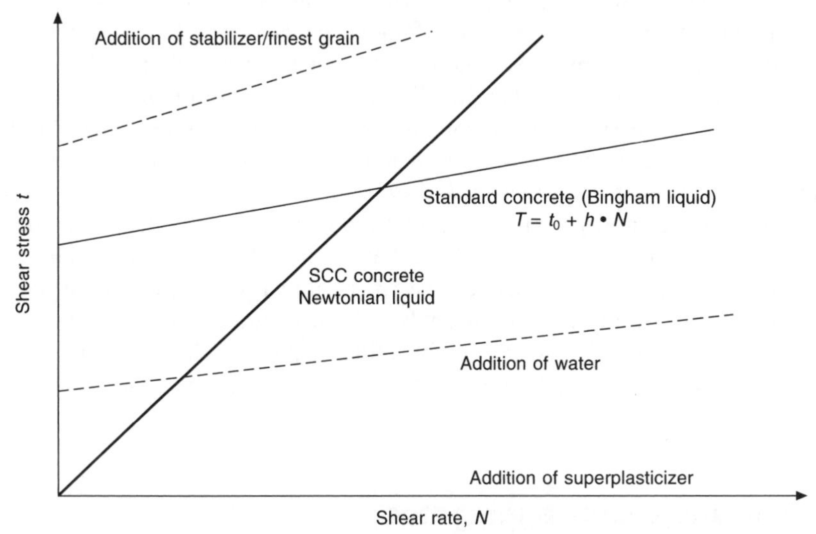


**Figure 9.1** Rheological properties of concrete. (Self compacting concrete, state of the art and future developments, in *Concrete Plant International*, pp. 50–56, 6 December 2000).

## 9.2.1 Cements and fine fillers

The number of cement types available within the UK and Europe has increased rapidly over the past decade. With the advent of EN 197-1:2000, the combinations of cementitious materials available for use in concrete has helped the development of concrete mixes for increasingly demanding circumstances (Domone and Chai, 1996). Many of the results have led to 'high-performance concretes' in the sense that many properties such as strength and durability have been enhanced.

The majority of research on SCC has been carried out using Portland cement (PC) as the binder. Pioneering work by Okamura in Japan and Bartos in the UK (Okamura and Ouchi, 1999; Bartos, 1999) focused primarily on the use of PC for SCC. In developing SCC, the total fines content of the mix is balanced against aggregate size and grading and, in general, the fines content is much higher than in conventional concrete for reasons of stability. The requirement for a high fines content leads to high cement contents, often in the range 450–500 kg/m$^3$.

The increasing number of cement types now available in the UK and across Europe has led to a variety of binder compositions being used by a number of workers. Fine particle binders such as PFA and GGBS have been used in SCC mixes (Domone and

Chai, 1996; Shindoh *et al.,* 1996; Nishibayashi *et al.,* 1996). These additions are known to increase the workability of fresh normal concrete mixes and the common beneficial effects of these materials, such as reduced PC contents and lower heat of hydration, have also been exploited in practical situations (Khayat *et al.,* 2000; Henderson, 2000).

Ultra-fine particle binders such as CSF have also been used in SCC. Such binders are often associated with reducing the workability and lowering slump, however, a number of workers (Petersson *et al.,* 1996; Khayat *et al.,* 2000) have used CSF in SCC, maintaining the desired workability by increasing the superplasticizer dosage.

The increased fines content requirement of SCC has also been offset by the use of fine filler materials. The most commonly used material is limestone powder as it is generally accepted as being economically feasible (Domone and Hsi-Wen, 1997; Petersson *et al.*, 1996; Sedran and DeLartarel 1996; Bartos and Grauers, 1999; Shindoh *et al.*, 1996). Limestone filler will help to maintain stability in a high workability mix although it will not contribute significantly to the compressive strength development of SCC (Domone and Chai, 1997). Limestone filler will also help to control the heat of hydration in mixes that have a high PC content (Nawa *et al.*, 1998).

A number of different filler types were tested in mortar mixes including a proprietary glass filler called Microfiller© and silica slurry (Petersson *et al.*, 1996). The work highlighted the importance of filler materials and showed that with various combinations of binders and fillers, the packing of materials strongly influences the plastic viscosity and yield shear stress. Lower cement replacement with a filler required higher total binder contents to achieve satisfactory rheology while maintaining segregation resistance (Petersson *et al.*, 1996).

## 9.2.2 Fine and coarse aggregates

The fine aggregate in SCC plays a major role in the workability and stability of the mix. The total fines content of the mix is a function of both the binder (and filler) content and the fine aggregate content with the grading of the fine aggregate being particularly important. The grading of fine aggregate in the mortar should be such that both workability and stability are simultaneously maintained. Standard concreting sands are suitable for use in SCC provided standard procedures are adhered to. Sands with fineness moduli of between 2.4 and 2.6 have been used in producing SCC (Goto *et al.*, 1996; Domone *et al.*, 1999). However, finer sands are recommended to ensure satisfactory segregation resistance.

SCC has been made with both gravels and crushed rock as coarse aggregate. The maximum aggregate size is commonly 20 mm but smaller aggregate sizes up to 10 mm have also been used (Domone and Chai, 1996). Coarse aggregate with a grading similar to that used in conventional concrete may be used.

## 9.2.3 Admixtures

The high level of consistence (workability) required from SCC while maintaining stability of the mix has led to the use of a number of admixtures within the concrete. High water contents to achieve workability were unfeasible as very high cement contents were often required to enable compressive strength requirements to be met (Figure 2). The advent of superplasticizers and further developments in admixture technology have meant the production of SCC has now become easier and a variety of materials may be used.

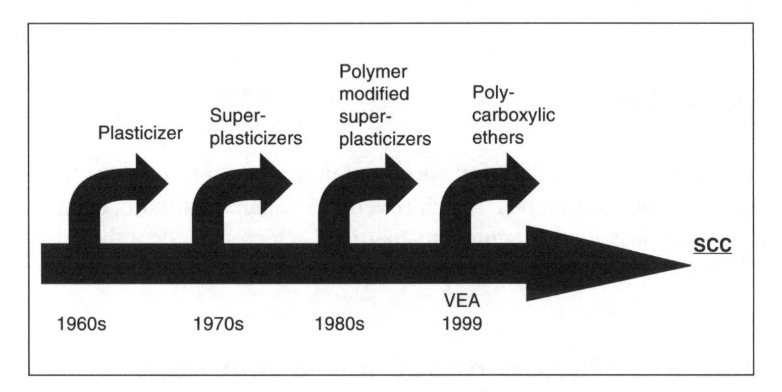


**Figure 9.2** Admixture development (Sika Ltd).

Advances in admixture technology have played a vital part in the development of SCC. Modern superplasticizers (based on polycarboxylic ethers) promote good workability retention and can be added at any stage of the batching cycle. They achieve this with a mechanism of electrostatic repulsion in combination with steric hindrance (Figure 9.3).

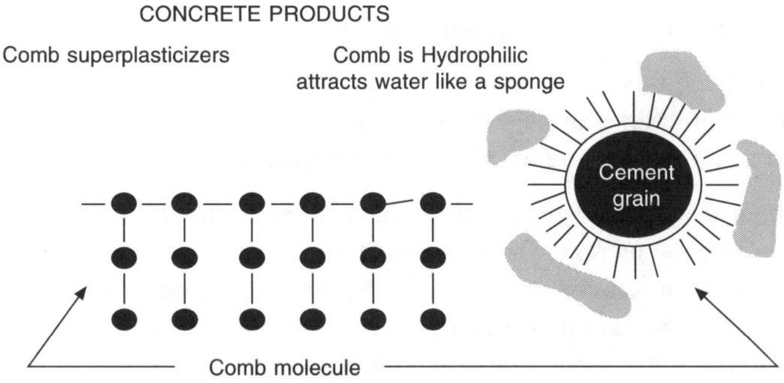

CONCRETE PRODUCTS
Comb superplasticizers
Comb is Hydrophilic
attracts water like a sponge
Cement
grain
Comb molecule

**Figure 9.3** The mechanism of polycarboxylic superplasticizers (Grace Construction Chemicals).

Viscosity modifiers (or segregation controlling admixtures) can be added to increase the resistance to segregation, while still maintaining high fluidity, allowing concrete to flow through narrow spaces.

Viscosity-modifying admixtures are typically high molecular weight soluble polymers, which in aqueous medium have increased viscosity because of their interaction with water (Ramachandran, 1995). These admixtures are effective in stabilizing the rheology of the fresh concrete and preventing segregation of the coarse aggregate from the other mix constituents (Okamura, 2000).

One of the most effective VMAs is welan gum, a natural polysaccharide which modifies the viscosity of the mixing water to maintain stability (Sakata *et al.*, 1996). Welan gum is a very expensive material which may prohibitively increase the overall cost of SCC (Rols *et al.,* 1999). Yurugi and co-workers used a glucose-based VMA with acceptable results (Yurugi *et al*., 1993). Other typical VMAs are based on cellulose ethers, alginates and polyethylene oxides, the latter increasing the viscosity of $H_2O$ by interacting through hydrogen bonding (Ramachandran, 1995).

## 9.3 Mix design

There are many different mix design theories but they all mainly separate SCC into a two phase design, namely, 'continuous' which covers the water, admixture, cement and fillers with a particle size less than 0.1 mm and 'particle' which considers the coarse and fine aggregate. Some of these theories are summarized below.

### 9.3.1 Influencing factors of SCC mix proportions

The first model for proportioning SCC mixes was developed in 1988 using constituent materials readily used in conventional concrete (Ozawa *et al.*, 1989). The model performed satisfactorily with regard to fresh properties and also long-term hardened properties such as shrinkage and density. The cementitious component of the model is much larger than conventional concrete with a subsequent reduction in the volume fraction of coarse aggregate.

Many workers involved in the development of SCC (Okamura *et al.*, 1989; Ozawa *et al.*, 1989; Yurugi *et al.*, 1993; Domone and Chai, 1996) recognized that three fundamental factors govern the fresh concrete properties:

(i) the fresh properties of the mortar
(ii) the volume of coarse aggregate in the mix
(iii) use of a superplasticizing admixture to enhance workability

The fresh properties of the paste and mortar have obvious effects on the concrete properties as with conventional concrete. The water/binder ratio heavily determines the workability of the mix, as will the proportion of sand in the mortar (Yurugi *et al.*, 1993). The volume content of coarse aggregate in the concrete will not only affect the workability but play a major role in the segregation resistance of the mix.

### 9.3.2 Properties of paste and mortar

Initial work on binder suitability for use in SCC (Domone *et al.*, 1999) showed that two factors could be used to determine the suitability for binder use in SCC:

(i) $\beta_P$ = *retained water ratio* – water adsorbed on the powder surface together with that required to fill the voids in the powder system and provide sufficient dispersal of powder.
(ii) $E_P$ = *deformation coefficient* – measure of the sensitivity of the fluidity characteristics to changes in water content.

An extensive investigation into a variety of binder types confirmed that these two factors could determine the suitability of cement combinations for use in SCC by using the slump flow test (Ouchi *et al.*, 1998; Domone and Chai, 1997). Typical values of retained water ratio and deformation coefficient for common binders are given in Table 9.1 (Domone *et al.*, 1999).

**Table 9.1** Retained water/powder ratios and deformation coefficients of powders (Domone *et al.*, 1999)

| | Portland cement | CGBS | PFA | Limestone powder |
|---|---|---|---|---|
| Retained water/powder ratio (by volume), $\beta_p$ | 0.96 | 1.17 | 0.69 | 0.77 |
| Deformation coefficient, $E_p$ | 0.044 | 0.043 | 0.030 | 0.037 |

The characteristics of the mortar within the mix are governed by the sand content as well as the paste characteristics. Provided there is sufficient workability within the paste, the sand content by volume should be approximately 40% (Okamura and Ozawa, 1995). To achieve self-compactability of the concrete, the deformability of the mortar should be high (Okamura, 2000).

## 9.3.3 Coarse aggregate content

The content of coarse aggregate in SCC is a vital parameter in ensuring that the mix has excellent flow characteristics and proper mechanical properties (Khayat *et al.*, 1999). A high coarse aggregate content can lead to a reduction in segregation resistance and also lead to blockage of the flow (Okamura *et al.*, 1998). The frequency of collision and contact between aggregate particles can increase as the relative distance between the particles reduces leading to an increase in internal stresses (Okamura, 2000). Limiting the coarse aggregate content reduces the potential for the formation of internal stresses leading to blockage of the mix.

Okamura and Ozawa (1994) proposed that the volume of coarse aggregate in SCC be kept to approximately 50% for the reasons mentioned above. Yarugi *et al.* proposed SCC could be obtained with coarse aggregate volumes lower than 33% providing the sand content in the mortar of the mix is in the region of 75% (Yarugi *et al.*, 1993). These limits are of course dependent on the maximum coarse aggregate size with coarse aggregate limits being slightly higher with lower maximum aggregate sizes (Domone *et al.*, 1999).

## 9.3.4 Common mix design methods for SCC

There is no unique mix design solution for the production of SCC and a wide variety of materials have been used (Domone *et al.*, 1999). Water/binder ratios are generally less than 0.5 and mixes have a lower coarse aggregate content and higher paste content than conventional concrete (Okamura and Ozawa, 1996). Admixtures and concrete additions such as PFA and GGBS contribute to both increases in workability and segregation resistance. An extensive survey of reported mix constituents and proportions from laboratory and *in situ* investigations showed that, although there were many variations in mix proportions, several factors were common to a majority of mixes (Domone and Chai, 1996); Table 9.2.

**Table 9.2** Summary of common factors from reported *in situ* and laboratory investigations

| Property | Comments |
|---|---|
| Water content | Typically 160–185 kg/m$^3$, similar to conventional concrete. Unclear whether this is due to SCC requirements or for durability purposes. |
| Admixtures | Superplasticizers: used to increase workability. Mainly naphthalene or melamine formaldehyde based.<br>Viscosity modifiers: used to control segregation in mixes with higher water/binder ratios. Cellulose or polysaccharide 'biopolymer'. |
| Binders | Typically in range 450–550 kg/m$^3$. PFA, GGBS, CSF commonly used to improve cohesion. Limestone filler also commonly used. |
| Water/binder ratio | From above, typical range is 0.30–0.36. |
| Aggregates | Both gravels and crushed rock used. Up to 20 mm nominal size is common. Lightweight SCC has also been produced (Hayakawa *et al.*, 1993). |
| Workability measurement | Numerous tests used to assess fresh properties. Common tests described in section 9.5. |

As with conventional concrete mix design, trial mixing is carried out to 'fine tune' mix proportions and make any adjustments as necessary, particularly when estimating the superplasticizer content and viscosity modifying admixture content (Khayat *et al.*, 1999). Superplasticizer dosage should be determined from mortar tests using the flow spread test and the V-funnel test, where the flow of a known volume of mortar through the apparatus is timed and expressed as an index (Ozawa *et al.*, 1994). These tests carried out on mortar minimize the need for trial mixing of concrete.

## 9.3.5 Rational mix design method – Okamura and Ozawa (1995)

A relatively simple mix design method was developed in Japan in 1995. The method was based on the fact that self-compactability of concrete can be affected by the characteristics of materials and mix proportions (Okamura and Ozawa, 1995).

The principle of the mix design method is that the coarse and fine aggregate content is fixed so that the self-compactability of the fresh concrete can be achieved by adjusting only the water/binder ratio and superplasticizer. The fundamental steps in the procedure are as follows:

(i) The coarse aggregate content is fixed at 50% of the solid volume of the concrete.
(ii) The fine aggregate content is fixed at 40% of the mortar volume.
(iii) The water/binder ratio is assumed to be 0.9–1.0% by volume depending on the properties of the binder(s).
(iv) The superplasticizer dosage and final water/binder ratio are determined so as to ensure self-compactability.

In mix proportioning conventional concrete, the water/binder ratio is a governing factor from the viewpoint of obtaining the required compressive strength. In SCC, the water/binder ratio is influential from the viewpoint of workability rather than obtaining compressive strength as, in the majority of cases the water/binder ratio is low enough to give the required strength for contemporary structures (Okamura, 2000).

## 9.3.6 Linear optimization mix proportioning – Domone *et al.* (1999)

An extensive laboratory investigation was undertaken to determine the suitability of materials available within the UK for use in SCC. The optimum mix design method is based on the rational mix design method of Okamura and Ozawa (1995) but modifies this by using the mathematical approach of linear optimization to produce an optimum mixture of water, powders and aggregate (Domone *et al.*, 1999).

Table 9.3 shows the proposed limits for mix proportions to achieve self-compactability. The limits are wider for coarse aggregate of nominal maximum size 10 mm as the segregation resistance is generally higher with smaller aggregate.

**Table 9.3** Limiting mix proportions for successful self-compacting concrete (Domone *et al.*, 1999)

| | Maximum aggregate size, 20 mm | | Maximum aggregate size, 10 mm | |
|---|---|---|---|---|
| Coarse aggregate content ($kg/m^3$) | 0.5 × dry rodded unit wt | | 0.5–0.54 × dry rodded unit wt | |
| Max water content ($kg/m^3$) | | 200 | | |
| Water/powder ratio by wt (w/p) | 0.28–0.40 | | 0.28–0.50 | |
| Water/(powder + fine aggregate) ratio by wt | 0.12–0.14 | | 0.12–0.17 | |
| Paste volume ($m^3/m^3$ concrete) | | 0.38–0.42 | | |
| | w/p | $v_{fa}/v_m$ | w/p | $v_{fa}/v_m$ |
| Volume sand/volume mortar $v_{fa}/v_m$ | <0.3 | 0.4 | <0.3 | 0.4 |
| | 0.30–0.34 | 0.40–0.45 | 0.30–0.34 | 0.40–0.45 |
| | 0.34–0.40 | 0.45–0.47 | 0.34–0.40 | 0.45–0.47 |
| | 0.40–0.50 | Do not use | 0.40–0.50 | > 0.45 |

The mix design procedure is as follows:

*Step 1:* A typical air content is chosen (1–1.5% for non-air-entrained mixes).
*Step 2:* The coarse aggregate content is fixed (Table 9.3).
*Step 3:* A binder composition is chosen using the guidelines in Table 9.3 or from past experience.
*Step 4:* The maximum water/binder ratio is chosen to ensure that the following three conditions are satisfied:

1. The paste has sufficient plastic viscosity (before addition of superplasticizer) to provide adequate segregation resistance. This is estimated by means of the specific gravity of the binder, retained water ratio ($\beta_p$) and deformation coefficient ($E_P$).
2. The concrete has sufficient compressive strength using the generalization of Feret's formula proposed by Sedran *et al.* (1996).
3. The durability requirements are met.

The minimum value of water/binder ratio from 1, 2 and 3 is chosen.

*Step 5:* The volume of the sand in the mortar is chosen.

*Step 6:* The paste content is calculated and adjusted if outside the limits (Table 9.3).
*Step 7:* The water and binder contents are calculated with the water being a limiting factor and set before Step 5 if greater than 200 kg/m$^3$.

The above steps may be calculated using a simple spreadsheet which allows the use of linear optimization to obtain the best solution within the overall constraints of the mix design (Domone *et al.*, 1999). A target minimum paste content has been found to be effective as this may give rise to a minimum cost mix as well as minimizing the effects of heat of hydration and shrinkage (Table 9.3).

*Step 8:* The superplasticizer dosage is then estimated from tests on the mortar component of the mix using the V-funnel test.
*Step 9:* The concrete mix is then made and tested for fresh concrete properties.

## 9.3.7 Model for self-compacting concrete – Petersson *et al.* (1996)

Petersson and co-workers developed a model for the mix design of SCC. The construction criteria are given by the unique demands that exist for each construction project such as strength and durability.

### *9.3.7.1 Void content*

The first stage is to consider the minimum paste volume required for the coarse and fine aggregate. This can be calculated by measuring the void content for different combinations of fine and coarse aggregate using the modified ASTM C 29/C29M method. The minimum paste volume should fill all voids between aggregate particles while covering all aggregate particle surfaces.

### *9.3.7.2 Blocking criteria*

The blocking criteria of the mix is based on the work carried out by Ozawa *et al.* (1992) on the mechanism of mortar flow and the role of mortar on blocking against reinforcement. Petersson *et al.* (1996) further developed this work to produce a model to calculate the limiting total aggregate content of a non-blocking concrete mix. The model is based on grading and maximum aggregate size.

### *9.3.7.3 Mortar proportions*

The optimum proportions of the mortar within the mix are determined by adjusting the water/binder ratio, superplasticizer and viscosity modifier content and sand content until the required yield shear stress and plastic viscosity are obtained. These can be measured by means of a viscometer (Petersson *et al.,* 1995).

### *9.3.7.4 Concrete proportions*

The concrete mix proportions are thus determined by means of the model developed for calculating the maximum total aggregate content of a mix without the risk of blocking.

The Petersson model for mix proportioning is based on the minimum paste content and could lead to low-cost mix designs due to the low cement content.

### 9.3.8 Solid suspension model mix design – Sedran *et al.* (1996)

The mix design method proposed by Sedran *et al.* is based on the Solid Suspension Model (De Larrard *et al.,* 1995). The principle of the Solid Suspension Model is that part of the water in concrete is used to fill the voids between the 'skeleton' (binder and coarse aggregate), the remainder is used to control the workability. By minimizing the voidage between the skeleton, the workability of a mix can be increased for the same water content. Essentially, the packing density of the binder and aggregate is improved. The Solid Suspension Model can predict packing densities of combined dry materials from their individual bulk density, grading, curve, packing density and mass proportion in a combination (De Larrard *et al.*, 1995).

In the development of the mix design method, a rheometer (BTRHEOM™, Sedran *et al.*, 1996) was also developed to measure the shear yield stress and plastic viscosity of the concrete and mortar.

The main steps in the mix design method are as follows:

*Stage 1:* Specifications for the concrete should be determined on the basis of slump flow or using the BTRHEOM. The plastic viscosity should be >1000 Pa but <200 Pa.
The 28-day strength is specified, as are the details of the most restricting confinement (e.g. narrowest space between reinforcement).

*Stage 2:* A combination of binders is fixed based on previous knowledge to satisfy compressive strength requirements and material availability.

*Stage 3:* The saturation level (the point at which no further water reduction occurs with increased dosage of superplasticizer for the binder is determined and tentatively half of this amount should be used as too near the saturation level will cause segregation (Sedran *et al.*, 1996).

*Stage 4:* The water demand of the binder combination with superplasticizer is determined with previous knowledge of material properties and water reduction effect.

*Stage 5:* The Solid Suspension Model is used to optimize the proportions of binder and aggregate. The water content is minimized and an arbitrary relative viscosity is chosen.

*Stage 6:* A sample of concrete is batched and the water content is adjusted to obtain the target viscosity.

*Stage 7:* The superplasticizer dosage is adjusted to achieve a suitable slump flow.

*Stage 8:* The potential compressive strength of the concrete may be calculated using the generalized Feret's formula as detailed in Sedran *et al.* (1996).

*Stage 9:* The fresh properties of the concrete, i.e. filling and passing ability are studied.

## 9.4 Plastic concrete

There are three main areas to be considered in the concretes plastic state, namely, filling ability, resistance to segregation and passing ability. These properties will be looked at in turn along with methods of assessment.

## 9.4.1 Filling ability

This property of the fresh concrete is related entirely to the mobility of the concrete. The concrete is required to change shape under its own weight and mould itself to the formwork in place.

To enable this to occur, the interparticle friction of the materials must be reduced. This can be achieved in two ways.

- First, surface tension can be reduced by the inclusion of superplasticizers.
- Second, optimizing the packing of fine particles can be achieved by the introduction of fillers or segregation controlling admixtures.

Measurement of the plastic properties can be achieved by the following tests:

- The *slump flow* test utilizes a British Standard slump cone (Figure 9.4), which is filled in one layer without compaction. The mean spread value in millimetres is recorded. Typical values lie between 650 and 800 mm. The test measures the mobility/deformity under a low rate of shear (self-weight). Assessment of segregation can be made subjectively but the test does not completely measure the filling capacity of the SCC in question. A further evaluation can be carried out at the same time. This is the $T_{50}$ value, which measures the time taken to reach a spread of 500 mm.

**Figure 9.4** Slump flow equipment.

  The value of slump flow results when viewed in isolation do not give a true indication of the concretes ability to self compact.
- The *BTRHEOM Rheometer:* The concrete is considered as a Bingham fluid and its behaviour is determined by the yield shear stress and the plastic viscosity. A low shear yield stress and a limited plastic viscosity value are required.

## 9.4.2 Resistance to segregation

SCC must be stable under mobile conditions (when concrete is moving) and two areas, therefore, need to be addressed:

- First, the amount of free water needs to be minimized to avoid bleeding. This can be achieved by the use of superplasticizers to reduce the water demand and a well-graded cohesive concrete to minimize segregation.
- Second, the liquid phase needs to be viscous to maintain the coarse particles in suspension, when mobile. This can be achieved by incorporating a high volume of fines in the mix and/or the introduction of a viscosity modifier.

Measurement of the plastic properties can be achieved by the following:

- The *GTM Stability Sieving* test (Cussigh, 1999), which measures the degree of separation of the coarse and mortar fractions. In this test 10 litres of fresh concrete are placed into a test container and allowed to settle over a 15 minute period. The coarse aggregate settles at the bottom and the upper part of the concrete in the container is then wet sieved and the volume of mortar calculated. The higher the value the more segregation has occurred.
- Visual inspection using the slumpflow method can also be used.

## 9.4.3 Passing ability

This is the ability of the concrete to pass round immovable objects in the formwork, such as reinforcement. The need for this ability will depend on the reinforcement arrangement.

Factors to be considered will be the space between reinforcement, which will influence the selection of the size and shape of the coarse aggregate and the volume of the mortar paste. The more congested the structure, the higher the volume of paste required compared with the amount of coarse aggregate.

Measurement of the plastic properties can be achieved by the following tests.

- The *L-Box* test is useful in assessing different parameters such as mobility, flow speed, passing ability and blocking behaviour. The apparatus (Figure 9.5) consists of a long rectangular section trough with a vertical column/hopper at one end. A gate is fitted to the base of the column allowing discharge of SCC into the trough. Adjacent to the gate is an arrangement of bars which permits assessment of blocking potential to be made. The flow speed can be measured by the time taken to pass a distance of 200 mm ($T_{20}$)

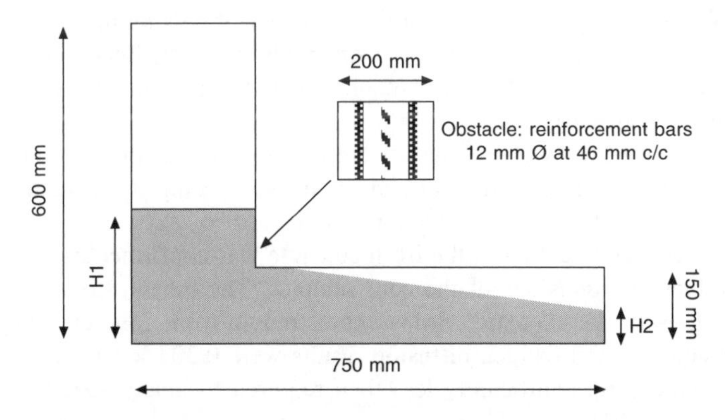


**Figure 9.5** L-box equipment (with rebars placed vertically downwards).

and 400 mm ($T_{40}$). Also the heights at either end of the trough ($H_1$ and $H_2$) can be measured to determine the levelling ability. The ratio of $H_1$ to $H_2$ approximates to 1 for SCC.

The test appears to be useful although there is no standardization of the principal dimensions of the equipment.

- A *J-Ring,* which is a 300 mm diameter steel ring holding vertical reinforcing bars at appropriate spacing, can be added to the slumpflow to assess the concrete's passing ability.

These tests for assessing the plastic properties of SCC are not a definitive list and are at present not recognized by any standards, but they are the most common in current use. The Advanced Concrete and Masonry Centre in Paisley is, however, coordinating a European working group investigating test methods for SCC.

## 9.5 Hardened concrete

### 9.5.1 Compressive strength

Due to the lower water/binder ratios associated with SCC, the compressive strength of SCC is usually higher than for conventional concrete.

When normal concrete is vibrated, water will tend to migrate to the surface of the coarser particles causing porous and weak interfacial zones to develop. If SCC has been well designed and produced it will be homogeneous, mobile, resistant to segregation and able to be placed into formwork without the need for compaction. This will encourage minimal interfacial zones to develop between the coarse aggregate and the mortar phase. Thus the microstructure of SCC can be expected to be improved, promoting strength, permeability, durability and, ultimately, a longer service life of the concrete. *In-situ* compressive strengths, determined using cores, have shown closer correlation to standard cube strength than conventional concrete (Gibbs and Zhu, 1999). Also, work has indicated that the reduction in compressive strength with increase in column height is less pronounced, showing good homogeneity of SCC (Gibbs and Zhu, 1999).

Trials were carried out at RMC Readymix Technical Centre to examine the hardened properties of SCC, using a total cementitious content of 480 kg/m$^3$ at a slumpflow of 700 mm, using a superplasticizer and a viscosity modifier (RMC Readymix Trials, 2001). The concrete was poured into a U-shaped mould as detailed in Figure 9.6, with obstructions placed in the unit (shaded).

Ultrasonic pulse velocity (UPV) tests were performed over the unit and cores were taken to determine the *in-situ* strength and the density within the structure. The cores were also tested for chloride and oxygen diffusion.

Satisfactory self-compaction of the fresh concrete was confirmed by the consistently high UPV values and densities of the core samples. The mean estimated *in-situ* cube strength was 81% of the standard 28-day cube strength from concrete sampled during casting. The chloride and oxygen diffusion results were $0.304 \times 10^{-12}$ and $1.44 \times 10^{-8}$, respectively, which are significantly less than required by many specifications (RMC Readymix Trials, 2001).

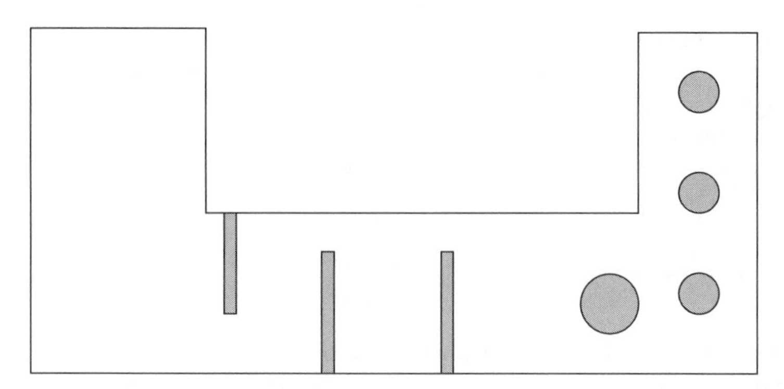

**Figure 9.6** Section of mould used for hardened property tests.

### 9.5.2 Frost resistance

Resistance to freeze/thaw is similar to conventional air entrained. Billberg reported that the freeze/thaw resistance of SCC was actually better than conventional concrete due to the refined microstructure associated with SCC (Billberg, 1999). Yurugi and co-workers stated that SCC was comparable to air entrained conventional concrete (Yurugi *et al.*, 1993).

## 9.6 Production and transportation

Owing to the need for the efficient dispersion of fine particles required to produce a homogeneous and stable mix, mixing time compared with normal concrete is increased. In addition, the need for an accurate total water content of the mix requires good knowledge of the properties of the materials being used. Consistency from the material supplier of moisture content and particle size distribution is critical. Sand grading and moisture content is particularly important.

SCC has been produced from different types of batching plant. The only difference being the size of the mixer units and the efficiency of the mixer. Different mixing techniques will impact on different mixing times. If the concrete has not been sufficiently mixed before transportation, there will be further dispersion of the superplasticizer through the concrete, leading to an increase in workability. These factors need to be considered after successful trial mixes have been established, owing to the nature of the controlled environment of the laboratory.

## 9.7 Placement

No special equipment is needed to be able to place SCC. The same pumps and skips can be used as for normal concrete.

Owing to the nature of SCC being used to reduce construction time, there will be no real advantage in placing the concrete by skip, as time is restricted to the amount of concrete the skip can hold. Generally for normal concrete, while the skip is returned to be filled, the compaction of the concrete is carried out by poker vibrators. This leads to SCC being pumped into place as the main option to save construction time. As with normal concrete, a well-designed SCC can be pumped considerable distances without any problems (Grauvers, 2000).

Pumping from the base of structures is feasible.

## 9.8 Formwork

In order to achieve the benefits of reduced construction time, SCC needs to be placed more rapidly than conventional concrete. With the need for vibration of the concrete eliminated, this can be achieved.

Reduced concrete placing time will usually result in an increased rate of rise of concrete within the structure. This will lead to an increase in hydrostatic pressure on the formwork which could require the formwork to be re-designed to accommodate the theoretical increase in pressure. However, one study confirms that SCC gives lower form pressures than normal vibrated concrete at the same rate of rise (Petersson, 2000). This is because, once the kinetic energy of the fresh concrete has dissipated, the concrete stiffens in a thixotropic manner and, therefore, no longer acts as a liquid (Petersson, 2000). More research is required on this subject.

At present, it is sensible to design formwork assuming full hydrostatic pressure.

## 9.9 Surface finish

In the UK, surface finish is one of the perceived key benefits of SCC leading to a whole range of architectural possibilities. There are several factors however, which interact to give the final surface finish.

- mix design
- workability
- formwork configuration
- formwork material
- mould release agent
- rate of rise
- method of placement

A series of trials were undertaken, at RMC's Technical Centre to examine the effect of different formwork materials together with different categories of mould release agent for the same SCC mix. The mix was designed with a total cementitious content of 500 kg/m$^3$, a free water/cement ratio of 0.36 and a polycarboxylate superplasticizer and VMA, at a slumpflow of 700 mm.

Units were constructed as detailed in Figure 9.7, which used eight different combinations of formwork and release agent. Steel and plywood were used as the formwork materials in conjunction with several categories of release agent, as detailed in Table 9.4.

The results of the trials are summarized in Table 9.4 and some examples shown in Figure 9.8. This gives the ratings (somewhat subjectively) of the combinations of type of

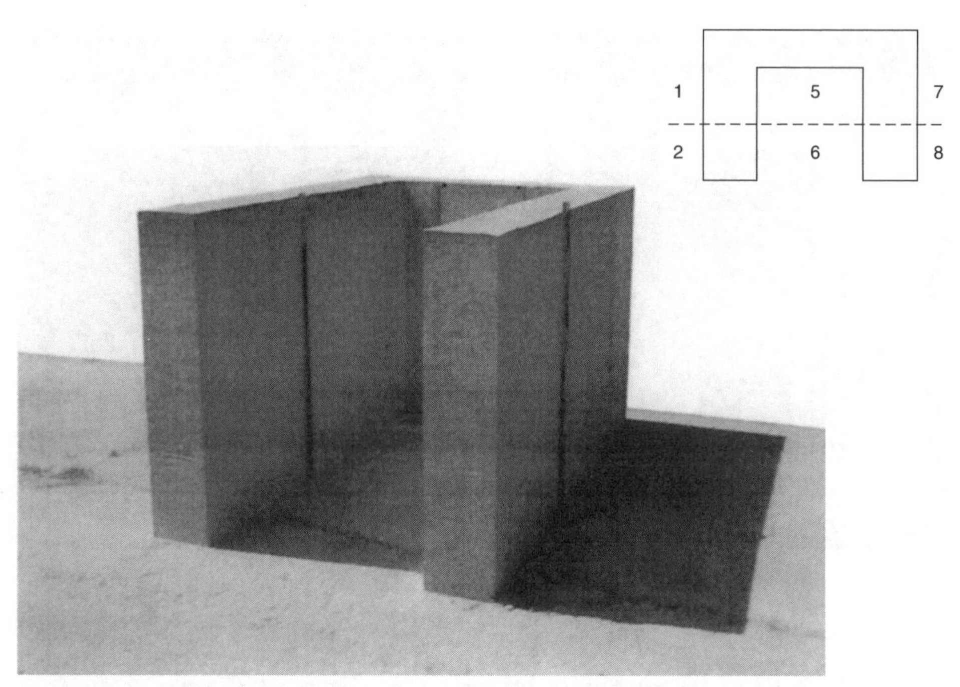


**Figure 9.7** Design of unit.

**Table 9.4** Results of trials

| Rating | Category of release agent* | Surface | Voids | | | |
|---|---|---|---|---|---|---|
| | | | 10 mm | 10–5 mm | 5–2 mm | <2 mm |
| Excellent | A | Plywood | – | – | – | – |
| Excellent/Good | B | Plywood | – | 3 | – | – |
| Excellent/Good | C | Plywood | – | 5 | – | – |
| Excellent/Good | D | Plywood | – | 5 | 10 | – |
| Good | E | Plywood | – | 5 | 20 | – |
| Good/Fair | B | Steel | – | 10 | 10 | 3 |
| Good/Fair | D | Steel | 4 | 10 | – | – |
| Fair | F | Plywood | 3 | 15 | >50 | – |
| Fair | E | Plywood | – | >50 | >50 | 3 |
| Fair | B | Steel | 2 | >50 | >50 | – |
| Fair | A | Steel | 2 | >50 | >50 | – |
| Less than fair | G | Plywood | 20 | >50 | – | – |
| Less than fair | H | Plywood | 20 | >50 | – | – |
| Fair/Poor | E | Steel | 2 | >50 | >50 | – |
| Fair/Poor | F | Steel | 5 | >50 | >50 | – |
| Unacceptable | H | Steel | 20 | >50 | >50 | >100 |
| Unacceptable | G | Steel | 20 | >50 | >50 | – |

*RMC categorization

release agent and formwork material, based on the general appearance and the number and size of voids present in an area of 0.06 m$^2$.

As expected, plywood provides a better surface finish than steel. It should also be noted that the type of mould release agent also plays an important role in the finished surface. Surprisingly, the release agents based on vegetable oil gave the poorest results.

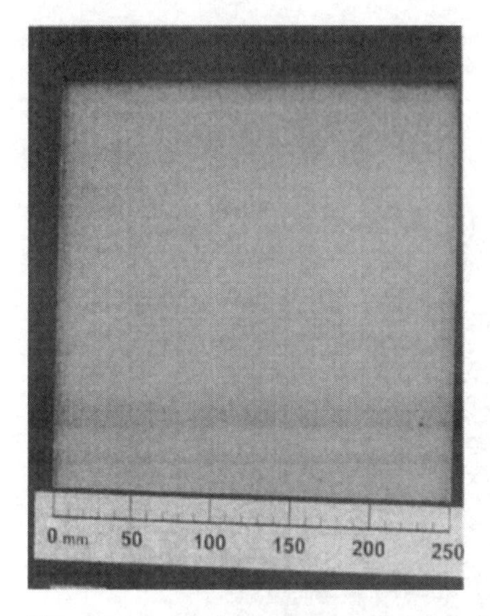


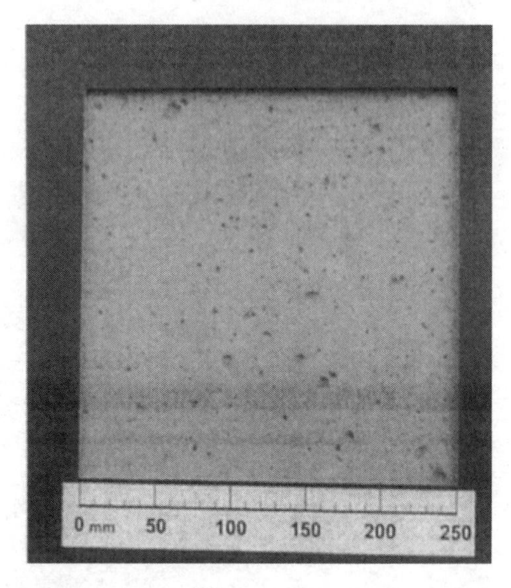


**Figure 9.8** Surface finish examples; same concrete, same formwork type, different release agent.

## 9.10 Mix design optimization – moving SCC to mainstream construction

Since the development of SCC, a total cementitious content of approximately 500–600 kg/m$^3$ has been used, typically achieving stengths in excess of 70 N/mm$^2$. Usually, such high strengths have not been a structural requirement.

One of the main drawbacks to mix designs in current use, however, is the increased cost attributable, in part, to the elevated cement contents required and the state of the art admixture technology.

The ability to reduce the total cementitious content of mixes and to incorporate additions would lower the strength and, more importantly, lower the cost, making SCC a more attractive and competitive proposition for mainstream construction. Research was undertaken by RMC Readymix in conjunction with the Building Research Establishment.

A series of laboratory trial mixes were carried out over a powder content range of 360–500 kg/m$^3$, with blend levels of 30% and 50% of limestone filler, using gravel, initially with only a superplasticizer. Figure 9.9 summarizes the performance of the different mixes. It should be stressed that the trials were investigating high-performance SCCs, with realistic slumpflows (700 mm).

The results showed that true self-compacting concrete could not be produced with just the addition of a superplasticizer below a binder content of 440 kg/m$^3$. Although the mixes were highly fluid they segregated. Instability was created by the excess water needed to achieve the desired workability in combination with the insufficient fines needed to maintain the viscosity. Significant strength reductions were obtained, as expected.

Further trials were then undertaken using a viscosity modifier at cement contents of 400 and 360 kg/m$^3$. The results showed that self-compacting concrete could be achieved in the laboratory with a total cementitious content of around 370 kg/m$^3$, using limestone filler. This is again illustrated in Figure 9.9.

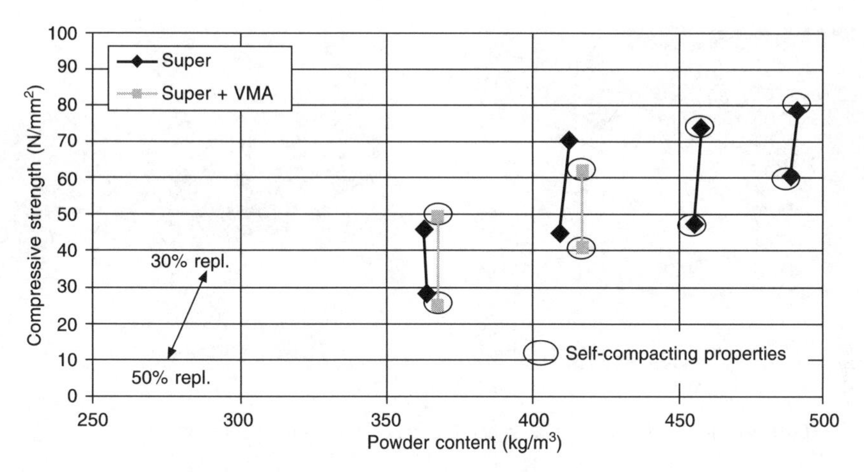


**Figure 9.9** 28-day compressive strength results for SCC containing various levels of limestone filler.

## 9.11 SCC applications

### 9.11.1 City environments

Within structural design, there is a general move towards slimmer elements (Marsh, 2002), particularly in building structures where the advantages are chiefly increased useable space and reduced self-weight. Slimmer elements, can lead to difficulty in vibration of the concrete because of congested reinforcement. This gives a great market opportunity for utilizing SCC.

Within a city location, environmental issues are very important. SCC leads to a reduction in noise levels for site neighbours due to the elimination of vibration equipment, thus also reducing the energy consumption. Material consumption will also be reduced due to less spillage and due to reduced cement consumption. Energy consumption and $CO_2$ emissions will also be reduced (Glavind, 2000).

Health and safety is an important factor on any site, but even more so within a city environment with more congested ground areas. Thus without the need to move pump hoses or handle vibrator equipment, the working environment will be significantly improved. Also, without using handheld pokers, which can cause blood circulation problems, there should be a reduction in reported injuries.

Quality of construction work is also vitally important and with the observed reduction in the number of skilled workers, SCC reduces the dependency of concrete quality on the workforce.

There have been many sites that have already used SCC in city location, including:

- John Doyle at HM Treasury, London
- Mann Construction of Moorgate, London
- Guys Hospital, London
- Midsummer Place, Milton Keynes
- Millennium Tower in Vienna
- Ares Tower in Vienna

Figure 9.10 Shows SCC being placed and a core from the structure

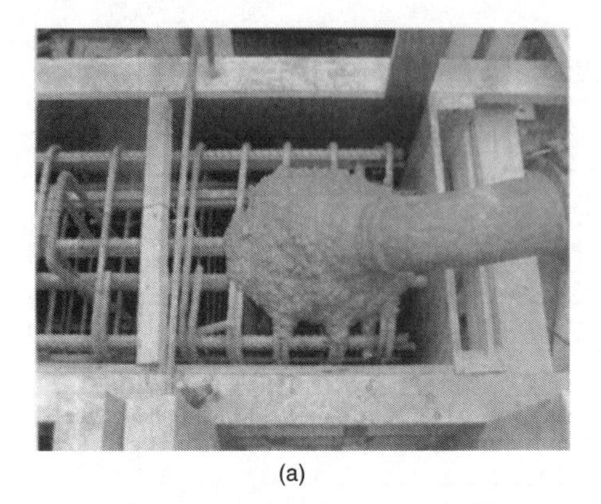

(a)

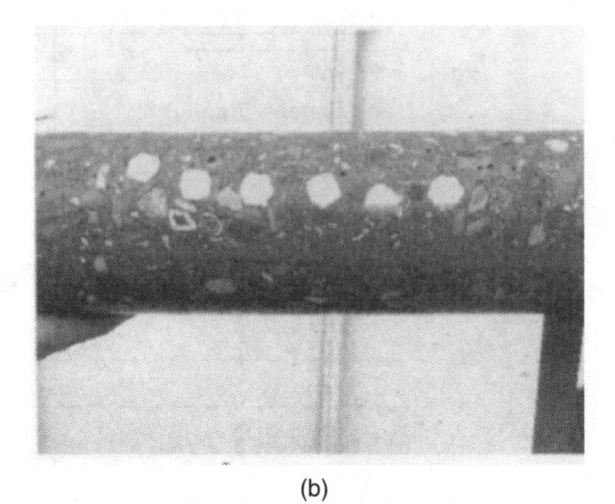

(b)

**Figure 9.10(a)** Overview of SCC being placed and (b) section of core from structure.

## 9.11.2 Precast

SCC used in the precast industry uses the same technology as cast *in-situ* SCC, but can have different requirements for early age strength development and workability life. Similar benefits can be experienced using SCC in the precast industry:

- complex congested moulds can be cast with more ease
- no skilled labour required for compaction
- no vibration leads to less damage to formwork and a reduction in exposure to 'white finger' risk
- no noise from vibration equipment gives a better working environment, improved productivity due to the potential for later working hours
- reduced health and safety risk.

# References


Bartos, P.J.M., Marrs, D.L. (1999) Development and testing of self-compacting grout for the production of SIFCON, *Proceedings of International Workshop on High Performance Fiber Reinforced Cement Composites,* Eds H.W. Reinhardt and, A.E. Maaman, Germany, May pp. 171–180.

Bartos, P.J.M., Grauers, M. (1999) Self-compacting concrete, *Concrete*, April, 9–13.

Billberg, P. (1999) Self compacting concrete for civil-engineering structures – the Swedish Experience. CIB Report 2:99, Swedish Cement and Concrete Research Institute.

Cussigh, F. (1999) Self compacting concrete stability, *Proceedings of First RILEM International Symposium on Self Compacting Concrete,* Stockholm, 13–15 September.

De Larrard, F., Hu C., Szitkar, J.C., Joly, M., Claux, F. and Sedran, T. (1995) A new rheometer for soft-to-fluid fresh concrete. LCPC internal report.

Domone, P.L., Chai, H.-W. (1997) Testing of binders for high performance concrete, *Cement and Concrete Research,* **27**, 1141–1147.

Domone, P.L., Chai, H.-W. (1996) Design and testing of self compacting concrete, *Production Methods and Workability of Concrete, RILEM International Conference,* Eds P.J.M. Bartos, D.L. Marrs and D.J. Cleland, June pp. 199–208.

Domone, P.L., Jin, J., Chai, H.W. (1999) Optimum mix proportioning of self compacting concrete, *Innovation in Concrete Structures: Design and Construction, Proceedings of Creating with Concrete,* University of Dundee, Dundee, September pp. 277–285.

Gibbs, J.C., Zhu, W. (1999) Strength of hardened self-compacting concrete, *Proceedings of First RILEM International Symposium on Self Compacting Concrete,* Stockholm, 13–15 September.

Glavind, M. (2000) How does self-compacting concrete contribute to implementation of sustainable/clean technologies in the construction industry? *Proceedings of Seminar of Self-compacting Concrete,* Malmö, November pp. 57–61.

Grauvers, M. (2000) From systems and surface properties, *Proceedings of Seminar of Self-compacting Concrete,* Malmö, November pp. 33–37.

Hayakawa, M., Kuroiwa, S., Matsuoka, Y. and Shindoh, T. (1993) Application of super workable concrete to construction of a 20-storey building, *High Performance Concrete in Severe Environments,* Ed. Paul Zia, ACI SP-140, pp. 147–161.

Henderson, N.A. (2000) Self-compacting concrete at Millennium Point, *Concrete,* April, 26–27.

Khayat, K.H. (1999a) Workability, testing, and performance of self-consolidating concrete, *ACI Materials Journal,* **96**, 346–353.

Khayat, K.H. (1999b) Effect of mixture proportioning on air void stability in self consolidating concrete, *Proceedings of First RILEM International Symposium on Self Compacting Concrete,* Stockholm, 13–15 September.

Khayat, K.H., Ghezal, A., Hadriche, M.S. (1999) Factorial design models for proportioning self-consolidating concrete, *Materials and Structures,* **32**, 679–686.

Khayat, K.H., Bickley, J., Lessard, M. (2000) Performance of self-consolidating concrete for casting basement and foundation walls, *ACI Materials Journal,* **97**, 374–380.

Marsh, B. (2002) Ove Arup. Personal communication.

Nawa, T., Izumi, T., Edamatsu, Y. (1998) State-of-the-art report on materials and design of self-compacting concrete, *Proceedings of International Workshop on Self Compacting Concrete* Kochi University, Japan, August.

Okamura, H., Ozawa, K. (1994) Self-compactable high performance concrete in Japan SP-169, American Concrete Institute, Detroit, pp. 31–44.

Okamura, H., Ouchi, M., Hibino, M., Ozawa, K. (1998) A rational mix-design method for mortar in self-compacting concrete, *The 6th East Asia-Pacific Conference on Structual Engineering & Construction,* Vol. 2, pp. 1307–1312.

Ozawa, K., Tagtermsirikul, S. and Maekawa, K. (1992) Role of materials on the filling capacity of fresh concrete, *Proceedings of 4th CANMET and ACI International Conference on Fly Ash, Silica Fume, Slag and Natural Pozzolans in Concrete,* Istanbul, pp. 212–237.

Ozawa, K., Maekawa, K., Kunishama, M., Okamura, H. (1989) Development of high performance concrete based on the durability design of concrete structures, *Proceedings of the 2nd East-Asia and Pacific Conference on Structural Engineering and Construction* (*EASEC-2*), Vol 1, pp. 445–450.

Petersson, Ö., Billberg, P., Van, B.K. (1996) A model for self compacting concrete. *Proceedings of RILEM International Conference on Production Methods and Workability of Fresh Concrete,* Paisley, June.

Petersson, Ö. (2000) Design of self-compacting concrete, properties of the fresh concrete, *Proceedings of Seminar of Self-compacting Concrete,* Malmö, November pp. 16–20.

Ramachandran, V.S. (1995) *Concrete Admixtures Properties, Science, and Technology,* 2nd edn. Noyes, Publications, Parkridge, NJ.

Rols, S., Ambroise, J., Pera, J. (1999) Effects of different viscosity agents on the properties of self-leveling concrete, *Cement and Concrete Research,* **29**, 261–266.

Sakata, N., Maruyama, K., Minami, M. (1996) Basic properties and effects of welan gum on self-consolidating concrete, *Proc. RILEM Int. Conf.* Production Methods and Workability of Concrete, Eds. P.J.M. Bartos, D.L. Marrs and D.J. Cleland 237–253 Scotland.

Sedran, T. and DeLarrard, F. (1996) Mix design of self-compacting concrete, *Proceedings of the International RILEM Conference on Production Methods and Workability of Concrete,* Paisley. Eds. P.J.M. Bartos, D.L. Marr, and D.J. Cleland, pp. 439–450.

Yurugi, M., Sakata, N., Iwai, M. and Sakai, G. (1993) Mix proportion for highly workable concrete, *Proceedings of the International Conference of Concrete 2000*, Dundee, UK.

Wallevik, O., Nielsson, I., (1998) Self compacting concrete – a rheological aproach, *Proceedings of International Workshop on Self Compacting Concrete,* Kochi University, Japan, August.

## Further reading

Bartos, P.J.M. (1978) Workability of flowing concrete *Concrete*, **10**, 28–30.

Bartos, P.M. (1998a) An appraisal of the Orimet Test as a method for on-site assessment of fresh SCC concrete, *Proceedings of International Workshop on Self Compacting Concrete,* Kochi University Japan, August.

Bartos, P.J.M. (1998b) Self compacting concrete, *Proceedings Betonarske Dny 1998*, Czech Society for Concrete and Masonry (CBZ), Pardubice, Dec 3–4 1998, pp. 219–229.

Bartos, P.J.M., Hoy, C. (1999) Interaction and packing of fibres: effects on the mixing process, *High Fibre Reinforced Cement Composites,* Eds H.W. Reinhardt and A.E. Naaman, RILEM Publications, Paris, May pp. 181–192.

Beaupre, D., Lacombe, P., Khayat, K.H. (1999) Laboratory investigation of rheological properties and scaling resistance of air entrained self-consolidating concrete, *Materials and Structures,* **32**, 235–240.

Building Research Establishment. *Practical Guide for Engineers Using* SCC.

Domone, P.L., Chai, H.-W. (1998) The slump flow test for high workability concrete, *Cement and Concrete Research,* **28,** 117–182.

Edamatsu, Y., Nishida, N., Ouchi, M. (1999) A rational mix design method for self compacting concrete considering interaction between coarse aggregate and mortar particles, *Proceedings of First RILEM International Symposium on Self Compacting Concrete,* Stockholm, 13–15 September.

Gaimster, R. (2000) Self-compacting concrete, *Concrete,* April, 23–25.

Hibino, M., Okuma, M., Ozawa, K. (1988) Role of viscosity agent in self-compactability of fresh concrete, *Proceedings of 6th East Asia Conference on Structural Engineering and Construction*, Taipei, Vol. 2, pp. 1313–1318.

Hu, C., Barcelo, L. (1998) Investigation on the shrinkage of self-compacting concrete for building construction, *Proceedings of International Workshop on Self Compacting Concrete,* Kochi University, Japan, August.

Kato, K., Hyun-Yang, S., Kunishima, M. (1993) A comparative study on the constructability of self placeable concrete, *Concrete 2000: International Congress*, September, University of Dundee, pp. 881–890.

Khayat, K.H., Roussel, Y. (2000) Testing and performance of fiber-reinforced, self consolidating concrete, *Materials and Structures,* **33**, 391–397.

Khayat, K.H., Ghezal, A., Hardriche, M.S. (2000) Utility of statistical models in proportioning self-consolidating concrete, *Materials and Structures,* **33**, 338–344.

Kitoh, M., Yamada, K., Nakamura, M., Ushijima, S., Matsuoka, S., Tanaka, H. (1998) Concept of super quality concrete, its properties and structural performance – the Association for the Development & Propagation of Super Quality Concrete Structures, *Proceedings of International Workshop on Self Compacting Concrete,* Kochi University, Japan, August.

Kwan, A.K.H. (2000) Use of condensed silica fume for making high strength, self consolidating concrete, *Canadian Journal of Civil Engineering,* 8.

Li, V.C., Kong, H.J., Chan, Y.-W. (1998) Development of self-compacting engineered cementitious composites *Proceedings of International Workshop on Self Compacting Concrete,* Kochi University, Japan, August.

Marrs, D.L., Bartos, P.J.M. (1996), Development and testing of self compacting low strength SIFCON slurries, *Production Methods and Workability of Concrete, RILEM International, Conference,* Eds P.J.M. Bartos, D.L. Marrs and D.J. Cleland, June, pp. 199–208.

Muroga, Y., Ohsuga, T., Date, S., Hirata, A. (1999) A flow analysis for self compacting concrete, *Proceedings of First RILEM International Symposium on Self Compacting Concrete, Stockholm, 13–15* September.

Nakahara, Y. (1998) Will self-compacting high-performance concrete be used more widely in Japan? *Journal of Materials in Civil Engineering,* **10**, 65.

Oh, S.G., Noguchi, T., Tomosawa, F. (1999) Mix Design for rheology of self compacting concrete design, *Proceedings of First RILEM International Symposium on Self Compacting Concrete,* Stockholm, 13–15 September.

Okamura, H., Ozawa, K., Ouchi, M. (2000) Self compacting concrete, *Structural Concrete,* **1**, 3–17.

Okamura, H. (1997) Self compacting high performance concrete, *Concrete International,* **19**, (7), 50–54.

Okamura, H. (1999) Self compacting concrete – development, present use and future, *Proceedings of First RILEM International Symposium on Self Compacting Concrete,* Stockholm, 13–15 September.

Okamura, H. *et al,* (1995) Mix design for self-compacting concrete, *Concrete Library of JSCE,* **25**, 107–120,

Ouchi, M. History of development and applications of self-compacting concrete in Japan *Proceedings of International Workshop on Self Compacting Concrete*, Kochi University, Japan, August.

Ozawa, K., Maekawa, K., Okamura, H. (1990) High performance concrete with high filling capacity, *Proceedings of RILEM International Symposium on Admixtures for Concrete*: Improvement of Properties, Ed. E Vasquez, Barcelona, May. pp. 51–62.

Petersson, Ö., Billberg, O., Norberg, J., Larsson, A. (1995) Effects of the second generation of superplasticizers on concrete properties.

Petersson, Ö., Billberg, P., Osterberg, T. (1998) Applications of self compacting concrete for Bridge castings, *Proceedings of International Workshop on Self Compacting Concrete,* Kochi University, Japan, August.

Sari, M., Prat, E., Labastire, J.F. (1999) High strength self-compacting concrete – Original solutions associating organic and inorganic admixtures, *Cement and Concrete Research,* **6**, 813–818.

Sedran, T., De Larrard, F. (1999) Optimization of self compacting concrete thanks to packing model, *Proceedings of First RILEM International Symposium on Self Compacting Concrete,* Stockholm, 13–15 September.

Skarendahl, A. (1998) Self-compacting concrete in Sweden. Research and application, *Proceedings of International Workshop on Self Compacting Concrete,* Kochi University, Japan, August.

Skarendahl, A. (2000) State-of-the-art of self-compacting concrete, *Proceedings of Seminar of Self-compacting Concrete,* Malmö, November, pp. 10–14.

Uno, Y. (1998) State-of-the-art-report on the concrete products made of self-compacting concrete, *Proceedings of International Workshop on Self Compacting Concrete,* Kochi University, Japan, August.

Walraven, J. (1998) The development of self-compacting concrete in the Netherlands, *Proceedings of International Workshop on Self Compacting Concrete,* Kochi Univeristy, Japan, August.

# Part 3

# Special processes and technology

# 10

# Sprayed concrete

*Graham Taylor*

## 10.1 Introduction

The process known as sprayed concrete was first used over a century ago. Since that time the uses to which it has been put have been many but have generally been where formwork and normal placing methods have not been possible; it has therefore held a niche market share of construction in concrete.

Once regarded as an uncontrolled method of placing, it was viewed with some scepticism by designers of more conventional structures using conventional concreting methods. However, in the last thirty years, respect for it has grown as more research has been undertaken, new materials have been developed and contractors, banding together in trade associations, have worked towards reassuring their potential clients of the integrity of sprayed concrete. But even so, some regard it as more of an art than a science.

The system involves projecting a concrete made with small-sized aggregate at a hard surface. Impelled by compressed air, the material is rapidly placed and compacted and, with a high cement content and a low w/c ratio, its potential strength is high.

The original material was patented as 'Gunite' and this term has been used generically in the UK; in America, the term 'shotcrete' is prevalent but to harmonize with the rest of Europe it is now called Sprayed Concrete.

## 10.2 History

In the late 1890s, Dr Carlton Akeley, Curator of the Chicago Field Museum of Natural Science, developed the original cement gun as a method of applying mortar over skeletal matrices to form replicas of prehistoric animals, having been dissatisfied with the results of trowel application.

Despite early failure, he persisted with pressurizing the mortar using his cement gun, which was a single chamber pressure vessel into which he placed a mixture of sand and cement and then applied compressed air. The rest of the process is similar to today's dry mix method, with the mixture travelling down a hose, passing through a spray of water and then exiting through a nozzle before shooting into place.

Having proved the theory with this rather crude machine, refinements were made, such as the addition of another compression chamber and a geared feed-wheel. The process patented in 1911 is, with some further refinements, still in use today in the dry process.

Gunite, as it was then known, came to Britain in the 1920s and healthy competition between rival firms gave rise to technical developments. During the Second World War, Gunite, like everything else, was used as an expedient. One notable use was that basements of bombed buildings were cleared of rubble and then lined with gunite, in very short time spans, to produce water tanks for fire fighting. Camouflaged Spitfire aeroplane hangers were produced by spraying gunite onto stretched heassian.

Post-war there was much to be repaired and gunite was used extensively as it gave a fast and effective answer to the problems of damaged and deteriorated jetties, warehouses, factories, grain silos, water towers, bridges, etc. It was also realized that gunite could be used as a strengthening medium, not merely as a means of repair.

In more recent times there have been developments in the type of plant used; progressing from the double chamber machine, which sends pulses of materials down the hose, to the almost constant feed-and-supply machines – the rotating barrel gun. In the 1960s the wet process was introduced, where cement, sand and water are mixed together before being put into the machine. Materials are now often supplied pre-bagged and include various carefully selected sands and admixtures which impart additional properties and make the process more reliable.

Probably the biggest change over the last half century has been the use of sprayed concrete for new structures. Constraints of straight line construction have given way to the production of structures with complex geometry such as hyperbolic paraboloid, doubly curved and domed roofs. This period has also seen the development of several codes of practice, specifications and methods of measurement, now largely superseded by the EFNARC specification, illustrated in Figure 10.1, which itself will ultimately be replaced by European standards.

## 10.3 Definition of sprayed concrete

Sprayed concrete is a mortar or concrete pneumatically projected at high velocity from a nozzle to produce a dense homogenous mass. Sprayed concrete normally incorporates admixtures and may also include additions or fibres or a combination of these.

The force of the jet impacting on the surface compacts the material, which normally has zero slump, in place and can support itself without sagging whether on a vertical, sloping or, within certain limits, an overhead surface.

Some rebounding material is inevitable when concrete is projected at a relatively hard surface at high speed. This material, which consists mainly of the larger aggregate particles, is known as rebound and should be discarded because it is of unknown grading.

**Figure 10.1** Cover of EFNARC Specification.

## 10.4 The wet process

### 10.4.1 Definition

In this technique, cement, aggregate and water are batched and mixed together prior to being fed into a purpose-made machine and conveyed through a pipeline to a nozzle where the mixture is pneumatically and continuously projected into place. This can be seen diagrammatically in Figure 10.2. The mixture normally incorporates admixtures (including almost certainly an accelerator) and may also include additions and/or fibres.

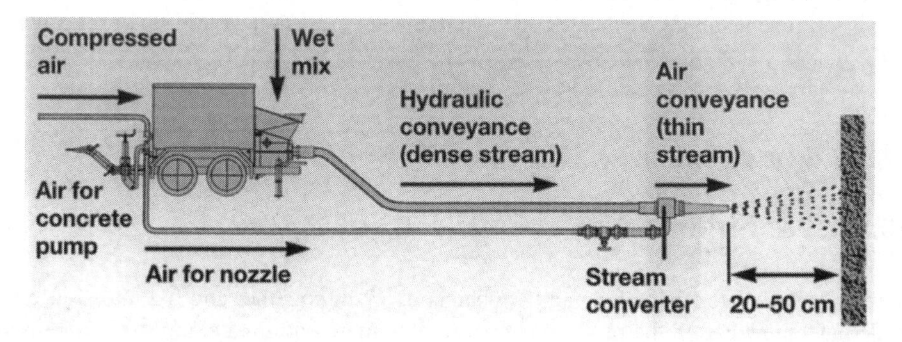


**Figure 10.2** Diagram of a typical wet process system.

### 10.4.2 General

Concrete for wet spraying can be supplied readymixed or from a site mixer, which may use individual constituent materials or blended and bagged dry material. Bagged material is often used for small quantity requirements such as repairs.

Water can be accurately controlled and, with the use of water-reducing plasticizers, a w/c ratio of less than 0.45 is easily achievable.

Sprayed concrete is a cement-rich material, although the surface area of the sand is high. Cement contents are normally in the range 350–450 kg/m$^3$ and 28-day strengths can be expected to be 30–60 MPa.

It is normal for the sprayed concrete contractor to design the mix so that he can take into account the performance of the wet concrete when it is being pumped and when it is in position as well as minimizing rebound; all in addition to the specified requirements of strength and durability.

Rebound is less in this process than in the dry process.

### 10.4.3 Equipment

The wet process uses either piston or worm pumps to deliver the concrete to the nozzle in a dense stream. Compressed air is introduced at the nozzle to project and compact the concrete onto the substrate. A set accelerator may also be added at the nozzle to give the in-place material an early mechanical strength, which can minimize sloughing off.

For low output requirements such as repairs (up to 4 m$^3$/h) a worm pump would be suitable and this can take a maximum aggregate particle size of 4 mm. This output may seem very low to the conventional concretor but it produces an acceptable output of sprayed concrete. For higher outputs, such as in tunnel construction (in the range 4–25 m$^3$/h), a double piston pump, capable of taking 20 mm aggregate, is required.

The entire spraying system should be able to deliver a constant stream of concrete, free from pulsations which can give rise to segregation of the mix and over-dosing of the admixture(s).

The lines, of either flexible hoses or steel pipes, should be laid as straight as possible. They should be of uniform diameter and well sealed. The actual size used is dependent on the maximum aggregate size or the inclusion of a fibre in the mix. Prior to concreting, the lines should be lubricated with a grout.

All equipment should be cleaned and maintained to prevent the build-up of set concrete in the delivery system.

## 10.5 Dry process

### 10.5.1 Definition

Here only the aggregate and cement are batched, mixed together and fed into a purpose-made machine wherein the mixture is pressurized, metered into a compressed air stream and conveyed through hoses or pipes to a nozzle, where water is introduced as a very fine spray to wet the mixture, which is then projected continuously into place, as shown in Figure 10.3. Additions and/or fibres may also be in the mixture.

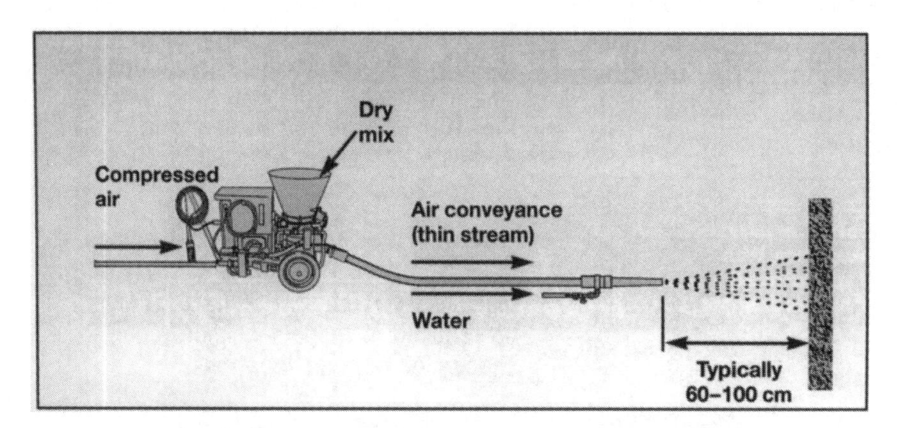


**Figure 10.3** Diagram of a typical dry process system.

## 10.5.2 General

Because the concrete in the hose does not have a workability requirement, which is normally required for transportation, and the compaction force is high, water added at the nozzle needs only to be sufficient to hydrate the cement and give some fluidity to the mix. This produces a mix with a w/c ratio potentially as low as about 0.35 and consequently zero slump, enabling it to be placed without admixtures on vertical and overhead surfaces.

Should they be required, admixtures can be added, either dry in the mix or wet at the nozzle. Fibres can, of course, be added to the dry ingredients.

The nozzleman controls the water addition and while this at first seems somewhat unprofessional, if too little water is added the amount of rebound increases and if too much is added, the concrete will not have sufficient rigidity to stay in place. In tunnels, remotely controlled robotic spraying is used and this reduces the hazards to the operators as well as reducing the need for access platforms to the higher levels.

When using a basic sand/cement mixture, rebound is higher than for the wet process; typically 10–15% when spraying downwards and as much as 50% when spraying overhead. Because this rebound is mainly the larger aggregate particles, the mix as-placed will be richer than the as-mixed materials, which would have an aggregate/cement ratio in the range 3.5–4.5 by weight. The resulting 28-day strength is typically 40–50 MPa. Pre-bagged materials, which contain cohesion-promoting admixtures, do not produce as much rebound.

However, dry process sprayed concrete can be the more variable. For example, because it is built up in layers and rebound is high, it can have variable density, especially at the interface between two layers. Rebound can also become trapped in corners, reducing the density. But because of its low w/c ratio, shrinkage is generally lower than for other similar but wetter concretes.

## 10.5.3 Equipment

The twin chamber machines referred to earlier have now virtually ceased to be used and have been replaced by the rotor type (shown in Figure 10.4), where the dry materials are fed into an open hopper, which feeds a rotating barrel. As the barrel rotates, the pockets are brought under a compressed air supply which projects the material down the hose. These machines, sometimes referred to as guns, have outputs up to 10 $m^3$/h.

**Figure 10.4** Typical dry spray machine (gun).

The length of hose is important. Short hoses will give rise to a pulsating effect in the material being delivered to the nozzle but this tends to even out in longer hoses. While pressure, and the resulting exit speed from the nozzle, reduces with distance from the gun, hose lengths of up to 600 m are possible. With moderate hose length, the exit speed can be up to 125 mph.

## 10.6 Constituent materials

All materials must comply with the standards in force at the time. Sprayed concretes should contain cements, cementitious additions and aggregates or sands that have been proved to be suitable for this type of work. Sprayed concrete contractors have found by experience that cements of particular types and from particular sources work best for them.

Preliminary tests are normally carried out to establish the suitability of the constituent materials and the final product. The term 'suitability' covers all aspects of the process as well as the final product's physical characteristics.

## 10.6.1 Cement

Most types of cement are used, as in any concrete work, depending on the requirement of the finished product. The cement to be used should be fresh and free from lumps.

## 10.6.2 Aggregates

Most sprayed concrete uses sand but aggregates up to 10 mm in size may be used where the section thickness allows and provided the machinery will accommodate them. It is also possible to spray larger size aggregate but this has no advantages and larger particles are more prone to rebounding. The grading of the sand is important and should be within the limits shown in the following table:

| ISO sieve (mm) | Max % passing | Min % passing |
|---|---|---|
| 16 | 100 | 100 |
| 8 | 100 | 90 |
| 4 | 100 | 73 |
| 2 | 90 | 55 |
| 1 | 72 | 37 |
| 0.5 | 50 | 22 |
| 0.25 | 26 | 11 |
| 0.125 | 12 | 4 |

The finer region is more suitable for dry mixes, although a high percentage of particles <0.25 mm can lead to dust problems. Sand for finishing coats (sometimes referred to as flash coats as they are flashed over the surface as a sort of render) may be finer than the grading limits shown. Finer sands will generally produce a concrete with greater drying shrinkage while coarser sands will give more rebound.

Aggregates need to be slightly damp to help suppress dust and to guard against segregation of the cement and sand in the hoses. The moisture content needs to be constant at around 5–6%. Too much moisture promotes clogging or blocking of the machine.

## 10.6.3 Mixing water

The requirements for mixing water are the same as for conventional concrete.

## 10.6.4 Admixtures

As will be seen later, specific admixtures are normally used, especially in the wet process. Like water, they must comply with the standards in force at the place of work.

## 10.6.5 Fibres

Both steel and synthetic fibres are used to satisfy a performance requirement of the finished concrete. Steel fibres, which may be included for abrasion resistance, have the ability to cause blockages in pipes and hoses, hence the general requirement to keep their length down to not more than 0.7 of the pipe/hose diameter until it has been proved that longer fibres can be sprayed without causing blockages. Synthetic fibres, used, for example, to control cracking, are flexible and, provided they are well-distributed, flow easily through the hose.

### 10.6.6 Cementitious additions

Silica fume, pfa and ggbs can all be used in sprayed concrete; generally up to the following percentages:

| Material | Max addition (% of Portland cement) |
|---|---|
| Silica fume | 15 |
| pfa | 30 |
| ggbs | 30 |

## 10.7 Spraying procedures

If a bond is required, as in the case of repairs or strengthening work, the substrate must be clean before spraying of concrete can commence. If the substrate surface is porous it may need wetting down to kill excessive suction from the concrete being sprayed. In dry spraying in particular, the w/c ratio of the sprayed concrete is already very low and suction from the interface surface might reduce it to below the critical level for complete hydration. However, there should be no free water left on the surface before spraying commences.

At start-up, the nozzleman should direct the spray away from the works until he judges that the concrete coming out of the nozzle is of an acceptable quality.

Layers are built up by making several passes of the nozzle over the area, generally in an overlapping oval configuration. The nozzle should be at right angles to the substrate and at the correct distance from it. This would be between 600 mm and 1 m for the dry process and between 250 and 500 mm for the wet process. The nozzleman will recognize the correct distance by the amount of rebound and the effectiveness of compaction.

Thickness of the section to be sprayed can be indicated by stretched 'piano' wires located at the required finished surface.

Application should start at the bottom of a vertical or sloping surface and from the shoulder to the crown for overhead working.

Figure 10.5 shows a nozzleman spraying concrete for the stabilization of a rock face.

Corners and sharp angles should be avoided as these will trap rebound material. This also applies to joints. The accepted technique is to slope the concrete at the end of a run then, while the concrete is still fresh, cut back in a straight line, about 50 mm into the concrete so that only a small right-angled cut-out section is produced. Work can then start up again at this point and all that will be seen is the straight line.

The nozzleman's assistant should remove any rebound that might become encased in the works and this rebound should be discarded. He may also be required to trim back the

**Figure 10.5** Nozzleman at work.

sprayed material to the line indicated by the stretched wires, although it is better practice to spray to a thickness less than that indicated and to finish off with a flash coat. Trowelling or screeding is best avoided as the action involved creates tension cracks in the sprayed concrete's surface.

Once a section is complete, it must receive curing, which can be a sprayed-on membrane (if no further coating is to be applied) or any of the other conventional curing techniques. Unfortunately, the large areas involved often do not receive any proper curing and the result can be premature drying out, excessive shrinkage or delamination.

## 10.8 Quality assurance

### 10.8.1 Introduction

Sprayed concrete is more dependent on those who place it than conventional concrete; a trained nozzleman is therefore imperative. This also makes it difficul to separate quality control from quality assurance. The nozzleman controls the water in the dry process and, as a result the *in-situ* mix proportions will be different from the batched proportions. If the requirement is that it should adhere to the substrate, it is the nozzleman and his assistant whose actions will control whether or not it does. The nozzleman's method of working may create variations in the quality of succeeding layers and therefore density (or perhaps even homogeneity), as well as strength, needs to be checked. Cores are used for this purpose and these can also help to check the thickness of the cover to the reinforcement.

### 10.8.2 Prequalification of nozzlemen

Each nozzleman should demonstrate his skill prior to commencement of spraying and this is usually done by spraying into test panels, usually about 750 × 750 × 100 mm thick, set at the orientation of the proposed works (vertical, sloping, horizontal or overhead). The test panels should replicate the type of concrete, i.e. reinforced (bar, fabric or fibres) or unreinforced.

Cores are subsequently taken to ensure that strength, density, freeze/thaw resistance, chemical resistance and filling behind reinforcement are all acceptable.

## 10.9 Quality control

Quality control must be continuous throughout the works, from drawings through to finished product and must be carried out by someone with experience of this type of work. The following should be included:

### 10.9.1 Drawing review

The method of placing the concrete must be considered while the drawings are being prepared. Reinforcement is a main consideration – bar sizes, locations and staggering of overlaps, methods of (rigid) support are all important.

### 10.9.2 Potential debonding and delamination

To detect the presence of debonding and delaminations, the concrete surface is tapped with a steel hammer some 2 to 3 weeks after placing. Sound concrete will ring, while debonded materials will give a dull sound. A hand placed on the surface during the tapping will help to detect the extent of any debonding.

### 10.9.3 Tensile bond strength

While this is not a perfect test and can only tell the strength at one particular location, it is widely used to test the interface bond strength or the tensile strength at the weakest point. The test may need interpretation.

To perform this test, a core bit is used to drill just into the underlying layer, which might be a substrate or a previously sprayed concrete layer. A steel disc is attached to the top of the core using epoxy resin and a pull-off load is applied to the disc through a loading rig and hydraulic ram, until failure occurs. Any deviation of the direction of the force from the perpendicular can induce bending which will give very misleading results. The failure plane is then examined to determine if it has occurred at the interface or within the body of the sprayed concrete, or even within the substrate.

## 10.9.4 Strength and density

Cores taken from the placed material can be tested as for normal concrete. Although the core diameter is usually small, the aggregate is also smaller than for conventional concrete and specimen size and proportion allowances can be made.

## 10.9.5 *In-situ* testing

There are various tests which can be carried out *in-situ.*

Entrapments of sand or rebound material, if suspected, can be verified by taking cores. Ultrasonic pulse velocity testing (Pundit) can be used to indicate strength and thermography or X-rays can be used (rarely) to evaluate the condition of the sprayed concrete.

## 10.9.6 Daily test panels

While these, shown in Figure 10.6, are normally specified, to test mainly for compressive strength and density, they may not represent the concrete in the works. The nozzleman obviously knows what test panels are and what they are used for, so he ensures that only the best concrete and workmanship go into them. The quality of concrete at start-up is frequently below the required standard.

**Figure 10.6** Test panel.

If cores from these daily test panels are found to be sub-standard, then cores should be taken from the *in-situ* concrete sprayed on that day and tested as required.

### 10.9.7 Conclusion

While the above may indicate difficulties in controlling the quality of the end product, the material has a higher than normal cement content and the spraying process will only allow good material to pass through the system. The problem, if any, is often associated with the skill of the nozzleman and it is therefore imperative that only well-trained, competent ones are used.

## 10.10 Applications of sprayed concrete

The Sprayed Concrete Association's book *An Introduction to Sprayed Concrete* starts with the words 'Sprayed concrete is a versatile and economic method of concrete placement, with endless possibilites . . .' and, yes, the list of actual and potential applications is almost limitless but it succeeds most notably where the use of rigid formwork is absent.

Some of the types of structures for which it can be used are set out below.

### 10.10.1 New construction

General structures (see Figure 10.7)
Shell roofs and domes
Barrel vaulted roofs
Retaining walls
Secant pile wall facing
Diaphragm walls
Silos
Caissons

**Figure 10.7** A church in France.

Bank vaults
Tunnel linings
Storage reservoirs
Sea and river walls
Reservoirs
Dams
Swimming pools
Canal linings
Water ditches
Towers (see Figure 10.8)

**Figure 10.8** Ventilation tower on a road tunnel.

## 10.10.2 Free-formed structures

Swimming pools (see Figure 10.9)
Landscaping
Climbing walls
Sculpture
Zoo structures
Leisure pursuits such as theme parks, water slalom courses and bobsleigh runs

**Figure 10.9** Swimming pool.

## 10.10.3 Strengthening and repair

Sea and river walls
Aqueducts
Water towers
Canal linings
Ditches
Concrete damaged by reinforcement corrosion
Fire damaged structures
Cooling towers
Bridges (see Figure 10.10)
Jetties and wharves
Brick arches and tunnels

**Figure 10.10** Viaduct renovation and strengthening.

## 10.10.4 Protective coatings

Fire protection of structural steelwork
Refractory linings
Pipeline encasement
Rock and soil stabilization (see Figure 10.11)

**Figure 10.11** Rock stabilization.

## Further reading

*An Introduction to Sprayed Concrete* (1999) The Sprayed Concrete Association.

EFNARC (1996) *European Specification for Sprayed Concrete*.

Austin and Robins (eds) (1995) *Sprayed Concrete – Properties, Design and Application*. Whittles Publishing Services.

Concrete Society (2002) Construction and repair with wet-process sprayed concrete and mortar, Technical Report 56.

# 11

# Underwater concrete

*Graham Taylor*

## 11.1 Introduction

For those used to concreting on dry land, concreting under water presents various challenges. Transporting, compacting, quality control, finishing and accuracy must all be carried out successfully in this different, and often difficult, environment.

There are, however, many common aspects, chief of which is that air is not required for the setting and hardening of concrete – it sets and hardens just as well, and often even better, under water – but it must be fluid enough to flow into position and be self-compacting as conventional vibration is not practicable under water.

During transporting and placing, conventional concrete and water must be kept apart and, when they inevitably do come into contact, rapid interface flow must be minimized or cement may be washed out to form a weak layer. Washout can be obviated by the use of an admixture to make the concrete non-dispersible but this comes at a cost and contractors unwilling to pay the additional expense involved often adhere to the more traditional methods of placing under water.

## 11.2 Conventional methods of placing

### 11.2.1 Tremie

The principle of this method is that concrete is poured down a pipe or tube from above the surface and is forced into the mass of concrete already in place by the weight of concrete in the tube. The tube is surmounted by a hopper ('tremie' in French) and the whole is suspended from a staging or frame, mounted so that it can be moved vertically when held by a crane. As the pour rises, sections of the tube can be removed to facilitate working. A convenient diameter for the tube is 8 to 16 times the maximum aggregate size and 250 mm is a common diameter. Figure 11.1 shows a diagrammatical representation of a tremie.

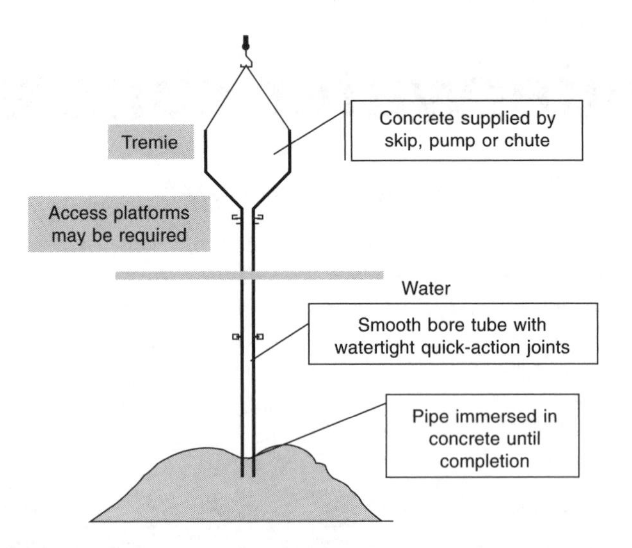


**Figure 11.1** Schematic of a tremie.

Before starting the pour, a plug is inserted into the tube to stop the concrete and water intermixing. This plug can be purpose-made (similar to a bath plug), a sponge rubber ball or exfoliated vermiculite, which is the most common method in the UK.

At start-up the bottom of the tube should be on or very close to the sea or river bed, sufficient to allow the water in the tube to escape and to force the first load of concrete to spread out horizontally into a mound shape. The concrete pouring should be continuous with the bottom of the tube always inside previously placed concrete. If this immersion depth, normally at least 0.5 m, is not sufficient, a breakthrough will occur and the pour will have to be abandoned for the day. Any air that is in the concrete being placed will pass through the previously placed concrete and bubble to the surface, disrupting the settled concrete as it goes.

The flow of concrete in the tube is governed by gravity and friction with the tube wall, so the tremie has to be moved up and down to regulate the flow. A crane driver with a good 'feel' for this is useful. The tube should be restrained from lateral movement whilst placing concrete. The placed concrete spreads out horizontally on the bed in a circle, with the top of the pour domed upwards.

Tremies are best used for thick pours of any area. For large area pours, multiple tremies are used, spaced at about 4–6 m apart, depending on the flatness required for the top level. The slope of the concrete surface from a tremie is likely to be in the range 1 in 9 for tremies close together to 1 in 6 for those spaced far apart as the slope increases with distance from the pipe.

## 11.2.2 Hydrovalve

This Dutch innovation is a refinement of the tremie and is shown in Figure 11.2. Instead of a solid pipe, it uses a collapsible fabric tube which is kept closed by the water pressure until the weight of concrete in the system overcomes the hydrostatic pressure and the skin friction. A plug of concrete then descends slowly and the tube is sealed behind it by the water pressure. The bottom section of the flexible tube is encased by a rigid tubular section the bottom of which, except at start-up, is at the desired surface level of the concrete.

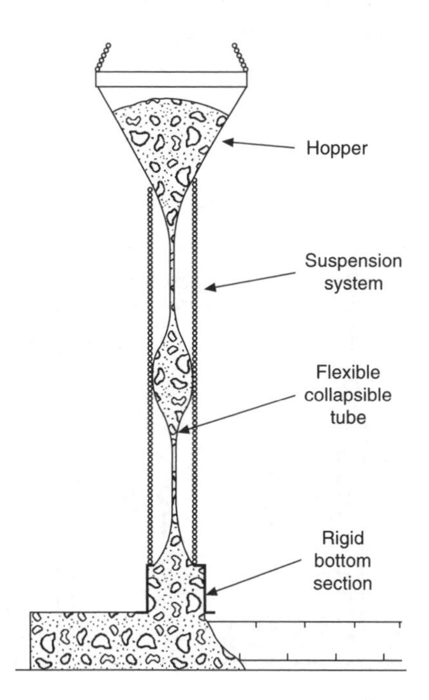


**Figure 11.2** Schematic of a hydrovalve.

The Hydrovalve is used for casting relatively thin sections of concrete in a to-and-fro pattern, with each run of concrete placed on the sloping exposed face of the previous run. As it is generally carried out in still water and each subsequent row of placements scours off any disturbed material on the sloping face, the end product is sound. The 'valve' sits on top of the pour so there is no pipe immersion, as there is with a tremied placement, and the valve levels off the top of the pour. Placements up to 750 mm thick through reinforcement are possible with this method.

Another Dutch refinement of the tremie is the 'hop dobber', where the tremie pipe is in two sections. The upper section is slightly smaller in diameter and fits into the top of the lower section, which has a surrounding flotation chamber to keep the bottom of the pipe at a constant level. Its method of operation is very similar to that of the hydrovalve.

## 11.2.3 Skips

Skips are more suitable for thin pours, although it is possible to bury the mouth of the skip in previously poured concrete to produce deeper pours. The skip should be fully charged in the dry and covered with a pair of flexible and overlapping covers; these prevent washout during lowering and also during discharge because they stay in contact with the top surface of the concrete as it flows out of the bottom.

The best type of skip has vertical sides to allow the concrete to fall freely and vertically downwards. Discharge is the critical part of the operation and to minimize intermixing with water, the skip must have a skirt. Some skip doors are hinged at the sides and open from the centre, which produces a skirt. A better type of skip, shown in Figure 11.3, has a skirt which encases the lower part of the skip and this descends as the central roller doors open.

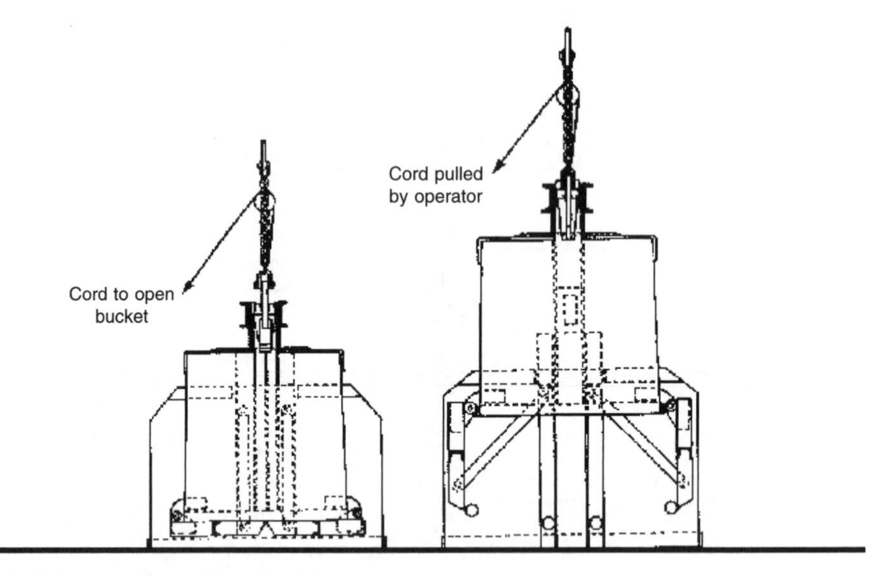


**Figure 11.3** Bottom-opening skip with skirt.

Skips normally trip open automatically on touching the base or previously placed concrete but any projections from the base will also trigger the mechanism, causing premature discharge.

## 11.2.4 Pumps

The constitution of concrete for underwater work is similar to that required for pumping, so pumps provide an ideal placing method. Both static and mobile pumps can be used. Water pressure helps to counteract gravity, which can be a problem when placing in the dry. Underwater pours are seldom at depths greater than 30 m, the depth limit when pumping in the dry.

As with tremie work, broken seals and air pockets can be a nuisance and the operator of a pump does not have the same 'feel' for the pour as a tremie operator. A pump pipeline is normally at an angle when used under water and is therefore not as controllable as a tremie tube; preventing it moving sideways is essential in preventing washout.

### 11.2.5 Toggle bags

Where small amounts of concrete are required, such as in repair work, the toggle bag is ideal. The waterproof bag is filled in the dry with wet concrete and the mouth is closed with a tie rope and toggle. At the placing location the concrete is squeezed out by a diver and rammed into place. The use of a diver adds to the cost of the operation.

### 11.2.6 Bagwork

The type of bags used here are normally made from an open-weave material such as hessian. They should be half-filled with plastic concrete, sealed and then taken under water and placed by a diver. Partial filling allows them to be moulded into shape and gives them good contact areas with adjacent bags. Grout from the mix seeps through the open-textured material allowing bond to be established with adjacent bags. For additional stability the bags can be spiked together with small-diameter reinforcing bars.

Divers prefer to handle bags of dry-mixed concrete and to grout up between bags. However, this system places too great a responsibility on the diver. The dry mix concrete is never fully wetted-out by water seeping in, the concrete cannot be fully compacted and contact surfaces are minimal.

Diver-handled bags are usually of 10 to 20 litres capacity but 1 $m^3$ bags can be placed using a crane.

### 11.2.7 Concrete packaged under water

Quilted revetment bags or mattresses are used as protection against erosion, scour and water seepage. They consist of a double skin of woven permeable fabric connected together by threads. Sections are zippered together for continuity, laid on banks below and above water level and then pumped full of sandy grout. Cable or rope reinforcement is threaded through them to ensure integrity after settlement and shrinkage cracking have taken place.

Filling gaps between structural units can be done by inserting collapsed plastic or nylon bags into which grout is then pumped until the gaps are filled.

### 11.2.8 Grouted aggregates

In order to obtain good results with this system, also known as pre-placed aggregate concrete, it requires considerable skill and experience in the application of the process. It must be undertaken by a skilled, specialist contractor.

The space to be concreted, often determined by formwork, is filled with single-sized aggregate, into which are inserted grout pipes, as shown in Figure 11.4. The aggregate represents about 65–70 per cent of the overall volume to be concreted. Grout is then pressure-pumped in to fill the voids left by the aggregate particles. One patented process uses a colloidal suspension, produced by mixing the cement, and possibly pfa, with water in a high-speed mixer, followed by the addition of sand and further mixing. Colloidal grout is stable and has good flow properties.

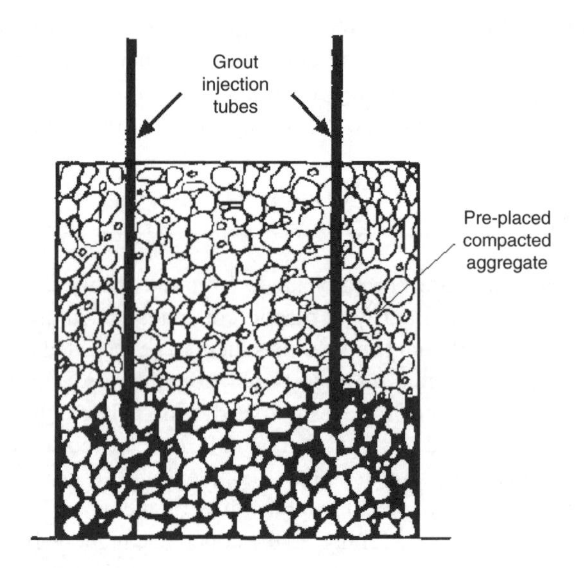


**Figure 11.4** Pre-placed aggregate.

The grouted aggregate technique is especially useful where the water is flowing and where undercuts which are inaccesible to tremies and skips need to be filled.

Stones placed in a formwork box reduce the flow of water through them, which allows the grout to penetrate rather than be washed out but, prior to grouting, it can cause sediment to be deposited on the aggregate and form a barrier between grout and stones if the water is silty. To minimize this the whole operation should be carried out on the same tide so that silt does not settle when the water is relatively still – at the top and bottom of a tide.

The grout pipes are rigid and spaced about 2 m apart, with their ends close to the bottom of the pour. As with tremie work, they are raised as grouting proceeds.

A typical grout consists of a blend of Portland cement and pozzolana (at a ratio of between 2.5:1 and 3.5:1 by mass) mixed with sand (at a ratio between 1:1 and 1:1.5 by mass) and with a w/c ratio of 0.42 to 0.5. An intrusion aid is added to improve the fluidity, suspending and cohesive qualities of the grout. The intrusion aid also delays stiffening of the grout and contains a small amount of aluminium powder, which causes a slight expansion before setting takes place.

The drying shrinkage of grouted aggregate concrete is lower than that of ordinary concrete, due to the contact between the large aggregate particles. This contact restrains the amount of shrinkage that can take place but shrinkage cracking can occasionally develop.

Strengths of about 40 MPa are usual but higher strengths are possible. Provided a strong aggregate such as flint is used, the resulting concrete will have excellent abrasion resistance.

## 11.3 Concrete properties

This section deals with concrete which does not include an anti-dispersant admixture.

### 11.3.1 Basic requirements

The requirements for the concrete to be free-flowing, cohesive, self-compacting and not prone to bleeding are generally achieved by aiming for a high workability (100–150 mm slump) and a high fines content (sand, cement and possibly a fine filler). These allow the free water/cement ratio to be kept moderate so that the concrete maintains its cohesiveness and is not prone to segregation. Free w/c ratios in the range 0.40 to 0.45 are common.

### 11.3.2 Cementitious materials

The cementitious material content is likely to be 375 ± 50 kg/m$^3$ in order to increase cohesion – not because some cement will be washed out. This will ensure a good strength but not the full potential as there is likely to be some residual voids as a result of trapped air and the higher than normal water content.

The flow of the concrete has been found to be enhanced by the inclusion of about 15 per cent of pozzolanic material.

In large pours, internal temperatures can reach 70–95°C and cracking can develop when the concrete cools, particularly in unreinforced concrete. To reduce this risk, the use of a cement blend consisting of about 16 per cent Portland cement, 78 per cent coarse-ground ggbs and 6 per cent silica fume has been suggested, with the concrete being pre-cooled to 4°C prior to placing.

### 11.3.3 Aggregates

Aggregates should be rounded in shape, such as Thames Valley Gravel, as these will have a low water demand and give better flow and self-compaction characteristics. Crushed rock or other angular material should be avoided if possible. Gradings should be continuous to reduce the susceptibilty to segregation. The maximum aggregate size is usually 20 mm, although this is often dictated by the placing method.

The sand content is normally high at 45 per cent-plus of the total aggregate; lower amounts would make the mix prone to washout.

## 11.3.4 Admixtures

There is normally sufficient cement and fine sand particles in underwater concrete to obviate the need for an admixture but should the concrete not be sufficiently cohesive a plasticizer could be included. For massive pours there is often a need to keep the concrete workable for extended periods and a retarder would be needed. Superplasticizers are available for high-performance concrete. Non-dispersible admixtures, which allow concrete to be in free fall through water without washout, are dealt with later in this chapter.

## 11.3.5 Strength

With a high cement content, strengths of 50 MPa are readily achievable, although designers often feel they have to allow for weak bands in the concrete and assume strengths of around 30 MPa in their designs. As noted above, because underwater concrete is not mechanically compacted and there is an excess of water in the mix, it will not reach its full potential strength.

Strengths of 40 MPa are readily achievable when using tremies, pumps, Hydrovalves and skips, although concrete placed with the latter is more prone to having loose layers. The other placement methods produce homogenous concrete provided they have been used properly.

## 11.3.6 Workability

The workability of the mix determines the degree of compaction underwater. As can be seen from Figure 11.5 with a slump of 150 mm, concrete which is not vibrated will

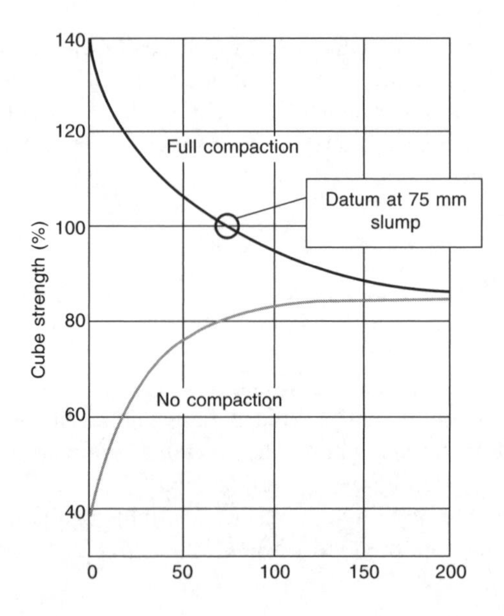


**Figure 5** Workability.

achieve almost as much strength (95 per cent) as the fully compacted concrete. There is not much benefit in having a slump greater than this as high-slump concrete will be more prone to washout. Even at 100 mm slump, the strength difference is relatively small, at about 88 per cent.

## 11.4 Non-dispersible concrete

Non-dispersible concretes (NDCs), containing anti-washout admixtures (AWAs), are formulated to produce cohesive concretes and grouts for underwater use.

### 11.4.1 The admixtures

AWAs contain thickening/viscosity-modifying agents which are effective in making the concrete flow when moved but highly viscous when at rest. They also contain a superplasticizer, preferrably of the sulfonated melamine formaldehyde type, and an anti-foam agent to minimize air-entrainment.

The group of thickening agents comprises a number of high molecular weight, water-soluble polymers which modify the rheology of cementitious compositions. Natural gums such as alginates and guar have been largely superseded by cellulose ethers (methyl, hydroxyethyl and hydroxypropyl celluloses) and polyethylene oxides.

Polyacrylamide-modified cement concrete (PMCC) is highly cohesive and, if based on cellulose ether, produces self-levelling concrete which can be discharged from a pump delivery pipe through about 1 m of water without significant loss of cement. Cellulose ethers can be used alone to produce underwater concrete but there are difficulties in obtaining sufficient workability for the concrete to be self-levelling and self-compacting.

A 10 per cent silica fume cement replacement has been successful in increasing cohesiveness and washout resistance in non-turbulent conditions but this would not be suitable for free-fall or turbulent underwater conditions. The inclusion in the concrete of a true AWA allows concrete to be dropped through water without washout.

Table 11.1 shows some typical non-dispersible underwater concrete mixes and their fresh properties.

### 11.4.2 Concrete handling

Non-dispersible concrete was developed to overcome some of the problems associated with tremie operations and whilst a conventional skip can be used, it is more convenient to use a pump, except that the pump lines are difficult to clean out at the end of the pour.

Concrete modified with an AWA is sufficiently cohesive and water-resistant to be undamaged by its passage through water. It tends to descend in discrete amounts rather than in a stream and when it reaches previously placed concrete it seals itself on, displacing the surrounding water. With good flow and self-levelling characteristics, it will flow around and encapsulate reinforcing bars.

**Table 11.1** Typical underwater concrete mixes

| | Control | Conventional UWC | | High performance UWC | |
|---|---|---|---|---|---|
| | | Mix 1 | Mix 2 | Mix 1 | Mix 2 |
| Water/cementitious material ratio | 0.50 | 0.43 | 0.43 | 0.43 | 0.47 |
| Cement (kg/m$^3$) | 390 | 450 | 450 | 346© | 278 |
| Pfa (kg/m$^3$) | – | – | – | 86 | – |
| Ggbs (kg/m$^3$) | – | – | – | – | 278 |
| Water (kg/m$^3$) | 195 | 194 | 194 | 186 | 258 |
| Coarse aggregate (kg/m$^3$) | 971 | 971 | 971 | 1045 | 827 |
| Sand (kg/m$^3$) | 794 | 794 | 794 | 764 | 706 |
| AWA (% mass of cem. material) | 0 | 0.50 | 0.60(b) | 0.20 | 1.65(f) |
| Superplasticizer (litres/100 kg cement) | 1.0 (a) | 1.21 (a) | 1.76 (a) | 3.9 (e) | 0.2 (a) |
| Slump (mm) | 160 | 215 | 220 | 225 | – |
| Flow (mm) | 265 | 330 | 340 | 370 | 510 |
| Washout % – Plunge test | 16.3 | 3.6 | 3.9 | 2.5 | 4.1 |
| Washout % – Spray test | 52.2 | 13.9 | 24.4 | – | – |

Key:
(a) Melamine-based SP
(b) Powdered cellulose AWA
(c) Silica fume cement (18 per cent superplasticizer)
(d) Welan gum
(e) Napthalene-based superplasticizer
(f) Liquid cellulosic AWA (litres/100 kg cementitious material)
*Source*: Sonebi, Paisley University, UK.

## 11.4.3 Testing

There is, as yet, no standard, accurate, reproducible and meaningful test for non-dispersible concrete, although several have been proposed and used. These include:

- Plunge test
- Spray test
- Stream test
- Drop test
- pH factor test.

In the **plunge test**, a small basket with 3 mm diameter holes is filled with the fresh concrete and weighed. It is plunged into a water tank three times and the resulting loss of mass is then determined.

For the **spray test**, a mould is filled with about 1 kg of fresh concrete and placed on a baseplate before the mould is removed. The baseplate and sample are suspended on an electronic balance and the sample is sprayed with water for 4 minutes. The loss in mass is then measured.

The **stream test** equipment consists of a half-round gutter, 100–150 mm in diameter and 2 m long, set at an angle of 15–20°, as shown in Figure 11.6. A sample of the fresh concrete is placed 300 mm from the raised end and a given quantity of water is poured down the gutter and over the sample. Washout is then visually assessed.

In the **drop test**, a small scoopful of fresh concrete, weighing between 300 and 500 grams, is dropped into a graduated cylinder filled with water. The resultant turbidity

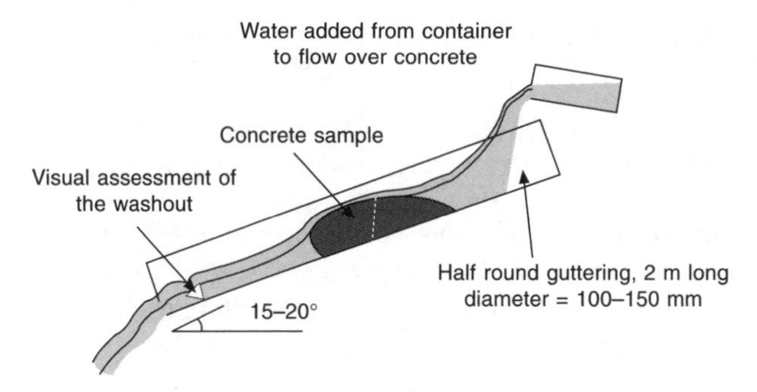


**Figure 11.6** Stream test.

is assessed visually or is measured by spectrophotometry, etc. This test is rapid, simple and qualitative.

For the **pH factor test**, a sample of fresh concrete is dropped into a beaker containing water. After 3 minutes a quantity of the supernatant solution is decanted into another beaker and its pH is measured. The higher the pH, the greater the washout.

Results from the plunge test and the spray test are shown in Table 11.1.

## 11.5 Formwork

Formwork needs to be robust so that it can withstand rough treatment by flowing water and heavy currents. It also needs to be simple so that divers can place and strike it easily. Accuracy of the formwork and of its final position are always problematic and this means that designers need to allow for variations in line and level.

It is normally prefabricated in the dry and lowered into position with a crane. Buoyancy has to be overcome if timber is being used and sandbags will help with this as well as sealing the bottom of the form onto the sea or river bed.

Flexible fabric formwork can be assembled easily into three-dimensional shapes by divers. There is a range of fabric types available, with varying strengths and permeabilities. As there is water pressure supporting the outside, small, simple shapes are able to resist the pressure of the concrete from within but larger and more complex shapes require external support.

A permeable fabric can allow the w/c ratio near the fabric to be reduced, giving an increase in strength at the concrete's surface. This is a benefit because although the fabric is left in place after concreting, it is often worn away and does not provide a protective layer.

## 11.6 Reinforcement

As with formwork, reinforcement details should be kept simple. It is best to use large-diameter bars widely spaced, securely held and with adequate cover. Cages must be detailed to allow for the placing technique being used; for example, when using a tremie, large flat areas of congested steel will hinder lateral flow.

Reinforcement cages should be prefabricated in the dry and then lowered into position.

## 11.7 Conclusion

The placing of concrete under water is a specialist operation, often working in remote and difficult conditions and in an unusual environment. If it is incorrectly carried out the consequences can be undetected and serious; it is therefore essential to pre-plan and to use personnel experienced in this particular field.

## Further reading

*Underwater Concrete*. Heron publication vol 19, 1973, No. 3 – Delft University and TNO Institute, the Netherlands.

*Underwater Concreting – TR35.* The Concrete Society.

# 12

# Grouts and grouting

*S.A. Jefferis*

## 12.1 Introduction

This chapter considers two general classes of grouts: cement-based structural grouts, which will be considered in detail, and geotechnical grouts, which may be based on clay, cement, additions (formerly known as cement replacement materials), sodium silicate or specialist chemical systems, which will be considered in less detail. Cement-based materials also may be used for the solidification/fixation of waste materials, especially low and intermediate level radioactive waste and hazardous toxic waste. These applications are not considered and for hazardous waste the reader is referred to specialist texts (Conner, 1990).

Grouting materials are some of the most complex systems that the engineer will encounter as it is necessary to control not only the fluid properties but also the set properties and to place the material into a void which may be of complex shape or into ground which may be heterogeneous. The void will be out of sight so that the operator who must rely on her/his experience to design and operate the grouting system and to assess what is actually being created. The use of grouting materials requires an understanding of:

- the intrinsic properties of the grout materials and health and safety requirements
- the preparation procedures that will be used on site
- the fluid properties of the grout
- the nature of the void/void system to be grouted
- the behaviour of the grout during and after injection into the voids
- the quality control methods to be used on site
- the properties of the hardened material

- the test methods for the hardened product
- the durability of the grout.

There is some guidance on grouts and grouting but the physical and chemical behaviour of this class of materials is still poorly researched and poorly understood by the majority of engineers. Experience is very important and the grouting specialist must use her/his experience and pragmatism to cover those parts where science has still failed to penetrate.

## 12.2 Health and safety

The grout system must have appropriate physical and chemical properties so that the health and safety of the grouting personnel is not put at risk either in the short term from acute effects or in the long term from chronic effects of exposure. Similarly, the grout and any material released from it prior to, during set or wash down of the grouting plant, or in service must not be damaging to the wider population or the environment.

It should be remembered that in many grouting operations, especially in geotechnical engineering, the quantity of grout used may be measurable in tens to hundreds of tons and that not all of this grout may remain at the location where it was injected. In particular it may be washed out by groundwater flow or spilled etc. Diluted grout materials may not react in the same way as the concentrated grout systems – indeed for some chemical grouts the setting reactions may not occur at all. This can be particularly significant if the initial reactants are more toxic than the final set grout – and this is often the case especially as the initial reactants will be more physically available as powders or liquids rather than a solid within a void. Great care must be taken when disposing of unreacted grout materials or grout-contaminated water, for example the water pumped from a tunnel during grouting works.

Grouts therefore have the potential to damage the environment and the issue of environmental damage is likely to become increasingly important. At the present time in the UK there appears to be no regulatory standard for, or specific regulatory controls on, what can be injected into the ground as a grout (licences are of course required for continuing discharges to surface or groundwater, for example, for industrial effluents). This is in marked distinction to the controls on the injection of waste waters etc. In other parts of Europe and particularly Germany there are already protocols for the injection of geotechnical grouts.

## 12.3 The void to be grouted

Very many different forms of void may require grouting and the void seldom will be a simple regular shape. For example in duct grouting in post-tensioned concrete the grout may have to penetrate the interstices of a tendon bundle and also fill the annular space between the tendons and the wall of the duct. The tendon bundle will not be central within the duct and its eccentricity will vary along the length of the duct which itself may be horizontal, vertical, inclined or undulating (see Figure 12.9). If full grouting is to be achieved careful attention must be paid to the flow behaviour within the duct/void and the positioning of injection points and vents. A fuller discussion of duct grouting is given in section 12.8.

For geotechnical grouting the grout may have to penetrate the interstices between soil grains without blocking or filtering. The voids often will be water filled and the grout must successfully displace this water without unacceptable mixing with it and dilution by it.

Thus in general the grout flow path or elements of it will be tortuous and three dimensional. The driving force causing the flow also must be considered. Normally grouts will be injected into a void under pressure. However, once in the void the grout flow will be influenced by gravity as well as the injection pressure and the effects of gravity (particularly gravity driven flow) will continue after injection has ceased. Failure to appreciate the effects of gravity has led to poor filling of pre-stressing ducts and failures in geotechnical and other grouting applications.

The void almost invariably will be invisible during and after grouting so that the grouting operator will have little feedback on the performance of the operation and whether the grouting has been successful. It is important to keep an open mind about the behaviour of the grout as it is injected into the void and the final disposition of the grout – it may not follow the text book behaviour as it will not have read the book! Examination of the grouted product can be invaluable in developing grouts and grouting procedures. Grout trials should be undertaken if new or complex void geometries are to be grouted and the resulting products should be subjected to destructive investigation. For a final structure inspection may be limited to non-destructive techniques such as ultrasonics, radar, radiography etc. and the degree of inspection that can be achieved may be rather limited.

## 12.4 Grouting theory and standards

Many countries have national standards for grouting work and there are also European standards, for example there are three standards relating to various aspects of the grouting of ducts in post-tensioned concrete (BS 1996a,b,c). There is also a revision of BS 3892, Part 2 which relates to the use of pfa in grouts. European standards are also in preparation on geotechnical grouting (prEN 1996a,b). However, these are mostly standards for the control, testing and use of grouts. Rather less information is available on the actual behaviour of grouts and the theory which underpins their design. Indeed it could be said that the best way to design a grout is to borrow a recipe from a previous successful job. Successful mixes are re-used and the recipes are adjusted to specific site conditions until problems occur and then designs are re-trenched. The grout designer like any artist has a palette of materials – Portland cement, pulverized-fuel ash (pfa), ground granulated blast furnace slag (microsilica and metakaolin are seldom used) sodium silicate and other chemical grouts, sand, aggregate and of course different mixer systems – though the key importance of the mixer is seldom recognized.

Grouting practice therefore is a blend of materials science, engineering and experience. Historically pragmatism has been a strong feature of many grouters but this may no longer find favour in our quality assured world.

## 12.5 Cement grouts

Cementitious grouts may be used for a very wide range of applications including:

structural applications such as grouting of tendons in pre-stressed concrete
grouted connections and grouted repairs for off-shore structures
geotechnical grouting for groundwater control and for soil strengthening
compensation grouting and slab jacking for the control of settlement
contact grouting behind tunnel linings during tunnel construction to fill the remaining voids and so improve the stress distribution on the lining and limit surface settlements
grouting of conductor pipes in oil and gas wells to prevent loss of hydrocarbon product
slurry trench cut-off walls for the containment of contaminants in the ground or the control of landfill gas migration
encapsulation/fixation of radioactive and toxic wastes and many other aspects of pollution control
*in-situ* stabilization of contaminants etc.

These applications involve many different engineering disciplines including structural, geotechnical, chemical and environmental engineering. Although all the applications may use what are basically cementitious grouts each has tended to develop its own form of specification and test procedures. As a result there is little consensus as to what is required of a grout, how it is to be specified and how its properties are to be measured and controlled.

From the client's/specifier's standpoint cement grouts are often assumed to be simply mixtures of cement and water which can be used to fill voids of any shape or size. In fact, all cementitious grouts are complex, time and shear history dependent, non-Newtonian fluids.

Failure to appreciate the potential complexity of grout systems and the importance of the form of the void to be grouted may lead to grouting failures in any of the above applications. In an overview such as this it is not possible to discuss all the applications of grouts and the special problems that may occur and the discussion is therefore limited to some of the fundamental features of grouts and especially the design of grout systems. The bibliography at the end of this chapter gives details of other works that may be consulted on specific topics.

## 12.5.1 Cementitious grout materials

The materials used in cementitious grouts may include:

Portland cements
calcium aluminate cements
pulverized-fuel ash
ground granulated blastfurnace slag
fillers such as sand or limestone flour
admixtures including plasticizers, superplasticizers
pre-set and post-set expansive agents etc.

An essential feature of all grout systems is that they must be sufficiently fluid to be injected into the void to be grouted and must set to a solid. To ensure satisfactory performance requires a very full understanding of the properties of the grout in the fluid state, during setting and in the hardened state. To begin to understand the behaviour of a cementitious grout system it is necessary to consider the mix proportions and in particular the volume fraction occupied by each of the mix components rather than just traditional weight ratios such as the water/cement ratio. As there is seldom any aggregate the W/C ratio is the free W/C ratio but grouters keep it simple and just refer to the water–cement ratio to add ‘free’ would confuse them.

In the past grouting was often undertaken with a normal Portland cement as it was felt there was no necessity to have a special grouting cement. However, BS 197 allows a rather wide range of compositions and the focus for construction cements is on strength – typically 28-day strength. There is no requirement to test for or maintain particular flow (rheological) behaviour in a grout produced with the cement. As a result Portland cements from different works and indeed different batches from the same works may show quite different behaviour if used as grouts, flow behaviour, bleeding, filtration etc. The properties of the grout also may be affected by the age of the cement if it has been subject to aeration (exposure to atmospheric humidity (Bensted, 1997)). For sensitive operations such as tendon grouting it can be preferable to use a Portland cement system that has been specifically formulated and quality assured for duct grouting (Concrete Society, 2002).

## 12.5.2 Mix proportions

In all cementitious grouts, even low water/solids systems, there is a substantial water filled porosity. This is essential for flow. As a result the solids in all cementitious grouts will have the potential to settle and express some water as bleed water (or as filter loss if the grout is contained within a leaky or otherwise permeable void). By way of example, Table 12.1 shows the water filled porosity of fresh grouts prepared at water/cement ratios in the range 0.25 to 0.5 and the potential for bleed/filtration (the volume of water that could be expressed from the grout as a fraction of the original volume of the grout) if the grouts were to settle/compact to a porosity of 40% – an unexceptional value for a particulate material settled through water.

**Table 12.1** Density, porosity, potential bleed, cement content and adiabatic temperature rise for Portland cement grouts as a function of water/cement ratio

| Water/cement ratio | Density ($kg/m^3$) | Initial porosity (%) | Potential bleed (%) | Cement content ($kg/m^3$) | Adiabatic temperature rise at 1 day (°C) |
|---|---|---|---|---|---|
| 0.25 | 2203 | 44.1 | 6.8 | 1762 | 133 |
| 0.30 | 2105 | 48.6 | 14.3 | 1620 | 120 |
| 0.35 | 2023 | 52.4 | 20.7 | 1498 | 109 |
| 0.40 | 1951 | 55.8 | 26.3 | 1394 | 100 |
| 0.45 | 1889 | 58.6 | 31.1 | 13.03 | 92 |
| 0.50 | 1835 | 61.2 | 35.3 | 1223 | 85 |

It can be seen that the potential for bleed/filtration is substantial even for a grout of water/cement ratio as low as 0.25 – a grout which would be quite unpumpable. Of course not all of this potential settlement may be expressed and the actual settlement may be much less than the potential settlement. The actual settlement may be influenced by:

- the level of shear during mixing of the grout
- the mixing time

- the addition of admixtures (see brief discussion in Domone and Jefferis, 1994)
- the cement type
- the fineness of the cement
- the set time
- the grout temperature and that of the duct/void
- the drainage path length(s)
- the form of the void to be grouted (shape, drainage, inclination to the vertical, obstructions etc.)

In small-scale tests on cylindrical grouts samples in the laboratory the initial porosity of the grout (and hence the water/cement ratio) and the set time may be the dominant parameters. In the field the drainage path length may be crucial.

In summary: pumpable cementitious grouts will always have a potential for considerable bleed/filtration. A grout that behaves satisfactorily in the laboratory or in one void system may show serious bleed, settlement etc. in the field or another void system. The mix proportions of a grout are fundamental design parameters and should not be changed without reference to the designer. A study of the literature shows that poor performance or durability of grouts, and especially chemical grouts used in geotechnical engineering, often can be traced to a failure to appreciate the importance of mix proportions.

## 12.5.3 Heat release

Table 12.1 also shows the cement content for the grouts and the estimated adiabatic temperature rise at one day from mixing (assuming that the heat released up to this time is 250 kJ/kg and that all this heat is available to raise the temperature of the grout). It can be seen that the cement content is very much higher than for concretes and grout designers must be aware that substantial temperature rise (which can result in temperatures above 100°C) may occur in large or insulated grout masses even if cured under water (Littlejohn and Hughes, 1988). With calcium aluminate grouts it may be difficult to avoid temperatures in excess of 100°C unless thin grout sections are used with water cooled metal formwork (Jefferis and Mangabhai, 1990). The temperature rise may be important in the laboratory as well as the field. For example if plastic or perspex formwork is used for laboratory models its insulating properties may lead to higher temperature rises than would occur in the field. Laboratory samples for strength measurement should be cast in metal moulds and covered (e.g. with a glass plate) after moulding and cured under water. If rate of strength development is an important issue temperature matched curing may be necessary.

Recent developments in bridge deck grouting in the UK suggest that the duct should provide the first line of defence to prevent corrosive agents such as air, water and de-icing salts reaching the tendons and that the ducting material should be plastic (Concrete Society, 1993; 2002). Thus the grout is a second line of defence for the protection of the tendons from corrosion. The first line of defence is the plastic ducting. This is an excellent approach but as the ducting is plastic the setting grout may reach rather higher temperatures than would occur in a metal duct. It is not suggested that this will be damaging but the grouter should be aware of it.

### 12.5.4 Grout mixing

The grout mixer and the mixing time can markedly change the behaviour of a grout. Often grout mixers are classed as high shear or low shear. However, there is no clear dividing line between these two classes of mixer. In the field, paddle mixers with a rotational speed of a few hundred rpm are likely to produce a low shear mix. Mixers with rotational speeds of 1400 to 2000 can produce high shear effects – low bleed, good internal cohesion etc.

Time of mixing is also important. Low shear mixing for extended times (perhaps in excess of 30 minutes) can produce results similar to high shear mixing for some grout systems and this has led to theories of mixing based on the total energy input to the grout. Total energy input per unit volume can be a useful parameter but it is not sufficient to describe mixing behaviour. Mixer geometry and mix energy per unit volume in the mixing zone are key parameters. Many mixers are effectively a reservoir linked to a mixing zone (the zone immediately around the rotating element) and it is the intensity of mixing in this zone and the number of passes an element of grout makes through this zone that controls mixer performance. It can be very difficult to simulate field mixing in the laboratory. In the laboratory a 1 kW capacity motor might be used to drive a 1 litre mixer. In the field the available motor power may be much lower at perhaps 0.015 kW/litre. This much lower power cannot be compensated by adjusting the mixing time as in the laboratory it may be necessary to mix for about 3 to 5 minutes to ensure that all the materials have been effectively wetted and incorporated in the mix. In the field the pressure to achieve rapid production will mean that mixing times are equally short. It therefore should be recognized that laboratory mixed samples may not be good indicators of flow and bleed behaviour etc. Jefferis (1985) gives an example of an extreme grout system comprising Portland cement with attapulgite clay and Dead Sea water. The set time of this grout when prepared in a low shear mixer was over 2 years but high shear mixing gave a set time of 24 hours. A further brief review of the effects of mixing is given in Domone and Jefferis (1994).

## 12.6 Cementitious grout testing concepts

As already noted there are national, European and international standards (see for example BS EN 445: 1996) on grout testing but unfortunately there still remains a wide range of non-standard test procedures and the underlying purpose of some tests may not be obvious. The possible purpose of tests include:

(a) quality control of materials on site
(b) quality control during grouting on site
(c) proving of the hardened properties of the grout
(d) investigation of materials in the laboratory
(e) laboratory-based research on the development of grout systems

Adoption of test procedures without recognizing their underlying purpose can lead to inappropriate controls being used. For example, a bleed test will be useless as an on-site quality control procedure during grouting as it will take a minimum of several hours. Control tests during grouting will be of use only if they are robust and give rapid results that can be acted on during grout injection as in most applications removing an unsatisfactory grout after set will be impossible without severe damage to the structure. However, bleed tests may be carried out prior to commencement of the grouting operation to prove the grout design, materials and mixing procedure.

Other properties such as long-term strength, permeability and durability cannot be determined by simple on-site tests and should be assured by laboratory investigations prior to commencement of grouting operations on site. During grouting work samples may be taken to confirm that the appropriate hardened properties are achieved – with the proviso that, if satisfactory properties are not achieved, the structure may have to be demolished. If particular hardened properties are critical then rapid on-site tests should be developed/specified to ensure that the grout, as used, will develop these properties.

In the past on-site testing tended to be focused on rheological properties as it was assumed that if the rheology was consistent then the grout would be consistent and that grouts of the same rheology would have similar other properties. Unfortunately this is not true – especially for cement-based grouts despite the fact that the achievement of consistent rheology is actually very difficult (unless a test procedure with a poor resolution is used, e.g. a test funnel with a large diameter spigot). Small variations in mixer type (including wear of the wetted components), mixing time, batch size, age of the grout at the time of test, age of the cement at the time of mixing and the cement chemistry and fineness may have much greater effect on the grout rheology than changes in more fundamental properties such as water/cement ratio. Modest changes in rheology, for example as a result of modest changes in batch size, may have quite modest effects on the performance of the grout whereas an undetected increase in water/cement ratio (undetected perhaps because the higher water/cement ratio grout was mixed for rather longer) could lead to substantial problems.

BS EN 445, 446 and 447 (BS 1996a,b,c) give details of test procedures and required properties for grouts for prestressing ducts. Specified properties include flow behaviour, bleeding, expansion and hardened strength. When studying these standards the reader should be aware that the international drafting committee considered several different national standards and the final documents were the inevitable compromise. As a result there are parallel test procedures for several grout properties. For example:

(a) There are 3 permitted sample dimensions for the bleed test sample. The test may be carried out in a 250 ml measuring cylinder, 25 mm in diameter with a 95 to 100 ml sample (i.e. a sample depth of 194 mm to 204 mm) or with a 150 mm deep sample in a 50 mm diameter transparent tube or with a 100 mm deep sample in 100 mm diameter can. For the first two tests the depth or volume of bleed water is measured at 3 hours and the bleed calculated as a percentage of the original depth or volume of grout. For the third test the height of the sample is measured immediately after placing and at 24 hours and the percentage change reported. This latter procedure also may be used to assess expansive grouts. It should be noted that the first two procedures use the volume of water expressed at 3 hours and the third procedure the change in solid volume at 24 hours. The change in solid volume may be sensibly equal to the volume of water expressed at early times such as 3 hours. By 24 hours all bleed water may have been re-absorbed and thus the change in solid volume (rather than the volume of bleed water) must be recorded for tests exceeding a few hours. In practice both the change in solid volume and the maximum volume of bleed water expelled are key parameters as both will lead to the formation of voids and the presence (even temporary) of free water can lead to density driven movement of the grout with the formation of fissures and cracks (see also section 12.7.3). As the specimens for all three tests have different depths the results obtained also will be slightly different – even if measured at a single time (see section 12.7.3.3).

(b) Flow properties may be measured with a flow cone or by a plunger immersion test (a test still new to the UK). It should be noted that due to lack of matching data for these two instruments a compromise could not be achieved, in the standards committee, as to the required flow properties for a grout immediately after mixing. As a result BS EN 447 (BS 1996c) requires that the grout immediately after mixing should have an immersion time of *greater* than 30 seconds or a flow cone time of *less* than 25 seconds. As thicker grouts will show an increase in time for both tests the effect is that if the immersion tests are used there is a minimum permissible rheology immediately after mixing but if the flow cone is used then there is a maximum. Furthermore the permitted times 30 minutes after mixing or at the end of the injection period are less than 25 and less than 80 seconds for the cone and immersion tests respectively, i.e. the immersion test time has been increased from 30 to 80 seconds but the cone time is constant at 25 seconds. The reason for this is that the immersion test is much more sensitive to the rheology of the grout (viscosity, yield value etc. than the cone), see Figure 12.5 (note the standard allows that all the cone times may be increased to 50 seconds for some grouts prepared in mixers with a high shearing action). The grout designer should be aware of the significance of the choice of test procedure.

(c) Strength may be measured on 100 mm diameter by 80 mm high cylinders (these may be prepared by cutting the ends of the can samples of the bleed/expansion test noted above), 40 × 40 × 160 mm prisms or with permission of the 'competent authority', on cubes of dimension not exceeding 100 mm. No interconversion factors for the effects of specimen dimensions are referenced.

In general, except for flow tests immediately after mixing (as discussed above), there is no direct conflict between the other 'parallel' test procedures (mainly because there are few specified test results). However, it should not be assumed that all the parallel procedures will rank all grouts equally – see, for example, Figure 12.1 which shows the effect of shear rate on the ranking of grout rheology. BS EN 445 (BS 1996a) provides a useful

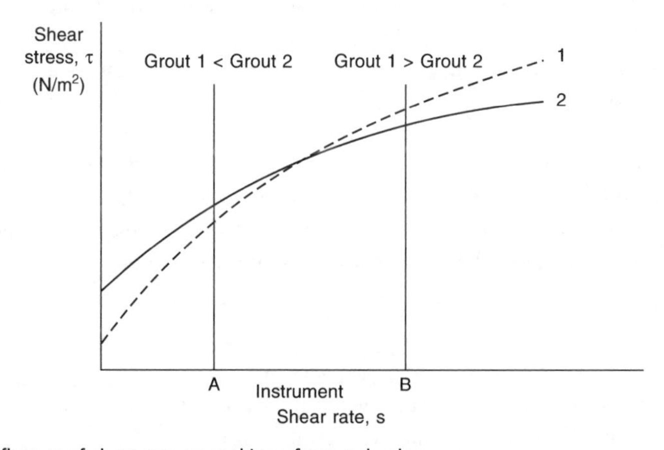


**Figure 12.1** Influence of shear rate on ranking of grout rheology.

baseline for grout investigation and it is to be hoped that refinements will be introduced as more information becomes available.

## 12.7 Grout properties

Typically the properties required of an ideal grout will include that it:

- is of low viscosity, yield value and/or gel strength so as to enable penetration of fine pores, fissures etc. (except for specialist grouts such as compensation grouts which may be designed so as not to flow significantly from the point of injection)
- completely fills the void to be grouted. This may be more of an issue of grouting procedure than grout materials but the grout may have to be tailored for the injection equipment, the pumping distance and the volume of grout to be injected etc.
- remains in the void after placement and is not washed out for example by ground water flow, thixotropic grouts may be advantageous
- is not subject to segregation of the mix constituents for example as a result of density differences causing settlement of solids, bleed and filtration during and after injection
- sets and hardens rapidly at a predictable time after injection thus minimizing the delays in commissioning structures etc.

Brief details of some test for flow properties are given in sections 12.7.1 to 12.7.5. Further details of procedures are given in Jefferis (1991) and Rogers (1963). It should be noted that several of the procedures were developed for testing oilwell drilling fluids and 'imported' into grouting work.

### 12.7.1 Flow properties – rheology

Rheology is important because of its influence on the injection process and on the behaviour of the grout in a void prior to set. Discussion of the theory of rheology – the flow of materials is beyond the scope of this chapter. However, a brief discussion of some of the key parameters is necessary.

For simple fluids such as water in laminar flow there is a linear relationship between shear stress and shear rate and the fluid flow can be described by a single parameter, the viscosity. These fluids are known as Newtonian fluids. Suspension grouts such as cement or clay-based systems, or chemical grouts as set approaches, show more complex shear stress–shear rate behaviour. These grouts can behave as solids at low shear stresses and flow only when the shear stress exceeds a threshold value – thixotropic behaviour is a special case in which the fluid properties are dependent on the shear history of the sample (as will be the shear stress–shear rate relationship itself). The simplest rheological model which may be used to describe grouts which show a threshold value is the Bingham model:

$$\tau = \tau_o + \eta_p \dot{\gamma} \tag{12.1}$$

where $\tau$ is the shear stress, $\dot{\gamma}$ the shear rate, $\eta_p$ the plastic viscosity and $\tau_o$ the yield value. In practice the model is a substantial oversimplification.

Figure 12.2 shows the shear stress–shear rate plot (the equivalent, for a fluid, of the

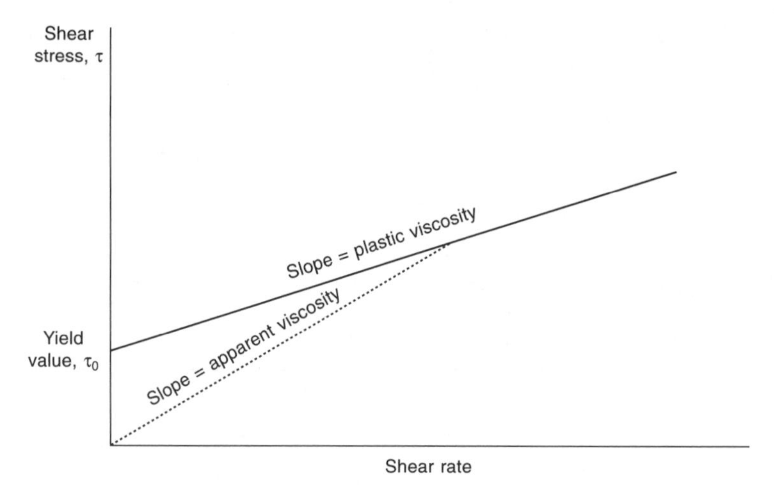


**Figure 12.2** Shear stress–shear rate plot for ideal Bingham fluid.

stress–strain plot for a solid) for the ideal Bingham fluid. The figure identifies the yield value $\tau_o$ and the plastic viscosity $\eta_p$ and also the apparent viscosity, which is the slope of the line joining any point on the plot to the origin. The apparent viscosity is the viscosity that would be determined at the given shear rate if the fluid were assumed to be Newtonian.

In practice the shear stress–shear rate plot is not a straight line as implied by equation (12.1) but a curve as shown in Figure 12.3(a). For any shear stress the tangent to the curve gives the equivalent Bingham plastic viscosity and the extrapolation of this tangent to the shear stress axis gives the equivalent Bingham yield value. Figure 12.3(b) shows the effect of time on the low shear rate section of Figure 12.3(a) for setting and thixotropic fluids.

When at rest, a grout may behave as a solid and the shear stress to break this solid is known as the gel strength. The gel strength is thus the yield point at an infinitesimally low shear rate. A further complication is that the rheology of suspensions grouts is time dependent and shear history dependent. A grout investigated after a period of high shear agitation will be found to be thinner (lower values of plastic viscosity, yield value and gel strength) than a grout which has been left quiescent. This may be due to setting and also for non-setting systems such as clay suspension due to development of a reversible structure in the system due to interaction of the particles and development of preferred alignments so that the system shows thixotropy.

Polymer solutions and thus polymer-based chemical grouts show rather different behaviour in that few show a true yield value, most are pseudoplastic for which the simplest rheological model is:

$$\tau = k\dot{\gamma}^n \tag{12.2}$$

where $k$ is a constant and is a measure of the consistency of the fluid, and the exponent $n$ is a function of the type of polymer and the shear rate, though it may be sensibly constant over a range of shear rates. For pseudoplastic materials $n$ will be less than 1. Materials with $n$ greater than 1 show dilatant or shear thickening behaviour.

The general shape of the pseudoplastic shear stress–shear rate curve, for $n < 1$, is shown in Figure 12.4. Extrapolation to zero shear rate from any point on the curve will

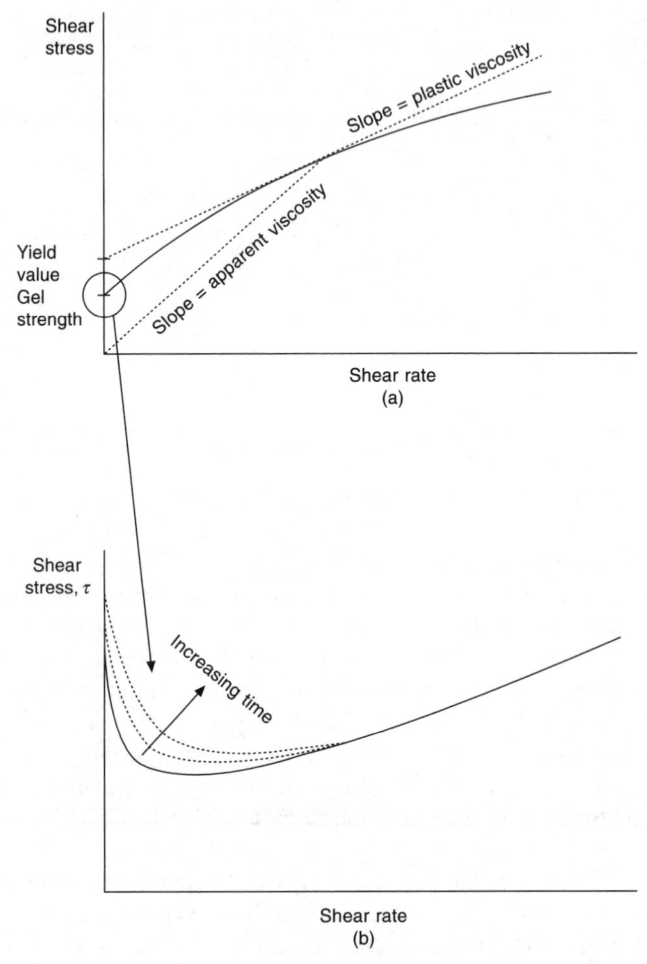


**Figure 12.3** (a) Shear stress–shear rate plot for general fluid: (b) effect of setting or thixotropic gel development.

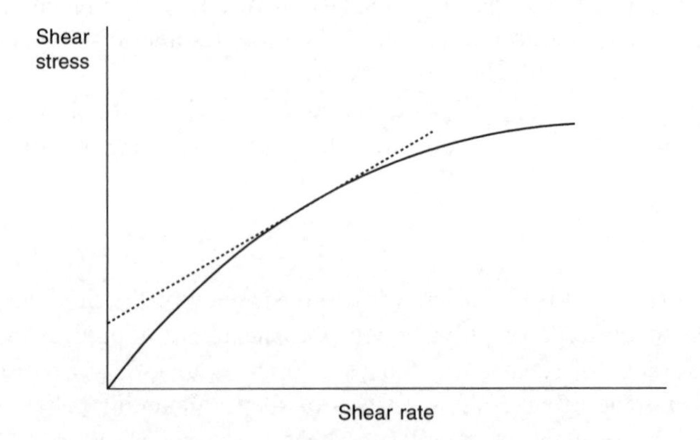


**Figure 12.4** Shear stress–shear rate plot for pseudoplastic fluid.

give an apparent yield value. However, the curve actually passes through the origin. Such materials therefore do not have a true yield value but show increasing viscosity with decreasing shear rate. This behaviour can be exploited to reduce settlement and bleed of suspension grouts – the high viscosity at low strain rates leading to reduced settlement rates for small solids (pseudoplastic polymers are used in domestic products to prevent the settling of solids in cleaning products and the segregation of oil and vinegar in vinaigrettes – an edible biopolymer xanthan gum often being used in the latter).

Apparent viscosity and gel strength (rather than yield value) are perhaps the most obvious and easily measured properties of a grout and therefore tend to be used for control purposes. However, there is little consensus as to what are acceptable values or even desirable values.

Furthermore there is no single standard test procedure that can be used to rank all the different grouts that may be used in the diverse applications mentioned in section 12.1. With the presently available instruments some grouts will be effectively so thin that no resolution can be obtained and others so thick that no flow occurs or readings are off the scale of the instrument. It is therefore difficult to obtain comparable rheological data on the grouts used in different applications. Instruments that are typically used to assess the fluid properites of cementitious grouts include:

flow cones
rotational viscometers
the plunger immersion test
flow troughs
the consistometer

For chemical grouts rheological measurements are rather seldom undertaken in the field as the grouts are usually formulated to be of low to moderate viscosity and test instruments with the necessary resolution tend to be fragile and/or very expensive. Furthermore viscosity is a poor control test for these grouts and attention should be focused on

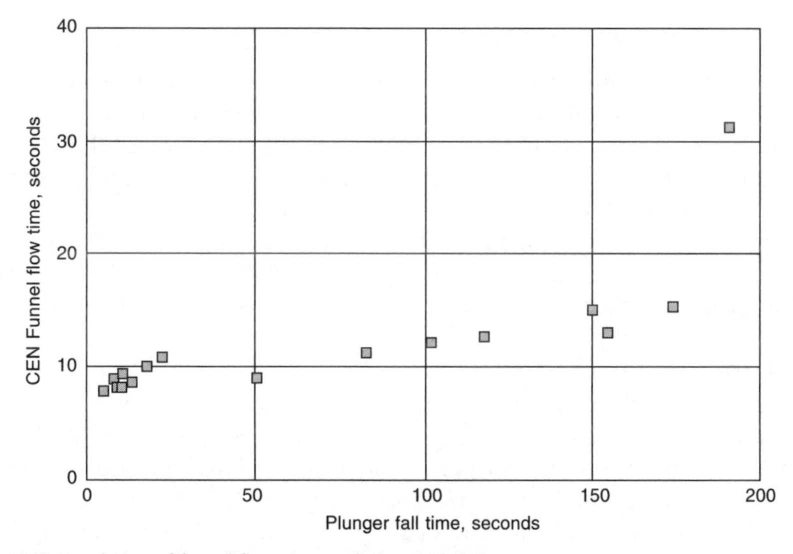


**Figure 12.5** Correlation of funnel flow time and plunger fall time.

accurate batching. Rheological measurements, if made, are likely to be undertaken in the laboratory as part of research on grout design, formulation etc. For this rotational viscometers such as the Fann viscometer as described in section 12.7.1.2 (Rogers, 1963), may be used though even this has rather limited resolution with a minimum scale division of 1 centiPoise (the viscosity of water).

#### 12.7.1.1 Flow cones

There is a wide range of cones including the Marsh cone, the FIP cone (FIP, 1990) and the CEN Cone (BS EN 445, BS 1996a). There is a tendency in the literature to refer to all cones as Marsh cones. This is incorrect and has led to much confusion. There is only one Marsh cone, which was developed for testing oilwell drilling fluids (Rogers, 1963) and because of its relatively small discharge spigot (3/16 inch diameter) it shows very extended flow times for most cementitious grouts and is unsuitable for them. The CEN and FIP cones have been designed for cement grouts. However, it should not be assumed that these cones will be suitable for testing all cement grouts. For some grouts the flow time may be so short that no useful resolution of flow properties can be obtained and for others the times may be very extended or flow may stop.

When quoting the results from any cone test it is important to specify the dimension of the discharge spigot of the cone, the volume of grout to be filled into the cone and the discharge volume. The flow time for water also ought to be quoted as a control. Unfortunately in much of the literature these data are not stated. For example, for the FIP cone, FIP (1990) gives neither a figure for the quantity of grout to be put into the cone nor the discharge volume and thus the operator is at liberty to choose his own figures.

It should be noted that although most specifications require that the volume of grout drained from the cone is less than that put into it, a few require that the cone is drained completely. This is unsatisfactory for cement grouts as much grout may remain adhering to the walls of the cone and the flow may reduce from a steady stream to a drip before the cone is empty. The time at which flow stops then may be several hours – not a satisfactory situation as the test then may take longer than the actual grouting operation.

#### 12.7.1.2 Rotational viscometers

There is no standard rotational viscometer for grout testing. The Fann viscometer, which was developed for testing oil well drilling fluids, is often quoted, in grout specifications, as the required test instrument (Jefferis, 1990; Rogers, 1963). However, it has a relatively narrow test range (this varies with the particular model of the instrument but is typically of order 0 to 300 centPoise (cP) with a minimum scale division of 1 cP). Low water/cement ratio grouts may give readings off the scale. Furthermore, for cementitious grouts, the small spacing between the stationary bob and rotating sleeve of the instrument (0.59 mm) makes it difficult to ensure that the test space is properly filled with grout.

#### 12.7.1.3 The immersion test

The plunger immersion test is relatively new to the UK and is best suited to superplasticized cementitious grouts though this is not explicitly stated in BS EN445 and BS EN447 (BS 1996a,c). It consists of a vertical tube of internal diameter 62 mm and height 900 mm and a torpedo shaped plunger of diameter 58.2 mm and overall length 300 mm (BS EN445). The tube is sealed at the lower end and for a test it is filled with approximately 1.9 litre of grout (so that the plunger when introduced into the tube can be submerged without overflow of the grout). The plunger is then submerged in the grout and allowed to fall through it. The time for the plunger to fall 0.5 m is recorded. BS EN445 (BS 1996a) requires that the test is repeated three times and that the average of the second and third times is reported. The first result is ignored, the test being to condition the grout so that more repeatable results can be obtained in the second and third tests. If simple water/cement grouts are used without a superplasticizer the test time may be very extended.

#### 12.7.1.4 Flow troughs

The basic principle of the flow trough is that a fixed volume of grout is filled into a wide-mouthed funnel mounted over a trough and then released by removing a stopper. The distance the grout flows along the trough is recorded. The flow trough is a simple and robust instrument capable of giving useful results for a wide range of grouts in both structural engineering, geotechnical applications and radioactive waste stabilization.

A flow trough is specified in BS 3892 Part 2 (BS, 1997) (in the past this particular trough has been referred to as the Colcrete flow trough). There are other flow troughs.

#### 12.7.1.5 The consistometer

The consistometer was developed for the oil well cementing industry. In this instrument a pot filled with grout is rotated at constant speed and the torque on a stirring frame inserted in the grout is measured. The grout container is mounted in a water bath so that the grout temperature can either be kept constant or can be controlled to follow a predetermined programme. There is a high-pressure version of the instrument so that grouts can be tested under the hydrothermal conditions that may occur in oil wells (unfortunately the high-temperature version has a different geometry to the low-temperature version and the results of the two instruments are not directly interconvertible). The instrument is very much an empirical device but is useful in that the grout is kept stirred and thus data can be obtained over extended times without segregation of the grout (in some rotational viscometers segregation of water from the solids can lead to serious reduction in the measured rheological parameters. Segregation may be both gravity driven and from centrifugal forces due to the viscometer rotation.

#### 12.7.1.6 Prediction of grout pipeline pressures

One of the main concerns with grouting operations is often that the pressure required to pump the grout will be excessive. In practice, pumping pressures are seldom a problem and well within the capacity of most standard pumping equipment unless unusually long pumping distances are involved. A considerable amount of work on pumping grouts was undertaken in relation to grouting operations for offshore oil rigs (Littlejohn and Hughes, 1988; Jefferis and Mangabhai, 1988). Jefferis and Mangabhai (1988) provide a simple procedure for the estimation of grout pumping pressures. Figure 12.6 shows a plot of wall shear stress against average flow velocity for a 0.3 w/c ratio grout. The wall shear stress is the shear stress exerted by the flowing grout on the wall of the pipe and thus the stress which is responsible for the pressure drop as the grout is pumped. The wall shear stress can be obtained from the pressure drop in the pipe using equation (12.3) and the average velocity is the grout flow rate divided by the cross-sectional area of the pipe. Figure 12.6 shows that for velocities up to about 4 m/s for this grout the wall shear stress increases rather gently but at higher velocities the stress and thus the pumping pressure increases rapidly. The Concrete Society's Technical Report 47 (Concrete Society, 2002) suggests a maximum grouting rate in a duct of 10 m/min. For a grout delivery pipe of diameter 25 mm supplying grout to a 100 mm diameter duct with a tendon bundle occupying 40% of the duct area, the ratio of grout velocity in the delivery pipe to that in the duct will be about 10. Thus 10 m/min in the duct would require 100 m/minute or 1.6 m/s in the duct which is in the region of Figure 12.6 where the wall shear stress is relatively independent of the grout velocity.

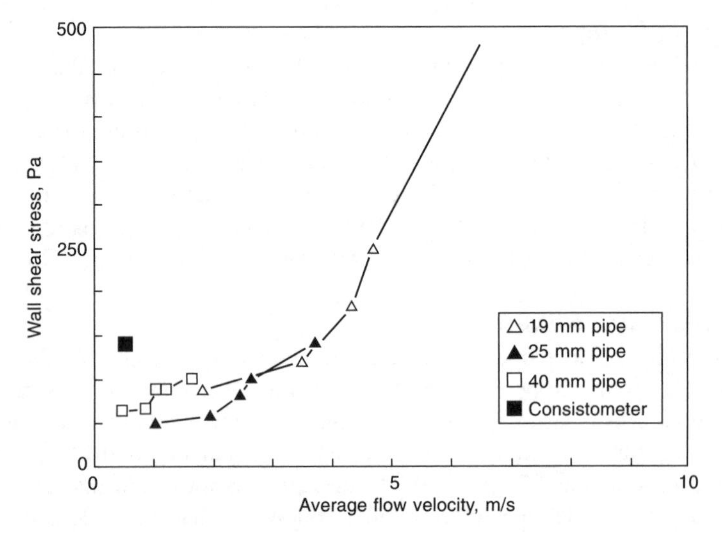


**Figure 12.6** Wall shear stress as a function of flow velocity.

To predict the wall shear stress a rotational viscometer may be used. The rotational speed of the viscometer should be selected so that the peripheral speed of the rotating element is equal to the planned average velocity in the pipe. The torque reading shown by the instrument then can be used estimate the shear stress.

## 12.7.2 Grout penetration

### *12.7.2.1 Effect of grout yield value*

For a grout properly to fulfil its function, for example to provide protection to a tendon, or provide shear transfer across an offshore grouted connection or increase the strength or reduce the permeability of a soil, it is important that the grout fills the void and all its interstices. However, the flow properties of the grout may delay or inhibit the flow of the grout. Important parameters will include the rheological properties of the grout (viscosity and yield value), capillary forces, the hydrostatic pressure of the grout at the location to be penetrated, the applied grouting pressure and the pressure within the void.

A grout which shows Newtonian or pseudoplastic behaviour, i.e. one that has no yield value ultimately will penetrate all the interstices provided a driving pressure is maintained for sufficient time – though the rate of penetration may be slow if the grout is very viscous and provided that capillary effects do not inhibit penetration.

However, for a non-Newtonian fluid which shows a yield value, full penetration may not be possible. A grout of yield value, $\tau$ will penetrate a distance $L$ in a tube of radius $r$

or between parallel plates a distance $r$ apart under a pressure $P$ (total hydrostatic pressure plus applied pressure less any pressure within the void – pressure can build in a sealed tendon bundle) where:

$$L = \frac{Pr}{2\tau} \tag{12.3}$$

At duct grouting pressures the effect may be small. For example a pressure of 500 kPa will force a grout of yield value 50 Pa (a potentially modest value) a distance of 5 m into a slot of width 1 mm or a tube of radius 1 mm (note 500 kPa is the pressure recommended in Clause 6 of the specification in the Concrete Society's Technical Report 47 (Concrete Society, 2002) and is equivalent to a 26 m head of grout of density 1.95 g/ml, i.e. water/cement ratio 0.4, see Table 12.1).

#### 12.7.2.2 Effect of interfacial capillary tension

If the grout wets the walls of the void to be penetrated then capillary forces will tend to pull liquid from the grout and potentially some or all of the solids in the grout into the void. The capillary pressure across an interface of radius $r$ is given by:

$$P = \frac{2\gamma}{r} \cos\theta \tag{12.4}$$

where $\gamma$ is the interfacial tension and $\theta$ the angle of contact between the liquid in the grout and the fluid filling the void (typically air but if the void has been pre-wetted then water may remain in the fine voids so that there is no interfacial tension and thus no tendency for the grout to be drawn into fine voids). In many situations such as soil grouting the fine voids are likely to be wet before grouting and thus capillary forces may be of little significance. However, the author is not aware that the effects of capillary forces have been considered when deciding whether or not to flush ducts with water before structural grouting. (Technical Report 47 discusses the general effects of water flushing but does not consider capillary tensions.)

Surface tension effects can add usefully to penetration if the grout fluid wets the surface to be penetrated. The penetration $L$ of a Bingham fluid of yield value $\tau$ due to surface tension in a horizontal tube or fissure with no applied pressure is given by:

$$L = \frac{\gamma \cos\theta}{\tau} \tag{12.5}$$

Note that this penetration is additional to that due to any applied or hydrostatic pressure and is independent of the void dimension (tube diameter, plate separation).

Pure water has surface tension of order 0.078 N/m and if there were full wetting (zero angle of contact) then for a 10 Pa yield value grout, the effect would add approximately 8 mm of further penetration. Full penetration of fine voids has been demonstrated using a wedge shaped glass mould (two sheets of glass clamped together along one edge but pre-wetting of the void has not been investigated would seem to be worthwhile.

If the fluid does not wet the void to be penetrated (i.e. the angle of contact is greater than 90°) then pressure will have to be applied to overcome the capillary effects. This may occur if surfaces are oil coated.

The effect of interfacial tension on grout penetration would appear to be an area where research could be useful.

## 12.7.3 Bleed

Bleed is a gravity driven process whereby the solids in a particulate grout settle because of their higher density than the surrounding pore fluid. Bleed is therefore analogous to self-weight consolidation of soils, but in grouts allowance must be made for set. Bleed may be stopped by stiffening the grout before the full consolidation that would occur in a comparable non-setting system has developed and thus the actual bleed may be less than the potential bleed (see Table 12.1 for examples of potential bleed).

From consolidation theory it can be shown that for a vertical column of grout at short times (strictly small values, less than about 0.08, of the consolidation time factor ($C_v t/h^2$) where $C_v$ is the coefficient of consolidation, $t$ the time of settlement and $h$ the depth of the grout mass, see Figure 12.7a) the rate of surface settlement (i.e. the rate of bleed) is independent of the depth of the grout mass (and depends only on the density of the grout, $\rho_g$ relative, to that of its pore fluid, $\rho_w$ and the permeability, $k$ of the grout to its pore fluid) and is of the form:

$$\mathrm{d}h/\mathrm{d}t = (\rho_g/\rho_w - 1) \cdot k \qquad (12.6)$$

Thus if settlement is stopped by set at small values of the time factor then the amount of bleed will be equal to the rate of settlement multiplied by the set time, $t_s$ that is:

$$\text{Bleed} = \mathrm{d}h/\mathrm{d}t \cdot t_s = (\rho_g/\rho_w - 1) \cdot k \cdot t_s \qquad (12.7)$$

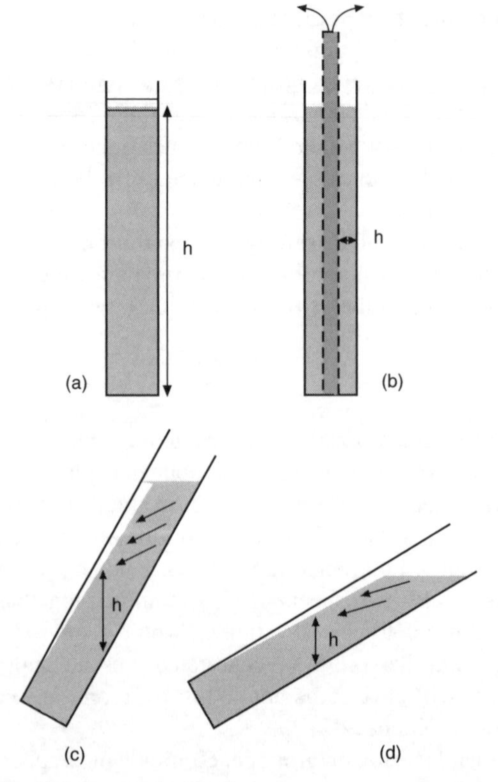


**Figure 12.7** Controlling length parameter for bleed.

and as the rate of settlement is independent of depth the amount of settlement will be independent of depth.

A check on equation (12.7) can be obtained by using it to predict the permeability of a grout. For example, if a 0.35 water/cement ratio grout shows a bleed of 2 mm at 2 hours then from equation (12.7) its permeability can be calculated as $3 \times 10^{-7}$ m/s which is a reasonable value for a silt sized material such as cement.

For larger values of time factor, for example if the set time is slow or the grout column is short (less than perhaps 0.5 m), bleed will be a function of grout depth. It follows that although there may be some justification for quoting bleed as a fraction of column height for short columns it is not helpful for tall columns. When reporting any bleed results the grout depth always should be quoted. It is therefore rather unfortunate that in the BS EN 445 standard (BS 1996a) the depth of the bleed test specimens is not explicitly fixed and common to all the test procedures.

In laboratory/research investigations of grout bleed several different test depths always should be used so that the effect of depth can be assessed.

A number of other keys issues follow from the analysis of bleed based on consolidation theory:

**Intermediate bleed lenses:** as bleed is independent of depth, if some imperfection in the duct, change of section, inclination etc. leads to the formation of an intermediate bleed lens, this lens may be of volume equal to that of the bleed water which collects at the top surface. Once formed, such a lens may not be stable and may tend to migrate up through the setting grout causing damage to a considerable length of duct. There will be no evidence of intermediate lenses at the top of the duct and the best protection against their formation is to use low bleed grouts and design to prevent bleed water accumulation.

**Inclined ducts:** analysis of bleed based on consolidation theory shows that inclined tubes will tend to show higher bleed than vertical tubes – a result observed by Cambefort in 1964 (Cambefort, 1964) but not explicable at that time (it is the vertical depth of grout, or strictly the distance to a drainage surface, that controls the rate of consolidation and in an inclined tube this will be less than the full height of the grout, see Figure 12.7c and d). A current French procedure for testing duct grouts involves the use of an inclined test section (see Concrete Society Technical Report 47 (Concrete Society, 2002), Appendix A7).

**Rate of bleed and amount of bleed:** in the consolidation time factor $C_v t/h^2$ the distance to a drainage path, $h$ appears to the power two. Thus its influence is particularly marked, for example, for a 10 m vertical prestressing duct, if sealed at the base and with no leakage to the tendon then $h^2 = 100\ m^2$. However, if the tendon/tendon bundle acts as a drain (water has often been reported spurting from tendons – driven by the density difference between the grout in the duct and the water in the tendon bundle, see section 7.6) then $h$ will be the annular thickness of grout around the tendon bundle – perhaps 10 mm (see Figure 12.7) and thus $h^2 = 10^{-4}\ m^2$. Thus a leaky tendon bundle allowing radial drainage may increase the time factor at any grout age by $10^6$ (strictly allowance should be made for radial consolidation but this simple analysis gives the order of magnitude). The increase in bleed as a result of the increase in the time factor will depend on the actual value of the time factor – the effect is strongly non-linear and depends on the precise drainage conditions. Typically 100% of the bleed may be achieved if the time factor reaches about 2 before set whereas 50% might occur at a time factor of 0.1.

In the field the effect can be dramatic, a duct with a sealed tendon bundle may show a few millimetres of bleed whereas with a leaky tendon bundle a substantial fraction of the potential bleed may occur (see Table 12.1). Thus of the order of 2 m of bleed could occur in a 10 m duct with a 0.35 water/cement ratio grout. The drainage to the tendon bundle may be so significant as to develop a substantial filter cake around the tendon (see section 12.7.4 and Domone and Jefferis, 1994).

In summary, bleed tests of the type specified in BS EN 445 (BS 1996a) are not useful indicators of the bleed that can be expected in a full scale grouting application such as a stressing duct. They are tests of the behaviour of a grout under particular test conditions and cannot be scaled to predict *in-situ* performance without a detailed understanding of consolidation processes.

### *12.7.3.1 Control of bleed*

Bleed is driven by the difference in density between the solid grains and the pore water (an exactly analogous situation occurs with the separation of cream from milk on standing except that the droplets of cream are of lower density than the continuous aqueous milk phase and rise through it). Bleed will be controlled by:

(a) Settlement of the grout solids and upward escape of porewater between the grains of the grout. Even in a fresh grout these may be linked by some bonding to form a weak solid which tends to settle as an intact mass rather than individual grains. Hence it is reasonable to consider the permeability of the grout to its pore fluid.
(b) The particle size distribution of the grout solids as this will affect the permeability. Lower permeability can be achieved by using finer cements, adding some fines or by longer or higher shear mixing.
(c) The viscosity of the pore fluid. Viscosifying the pore fluid can reduce bleed but may excessively thicken the grout – unless shear thinning viscosifying agents are used (see discussion on pseudoplastic fluids in section 12.7.1).
(d) The amount of water that actually can be squeezed from the grout. This will depend on the potential bleed of the grout (see Table 12.1), the nature and strength of bonding between the solid grains in the grout and the effective stress acting on any element of the grout (the effective stress at any point is the difference between the total stress exerted by the grout and the local fluid pore pressure – it is the stress carried by the solid grains). The grout at the base of a tall duct therefore may have a higher potential bleed than that nearer the surface or in a short duct – though this may not be expressed, because of the long drainage path, if the only drainage is at the top surface of the duct. It follows that any leaks at the base of a duct may substantially increase bleed.
(e) The stiffening time of the grout, i.e. the time at which the grout has sufficient strength to stop settlement of the solids. Again this may depend on the effective stress acting on the solids, i.e. movement may stop later at the base of tall ducts. Adding an accelerating admixture therefore will reduce the bleeding time.

### *12.7.3.2 Channelled bleed*

Grouts showing more than about 2% bleed in small-scale tests are unlikely to be suitable for structural applications not only because the void will not be properly filled but also because of the risk of a more serious type of bleed – channelled bleed. This can occur in unstable grouts (e.g. high water/cement ratio systems). Channelled bleed will occur if the pressure of the escaping pore fluid is sufficient to fracture the grout mass and tend to force a flow channel through it (actually this is the incipient condition in all freshly mixed grouts) and the quantity of water is sufficient to develop the flow channel. Once formed a channel will acts as a drain and allow rapid escape of pore fluid thus exacerbating the bleed. Fines may be carried in the water escaping along the channels to be deposited as small 'volcano' like mounds at the grout surface. The channels may develop at discrete points perhaps associated with some discontinuity in the grout mass which encourages the localized collection of bleed water. This could be under a horizontal section of tendon or at any significant change in direction of the duct.

In high water/cement ratio grouts so many bleed channels may develop that the surface 'volcanoes' coalesce to produce a roughened grout surface which may not be immediately recognizable as the result of channelled bleed.

Bleed channels will remain once the grout has set and may allow subsequent ingress of atmospheric carbon doxide, water and aggressive agents etc.

Channelling also can occur if water bleeding from one area of the grout can be conducted to another area of the where the local pore pressure is lower. For example if a leaky tendon bundle acts as a pipe for bleed water. This is considered in section 12.7.6.

### 12.7.3.3 Measurement of bleed

A typical simple test for bleed involves the use of a measuring cylinder (volume perhaps 100 to 1000 ml) filled with the grout and left undisturbed for up to 24 hours. The volume of water released or the final volume of set grout may be recorded and the bleed expressed as the volume of water or loss of volume of solid grout as a percentage of the initial grout volume. As the hydration reaction can lead to the re-absorption of some or very often all the bleed water by about 24 hours, the use of the volume of water expelled is likely to underestimate the bleed after about three hours.

BS EN 445 (BS 1996a) sets out some codified procedures for bleed measurement. Criticisms of these procedures are:

- only small samples are tested
- the effect of drainage to a tendon or tendon bundle is not considered
- the effect of effective stress level in the grout is not considered.

The procedures in BS EN 445 (BS 1996a) involve relatively small drainage path lengths and thus relatively full bleed is likely to occur though at the low effective stress levels of a shallow sample. Concrete Society Technical Report 47 (Concrete Society, 2002) proposes the use of a 1.5 m high grout specimen with a tendon. The pressure driving drainage is thus significantly higher than in a sample of height about 10 cm. However, the stress level still may be much smaller than will occur in a tall vertical duct or an undulating duct with a crown to trough height difference of much greater than 1.5 m. Thus although the test may be an improvement on small sample tests it will not provide conservative results for grouts involving higher stress levels. Technical Report 47 also gives details of a test developed in France which uses an inclined, pressurized duct. The test is complex and relatively expensive but usefully investigates the effects of filtration and the possible movement of grout as a result of bleed in an inclined duct – though this could be inhibited by the development of a filter cake around the tendon bundle. Wisely, Technical Report 47 also recommends that full-scale trials are carried out.

## 12.7.4 Filtration

Bleed occurs as a result of the difference in density between the cement grains and the mix water and thus is the result of a gravitationally driven separation of solid and liquid. The two phases also may be separated by pressure filtration.

If there is an escape path (e.g. a leak in a duct, a leaky tendon bundle or at a stressing head) water may be forced from a grout by the hydrostatic pressure of the grout itself and/or an externally applied pressure (for example if the duct is pressurized at the end of grouting). Filtration will lead to the loss of water and the deposition of a filter cake. Initially the rate of water loss will be high but as the cake develops the rate will decline. For a simple incompressible filter cake the rate of loss of water will be proportional to $k_c \cdot t^{-1/2}$ where $k_c$ is the permeability of the filter cake and $t$ is the time since cake formation started. The thickness of the resulting cake will be proportional to $t^{1/2}$.

Filtration differs from bleed in that the rate of escape of water is a function of the permeability $k_c$ of the developing cake rather than that of the grout mass to its pore fluid.

In practice it can be difficult to distinguish between bleed and filtration and they can be regarded as the simple extremes of a more complicated process which actually will involve a combination of the two mechanisms. At high effective stresses (compared with the gel strength of the grout) filtration will dominate. At low effective stresses bleed will be the main process. The properties of the grout, the *in-situ* effective stresses and the drainage conditions will determine which of the processes occur/dominate.

It is important to note that if either filtration or bleed occurs the volume of the remaining grout will be reduced. Thus if water drips from the tendon bundle at the anchorage of a stressing duct then a void must be forming somewhere within the duct unless further grout is being pumped into the duct to make good this loss (the volume loss must be made good before the grout begins to stiffen as otherwise its integrity may be compromised).

Filtration is not intrinsically undesirable as it can lead to a tight filter cake around a tendon bundle or other permeable feature thus providing very good protection. However, if the volume loss is not made good the effects may be very damaging as large voids can develop – the author has observed filtration voids of the order of one metre in length in laboratory tests on stressing ducts.

## 12.7.5 Combined bleed and filtration

In a duct during grouting there may be water loss by both bleed and filtration. The test procedures for bleed given in BS EN 445 (BS 1996a) examine only vertical settlement in small samples. As discussed in section 12.7.3.3, it has been suggested that a more useful procedure is to measure the volume loss of a sample which includes a length of pre-stressing tendon to act as a vertical drain. This may give a more realistic result, however, care is necessary. The balance between bleed and filtration will depend on the effective stress and the properties of the grout. If the effective stress (the differential between grout pressure and pore pressure in the duct at any point) is less than the pressure necessary to mobilize movement of the grout solids (or more simply less than some pressure comparable to the gel strength/yield value of the grout) then filtration may not occur and the measured volume loss will be less than that which occurs in the field.

## 12.7.6 Escape of water

It is also appropriate to consider whether in a full-scale duct water moving into the tendon bundle should be allowed to escape. Most standards now require that the duct and the ends of the tendon bundle are sealed. If water cannot escape from the tendon at the ends it may escape within the duct at any point where the external grout pressure is less than the internal water pressure in the tendon boundle. In principle in a duct (whether vertical or inclined), water escape may occur at any point above the point of water entry into the tendon as is shown by the following analysis.

Consider the situation shown in Figure 12.8 with a duct inclined at an angle $\theta$ to the horizontal. The pressure, $P_{gH}$ at a distance $H$ from the upper end of the duct (measured along the duct) exerted by a fluid grout (i.e. before any stiffening has occurred) is given by:

$$P_{gH} = \rho_g H g \sin \theta \tag{12.8}$$

where $\rho_g$ is the density of the grout and $g$ is the acceleration due to gravity.

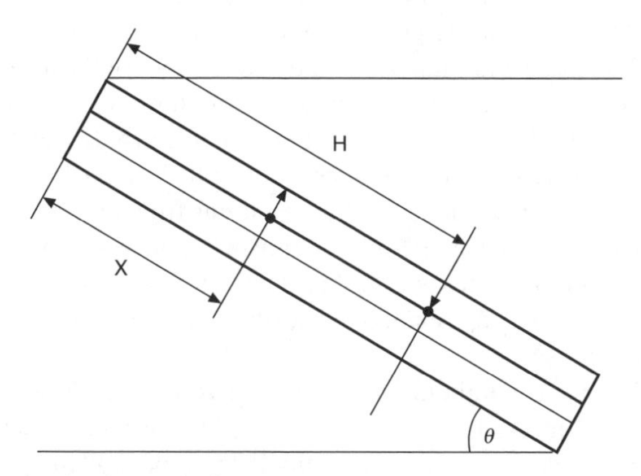


**Figure 12.8** Section of inclined grouting duct.

If water enters the tendon bundle at this distance $H$ and there is no leakage from the bundle then the pressure in the tendon bundle at this point will be equal to $P_{gH}$.

Now consider the situation if water tends to escape from the tendon bundle at a distance $x$ from the upper end of the duct. The water pressure in the tendon at this point, $P_{wx}$ will be the entry pressure less hydrostatic pressure loss in the water:

$$P_{wx} = P_{gH} - \rho_f g(H - x)g \sin \theta \tag{12.9}$$

where $\rho_f$ is the density of the fluid within the tendon bundle (the density of water if the fluid is water but see below). However, the pressure of the grout outside the duct at $x$ will be:

$$P_{gx} = \rho_g x g \sin \theta \tag{12.10}$$

Hence the differential pressure between the fluid in the tendon bundle and the grout outside is:

$$P_{wx} - P_{gx} = (\rho_g - \rho_f)(H - x)g \sin \theta \tag{12.11}$$

and thus as $\rho_g > \rho_f$ the water pressure in the tendon bundle will be greater than the external grout pressure for all $\theta > 0$ and $x < H$.

Thus water entering a tendon bundle will tend to escape at points higher up the bundle and in doing so may form a lens or void within the grout and the author has observed a 'blow hole' originating from the tendon bundle at the crown of a model duct used to simulate bridge deck grouting (similar in shape to that shown in Figure 12.9(a)).

Initially the interstices of a tendon bundle are likely to be filled with air and some water inflow will be necessary to pressure the space within the bundle. However, once the space has been pressurized the differential pressure as given by equation (12.11) may be higher than for a purely water filled space as the average density of the water and air in the duct, $\rho_f$ will be less than that for pure water $\rho_w$.

The above analysis is a demonstration of a more general concept. If any discontinuity in the duct or grout promotes segregation of water and grout with migration of the water then the pressure of this water may be sufficient to fracture the grout with the formation of a water filled void or lens. This is a demonstration of the unfortunate truth that complete filling of a duct with grout will not ensure that there are no voids. Rearrangement of the grout solids and liquid within the duct, after injection, can lead to voids. The ability of a grout to resist fracture by water pressure will increase with time as the grout takes on set. Bleed also will increase with time and the balance between the development of bleed and the development of set can be very important. Anything which changes this balance may affect the behaviour of a grout in a duct. For example grouts may behave differently in cold weather and hot weather (low temperatures will slow set and so increase bleed).

As already noted voids also will occur if water drains from the duct, for example at the ends of the tendons after grout injection has finished. All water loss from a duct after grouting produces a void or voids within the duct. The escaping water also may carry away some of the cement but for small leaks a filter cake is likely to form and clog the leak so that only water escapes.

Finally, any void or lens within the duct can lead to movement of grout after the end of injection. In inclined ducts such movement can lead to particularly severe damage. The bleed void tends to migrate upwards and eventually the grout collapses to fill the void (as shown by the arrows in Figure 12.7c and d). Slumping flow of a partially stiffened grout may lead to tearing and damage which is not healed by the continuing hydration. Similar severe damage can occur if bleed lenses develop within a vertical duct and then begin to migrate upwards through the stiffening grout.

## 12.7.7 Injectable time

The time for which a grout is injectable will depend on the properties of the grout itself and geometry of the void to be injected, the rheology of the grout and the injection equipment and whether or not the grout is kept agitated after mixing and before injection. For most applications it will be important that the injection is complete before the first injected grout at the leading edge of the flow begins to set.

Set time may be determined by penetration tests such as the Vicat needle (but at the design water content for the grout and not at the water content for standard consistence of the cement). The results of such tests may be of rather little use when considering grout injection as pumpability may be significantly impaired long before set. However, set time data may be useful when considering the stability of a grout in the void from bleeding or the wash-out of a geotechnical grout by flowing groundwater.

Many cement-based grouts, especially those with high levels of ground granulated blastfurnace slag replacement may be held for several hours, after mixing and prior to use, provided that they are kept agitated. However, they may set very rapidly once agitation ceases and this can lead to setting of the grout in the equipment if it stops for any reason such as a power failure or other malfunction. While slag-based grouts are not normally used for duct grouting they are often used in radioactive waste cementation.

### 12.7.8 Set time

Ideally, at a controllable time the grout should convert from a free-flowing liquid to a hardened solid without loss of volume or if required with a controlled increase in volume. However, few grouts achieve this ideal and cementitious grouts are not among these few.

For some grout systems and especially the chemical grouts used in geotechnical engineering the setting is sufficiently sharp that a set time may be defined simply as the time at which the grout is no longer pourable from a test tube or beaker etc. However, for cement-based grouts the determination of a limit of pourability would be very subjective and operator dependent. For these grouts a set time may be defined by a penetration test such as the Vicat needle. As already noted the results from such a test may be of little use when considering grout injection but may be useful when considering the stability of a grout in a void, e.g. from bleeding or wash-out by flowing groundwater. If penetration tests are used it is important that the grout is tested at the water/cement ratio to be used in the field and not at the water/cement ratio for standard consistence.

In principle, for all grouts, injection time and set time could be defined by the time to reach specific viscosities. Occasionally this may prove useful but most grouts are or become non-Newtonian fluids as setting approaches and the viscosity of such liquids is sensitive to shear rate, shear history and the viscometer test geometry. Also the viscometer ultimately must break up the setting gel or else it will stall. Once the setting grout has been broken up its viscosity will be meaningless as an indicator of grout behaviour. The viscosity at which break-up occurs will depend on both the test parameters and the grout. In some systems and with some grouts, very high viscosities may be achieved (especially if low shear rates are used). In other systems the limiting viscosity may be much lower. Thus if viscosity is to be used as a control parameter for the injectable time or set time then the test parameters (shear rate, shear history and test geometry) must be defined and confirmed, by prior testing, as appropriate to the grout under consideration – and procedures may have to be changed if the grout is design changed.

In summary, for cementitious grouts there is a more or less gradual transition from liquid to set material to hardened solid. It is not possible to identify unique injection, setting or hardening times. Arbitrary tests may be used to define these times but must be appropriate to the grout and the application.

## 12.8 Grout injection

There are a few basic rules that must be observed for grout injection. These include:

(a) Grouting plant should be carefully cleaned before and after use. Flakes of hardened grout from a previous mix, if dislodged from the walls of a mixer or a grout line, may accumulate in line especially at any change of diameter and cause a blockage.
(b) In-line strainers should be used to remove clogging materials/lumps in the grout. However, if used, they must be of sufficient surface area that they themselves do not clog too rapidly.
(c) Leaks in a grout line will lead to loss of pore water and deposition of a filter cake within the line. This cake may build to clog the line or if dislodged by the grout flow may block the line elsewhere.
(d) The void to be grouted should be watertight to avoid volume loss by filtration or grouting should continue until water loss by filtration has stopped. This latter procedure is difficult to apply but will build a tight filter cake around any leaks.
(e) Grout, if disturbed after taking on any initial set, will be 'cracked'. The cracks will persist in the hardened grout unless the grout is completely re-mixed after any disturbance. A grout mass may be cracked if the separation of bleed water allows the grout to slump after placement.
(f) The volume of the void to be grouted should be calculated prior to grout injection and the injected volume should be compared with this theoretical volume.
(g) Generally, flexible hoses will be used for grout lines. These lines will expand under grout pressure especially if there is a blockage in the line. Undoing a pressurized grout line will release a significant volume of grout under substantial pressure – quite enough to cause serious injury, loss of eyes, etc. Never do this or look down the end of a grout line.
(h) The duct or other void to be grouted must have vents to permit the escape of displaced air (or sometimes water). Very often it will be necessary to have more than one vent. For example, in an undulating duct it can be necessary to have a vent at each high point and at a point a little beyond it – measured in the direction of grouting.
(i) Grout will not flow back down a vent pipe. In some grouting designs it is assumed that excess grout in a vent pipe or other reservoir will flow back down into the duct to displace air voids or bleed water. This does not happen! The author has often observed ungrouted voids at the foot of vent pipes which are themselves full of grout.
(j) Cement grouts are inexpensive materials and sufficient grout should be wasted at each vent to ensure that all air is displaced from the duct. Concrete Society Technical Report 47 (Concrete Society, 2002) requires that at least 5 litres of grout shall be allowed to flow from each vent after sufficient has vented that the fluidity of the grout is, by visual assessment, equivalent to that of the grout injected. It is good practice to calculate the void volume of the duct per metre length so that the volume of grout vented can be related to the volume of duct that has been purged – venting 5 litres of grout may purge only a rather short length of duct and be insufficient to purge trapped air at any distance from the vent.
(k) The grout must not contain material with a size comparable to that of the voids to be injected (indeed the maximum grout particle size should be perhaps no more than 25% of the void size though this will depend on the application and the amount of grout to be injected at each injection point). Oversize material will not penetrate but will collect at the injection surface and clog it. Only a very small amount of oversize material is necessary to clog all the openings at a surface. For very fine voids such as fine rock in deep mine shafts treated with pure solution chemical grouts it may be necessary to centrifuge the grout to remove very fine solids.

## 12.8.1 Duct grouting

For duct grouting there are some more specific issues and it is necessary to understand how a grout can flow in a duct. At the extremes, two types of grout could be considered to fill a duct.

(1) A grout which is so stiff that it flows like toothpaste from a tube advancing along the duct on a uniform front and sweeping all air and any water in the duct before it.
(2) A very fluid grout which behaves like water. When flowing upwards it will fill all voids and displace air. However, when flowing downwards it will take the path of least resistance and will not tend to spread across the duct to fill it.

Unfortunately a grout of type 1 cannot be achieved and will not work. In a stressing duct the tendon will not be central in the duct over the length of the duct. Grout and indeed any viscous fluid will flow more rapidly in more open areas where the tendon is far from the wall (for example above the tendon near the crown of a duct where the tendon is pulled down by the stressing force) and more slowly in constricted regions (e.g. beneath the tendon near the crown). As a result flow in the open areas will overtake that in the tighter areas and may seal off sections of ungrouted duct. Grouts therefore should be designed generally of fluid rheology so as to promote filling of interstices. However, fluid rheology will cause problems in any descending sections of a grout duct (descending in the direction of grouting).

Figure 12.9(a) shows a section of a trial duct of diameter 60 mm and length 8.3 m containing three 15 mm diameter stressing strands (Jefferis, 1991b). The duct was not vented and was grouted from the left-hand end as shown in the figure using a 0.4 water/cement ratio grout. It can be seen that some air has accumulated at the injection end and has not been swept out by the grouting process despite the fact that 5 litres of grout were wasted at the discharge end of the duct. Some air has been swept along the duct as seen at cross-section 1. Air also has been trapped at the crown and in the descending limb (cross-sections 2 and 3). This is the result of grout runaway down the descending limb. As soon as the grout reaches the crown it starts to run down the descending limb eventually filling and so sealing off the second trough in the duct and leaving a trapped void in the limb above. The more fluid the grout the greater the tendency for runaway as is shown by Figure 12.9(b) which relates to a much more fluid 0.46 water/cement ratio grout. However, in the trials it was found that even the stiffest grouts injectable left some trapped air at cross-sections 2 and 3 and thus the Type 1 stiff grout approach did not work.

The only way to avoid trapping air at a crown in a duct is to vent it. Unfortunately if a single vent is used at a crown then some air will remain trapped at the crown. To achieve full grouting it is necessary to vent both the crown and the descending limb – with a vent about 400 mm after the crown. If only a single vent can be used it may be most effective if it is just after the crown as in Figure 12.9(c).

The use of gas expansive admixtures (admixtures which release gas into the grout) has been suggested as a way of improving duct filling. However, these admixtures cannot completely eliminate voids due to trapped air. The gas expansion, if it pressurizes the duct, may cause these voids to reduce in volume but the pressures are unlikely to be high enough to effectively eliminate such voids. Furthermore it should not be assumed that all the gas will remain in the grout – it may escape from the grout and form new voids (clearly admixture design must address this). Figure 12.9(d) shows a duct grouted with

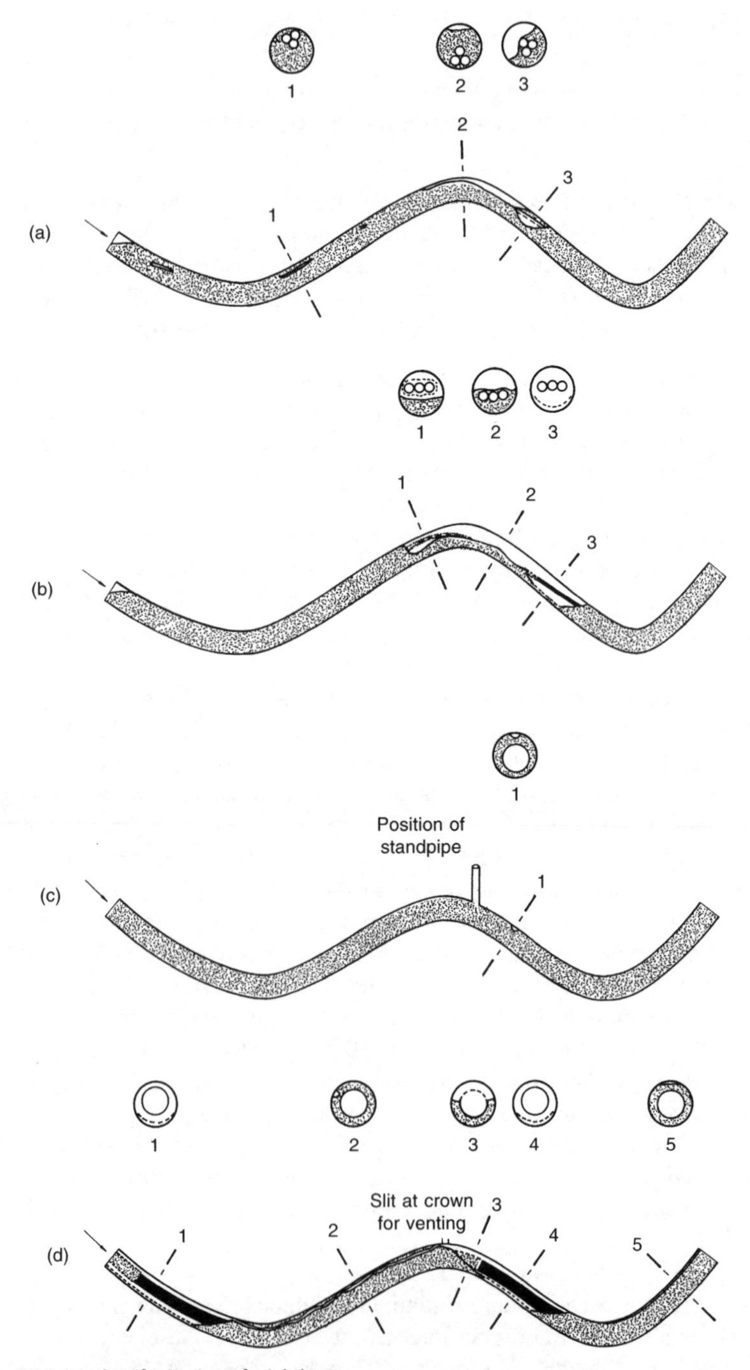


**Figure 12.9** Results of injection of trial ducts.

0.5% of an aluminium-based gas expansive admixture. There has been some escape of gas near the inlet after the completion of grouting and a channel has been blown through the grout and large voids remain in both descending limbs.

## 12.9 Volume change

As already discussed, a grout may lose solid volume by bleed and filtration prior to set. However, there also can be volume changes as a grout transforms from liquid to solid. These volume changes may result from several different mechanisms and can have a profound effect on the performance of the grout and its durability. Figure 12.10 shows various types of volume change that can occur in a grout from the fresh, just mixed state via setting to the in service state.

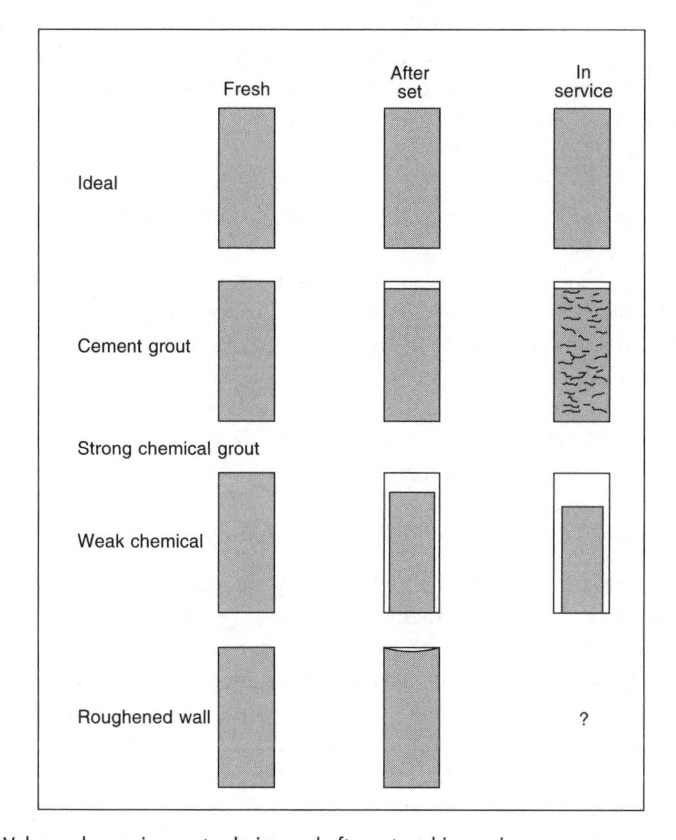


**Figure 12.10** Volume change in grouts, during and after set and in service.

For cementitious grouts there may be some loss of water via bleed and filtration prior to set so that the volume just after set is less than the as-mixed volume. Continued hydration of the cement will then lead to some further loss of volume as follows.

### 12.9.1 Hydration of Portland cement grouts

The overall volume of grout $V_G$ (i.e. that which could be determined from the external dimensions of the grout mass) will be the sum of two volumes. Thus:

$$V_G = V_S + V_P \qquad (12.12)$$

where $V_S$ is the volume of the solids and $V_P$ the volume of pores within the grout mass.

For cementitious grouts, $V_G$, $V_S$ and $V_P$ all change during the setting and hardening process. At the time setting starts $V_G$ will be equal to the original volume of liquid grout less any bleed or filtration water etc. that has been lost prior to the onset of set (for the purpose of this analysis, setting is considered to be the time at which the grout begins to behave as a solid and thus $V_G$ is definable as the volume of a solid mass). During hydration of the cement, $V_S$ will increase as a result of the reaction of cement and water and the formation of solid hydrates. On full hydration, Portland cement grains may increase in volume by the order of 50%. $V_P$ will decrease as the reaction progresses and water is removed and converted to solid hydrate products. However, the increase in solid volume is less than the volume of liquid consumed by reaction (simplistically the water taken into the solid may be considered to be in a more ordered form than in the liquid and of a higher effective density). Thus, in principle, the hydration reaction will lead to a reduction in the overall volume $V_G$. During the first stages of setting while the grout is still rather weak there may be an actual reduction in $V_G$. However, at later stages any reduction in volume will be resisted by the developing strength and the external dimensions of the grout will become effectively fixed. Thereafter there may be small changes in volume as a result of thermal expansion from the heat of hydration. There also is often a small expansion after any bleeding has ceased. This appears to be due to the increase in the solid volume tending to push the hydrating cement grains apart irrespective of the fact that the overall hydration volume change is negative (though most of the expansion will be into adjacent pore space and only a fraction of it will be manifested as overall expansion).

As the volume of water removed by hydration is greater than the swelling of the cement grains the net effect is that there will be insufficient water within the grout mass to fill the remaining pore volume, $V_P$ and as a result the pore pressure within the hydrating grout mass will drop. This reduction in pore pressure will be offset by:

(a) Reduction in $V_G$. However, as noted above as the grout gains strength changes are likely to be very small and actually there may be a slight increase in $V_G$.
(b) Water drawn in from the surroundings (if water is available). In large grout masses the rate at which water is removed by the hydration reactions may be large compared with the rate at which it can be drawn in (this will be controlled by the permeability of the set material, the flow path length and the pore pressure difference).
(c) The expansion of any gases trapped within the grout mass. For example, some air may have been entrained during the grout mixing and any gas initially sorbed on the surface of the dry cement grains may be released on wetting but remain in suspension (small gas bubbles will rise only slowly through viscous grouts and will remain trapped in grouts which show a yield value; see also the discussion of the effects of trapped air on the ultrasonic pulse velocity in fresh cement grouts given in Domone and Jefferis, 1994).
(d) The de-gassing of gases dissolved in the mix water.
(e) Air drawn in from the surroundings, though before any air can be drawn in the necessary capillary pressure to draw air into previously water filled pores or cracks must be overcome. This pressure may be substantial.
(f) The formation of water vapour within the grout mass (in a grout mass the pore pressure can easily reduce to and indeed fall below the vapour pressure of water).

The reduction in pore pressure may be sufficient to cause cracking of the grout. This, and the potential for the grout to shrink as it sets, is of particular significance in the grouting of gas wells. At the time of setting the reduced hydrostatic pressure and the tendency for the grout to shrink may allow the formation of gas flow channels which will not heal as set progresses. It is important to note that if cracks form in a grout as a result of negative pore pressure or slumping of the grout to fill voids caused by bleeding or filtration they will not be healed by hydration and will persist in the final set material.

### *12.9.1.1 Pore pressure in a setting grout*

Figure 12.11 shows the pore pressure in a small sample of grout. The initial reduction in pore pressure for this sample up to about 1 hour is as a result of consolidation of the sample with settlement of the solids and expulsion of water until the solids begin to come into contact and support some of the weight of material above. Thereafter there is a sleep reduction in pore pressure as a result of the volumetric shrinkage on hydration. The pore pressure can drop to about the vapour pressure of water (i.e. about –1 atmosphere gauge pressure or lower if the water in the void does not cavitate and 'boil'). In practice cracking of the grout often prevents the pressure dropping to this level, cracking can be seen as sharp rises in pore pressure often followed by a further reduction in pressure as hydration continues – it would seem that although cracks can relieve the negative pressures not all such cracks reach a free surface and maintain a connection to it. This is particularly the case for locations deeper within a grout mass.

In environments where the setting grout is under a high hydrostatic pressure such as

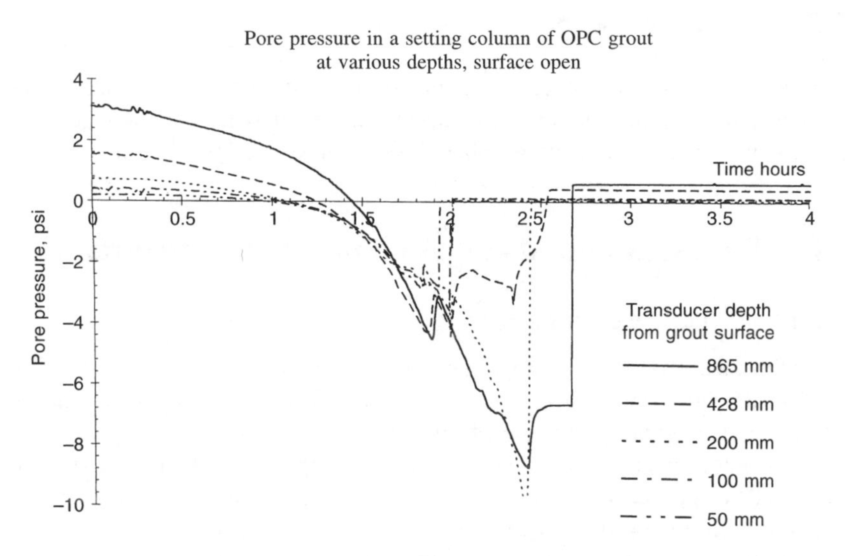


**Figure 12.11** Pore pressure development in an unconfined sample of setting grout, from Holmes, 1995.

deep in the grout column when grouting an oil well, the differential pressure can be much higher. Figure 12.12 shows the pore pressure for three grout samples subject to an initial confining pressure of 200 psi. It can be seen that the pore pressure drops to about –1 atmosphere gauge but the differential pressure will now be about 201 psi whereas in an unconfined grout the maximum pressure differential would be about 1 atmosphere (14.7 psi) and thus with much reduced potential for cracking.

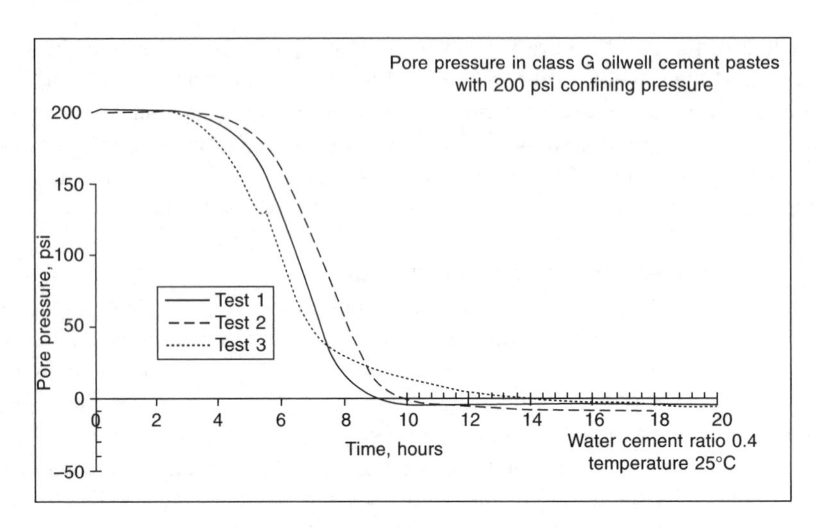


**Figure 12.12** Pore pressure development in a confined grout sample, from Holmes (1995).

### 12.9.2 Drying shrinkage

In addition to hydration shrinkage (and in chemical grouts syneresis, see section 12.13), grouts may show significant shrinkage on drying and this may lead to cracking. Cracking should be considered in applications where a grout may be allowed to dry.

## 12.10 The hardened state of cementitious grouts

In the hardened state the grout should:

(a) Have the necessary physical properties, e.g. low permeability, high strength, stiffness etc. The development of these properties may take from hours to weeks from the time of mixing depending on the grout formulation.
(b) Have chemical and physical properties that are compatible with the void system to be grouted (for example the chemistry of a soil) or have appropriate chemistry to protect material within the grout, for example pre-stressing tendons.
(c) Endure for the design life of the grouted structure without loss of performance, reduction of solid volume by syneresis (see section 12.13), shrinkage or chemical reactions or unacceptable expansion.

The relative importance of each of the above properties will depend on the particular application of the grout and, as indicated in the introduction to this chapter, these are very diverse.

The hardened properties of cementitious grouts have been the subject of much research (Domone and Jefferis, 1994).

## 12.11 Durability

Durability must be expected to become an increasingly important issue especially where grouts are used to solidify toxic or hazardous wastes or to contain contaminated sites. However, it should not be forgotten that water is a very powerful solvent and the dissolving power of water for some grout constituents may significantly exceed the reaction power of many contaminants in the water. Also the contaminants that are most damaging to human health, the environment etc. may not be the most damaging to a grout. Furthermore the concentrations of many contaminants in toxic materials may be so low as to be insignificant – to the grout, though not to man and the environment.

Immersion tests (immersion of specimens of the grout in aggressive solutions) can be very useful in identifying severely damaging reactions such as sulphate attack and results may be available after only a few weeks to months of immersion. If there is a reaction between the grout and the immersion fluid it will be necessary to replace this fluid at regular intervals (for example, one litre of a fully dissociated acid of pH 3 can be neutralized by 37 mg of lime, calcium hydroxide, from the cement in a grout. A single grout specimen may contain tens of grams of lime and thus be capable of neutralizing hundreds of litres of such acid).

However, it should be remembered that in such tests the sample will not be subject to any confining stress. For low strength chemical and cementitious grout systems this may lead to an overestimation of the effects of contaminants etc. To obtain a fuller understanding of the grout behaviour it may be necessary to permeate the grout with the test solution under confined conditions. However, short-term permeation tests will not suffice, sufficient of the test solution must be permeated through the grout to ensure that full reaction can take place between the grout and test fluid. Unfortunately with low permeability grouts this can take years or even hundreds of years – in which case it may be simpler to demonstrate that no unacceptable damage will occur within the lifetime of the grouted element.

Finally it is important to recognize that the injection of a grout can change the local chemical environment, for example by increasing the pH which is an important benefit in preventing tendon corrosion in structural grouting but can cause problems in geotechnical grouting, for example the release of ammonia from nitrogenous soils (Bracegirdle and Jefferis, 2001). Geotechnical grouting also can prevent air ingress to soil so that an environment that may have been previously aerobic becomes anaerobic (Jefferis, 2003).

## 12.12 Chemical grout systems – materials issues

The setting of grouts may be based on many different chemical reactions involving materials such as Portland cements, sodium silicate, sodium aluminate, polyacrylamides, polyacrylates, polyurethanes and a wide range of resins. Fundamentally, any system which is initially liquid and sets to a solid could be employed as a grout. However, as grouts generally will be used in large quantities the balance of cost and performance can be a prime consideration (though too often just raw material cost is considered and not performance). For any grout mix the grout designer must seek to ensure that:

(a) The grout mix proportions are reasonable. For cement grouts it is well known that careful selection of water/cement ratio is crucial for effective grouting. However, it seems less well appreciated that other grout systems must be designed and that concentrations of hardeners etc. cannot be varied at will. For example in some recent papers on silicate–ester grouts the degree of neutralization of the sodium hydroxide in the sodium silicate appears not to have been considered and some quite inappropriate formulations have been tested and not surprisingly have been found to be of poor durability (note sodium silicate can be regarded as a solution of silica in sodium hydroxide and the silica will precipitate if the sodium hydroxide is neutralized).

(b) The chemistry of the grout is compatible with that of the void to be grouted both for the fluid grout and the setting and hardened products. In normal soils problems of chemical incompatibility are rather rare but may occur in saline soils or other chemically active systems. Compatibility may be much more of an issue in contaminated soils, which are becoming increasingly a focus of attention.

(c) The durability of the grout is satisfactory. Many chemical grouts are used in short-term or temporary works and long-term mechanical performance is not an issue.

(d) The grout during preparation and injection does not pose health and safety issues for the grouting personnel or any others who may be affected by the works. The grout also must not pose unacceptable risks to the environment and especially groundwater.

(e) The conditions noted in (d) must apply not only during injection but also in service and to any degradation products from the grout or reaction by-products from the hardening process etc.

## 12.12.1 Fluid properties

Most chemical grouts will be placed by injection and thus viscosity and yield value (if the grout is non-Newtonian) will be important design parameters. The development of the grout rheology with time also will be important – if large volumes are to be placed the injection process may take some time. Ideally the grout must:

- be of low viscosity and low yield value so as to enable penetration of fine pores fissures etc. (except for specialist grouts such as compensation grouts which may be designed so as not to flow significantly from the point of injection)
- be of unchanging rheology during the injection period
- completely fill the void to be grouted. This may be more of an issue of grouting procedure than grout materials but the grout may have to be tailored for the injection equipment, the pumping distance and the volume of grout to be injected etc.
- remain in the void after placement and not be washed out, for example by groundwater flow or tend to migrate downward under gravity. Thixotropic grouts which thicken as the shear rate reduces (independently of set) may be advantageous
- not be subject to segregation of the mix constituents as a result of density differences (e.g. settlement of solids, bleed or the separation of immiscible liquid phases)
- not be subject to separation as a result of differential sorption of the chemical constituents on the solid surfaces (this can happen with silicate grouts hardened with esters as the components are immiscible and surface active agents may need to be incorporated in the formulation to reduce the potential for separation).

The addition of surface active agents (e.g. detergent type materials which markedly reduce the surface tension of the mix water) also may enhance the penetration of fine interstices and improve the adhesion of the grout to solid surfaces.

## 12.12.2 Setting

The grout should set at a defined time which should be adjustable by changing the mix formulation or by the addition of an accelerating or retarding admixture. Ideally at the set time the grout should rapidly convert from a liquid to a hardened solid without loss of volume or if required with a controlled increase in volume. However, for most practical grouts there is a gradual transition from liquid to set material to hardened solid and it is not possible to identify unique setting or hardening times.

## 12.12.3 Measurement of set time

As discussed for cement grouts, set time may be determined simply as the time at which the grout is no longer pourable from a test tube or beaker etc. For some chemical grouts this is quite sufficient and the measured time will be little influenced by the dimensions of the tube, beaker, etc. For cement-based grouts the determination of a limit of pourability would be very subjective and operator dependent.

Sometimes set time is specified as the time to reach a given viscosity, e.g. 100 cP (i.e. 100 times that of water). However, it is not always appreciated that a set time defined in this way will be sensitive to the type of viscometer and the shear rate used for the test. For example, in a rotational viscometer the developing grout structure may be substantially destroyed by the shearing so that a viscosity of 100 cP is never achieved even after set as the material becomes a fragmented mush and does not develop into a continuous gel system.

An empirical test for the set of silicate–aluminate grouts involves drawing a wire across the surface. The set time may be defined as the time at which the furrow ploughed by the wire persists for just a moment. This marks the time at which the grout transforms from an effectively Newtonian liquid to a gel. The test cannot be used with silicate–ester systems as the liquids are immiscible and the ester will float on the surface and obscure the setting silicate.

In summary it is important to realize that set is not a uniquely definable condition. The measured set time will depend on the procedure used and the nature of the setting reaction.

## 12.12.4 Injectable time

In addition to setting and hardening times it also may be necessary to consider the time for which the grout is injectable, as a grout may become uninjectable prior to set. The injection time will depend not only on the viscosity-time relationship (or strictly the change of rheology with time) for the grout but also on the size of the voids to be grouted.

The quantity of grout to be injected also will be an issue (as for cement grouts, in many applications it will be desirable that the injection is complete before the first injected grout at the leading edge of the flow begins to set or becomes uninjectable).

## 12.12.5 Injected quantity

For geotechnical grouting it is common practice to use multiple stages of grouting. The distance betwen grout injection points must be related to the penetration that can be achieved with the grout. However, a continuous grout mass is seldom achieved with a single set of injection holes. Typically secondary points must be injected between the primaries and tertiaries between the secondaries if a grouted barrier is to be developed. With suspension grouts such as cement grouts or clay–cement grouts at each stage of grouting the injection may be continued until rejection; that is no further grout penetration can be achieved at the design grouting pressure (this will depend on grouting procedure and potential for hydrofracture etc.). With low viscosity chemical grouts grout rejection is unlikely to occur and grouting must be controlled by volume. At each grouting point a calculated volume of grout must be injected.

## 12.12.6 Geotechnical grout penetration

The penetration of grouts into soils is a fundamental issue and is influenced by parameters including:

- the pore size distribution of the soil
- the spacing of the grout injection points
- the rheology and set time of the grout at the mixing and ground temperature
- the injection pressure that can be applied (high pressures may lead to hydraulic fracture, especially for shallow injection points)
- the grouting plant, procedure etc.

As a general guideline, the following soil types may be injected with grouts (Karol, 1990):

| **Grout** | **Dominant soil type** | **Dominant particle size** range |
|---|---|---|
| Cement grouts | Coarse sand | (2 mm to 6 mm) |
| Silicates (high concentration) | Medium sand | (600 microns to 2 mm) |
| Polyurethanes | Medium sand | (600 microns to 2 mm) |
| Silicates (low concentration) | Fine sand | (200–600 microns) |
| Acrylates | Coarse silt | (60–200 microns) |
| Acrylamides | Medium silt | (20–60 microns) |

Note that the dominant soil size is typically taken as the $D_{10}$ size, that is the diameter below which 10% of the soil mass lies.

## 12.13 Volume change in chemical grouts

The change of state as a grout transforms from liquid to solid may involve a change in volume. Most grouts, when set, will still contain most of the mix water and for the set product to be a solid this water must have been trapped in or become part of the solid phase. Thus, in the solid, at least some of the water will be in a more ordered state than in the liquid phase and thus possibly of a higher density (i.e. lower volume). The mechanics of the volume change on setting and the theoretical procedures required to analyse the change will depend on the nature of the setting reaction.

It is not possible to consider all the volume change mechanisms that may occur in chemical grouts. However, one process, syneresis merits comment. Syneresis is a general term for processes, other than gravity driven separation of solids, occurring within the grout mass which cause a reduction in the solid volume. The processes may be chemical or physical or a combination of both and may lead to shrinkage of the external volume of the grout or the opening of fissures within the grout. The separation of curds and whey in the first stages of cheese making could be described as syneresis. Depending on the chemical system, syneresis may be complete in few days or continue for many years and in either case can lead to a very substantial loss of solid volume. Figure 12.10 shows some possible volume change scenarios for different grout types.

For a chemical grout the volume of grout solid $V_S$ and the volume of fluid filled pores is $V_P$ can be related to the initial volume of the grout mix, $V_M$ as follows:

$$V_M = V_S + V_P \tag{12.13}$$

In practice, for chemical grouts, the distinction between the volume of solids $V_S$ and the volume of pores $V_P$ may be difficult. Thus it may be necessary to treat the grout as a homogeneous system of volume $V_G$ despite the fact that most chemical grouts contain much more liquid (almost always water though this may contain some dissolved salts) than solids so that:

$$V_P >> V_S \tag{12.14}$$

The solids are likely to be present as three-dimensional molecular structures involving linkages such as silicon–oxygen bonds. There also may be some larger water filled voids between molecular units. Water is fundamental to the structure of many chemical grout gels and if it is lost by drying the structure will collapse and may not swell again on re-wetting. It is therefore more convenient to consider the combined volume $V_G$ of grout solids and pore fluid:

$$V_G = V_S + V_P \tag{12.15}$$

The volume of grout $V_G$ may not be equal to the initial volume of mix $V_M$ as a result of syneresis during and after setting. For the analysis of syneresis it is necessary to introduce a volume $V_{E,t}$ which is the time-dependent volume of fluid exuded from the grout by syneresis (this parameter is not needed for cement grouts, see section 12.9.1), so that:

$$V_{G,t} = V_M - V_{E,t} \tag{12.16}$$

The volume of the set grout is therefore also time-dependent. Syneresis is the tightening of the structure with time, independent of any gravitational settlement of the grout solids or any drying. For some chemical grouts (e.g. dilute silicate grouts hardened with an ester) the syneresis shrinkage may occur during or soon after the setting so that at the time of set the volume of solid grout $V_{G,tset}$ may be less than 50% of the original volume of the mix constituents $V_M = V_{G,t=0}$ (where $t = 0$ is the time of mixing of the grout components). The solids may not settle to the bottom of the mould but may stick to parts of the solid surface of the mould so that the grout mass has a complex three-dimensional shape.

If there is a large initial volume loss, further syneresis may be quite small and for grouts showing such a loss $V_{G,t>tset}$ may be relatively insensitive to time, $t$ after set (i.e. the change in $V_{E,t>tset}$ may be small).

In contrast in some other grouts for example, silicate–aluminate grouts after setting the volume of gel is equal to the initial volume of the grout reactants, i.e. $V_{G,tset} = V_M$. However, syneresis develops with time and these grouts can show a shrinkage of more than 50% at age 5 years (the increase in syneresis being approximately linear with the logarithm of time). The volume loss is a three-dimensional shrinkage (with exudation of a matching volume of liquid). However, if tall samples are tested, e.g. in a 250 ml measuring cylinder, as the gels are soft they may slump so that only the near surface region shows the three-dimensional shrinkage (manifested as a tapering top to the grout column (Sheikhbahai, 1989). The relationship between the volume $V_{G,t}$ of the gel and the syneresis fluid $V_{E,t}$ at any time $t$ will be:

$$V_{G,t} = V_M - V_{E,t} \tag{12.17}$$

For grouts showing syneresis $V_M$ is effectively constant and is equal to the volume of grout at zero time, i.e. at the time of mixing of the ingredients. For cement grouts it is $V_G$ that is effectively constant after set.

In large test samples, the gel, if subject to syneresis, usually shrinks as a single mass and exudes excess fluid. However, as discussed below, if the shrinkage is inhibited it is possible that the syneresis fluid may remain within the grout mass so that there is no change in external volume or that the syneresis itself may be totally inhibited.

The above comments relate to relatively large samples of grout (of order 100 ml to 1 litre) tested as bulk materials. However, in geotechnical practice, the grout will be distributed throughout the pores of the soil system and the maximum dimension of any element of grout within a soil pore is likely to be in the range from a few tens of microns to a few millimetres (chemical grouts are unlikely to be used in soils coarser than sands for which cement or clay–cement grouts will be used).

As already noted, in some chemical grouts such as silicate–ester systems the bulk of any syneresis may be apparently complete within a few minutes/hours from mixing. However, in silicate–aluminate grouts syneresis continues for a much longer time. Clearly syneresis occurring as it does during and after grout set could substantially damage the strength and permeability of a grouted soil. It is therefore most important to know whether syneresis will occur in the small voids of a chemically grouted soil (chemical grouts generally will be used only in finer soils). It seems likely that the reduction of volume as a result of syneresis is proportional to the initial volume of grout. Thus grout samples of different sizes should all show the same percentage of syneresis at the same time after mixing. For example if the syneresis measured on a large sample in the laboratory is 20% then 20% syneresis might be expected in the pores of a soil. However, the boundary conditions also must be considered. In the laboratory, the long-term syneresis of approximately 0.5 litre samples of silicate–aluminate grouts was inhibited for 5 years (readings were stopped after 5 years) by gluing a coating of sand to the walls of the grout container prior to filling it with the grout (see Figure 12.10). It would thus appear that syneresis can be inhibited by bonding (this may be chemical or physical). If bonding can inhibit syneresis in large samples it should be even more effective in small samples as the surface area (bonding area) to volume (shrinkage generating) ratio is proportional to 1/(pore diameter). However, the situation may be yet more complex (Sheikhbahai and Jefferis, 1995). Bonding may inhibit the reduction in external volume but internal re-arrangements still may occur. In particular the gel may become locally densified with compensating regions of pure pore fluid remaining within the gel so that the external dimensions remain unchanged but internally it begins to open up. This could lead to a reduction in strength, an increase in permeability and perhaps a greater sensitivity to chemical attack. This is an area that would seem to need further work if chemical grouts are to be accepted for long-term soil modification.

### 12.13.1 Drying shrinkage

In addition to hydration shrinkage and syneresis chemical grouts will show shrinkage if subjected to drying conditions. This may be rather unlikely to occur in the ground but can be an issue for grouts at or near the soil surface and in laboratory tests samples. In general the grout, as it dries, will shrink back towards the soil–grain contacts. There will be little shrinkage of the soil skeleton as prior to grouting the grains will have been in contact and so will have little potential for further tightening. Drying of the grout phase will lead to an increase in the strength of the residual grout material and as this will be at the grain contacts the effect of drying is likely to be an increase in strength of the grouted soil. Grouted soil samples therefore should be tested for strength when wet and should not be allowed to dry before testing.

## 12.14 The hardened state for chemical grouts

The parameters listed in section 12.8 on cementitious grouts also need to be considered for geotechnical grouts. Thus the grout should:

- have the necessary physical properties, e.g. permeability, strength, stiffness etc. The development of these properties may take from hours to weeks from the time of mixing depending on the chemistry of the grout system
- have chemical and physical properties that are compatible with the void system to be grouted (for example the chemistry of a soil) or have appropriate chemistry to protect or encapsulate the material within the grout for example solidified toxic waste
- endure for the design life of the grouted system without loss of performance, reduction of solid volume by syneresis, shrinkage or chemical reactions or expansion (NB design life can be very difficult to establish)

The relative importance of each of the above properties will depend on the particular application and these are very diverse. The majority of applications of chemical grouts will be in geotechnical grouting for strengthening and/or water tightening soils. They also may be used as part of compensation grouting schemes (the injection of grouts to compensate for ground movements/settlements, e.g. those caused by tunnelling works). Chemical grouts can be used for water tightening masonry and stone in buildings where there is no damp-proof membrane or the membrane has failed. Solidification/stabilization with cement and chemical grants can be used to reduce the risks associated with toxic, hazardous and radioactive wastes.

## 12.14.1 Strength

The intrinsic strength of a set grout (i.e. the strength of the grout alone without any soil) may range from a few kPa for silicate–aluminate grouts to over 100 MPa for some cement grouts. For low strength grouts the strength of grouted soil is usually higher than that of the grout alone. Strength measurements for grouted soils are time consuming as it will normally be necessary to prepare samples by injection into specially prepared moulds. Simple mixing and compaction of grout and soil into moulds is seldom entirely satisfactory.

It follows that it is useful to have a guide to the strength of grouted soils and this can be obtained from measurements of the strength of the pure grout (perhaps with a hand vane) and then assuming that the strength of the chemically grouted soil is directly related to the strength of the grout. Thus:

$$f_S = B \cdot f_G \tag{12.18}$$

where $f_S$ is the strength of the grouted soil and $f_G$ that of the grout alone. Note this procedure is not appropriate for high-strength cement grouts where the matrix may have only modest influence on the properties of the grouted material.

$B$ will be a function of the fineness of the soil (generally it will be higher for finer soils) and the density of the soil (denser soils generally will give higher strengths). The strength of the grout itself, and the shape, surface roughness and surface chemistry of the soil particles are also likely to influence $B$. Values of $B$ may be in the range perhaps 5 to 50. The above equation should be applied with caution and regarded merely as an indicator and not a predictor of grout behaviour. $B$ also may vary with time. For better estimates of strength, $B$ can be calibrated for individual soils and grouts.

It should be noted that rather little work has been done on the effects of water in the voids of a soil. Much of the reported laboratory work having been done with grouts injected into dry soils. In the field most soils will be water saturated or at least water wet. More *in-situ* data on strength and stiffness of grouted materials would be useful.

## 12.14.2 Permeability

Permeability will be important for two reasons:

- the purpose of the grout may be to reduce the permeability of the grouted soils or rock etc.
- the flow of groundwater through a grout may leach materials from the grout and slowly destroy it

Pure clay–cement grouts may have a permeability of perhaps $10^{-11}$ to $10^{-7}$ m/s and pure cement grouts may achieve permeabilities below $10^{-12}$ m/s. However, grouted soils are likely to be much more permeable due to imperfections in the grouting of the soil and it should not be assumed, unless justified by field tests, that *in-situ* permeabilities less than about $10^{-7}$ m/s can be achieved even if laboratory tests on grouted soils show much lower values.

Rather little work has been done on the permeability of grouted soils. This may be because there are severe practical problems with permeability tests. In a well grouted laboratory soil sample the permeability may be effectively zero (perhaps $< 10^{-10}$ m/s). However, a small defect in the grouting may lead to a permeability increase of many orders of magnitude and it is difficult to simulate the imperfections in the grouting that occur in the field in the laboratory. More data on the *in-situ* performance of grouts is needed but it is difficult to obtain.

Syneresis and shrinkage may affect the permeability of a grouted soil. However, for some grouts, it may be many years before the effects of syneresis are manifested and short-term permeability testing may identify problems only in grouts which show high initial syneresis.

## 12.14.3 Durability

The grout should be physically stable, i.e. there should be no significant loss of volume, strength etc. during the service life.

The grout also should be compatible with the chemical environment of the void (see the discussion on the durability of cement grouts at section 12.11). Very often the chemistry of the groundwater will be relatively unimportant but the leaching action of any flowing groundwater may be very damaging. Permeable gels or poorly grouted soils may lead to rapid erosion. For example there are literature reports of large quantities of lime being leached from cement and clay–cement grouts, presumably with much damage to the grout.

Long-term tests are necessary to investigate the effects of leaching or reactive groundwaters on grout properties. In general the grout will be exposed to the flow of water and chemicals only from one face and so a leaching/reaction front will move through the grout changing its properties as leaching/reaction occurs. For a grouted barrier there may be little observable effect on the overall permeability of the system until this front has passed through almost the entire length of the flow path through the grout. For example if a very damaging reaction front has penetrated halfway through a grouted zone so that the first half of the material is effectively of infinite permeability then the overall permeability of the zone will have increased only by a factor of 2 – an increase which could pass unnoticed if the unreacted part of the grout showed a slight reduction of permeability with time as often occurs. A full analysis of grout interactions requires advection-diffusion-reaction modelling (see Jefferis 2000, 2003) for a brief introduction.

In practice there are few data on the long-term behaviour of grouts and once again it would be most useful if more *in-situ* data could be obtained – especially for grouts which have been in the ground for many years. Such data could be obtained when extending/ repairing existing grouted structures at relative modest cost if programmed into the works.

## 12.15 Chemical and geotechnical grout materials

The discussion thus far has distinguished two types of grout: cement-based grouts typically used for structural applications such as duct grouting or offshore work and chemical grouts which are mainly used in geotechnical grouting. However, there is a large class of grouts which lies intermediate between these two types – clay and clay–cement grouts. Pure clay grouts are seldom used as they are non-setting but if used may behave rather as chemical grouts (for example in respect of strength and syneresis). Clay–cement grouts are very widely used in geotechnical engineering as they are of relatively low cost and simple to prepare. Their properties will typically follow those of pure cement grouts but they are generally of much lower strength (typically of the order a few hundred kPa to a few MPa).

### 12.15.1 Active clays

Simple suspensions of active clays such as sodium smectite (bentonite) in water may be used as grouts. There is no setting reaction but as the suspensions are thixotropic there is some increase of strength once injection is complete and flow has stopped. However, the strength of the grout is unlikely to exceed a few tens of Pa. Furthermore the suspensions will be dilute, perhaps 5 to 10% solids, and thus easily eroded. Ion exchange with cations present in the groundwater may also lead to changes of interparticle bonding for the clay and loss of structural integrity. Clay grouts are therefore unlikely to be used except as temporary systems and so find little application. However, clays find considerable application as fillers and bleed control agents in mixed grout systems, e.g. clay–cement and clay–cement–silicate grouts where they will also be involved in clay-alkali reactions.

### 12.15.2 Cement grouts

Cement grouts for geotechnical grouting are often used at much higher water/cement ratios than those for structural grouting. Indeed there was a fashion for the use of little more than dirty water with water/cement ratios of up to 20. The theory being that the water would wash the cement into the pores to be grouted, particularly rock fissures and that once the pores had been blocked a tighter cement system would be deposited by filtration. Houlsby (Houlsby 1982; Houlsby 1990) discusses the use of high water/cement ratio grouts. Such grouts are likely to be markedly less durable than low water/cement structural grouts.

### 12.15.3 Clay–cement grouts

In principle any clay may be used in a clay–cement grout though it may be difficult to achieve both low viscosity and good bleed control with native clays, and design trials almost always will be necessary when using such clays. As a result, in practice, processed sodium bentonite is the most widely used clay in grouts though other clays have been used where convenient deposits of native materials are available.

The particular feature of sodium bentonites is that they disperse to fine particle size in water. The fine size gives suspensions of the clay substantial colloidal activity and also thixotropy (developing a reversible gel on standing and showing shear thinning with increasing shear rate). At suspension concentrations greater than about 3% (though this will depend on the source of the clay) the suspensions are non-settling.

Dispersion of bentonite in water can take some time. On contact with water the particles swell and once sufficiently swollen the clusters of particles break up. The dispersion appears to develop, and continue to develop, with time. Thus a time to full dispersion (hydration) cannot be identified – dispersion is an asymptotic process see (Jefferis, in Hayward Baker, 1982). Typically hydration times of a few hours will be sufficient. Addition of cement to a sodium bentonite adds calcium ions to the system which can exchange with the sodium ions on the clay. This exchange leads to flocculation and aggregation of the clay (the inverse of dispersion) and thus limits further dispersion.

The behaviour of bentonite cement grouts will be dependent on the mix proportions and the mixing procedure. If clay and cement are added to the mixer at the same time there may be little dispersion of the clay (even if high shear mixing or chemical additives are employed). Thus relatively high solids contents will be necessary to produce stable non-bleeding suspensions. Typically the design procedure seems to be to select a bentonite/cement ratio (this may vary from bentonite with a little cement to cement with a few per cent bentonite) and then to add sufficient water to achieve an acceptable viscosity. Often mixes of an acceptable viscosity will be unstable and show much bleed – especially with low shear mixes. However, this may pass unnoticed on site if the grout is immediately injected into the ground and no test samples are taken.

If hydrated bentonite suspensions are used (at least a few hours hydration), the clay content typically will be in the range 3 to 10% by weight of water and the cement content perhaps 100 to 350 kg/m$^3$ of grout. Additions such as pulverized-fuel ash (pfa) and ground granulated blastfurnace slag may be included in the grout; high proportions of the latter are commonly used in cut-off wall slurries (ICE, 1999).

For void filling – for instance old mine workings – pfa-cement grouts may be used with little or no clay and pfa/cement ratios of up to about 20.

## 12.15.4 Microfine cement grouts

Microfine cement grouts are a relatively recent addition to the range of grouting materials. Microfine cement typically will have almost 100% of particles finer than about 15 microns. They may be produced as microfine powders by special milling processes during manufacture or by wet milling cement/water mixes prior to use – the latter process requiring specialist plant on site. Microfine cements are likely to be used at water cement ratios of perhaps 1 to 2. The use of a retarding admixture may be essential. Microfine cement grouts will have strengths comparable to those of the conventional cement grouts used in geotechnical engineering but the injectability of chemical grouts.

## 12.15.5 Chemical grouts

There is a very wide range of chemical grouts available and it is not possible to review the features of all the potential systems and thus only some of the more widely used systems are presented. In terms of quantity of grout used silicate-based systems are the most widely used chemical grouts.

#### 12.15.5.1 Silicate–ester grouts

Sodium silicate is effectively a solution of silica in sodium hydroxide. If the sodium hydroxide is neutralized with an acidic material the silica will precipitate. The form of the precipitated silica is influenced by the nature of the neutralizing material. Silica also may be precipitated by the addition of strong electrolytes (e.g. concentrated solutions of calcium chloride). Indeed, the first chemical grouting process, the Joosten process developed in 1925, involved the separate injection of sodium silicate and calcium chloride so that they mixed and gelled in the ground. Separate injection is costly and may leave much silicate ungelled. Modern grouts are therefore designed to have a delayed set so that only single injection points are necessary. For example, ethyl and methyl esters of adipic, glutaric and succinic acids which are slowly hydrolyzed by water to release acid and an alcohol are convenient materials to slowly neutralize the sodium hydroxide in sodium silicate and thus give a delayed set. The quantity of ester required is fixed by the degree of neutralization required (typically for a durable grout at least 60% of the sodium hydroxide in the sodium silicate must be neutralized). The rate of setting is controlled by selecting an appropriate ester (or blend of esters). There is a common misconception, reinforced by some current literature, that the rate of setting can be controlled by varying the sodium silicate/ester ratio. The rate of reaction will be somewhat sensitive to the concentration of the reactants but the production of a good quality hardened grout requires that the reactants are used in the correct proportions. Thus varying the set rate of a silicate–ester grout by varying the relative proportions of silicate and ester may either produce a poorly durable product due to insufficient neutralization of alkali or be wasteful of the ester (which is the more expensive component). Sodium silicate–ester grouts can give strengths from a few hundred kPa to a few MPa with set times from about 10 minutes to 2 hours. Initial grout solution viscosities may be of order 5 to 40 cP. Strength and viscosity will be very sensitive to the concentration of silicate in the grout. It is normal practice to dilute the sodium silicate with perhaps 30 to 60% water (as supplied the sodium silicate could have a silica content of 26 to 30% with a silica to soda ratio of 2.9 to 3).

#### 12.15.5.2 Silicate–aluminate grouts

Sodium silicate solution also may be set by the addition of small amounts of sodium aluminate. However, only dilute solutions can be used. If concentrated silicate is added to aluminate or vice versa there is an immediate deposition of gel and the product is uninjectable. However, if the silicate and aluminate are diluted and continuously stirred while intermixed, a delayed set can be achieved with set times controllable from a few minutes to perhaps 2 hours. However, the gels are weak with strengths up to a few hundred kPa and are best suited to water tightening rather than strengthening soils.

#### 12.15.5.3 Acrylamide and acrylate grouts

In the 1950s a grout based on a mixture of 95% acrylamide and 5% methylene bis-acrylamide was developed. This grout had a viscosity comparable to that of water and an almost flat viscosity–time characteristic up to the point of set. However, there were concerns that unpolymerized acrylamides could be neurotoxic and manufacture was discontinued. Subsequently acrylate-based systems have been developed of lower or minimal toxicity.

#### *12.15.5.4 Polyurethanes*

Polyurethane-based grouts are a distinct class of grouts requiring different injection procedures as they are water reactive rather than water based. The basic reaction involves a polyisocyanate and a polyether or polyester etc. The grouts are relatively viscous and thus they should have poor penetration. However, the reaction with water produces carbon dioxide and pressures up to 400 psi have been reported and this leads to substantial secondary penetration of the grout and it is reported that the grouts can be used for soils down to silt size. Polyurethanes are flammable and this has reduced their application in tunnelling and other sensitive environments.

#### *12.15.5.5 Other grout systems*

Other grout systems include:

- resins (epoxy, polyester, furaic and phenolic)
- bitumen emulsions (stabilized suspension of colloidal sized droplets of bitumen in water). Emulsions will have a lower viscosity than the parent bitumen. The emulsion will break if the water is lost, e.g. to a soil to leave the parent bitumen
- latex emulsions
- lignin and lignin derivatives

Details of these systems can be found in texts on chemical grouting such as Karol (1990).

## 12.16 Conclusions

This chapter is intended to be an introduction to some of the major features of cement and chemical grouts and in particular to demonstrate that they are neither just fluids that will set to fill any void into which they have been injected nor concrete without the aggregate. In particular, the volume changes that occur prior to and during setting and hardening of the grout are complex and their significance will depend on the form of the void to be grouted.

At the present time there is a lamentable lack of advice on the design of grouts and there are many different empirical test methods. This makes the exchange of test data etc. difficult.

Further work is needed on the *in-situ* behaviour of grouts and particularly their durability in contaminated soils and groundwater.

Finally grout specifications must be appropriate to the grout material. A specification drafted for a chemical grout may be wholly inappropriate to a cement grout.

## References

Bell, A. (ed) (1994) Grouting in the Ground, Proceedings of the Institution of Civil Engineers Conference, London.

Bensted, J. (1997) Special Cements in *Lea's Chemistry of Cement and Concrete,* Hewlett, P.C. (ed.), Arnold, London.

Borden, R.H., Holtz, R.D. and Juran, I. (1992) Grouting, soil improvement and geosynthetics, Proceedings of the American Society of Civil Engineers Specialty Conference, New Orleans.

Bowen, R. (1981) Grouting in engineering practice, 2 Ed, Applied Science Publishers.

Bracegirdle, A. and Jefferis, S.A., (2001) Heat and ammonia associated with jet-grouting in marine clay, Third British Geotechnical Association Conference on Environmental Geotechnics, Edinburgh.

BS 3892, Part 3 BSI (1997) Specification for pulverized fuel ash for use in cementitious grouts.

BS EN 445 : BSI (1996a) Grout for prestressing tendons – Test methods.

BS EN 446 : BSI (1996b) Grout for prestressing tendons – Grouting procedures.

BS EN 447 : BSI (1996c) Grout for prestressing tendons – Specification for common grout.

Cambefort, H. (1964) *Injection des Sols, Tome 1, Principes et Methodes,* Editions Eyrolles, Paris.

CIRIA (1997a) Project Report 60, Geotechnical Grouting: a bibliography, September.

CIRIA (1997b) Project Report 61, Glossary of terms and definitions used in grouting: proposed definitions and preferred usage, September.

CIRIA (1997c) Project Report 62, Fundamental basis of grout injection for ground treatment, September.

Concrete Society (1993) Grouting specifications, Concrete, July/August pp. 27–32.

Concrete Society (1995) Interim technical report – Durable bonded post-tensioned concrete bridges. Report CS 111.

Concrete Society (2002) Technical report 47, Durable bonded post-tensioned concrete bridges, second edition.

Conner, J.R. (1990) *Chemical Fixation and Solidification of Hazardous Wastes,* Van Nostrand Rheinhold.

Domone, P.L.J. and Jefferis, S.A. (1994) *Structural Grouts,* Blackie Academic and Professional Publishers

Ewert, F.-K. (1985) *Rock Grouting with Emphasis on Dam Sites,* Springer-Verlag, Berlin.

FIP (1990) *Grouting of tendons in prestressed concrete, FIP Guide to good practice,* Thomas Telford.

Hayward Baker, W. (ed.) (1985) *Issues in Dam Grouting,* American Society of Civil Engineers.

Hayward Baker, W. (1982) Proceedings of American Society of Civil Engineers Specialty Conference, Grouting in geotechnical engineering, New Orleans.

Holmes, G. (1995) Early age volume change and pore pressure development in cement paste, University of London, PhD thesis.

Houlsby, A.C. (1982) Cement grouting for dams, Proceedings of the American Society of Civil Engineers Specialty Conference, Grouting in geotechnical engineering, New Orleans, Hayward Baker, W. (ed.)

Houlsby, A.C. (1990) *Construction and Design of Cement Grouting,* John Wiley

ICE (1999) UK National Specification for the construction of slurry trench cut-off walls as barriers to pollution migration, Institution of Civil Engineers, October.

Jefferis, S.A and Mangabhai, R.M. (1988) Laboratory measurement of grout pumping pressures, in Grouts and Grouting for the Construction and Repair of off-shore structures, Department of Energy off-shore Technology report No. OTH-88-289, HMSO, London.

Jefferis, S.A. and Mangabhai, R.J. (1990) Effect of temperature rise on the properties of high alumina cement grouts, International Symposium on Calcium Aluminate Cements, University of London.

Jefferis, S.A. (1985) Discussion on the Jordan Arab Potash Project, *Proceedings of the Institution of Civil Engineers,* **78**, 641–646.

Jefferis, S.A. (1991a) Grouts and slurries, in *The Construction Materials Reference Book,* Doran, D.K. (ed.) Butterworths.

Jefferis, S.A. (1991b) Development of site tests for grout, Final report to the Transport and Road Research Laboratory.

Jefferis, S.A. (2003) Proceedings of the Third International Conference, Granting and Ground Treatment, American Society of Civil Engineers, New Orleans.

Jefferis, S.A. and Fernandez, A. (2000) Spanish dyke failure leads to developments in cut-off wall design, *International Conference on Geotechnical and Geological Engineering,* Melbourne, Australia, November.
Karol, R.H. (1990) *Chemical Grouting,* 2 ed, Marcel Dekker, Inc.
Littlejohn, G.S. and Hughes, D.C. (1988) Thermal behaviour of grouted supports for pipelines, in Grouts and Grouting for the Construction and Repair of off-shore structures, Department of Energy off-shore Technology report No. OTH-88-289, HMSO, London.
Moxon, S. and Jefferis, S.A. (1996) Selection of grouts for offshore applications, International Congress, Concrete in the service of mankind, University of Dundee.
Nonveiller, E. (1989) *Grouting theory and practice*, Elsevier.
prEN288006 (1996a) Geotechnical works, Grouting
prEN288007 (1996b) Geotechnical work, Jet grouting
Rogers, W.F. (1963) *Composition and Properties of Oil Well Drilling Fluids,* 3rd edn, Gulf Publishing.
Sheikhbahai, A. (1989) Grouting with special emphasis on silicate grants. University of London, PhD thesis.
Sheikhbahai, A. and Jefferis, S.A. (1995) Investigation of syneresis in silicate-aluminate grouts, Geotechnique, The Institution of Civil Engineers, London.
Shroff, V.A. and Shah, D.L. (1993) *Grouting Technology in Tunnelling and Dam Construction,* A.A. Balkema, Rotterdam
Verfel, J. (1989) *Rock Grouting and Diaphragm Wall Construction,* Elsevier.
Weaver, K. (1991) *Dam Foundation Grouting,* ASCE.
Widman, R. (ed.) (1993) *Grouting in Rock and Concrete,* A.A. Balkema, Rotterdam.
Woodward, R.J. and Williams, W.F. (1988) Collapse of Ynys-y-Gwas Bridge, West Glamorgan, *Proceedings of the Institution of Civil Engineers,* Part 1, **88**, 635–699. See also the written discussion on the meeting.
Xanthakos, P., Abramson, L.W. and Bruce, D. (1994) *Ground Control and Improvement,* Wiley.
(2000) Grouting for the Hallendsas tunnel, Sweden, Ground Engineering.

# 13

# Concreting large-volume (mass) pours

*Phil Bamforth*

## 13.1 Introduction

This chapter provides guidance for the designer, the contractor and the concrete supplier, on the construction of large-volume pours. The advancing knowledge of the performance of fresh and hardened concrete and the wider use of composite cements and chemical admixtures has led to increased confidence in casting large-volume pours or massive sections. In the context of this chapter a large-volume pour and a mass pour are the same.

### 13.1.1 Definition of a mass pour

Concrete Society Digest No. 2 (Bamforth, 1984a) defines **a mass pour** as one of sufficient size to demand special attention to be given to logistical and technical considerations such as:

- Concrete supply
- Casting sequence
- Cold joints
- Plastic settlement

---

This chapter is based on Publication R135, 'Construction of deep lifts and large volume pours', P.B. Bamforth and W.F. Price, CIRIA, London, 1995, UK.

- Heat of hydration
- Early age thermal cracking

The American Concrete Institute Report 116T(1984a) defines mass concrete as 'Any volume of concrete large enough to require measures to be taken to cope with the generation of heat and attendant volume change to minimize cracking'. It is clearly inappropriate, therefore, to give a specific quantitative definition of a 'large-volume' pour, as this will be relative to the experience of the construction team and the scale of operations on a particular site.

## 13.1.2 Benefits

The principal benefits of large-volume pours are the savings in cost and timescale resulting from the reduction in the number of joints. These are expensive to form, requiring stop-ends when vertical and careful preparation of the concrete surface, regardless of orientation, when new concrete is cast against old. Furthermore, the current trend towards the imposition of very high stresses on massive concrete sections (e.g. heavily reinforced foundations to core-shaft tower blocks) has tended to create situations where the formation of joints or temporary stop-ends is sometimes virtually impossible, because of very congested reinforcement.

The disadvantages of cracks that might occur in situations where construction joints are not used appear to be comparatively minor. The importance of such cracks should be considered in relation to the structural requirements of the section, together with considerations of durability. For the protection of reinforcement, a maximum crack width of 0.30 mm has been considered acceptable. In addition, the elimination of joints removes potential cracks and zones of weakness, which are often highlighted in structures that are required to be water-retaining.

## 13.1.3 Technical considerations

While there can be significant benefits of scale resulting from the programmed use of large-volume pours, their successful execution demands an awareness of the specific technical requirements of the concrete in relation to the geometry of the element and the proposed methods for transportation, placing and compaction. The geometry has a significant influence on the materials and method that must be employed to achieve a monolithic and homogenous element. Particular attention must be given to:

- Planning to achieve continuity of concrete supply and compatibility of rates of delivery, distribution, compaction and finishing.
- Avoidance of cold joints. Here the rate and sequence of placing must be considered in relation to the stiffening time of the concrete.
- Plastic settlement. Large-volume pours are often deep slabs, e.g. raft foundations, and plastic settlement may result in cracking, characteristically above the top mat of steel.
- Early age thermal cracking. Excessive thermal gradients or restraint to bulk thermal contraction can cause cracking within days, or weeks, of casting. Cracking can be controlled by the use of reinforcement, and the risk of cracking can be minimized by appropriate materials selection and mix design, control of the mix temperature, insulation or cooling of the cast element and planning the sequence and timescale of construction.

In general, no significant adverse effects have been caused by the placing of concrete in large pours for reinforced concrete structures. Where difficulties do arise, this is usually due to a lack of awareness of those aspects of the concrete and the construction method which demand special attention, described above.

## 13.2 The designer's role

### 13.2.1 Approaches to the specification

There are two basic approaches to specification – **performance** and **prescriptive**. In practice, specifications are likely to be a combination of the two. In the former case the designer specifies performance limits and allows the contractor the flexibility to meet these limits. **The designer must ensure, therefore, that all aspects of the specification are compatible**. For example, a minimum cement content for durability may be inconsistent with low heat of hydration. The designer should also define the assumptions which have been made. This is particularly important in relation to limits on early age thermal cracking, where assumptions are made regarding the following:

- Temperatures and temperature differentials
- Coefficient of thermal expansion of the concrete
- Strain capacity of the concrete
- Restraint to thermal movement at critical locations (which may also involve assumptions about pour sizes and casting sequences)

This performance approach will require close cooperation between designer and contractor.

The prescriptive approach involves the designer in more detailed specification of the concrete mix type, maximum pour sizes and the construction sequence. In this case, the designer has responsibility for the completed structure, provided that accepted good practice is exercised during the construction process.

As in all construction projects, the designer can have significant impact on the construction process and it is, therefore, essential either that he makes himself aware of the construction process or that he works closely with the contractor. This is particularly important with regard to design for control of cracking. In the case of large-volume pours, the designer can influence the construction process in the following ways by:

1 Limiting the temperatures and temperature differentials resulting from heat of hydration.
2 Limiting pour sizes and configuration and the location of joints.
3 Defining the sequence of construction and the timescale (e.g. delay between adjacent elements).
4 Prescribing the concrete mix constituents and placing limits on the mix proportions.
5 Detailing steel which may lead to construction difficulties.

The designer must be aware of the impact of the design on the construction process, for example in heavily congested areas. Where such situations are unavoidable, due to some specific serviceability requirement, the designer should highlight these in the contract documents. Limits on pour sizes or the use of special concrete mixes should also be clearly defined to ensure that the contractor can take appropriate measures.

With regard to the concrete mix, it is important to achieve a balance between specified strength, durability, heat of hydration and the requirements for placing and compaction. In particular the strength and minimum cement content should not be overspecified.

### 13.2.2 Response to contractors' proposals

Where the contractor proposes large-volume pours, either at the tender stage or after the start of construction, the designer's role will be to review the implications for the structure in two respects:

1 Does the contractor have the experience and expertise to complete works successfully?
2 Will the functional requirements of the structure still be met?

The designer may judge the expertise of the contractor based on the quality of the proposal and his experience of similar construction methods. The assessment of the performance of the structure will require comparison between the design assumptions, e.g. regarding pour sizes, temperature rise, restraint, etc. and revised estimates taking account of the proposed changes. Once again, close liaison between designer and contractor is required. Approval is likely to be achieved more rapidly if the contractor presents a comprehensive technical proposal, based not only on the practicalities of construction but also on those matters which are likely to be of greatest concern to the designer, and in particular the risk of cracking.

### 13.2.3 Early age thermal cracking

**Reinforced concrete is designed to crack.** Where cracking is totally unacceptable then prestressing should be employed. Hence the designer's role is to control the location and the size of cracks to levels which are consistent with:

- Maintenance of structural integrity
- Durability
- Serviceability
- Visual appearance

CIRIA Report 91, Early age Thermal Crack Control in Concrete (Harrison, 1992) provides guidance on the significance of cracking under the above headings as shown in Table 13.1. Furthermore, the designer must consider whether early age thermal cracks will be additive to cracking which occurs due to service loads and, in the longer term, drying shrinkage. This will depend on the type of element, and its in-service condition.

**Table 13.1** Limiting crack widths

| Limit state | Limiting crack width (mm) | Comments |
|---|---|---|
| Structural integrity | Up to 0.5 | According to loading condition |
| Durability | 0.1–0.4 | According to environment |
| Serviceability (in water-retaining structures) | <0.2 | For self-healing |
| Appearance | Up to 0.4 | Depends on viewing distance |

As a general rule,

**Surface cracks** which result from **internal restraint** (i.e. temperature differentials within a pour) will tend to **close with time** as the stresses are redistributed, and these early stresses **need not be included in the structural design.** However,

**Through-cracks** which are caused by **external restraint** will not close, and **may increase in width** in the longer term, and the associated stresses will be **additional to load-induced stress**.

## 13.2.4 Limitation of cracking

In accordance with CIRIA Report 91, the maximum crack spacing, $S_{max}$, and the crack width, $w$, can be estimated using the following equations:

$$S_{max} = \frac{f_{ct}}{f_b}\frac{\Phi}{2\rho} \tag{13.1}$$

$$w = S_{max}\left[R(e_{th} + e_{sh}) - \frac{\varepsilon_{tsc}}{2}\right] \tag{13.2}$$

where:

$f_{ct}$ = tensile strength of the concrete
$f_b$ = bond strength of concrete to reinforcement
$\Phi$ = bar diameter
$\rho$ = percentage of steel
$e_{th}$ = thermal strain = $\alpha_c T_1$
$\alpha_c$ = thermal expansion coefficient
$R$ = restraint factor
$\varepsilon_{tsc}$ = tensile strain capacity
$T_1$ = difference between the centreline peak and the ambient mean temperature
$e_{sh}$ = drying shrinkage strain

The calculated crack width using this equation is the maximum 'average' crack width. However, recognizing the *in-situ* variability of concrete, there is a probability that some individual cracks will be greater than the calculated value. **Conformance must, therefore, be based on an average value taken over the full length of a particular pour.**

The contractor will have little influence over many of the above factors but, in a performance specification, will select the concrete mix within the requirements for strength, durability and early age thermal behaviour. To control the extent of cracking, it is common to specify allowable limits on the centreline peak temperature, $T_p$, and on temperature differentials $\Delta T_{max}$ during the post-construction period. Typical limits may be specified as follows:

- The max. temperature at any point within the pour shall not exceed . . . [commonly 70°C]
- The max. temperature differential within a single pour shall not exceed . . . [commonly 20°C]
- The max. value of mean temperatures between adjacent elements cast at the same time shall not exceed . . . [commonly 20°C]

- The max. value of mean temperatures between adjacent elements cast at different times shall not exceed . . . [commonly 15°C].

This is a simplistic approach, as the object is to limit restrained (locked-in) thermal strain, $e_r$, and the associated stresses that may lead to cracking. Temperature measurements are easy to obtain and to interpret, while strain measurements are much more complex in both respects. As the acceptable temperature limits are used to imply limits on strain, they should, therefore, be variable according to the assumed coefficient of thermal expansion of the concrete, $\alpha_c$, and the restraint to thermal movement, $R$. The relationship between the factors is demonstrated in the simple equation for evaluating crack risk proposed by Bamforth (1982):

$$e_r = K\,\alpha_c\,\Delta T\,R \tag{13.3}$$

and for no cracking $\quad e_r < \varepsilon_{tsc}$

where:

$\varepsilon_{tsc}$ = strain capacity under short-term loading
$\alpha_c$ = coefficient of thermal expansion of the concrete
$\Delta T$ = temperature change
$R$ = restraint factor (0 = unrestrained; 1 = full restraint)
$K$ = modification factor, 0.8, for sustained loading and creep

Clearly the allowable value of $\Delta T$ is inversely related to both $\alpha_c$ and $R$.

This approach, based on limiting the restrained strain, has also been adopted in CIRIA Report 91, which assumes a value of restraint of 1.0 at the interfaces between new and old concrete, and a modification factor of 0.5. This is consistent with BS 8007 (1987) for water-retaining structures, which assumes a restraint factor of '0.5 for immature concrete with rigid end restraints, after allowing for the internal creep of the concrete'.

Values of $\alpha_c$ may vary from as low as $7 \times 10^{-6}$ mm/mm°C for some lightweight concrete mixes, to higher than $12 \times 10^{-6}$ mm/mm°C for concretes using siliceous gravel aggregate. Furthermore, the aggregate also has an effect on the strain capacity, $\varepsilon_{tsc}$ (or resistance to cracking) of the concrete, with high values of $\varepsilon_{tsc}$ being associated with lower values of $\alpha_c$. Table 13.2, from Concrete Society Digest No. 2 (Bamforth, 1984a), gives estimated values of $\alpha_c$ and $\varepsilon_{tsc}$ for concretes using different aggregate types, together with limiting values for temperature drop and temperature differential.

The commonly used value of 20°C as a maximum temperature differential $\Delta T_{max}$, applies to gravel aggregate mixes which exhibit high $\alpha_c$ and low $\varepsilon_{tsc}$ relative to concretes using other aggregate types. With the use of a limestone aggregate, for example, which may yield concrete with an $\alpha_c$ as low as $8 \times 10^{-6}$ mm/mm°C, higher values of maximum temperature differential may be acceptable. In specifying, $\Delta T_{max}$, therefore, the assumed value of $\alpha_c$ should also be stated, hence defining the limit on differential strain used in the calculation of crack widths and providing a basis for accommodating the use of alternative aggregates. **The values in Table 13.2 are for guidance only.** Where data are available for a particular mix, the limiting temperature change can be calculated using the equation:

$$\Delta T = \frac{\varepsilon_{tsc}}{K \alpha_c R} \tag{13.4}$$

The limiting temperature differential may be derived using the above equation with an assumed restraint factor of 0.36 (Bamforth, 1982).

**Table 13.2** Limiting temperature changes and differentials to avoid cracking, based on assumed typical values of $\alpha_c$ and $\varepsilon_{tsc}$ as affected by aggregate type

| Aggregate type | Gravel | Granite | Limestone | Lightweight |
|---|---|---|---|---|
| Thermal expansion coefficient $\times 10^{-6}$/°C | 12.0 | 10.0 | 8.0 | 7.0 |
| Tensile strain capacity $\times 10^{-6}$ | 70 | 80 | 90 | 110 |
| Limiting temperature change in °C for different restraint factors: | | | | |
| 1.0 | 7 | 10 | 16 | 20 |
| 0.75 | 10 | 13 | 19 | 26 |
| 0.50 | 15 | 20 | 32 | 39 |
| 0.25 | 29 | 40 | 64 | 78 |
| Limiting temperature differential (°C) | 20 | 28 | 39 | 55 |

Restraint may also vary significantly and the designer must make some assumptions in his calculation which reflect likely restraints during construction. These will be affected by the selected pour sizes (length and depth), the time between adjacent casts and the sequence of construction. Guidance on restraint factors is given in CIRIA Report 91, together with a method for designing crack control steel. However, this generally assumes a restraint factor at the joint between new and old concrete of 1.0. No account is taken of the inherent stiffness of the new pour in relation to its immediate surroundings except in the modification factor *K*, which also takes account of creep and sustained loading effects. ACI Report 207.2R-73 (American Concrete Institute, 1984b) gives a more detailed approach for estimating restraint factors in relation to the length/height ratio of a pour as shown in Figure 13.1. The restraint at any point is determined by multiplying the restraint at the joints, calculated using equation (13.5), by the relative restraint at the appropriate proportional distance from the joint, obtained from Figure 13.1.

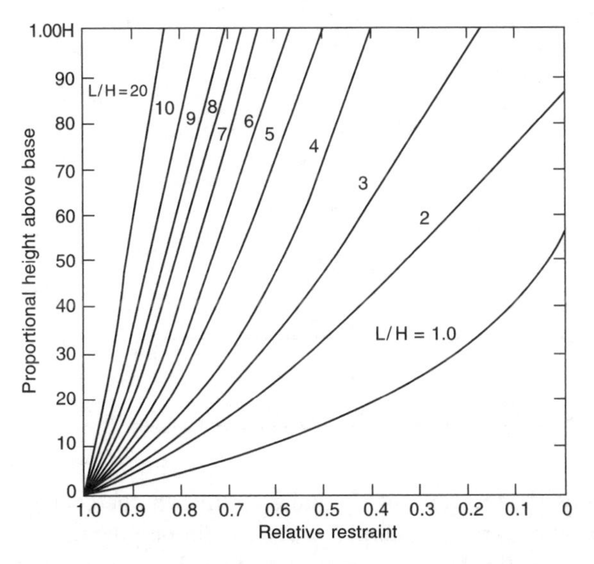


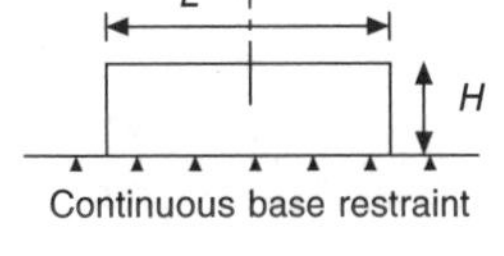


**Figure 13.1** Restraint factors for elements with continuous base restraint (American Concrete Institute, 1984b).

$$\text{Restraint at the joint} = \frac{1}{1 + \dfrac{A_n E_n}{A_o E_o}} \tag{5}$$

where $A_n$ = c.s.a. of new pour
$A_o$ = c.s.a. of old concrete
$E_n$ = modulus of elasticity of the new pour concrete
$E_o$ = modulus of elasticity of the old concrete

A comparison between the measured restraint through the height of a bridge pier cast on to a strip footing and values predicted using the ACI method is shown in Figure 13.2 (Bamforth and Grace, 1988), indicating that, provided the assumptions about the relative stiffness of the old and new concrete are appropriate, the method is reasonably accurate. Based on limited measured values of the early age modulus of heat-cycled concrete, and the estimated time for the new element to cool, the ratio of $E_n : E_o$ is likely to be in the range 0.7–0.8 (Bamforth, 1982) as cooldown occurs. The results in Figure 13.2 were obtained on the centreline of an 800 mm thick, 6.2 m high, 12 m long bridge pier, cast onto a 1 m deep by 2.85 m wide footing:

$$\frac{1}{1 + \dfrac{A_n}{A_o}\dfrac{E_n}{E_o}} = \frac{1}{1 + \dfrac{4.96}{2.85} = \dfrac{0.8}{1}} = 0.42$$

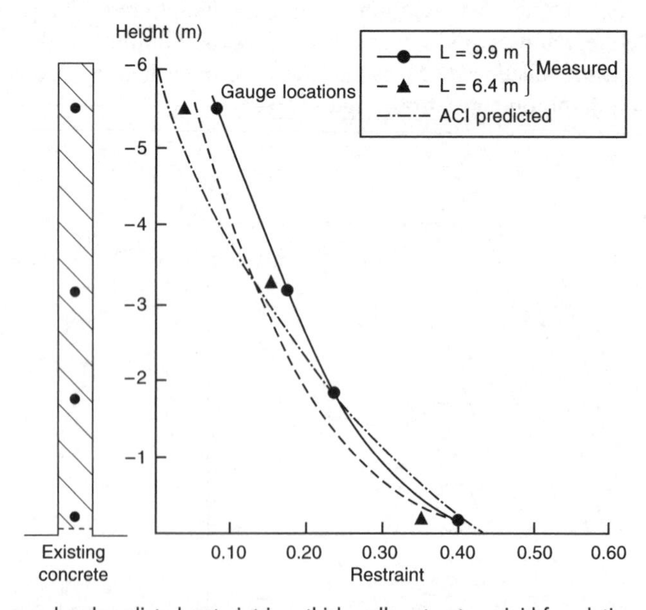


**Figure 13.2** Measured and predicted restraint in a thick wall cast onto a rigid foundation.

The reducing restraint towards the top free surface indicates that the percentage of steel may be reduced with height for early age thermal crack control purposes.

In some cases, say, when a high wall is cast onto an existing slab, the designer will be required to exercise judgement with regard to the effective cross-sectional areas (c.s.a.) of new and old concrete used in the calculation. The following rules of thumb may be applied therefore:

- When a wall is cast at the edge of a slab, the relative effective areas may be assumed to be in proportion to the relative thicknesses of the wall and the slab.
- When a wall is cast remote from the edge of a slab, the relative areas may be assumed to be in proportion to the ratio of the wall thickness to twice the slab thickness.

More complex geometries may require more detailed analysis. Hence, the designer should define within the specification, the following assumptions:

- The allowable temperatures in terms of the maximum value and differentials.
- The coefficient of thermal expansion of the concrete.
- The restraint factors at critical locations. (Where these are based on restrictions on pour sizes, this must also be stated.)
- The tensile strain capacity of the concrete.
- Acceptable crack widths, as measured at the surface.

The designer must also consider what actions should be taken in the event of the following:

1 Unacceptable cracking which occurs within the allowable temperature limits
2 Non-conformance with the temperature limits, but cracking within specified limits
3 Non-conformance with the temperature limits and excessive cracking

As design codes tend to be conservative, scenario 1 is unlikely, and scenario 3 is clearly the responsibility of the contractor. When scenario 2 occurs, this simply demonstrates conservatism in the design assumption and as experience is gained on a contract, the limits could be adjusted to reflect this.

It is becoming increasingly common, on large civil engineering structures, to specify full-scale trials to obtain performance data on the concrete which can be used to define limits on temperature differentials for use in construction. Where such trials are carried out, care must be taken to ensure that restraints are realistic, particularly in respect of walls cast onto rigid foundation, or slabs which link stiffer elements.

Complex computer models are also available, which enable scoping studies to be carried out to investigate the effects of mix type, pour geometry and environmental conditions (Emborg, 1989; Danish Concrete and Structure Research Institute, 1987) and these are occasionally used for critical structures or elements. However, the value of the output is often limited in absolute terms by the assumptions which have to be made regarding the early age properties of the concrete and their relationship with the temperature history or maturity of the concrete. Validation is also difficult without *in-situ* measurements of temperature, strain and stress, but trials may often have significant programme implications. This is an area which could benefit from further research.

## 13.2.5 Performance of construction joints

Many large-volume pours will be cast as part of much larger elements and this will necessitate the inclusion of construction joints. CIRIA Report 91 states that 'with full steel continuity, and assuming that the current cracking control theory for thermal cracking is correct, the crack widths at the joint will be no wider than elsewhere in the section'.

Hence, if the approach to cracking is to control the crack width and spacing by the design of reinforcement, the location of construction joints is unlikely to be critical. When other means are being employed to limit cracking, the performance of the joint can be assessed by considering the joint to represent the centreline of a pour which is double the length of the pour being cast.

## 13.2.6 Materials specification

The designer is responsible for specifying materials that are appropriate for producing a safe durable structure that is fit for its intended purpose. This is usually achieved by placing limits on the properties of the constituent materials and certain mix proportions (see section 13.4). In most cases, this will be accomplished by reference to Standards and Codes of Practice. However, the designer should avoid being overrestrictive in the choice of materials or limits on mix proportions as, in some cases, this can lead to difficulties with incompatibility between the specified requirements and the practicalities of construction. An example would be specification of an unnecessarily high minimum cement content (on durability grounds) together with stringent limits on the allowable heat build up within a large-volume pour. Consideration of the probable construction techniques should also be an important factor in specification.

## 13.2.7 Testing for conformance

On large contracts or for critical structures, consideration should be given to specifying the construction of full-scale trial mock-ups of representative sections, both to assess the effectiveness of the proposed concrete mix and placing techniques and to enable direct measurements of the *in-situ* concrete properties. In the event of inadequate performance, changes in concrete mix design or construction techniques would be required.

Inspection can be divided into three distinct phases, i.e. pre-, during and post-concrete placing. Where large-volume pours are to be cast, the designer should specify that the contractor's method statement makes provision for measurement or inspection as follows:

*Pre-placing*

- Section dimensions
- Reinforcement location (especially cover depth) and bar sizes
- Formwork treatment (adequate and appropriate release agent applied)
- Location of box-outs etc.
- Location of embedded inserts
- Location of embedded thermocouples or strain gauges
- Cleanliness

During placing

- Concrete workability
- Concrete temperature
- Rate of concreting

- Open time of live faces
- Effective curing

*Post-placing*

- Standard of finish (i.e. blowholes or sand runs on deep lifts)
- Monitoring of temperatures/strains
- Examination of cracking (both crack spacing and crack widths)
- Achievement of cover

The specification should include both the type and frequency of inspection and the method of analysis and interpretation.

## 13.2.8 Specification for monitoring of temperatures

In large-volume pours, the avoidance of excessive temperature rises and differentials is essential to minimize problems of thermal cracking. The specification of temperature limits is important in this respect and the following limits should be considered:

- Maximum placing temperatures
- Maximum temperature achieved within the section
- Maximum temperature differential between the centre and surface of the section
- Maximum temperature difference between the centre of the pour and ambient temperature
- Maximum temperature differential between successive pours or lifts.

The quantitative limits placed on these parameters should reflect the consequences of any cracking and the properties of the concrete actually used during construction (Figure 13.3). For example, the type of aggregate used in the concrete has a significant effect on the temperature differential that can be sustained without cracking (some commonly used values are given in section 13.2.4) and the effects of peak temperatures on *in-situ* concrete properties are much less deleterious when composite cements are used.

**Figure 13.3** BP Harding gravity base structure used lightweight aggregate concrete with temperature limits shown in the example specification. Note that the values are much higher than normally specified. No special thermal curing procedures were necessary and there was no evidence of early thermal cracking.

If temperature limits are specified, temperature monitoring is normally required by the specification. Embedded thermocouples are an appropriate technique and may usefully be combined with embedded strain gauges (e.g. vibrating wire gauges). The minimum number of thermocouples together with appropriate locations should be specified. Normally this would consist of a thermocouple at the centre of the pour together with additional thermocouples close to the top, bottom and side surfaces of the section. In view of the application, the thermocouples need only be accurate to the order of ± 1°C and reference calibration is not necessary. It is useful to specify detailed monitoring of the first few pours, relaxing the requirements once a pattern of behaviour has been established (perhaps just monitoring a smaller number of selected locations).

## 13.2.9 Action in event of non-conformance

Clear specification of the action to be taken in the event of any non-conformance with the specification is essential. This should include:

- Test plan for NDT survey and retrieval and testing of cores from suspect areas
- Assessment of consequences of non-conformance
- Action to be taken
  - Qualified acceptance
  - Local repairs (e.g. sealing, waterproofing, crack injection)
  - Removal of suspect concrete

Additional strengthening measures are as follows.

**Example specification – Control of concrete temperatures during hardening**

To control cracking as a result of early thermal movement, the Contractor shall take measures to ensure that the following requirements are satisfied. The Contractor shall describe in a Method Statement to be submitted to the Engineer for his approval, and before concrete operations are commenced, the measures to be taken to achieve these requirements.

A. The maximum temperature of the concrete during its early age heat cycle shall not exceed [80°C] unless it can be demonstrated by test that the compressive strength, after [56] days, of concrete heated to a higher temperature, exceeds the required design strength by a margin to be agreed with the Engineer.
B. In elements, or parts of elements, which are not restrained by existing concrete, the difference between the maximum temperature and the surface temperature (measured within 10 mm of the surface) shall not exceed [50°C] unless it can be demonstrated by test that greater differentials can be tolerated without cracking.
C. When a new pour is cast against existing concrete of the same thickness, e.g. adjacent pours for a thick slab, the difference between the mean temperatures in the existing and the newly cast sections shall not exceed [45°C]. The mean temperature shall be determined at a distance of 1 m from the interface between old and new concrete and shall be calculated using the equation:

$$T_m = 1/6(4T_p + T_{s1} + T_{s2})$$

where:

$T_m$ is the mean temperature
$T_p$ is the peak temperature at the centre of the section
$T_{s1}$, $T_{s2}$ are the surface temperatures on opposite faces.

D. When a new pour is cast against existing concrete of a different thickness, account shall be taken of the difference in the restraint excercised by the old section on the new, when deriving the mean temperature differentials. The Contractor shall include in his Method Statement, the temperature limits for the elements to be cast, together with the calculations supporting the limits, for approval by the Engineer before concreting operations are commenced.

E. The limits given in clauses B were calculated using the equation:

$$\Delta T = \frac{\varepsilon_{tsc}}{K \alpha_c R}$$

The limiting temperature differential has been derived assuming a restraint factor of 0.36.
The thermal expansion coefficient, $\alpha_c$ = [8.3] microstrain/°C
The tensile strain capacity of the concrete under short-term loading, $\varepsilon_{tsc}$, = [160] microstrain.
If measurements demostrate that values of $\alpha_c$ and $\varepsilon_{tsc}$ are different from those assumed then the proposed limits must be revised according to the equation above. The results for site trials may also be used to revise the proposed limits. However, trials must be of sufficient scale to ensure that the levels of restraint are the same as those that occur during construction.
Results from *in-situ* monitoring can provide values of actual concrete temperature and strain. By comparison with thermal strain in unrestrained blocks, actual *in-situ* restraint may be calculated. Such data may be used to support changes in bay size and layout as construction proceeds.

## 13.3 Planning

This section provides a detailed checklist for the contractor at the planning stage of construction of large-volume pours. The successful completion of a large-volume pour depends on cooperation between the designer, the contractor and the concrete producer. In addition, it is essential that planning begins at an early stage in order to forestall potential problems. The following points need to be resolved at least 2 to 3 weeks before the pour is to take place:

- Establishing authority, responsibility, planned activities and communications for each party concerned
- Agreement on approved material sources and mix details
- Provision for standby equipment
- Access to the site
- Setting a 'fixed' start date

Where deep lifts also involve large volumes, then all of the above still apply. However, the essential differences between large-volume pours are:

- the much higher rate of casting normally associated with the large-volume pour
- the restricted access for concrete delivery and compaction in the deep lift.

**The most important point is to ensure that there is adequate back-up plant available to maintain continuity in all stages of delivery, distribution, compaction and finishing of the concrete.**

## 13.3.1 Capability

Although every site is unique, and every pour will have its own special problems, previous experience in planning and undertaking the placing of concrete in large-volume pours is invaluable. The key to success is achieving continuity of casting at a rate which avoids cold joints and other construction defects, using methods which are consistent with recognized good practice. Hence, all those involved in the casting must be made aware of the importance of their link in the chain of:

- Production
- Delivery
- Distribution
- Compaction
- Curing

Other trades, e.g. steel fixers, carpenters, must also be made aware of the 'fixed' start time, as substantial plant and manpower has to be committed to the pour itself, much of which may be sub contracted.

The contractor can demonstrate his capability by providing a work procedure defining the management, method, programme and inspection and testing regime, which takes account of the unique features of the pour to be cast.

## 13.3.2 Planning meeting

For the benefit of those on a team who have not previously experienced continuous concreting beyond normal working hours, or the special procedures involved in large pours, there should be formal meetings between all involved at all levels, both to inform and to advise, the objective being to highlight critical aspects of the construction process, and in so doing to forestall potential problems.

## 13.3.3 Plant

In **large-volume construction** the main considerations are:

- Maintenance of a high rate of placement
- Distribution over a large area

As indicated in Table 13.3, most large-volume pours rely on pumps to achieve high rates of concrete placement (Laning, 1990; Randall, 1989; Anon., 1988, 1990; Grace and Hepplewhite, 1993; Fitzgibbon, 1977). In Germany in 1988, a record 17 000 $m^3$ pour was cast continuously using four pumps over a period of 78 hours (Anon., 1988). The average placing rate per pump was about 55 $m^3/h$, and 90 readymix trucks were used, operating from six batching plants. Higher placing rates of 100 $m^3/h$ per pump were achieved in the USA in 1990, where a 5700 $m^3$ pour was cast in only $11^1/_2$ hours (Laning, 1990).

**Table 13.3** Some examples of large-volume pours

| Ref no. | Location/ date | Pour details | | Time taken (hr) | Average rate of placing ($m^3/h$) | Plant used | Average rate per pump ($m^3/h$) | Concrete mix details |
|---|---|---|---|---|---|---|---|---|
| | | Vol. ($m^3$) | Thickness (m) | | | | | |
| 15 | Frankfurt, Germany, 1988 | 17 000 | 8.5 | 78 | 218 | 4 pumps | 55 | Not given |
| 10 | California, USA, 1990 | 5700 | 1.7–3.8 | $11^1/_2$ | 496 | 5 pumps | 99 | Grade 20 pfa |
| 11 | Seattle, USA, 1989 | 8230 | 4.7 | $13^1/_2$ | 610 | 9 pumps | 68 | Grade 40, 30% pfa |
| 12 | Jamshedpur, India, 1990 | 3600 | 4.3 | 44 | 82 | 6 pumps | 15 | Grade 25, BFS Cement, w/c = 0.42 |
| 13 | Thames Barrier, 1982 | 6600 | 5.0 | 72 | 92 | Tremie | – | Grade 30 50% pfa |
| 14 | Sheffield, 1976 | 3000 | 2.0 | 22 | 136 | 2 pumps | 68 | Not given |

The number of pumps needed will depend not only on the placing rate to be achieved but also on the plan area of the pour. Mobile pumps have folding distribution booms which vary in length from 15 m to over 50 m (Fitzgibbon, 1977). Hence boom pumps are particularly suitable for large-area, large-volume pours.

Typically, pumps can achieve practical placing rates of between 30 and 100 $m^3/h$, compared with average rates of the order of 8–12 $m^3/h$ with crane and skip (Cooke, 1990) and are, therefore, much more suited to large-volume concrete construction. Other placing methods may be appropriate when the plan area is smaller. For example, chutes provide a rapid method of distribution over relatively short distances and are most appropriate when high-workability concrete is used. Conveyors are more suitable for use with low-workability mixes. When night working is expected, daylight standard lighting must be provided, in particular, for reasons of safety.

**In view of the importance of maintaining continuity of placement, back-up must be available for all essential items of plant.**

## 13.3.4 Labour

It is essential that sufficient labour is allocated to each stage of the casting process to ensure that construction proceeds without interruption. Agreement must be obtained regarding lengths of shifts and changeover times and, on long shifts, back-up teams must be available to accommodate work breaks. It is preferable that experienced teams are used, and, where appropriate, additional training must be given.

### 13.3.5 Material supply

As continuity is an essential feature of the successful construction of large pours, the benefits of advanced planning in relation to all materials suppliers cannot be overstated. If readymix concrete is to be used, the supplier, who will also have to meet the requirements of other customers, must be given sufficient advanced warning when high placing rates are required. When the casting rate necessitates the use of more than one batching plant, this requires a high level of coordination by the concrete supplier. However, even when the casting rate is well within the capacity of a single plant, it is still advisable to utilize more than one, to ensure continuity in the event of one plant being unable to maintain supply.

When very large volumes are to be placed, or very high placing rates are required to meet the programme, it may not be appropriate to rely solely on site production, except on very large sites which have more than one batching and mixing plant. However, site-batching plant can be used in conjunction with a readymix supply. In this case, the contractor must advise all material suppliers well in advance and obtain assurances that the necessary rates of supply will be met.

### 13.3.6 Access

Access is important in a number of respects:

- Access for the concrete supplier to the site
- Access within the site to the pour
- Access to the point of delivery within the pour

Access to the site will depend not only on geography but also on the time of day, day of the week, and possibly the time of year. Consideration must be given at the planning stage to possible delays due to heavy traffic, road works, etc. and the benefits which may be achieved by working non-standard hours. For example, in some locations casting during the night or at weekends may allow easier access while in other cases the opposite may be true.

On-site, a temporary parking area is required for the readymix trucks, routes must be carefully coordinated, particularly if there are several delivery points, and an area must be allocated for the trucks to wash out. Access to the point of delivery within the pour is determined by the selection of plant.

### 13.3.7 Timing

The timing of the pour may be significant in a number of respects.

- It will determine ease of access of the site.
- It will determine availability of plant and concrete supply from subcontractors with many other customers.
- It will have an effect on the workability of the concrete and subsequent performance *in-situ*.

Availability of plant will be determined by the timing of the pour. For example, at night-time and at weekends there is less likely to be a high demand on concrete supply and it may be possible, therefore, to dedicate particular plants. This also helps to achieve consistency of supply.

Two examples of very large volume pours in the USA, cast at high placing rates over relatively short periods – 5700 $m^3$ in $11^1/_2$ hours (Laning, 1990) and 8230 $m^3$ in $13^1/_2$ hours (Randall, 1989) – involved night-time starts at 10.00 pm and midnight respectively. Technical advantages can also be gained by night-time concreting associated with lower ambient temperatures and the lower concrete mix temperatures than those occurring during the day. Particular benefits include:

- Slower rate of workability loss of the fresh concrete
- Longer stiffening time, reducing the risk of cold joints
- Slower rate of generation of heat of hydration, from a lower start point, thus reducing the risk of early age thermal cracking.

## 13.3.8 Quality assurance

A quality assurance system is designed to prevent mistakes occurring during construction and also to identify and remedy any problems and non-conformances that do occur and is therefore of great benefit in ensuring successful construction. Such systems (preferably complying with the requirements of BS 5750) should encompass all measures necessary to maintain and regulate the quality of the construction process.

A quality system should include inspection and testing of materials and fresh and hardened concrete as well as documented procedures for production, placing and curing of concrete. The testing plan should also form part of the quality assurance system. Close cooperation between all parties involved is a prerequisite of operating an overall quality system and for successful construction as a whole.

## 13.3.9 Monitoring

Where special monitoring is required, in addition to the normal inspection by the engineer prior to casting, and control of the concrete during casting, this should be clearly stated in the Specification. The most common 'additional' requirements for thick walls or large-volume pours are for temperature measurements. In such cases, the following should be agreed.

- The precise locations of the measuring points in relation to the limits given in the Specification (see section 13.2)
- The protection of the measuring points prior to and during concreting

- Calibration standards and accuracy
- The frequency of measurement, and for how long measurement must continue
- The form in which the results are to be presented
- The way in which the results are to be interpreted
- Action limits, and the procedures to be initiated when these limits are reached
- Actions to be taken in the event of non-compliance with the specified procedures or non-conformance with specified performance limits.

## 13.3.10 Contingency plan for loss of concrete supply

Although unlikely, there is always a finite risk of loss of materials supply, however well the project is planned and however much back-up is provided. The contractor is therefore advised to have a contingency plan for stopping the pour at any stage and incorporating joints. The contingency plan should include:

- preferred locations of joints, with advice from the designer
- the method for forming joints

This plan should be prepared in conjunction with the planning of the concreting sequence.

## 13.3.11 Responsibilities

The responsibilities for ensuring that the concrete is capable of producing a safe durable structure, fit for its intended purpose, is shared between the designer, the contractor and the concrete producer. The designer is responsible for specifying those properties that are required to ensure the safe structural performance and durability of the structure. These are typically limited to the characteristic strength, durability parameters such as a maximum water/cement ratio or minimum cement content, or cement type (often by reference to standards or Codes of Practice) and limitations on maximum temperatures or temperature differentials as a means of limiting crack widths. As already highlighted in section 13.2, it is important that the designer recognizes that on overconservative limitation on minimum cement content, which is generally perceived to increase durability, may actually have the opposite effect. This occurs in cases where heat of hydration is likely to lead to problems such as early age thermal cracking or excessively high-temperature build-up on the surface.

**Contractors' checklist**

In the planning stage, the contractor should pay particular attention to the following:

1 Advanced warning to all parties involved, and in particular, the material supplier. The local community, police, etc. should also be notified.
2 Concrete production – if readymixed concrete is to be used the contractor must check:
   - Location of the plant in relation to transit time risk of delays (e.g. in rush-hours)

- Throughput as determined by storage capacity, mixing rate, number of delivery trucks
- Quality control in relation to risk of rejecting non-compliant mixes
- Back-up plants, or use of several plants.

3 Access to the site – separating incoming and outgoing mixer trucks, provision of on-site space for queuing and washdown.
4 Concrete distribution – selection of plant for the required placing rate. Pumps are most common for very large-volume placements, but skips, chutes and conveyors are also used. Arrangement of back-up plant in case of breakdown. Lighting for night-time concreting or deep lifts.
5 Concrete placing sequence – planning sequence of placing from one or more points, layer thickness.
6 Placing and compaction – ensure sufficient labour and plant for thorough compaction.
7 Finishing – this may be delayed if the mix has a slow set.
8 Condition for non-standard working hours – check local by-laws for night and weekend working. Agree conditions with labour force, e.g. shift working, cover during breaks.
9 Quality control – rate of sampling of concrete; storage of test specimens; actions in the event of non-conformance. In the case of concrete subject to third-party certification, the quality assurance scheme of the concrete producer must be examined and approved by both contractor and designer.
10 Temperature monitoring – location of measuring points; method of recording temperature; agreed action limits; actions in the event of non-conformance.
11 Achievement of specified temperature limits – mix selection, trials to measure temperature rise or heat of hydration, cooling of the fresh concrete, *in-situ* cooling with embedded pipes, thermal curing.
12 Contingency plans for
- Delays or failure of concrete delivery
- Breakdown of plant for concrete distribution and compaction
- Extension beyond anticipated pour time (particularly if running beyond normal working hours)
- Severe weather, especially heavy rain.

## 13.4 Concrete mix design

This section provides guidance on the selection of concreting materials and mix design to achieve performance characteristics required for the successful completion of large-volume pours.

Good concrete mix design is an essential part of any successful concrete construction project, but placing concrete in large-volume pours requires certain additional considerations. The final concrete mix design must be capable of satisfying both the requirements of the designer in terms of structural and durability properties and the needs of the contractor for production and placing the concrete in large-volume pours.

There is no evidence that increased cement contents, *per se*, as opposed to their effect on reducing w/c ratio, are of benefit for durability, and hence, within reason minimum cement content should not be specified for durability requirements. In general the concrete specification, in accordance with BS EN 206–1: 2000 and the complementary BS 8500–1: 2002, will fall into one of the following categories:

- Designated mix
- Designed mix
- Prescribed mix
- Standard prescribed mix
- Proprietary concrete

Considering the number of factors that need to be taken into account for successful construction in large-volume pours, it is only really appropriate to specify designed mixes.

The contractor also has specific requirements for distribution and compaction of concrete. These may not always be compatible with other requirements, however. For example, a low cement content is desirable for low-temperature rise and low risk of cracking, but if the concrete is to be pumped, the cement content would normally be expected to be not less than about 290 kg/m$^3$ (British Concrete Pumping Association, 1988).

The designer must, therefore, be aware of possible constraints on mix design when specifying an allowable maximum temperature. As a maximum value of 70°C is often specified, this problem is unlikely to arise except in extreme conditions of summer concreting with a very rich mix. Nevertheless, the designer must be aware of the possible constraints which may arise when designing for the use of high-grade concretes in thick sections. Furthermore, the designer must also state the assumptions which have been made in the design process. These will include:

- The materials safety factor applied to the design strength.
- The aggregate type as it affects the thermal expansion coefficient and tensile strain capacity of the concrete, and hence the risk of early age thermal cracking.
- The temperature drop used in the calculation of crack widths.
- Restraints, as affected by pour sizes and construction sequence.

As some of these assumptions may differ from practice once a contract is underway, it is essential that they are stated in the contract documents to enable a sound technical dialogue between the designer and the contractor when discussing the consequences of changes, and considering the suitability of mixes in relation to the particular construction programme.

The detailed mix design will be produced by the contractor or concrete producer and will combine the specifier's requirements with the practical requirements for successful placing of the concrete.

Test results confirming the properties of the proposed mix (usually limited to compressive strength and workability) are often submitted for approval together with the mix proportions.

## 13.4.1 Properties to be considered

When designing a concrete mix for use in large-volume pours, the following properties should be considered:

(a) **Specified properties**
- Characteristic compressive cube strength

- Minimum/maximum cement content
- Durability requirements – resistance to sulphates, chlorides, freeze/thaw, ASR
- Limiting temperatures and temperature differentials

(b) **Construction properties**

- Workability and workability retention – to enable the concrete to be transported, placed and compacted. This must take account of the selected method of distribution, e.g. pump, skip, tremie
- Stiffening time – as it affects formwork pressures and cold joints
- Cohesiveness and bleed – particularly important in deep lift construction
- Strength development – as it affects formwork striking times
- Heat evolution – in relation both to the risk of cracking in large-volume pours and to the resulting *in-situ* properties.
- Finishing characteristics

Some of these properties are not covered by standard test methods, but quantitative assessments can be derived from site trials. The general principles of mix design are covered elsewhere (Teychenne *et al.*,1988; American Concrete Institute, 1984c).

## 13.4.2 Factors influencing key properties

Given that all the specified properties of the concrete mix have been achieved, construction in large-volume pours requires the following considerations.

### *Section dimensions*

The section dimensions influence the mix design in a number of ways. The minimum dimension influences both the selection of maximum aggregate size and cement type and content.

To avoid problems with aggregate arching, BS 8110 (1985) recommends that the nominal maximum size of coarse aggregate should be not greater than one quarter of the minimum section thickness. The spacing of the reinforcement must also be taken into consideration and it is generally accepted that the maximum aggregate size should be less than three quarters of the minimum clear spacing between reinforcing bars, bundles of bars or prestressing strands.

The minimum section thickness also governs the rate at which heat generated by cement hydration can be dissipated. Some specific figures for use in design are given in CIRIA Report 91. Thicker sections have the potential for considerable internal temperature rises (see section 13.5) and the use of concrete with either lower total cement contents or composite cements may be appropriate (Bamforth, 1984a). Narrow sections with congested reinforcement require concrete to have a high workability if placing without honeycombing or excess voids is to be achieved.

Concreting deep sections often involves placing concrete from the top of the form, and in order to minimize segregation as the concrete falls through the formwork and reinforcement cage, the mix must be cohesive. This influences the proportions of fine to coarse aggregate as well as the consistency of the cement paste.

The effects of bleed and settlement are of greater significance in very deep sections and again point to the need to achieve a cohesive concrete mix.

## Access and placing techniques

The technique proposed for placing the concrete and the ease of access to the structure can both influence the requirements of the mix design. When access is unrestricted, concrete may be placed by skip or directly from a readymix truck. In more restricted situations, pumping or tremie placement is often preferred. The transport time between mixing and placing will also influence the detailed mix design.

Concrete placed by skip, or by chute directly from the readymix truck frequently needs no additional considerations over and above the usual requirements for adequate workability. Successful pumping or tremie placing, however, both require cohesive concrete of medium/high workability if blockages are to be avoided. For pumping, a minimum fines content (cement + fine sand) of 400 kg/m$^3$ and slump of 75–100 mm is often recommended (British Concrete Pumping Association, 1988). For concrete to be placed by tremie a slump of 150 mm is commonly used.

If concrete is to be transported (or pumped) over long distances, retention of workability must also be considered. If this is due to loss of water by evaporation, it may be acceptable to add water prior to discharge. Where long haulage times are expected, an acceptable method of adding small quantities of water, to restore workability, should be agreed in advance. Alternatively, retarding admixtures and/or the selection of cements with longer setting times, e.g. composite cements with pulverized-fuel ash (pfa) or ground granulated blastfurnace slag (ggbs) may be used.

## Volume of pour and rate of placing

The volume of pour in itself does not influence the mix design, although the minimum dimension does. However, the rate at which concrete is to be placed will control the acceptable workability loss between batching, arrival at site and placing. For information on concrete properties in relation to transportation and handling, reference should be made to CIRIA Report 165 (Masterson and Wilson, 1997).

The stiffening time of the concrete must be compatible with the rate of placing. If the concrete sets too rapidly (relative to the placing rate), in addition to the formation of cold joints, there is also a risk that vibration may damage the lower levels of the lift. In such cases, consideration should be given to the use of slow stiffening concretes, i.e. concretes containing composite cements (Concrete Society, 1991) or the incorporation of retarding admixtures (Rixom and Mailvaganam, 1986). This will be of most significance in large-volume pours.

However, for deep lifts the effects of set retardation on formwork pressures are likely to be of greater significance. This is addressed in CIRIA Report 108 (Clear and Harrison, 1985).

## Workability

Workability is primarily controlled by the water content of the mix, hence for mixes without admixtures, higher workability requires more water and thus more cement to achieve a specified strength grade. Concrete must be workable in order that it can be properly compacted. The introduction of water-reducing admixtures (plasticizers and superplasticizers) has enabled high slump concretes to be produced at high strengths and without the need for excessive cement contents (Clear and Harrison, 1985). Workability is also improved by selecting the maximum aggregate size compatible with the section dimensions and reinforcement spacing and ensuring that both coarse and especially fine aggregates are properly graded (Teychenne *et al.*, 1988).

The fines content must also be high enough to avoid harshness. It should be noted that concretes containing pfa to BS 3892: Part 1 (1982) or to BS EN 450 with a loss-on-ignition of not more than 7%, have improved workability at a given water content, due to both the spherical shape and the grading of the pfa particles and its lower density which results in a higher volume of paste for a given weight of composite cement (Concrete Society, 1991). Modest improvements can also be achieved using ggbs to BS 6699 (1986).

It should be noted that when high quantities of admixture are used to produce workable concretes at low w/c ratios, there may be a tendency for the concrete to be very 'sticky' and resistant to flow under the influence of vibration even though the slump is suitably high. Excessive levels of plasticizing admixtures may also lead to segregation and retardation.

## Cohesiveness and bleed

Concrete placed in deep lifts is often poured from the top of the forms, and even when other placing techniques such as pumping and tremie placing are used, there is always the risk of the concrete segregating as it strikes the formwork and reinforcement. This can cause separation of aggregate from the mortar as the latter adheres to the reinforcement, and consequent honeycombing. Increasing the cohesiveness of the mix will minimize the risk of separation.

Cohesive concretes are characterized by a higher than average proportion of fine aggregate and a high cement content. Addition of pfa to the concrete, which results in a much lower water/composite cement ratio by volume, also improves cohesiveness. The use of fine sands tends to produce more cohesiveness than equivalent quantities of coarse sands, but this would need to be associated with increased admixture doses to maintain workability.

A well-proportioned pump mix is often suitable for use in deep lifts. This combines the cohesiveness and workability needed to avoid blockages in the pipeline. A rule of thumb (British Concrete Pumping Association, 1988) is to add an additional 4 per cent of fine aggregate over the optimum content for a normal mix (Teychenne, *et al.*, 1988) of given workability.

Closely allied to cohesiveness is reduced bleed and settlement. Bleed is the upward separation of water from the concrete under the influence of gravity. The results of bleed (and associated settlement of the concrete) are a general increase in the water content in the upper parts of a lift, together with water-filled voids under reinforcing bars and cracking in areas where settlement is restrained (Concrete Society, 1992).

Deep lifts are particularly prone to the effects of bleed, as the amount of settlement is generally proportional to the depth of the section.

As a rule of thumb, bleed may be excessive if the aggregates are very clean (i.e. the silt content is low) if less than 20 per cent passes the 300 micron sieve, and if the material tends to be single sized. Hence, bleed can be minimized by ensuring that the aggregates are properly graded and in particular, by including a sufficient quantity of fine material. This may be achieved by increasing the proportion of fine aggregate, by the use of a sand with a high proportion of material finer than 150 $\mu$, by using a more finely ground cement or by incorporation of an increased volume of cementitious materials such as pfa or by the addition of a much finer material such as microsilica (Concrete Society, 1993). While pfa retards the set, and the period or bleed is extended, the rate of bleed is usually reduced sufficiently to more than offset this (Concrete Society, 1991). Microsilica concretes in particular may exhibit very low bleed to the extent that exposed faces are very sensitive to curing and horizontal surfaces are difficult to finish. Concretes containing a high percentage of ggbs are sometimes more prone to bleed (Concrete Society, 1991).

There is an optimum fine aggregate grading and content leading to a minimum void content and reduced bleed (Figure 13.3) and if this is exceeded, the bleed may actually increase. The optimum sand content can be estimated simply, by weighting the combined aggregate in a bucket of fixed volume. As the sand content is varied, the weight (or bulk density) will also change and the maximum weight indicates the minimum voids (Figure 13.4). These aggregate proportions would then be used in subsequent trial mixes.

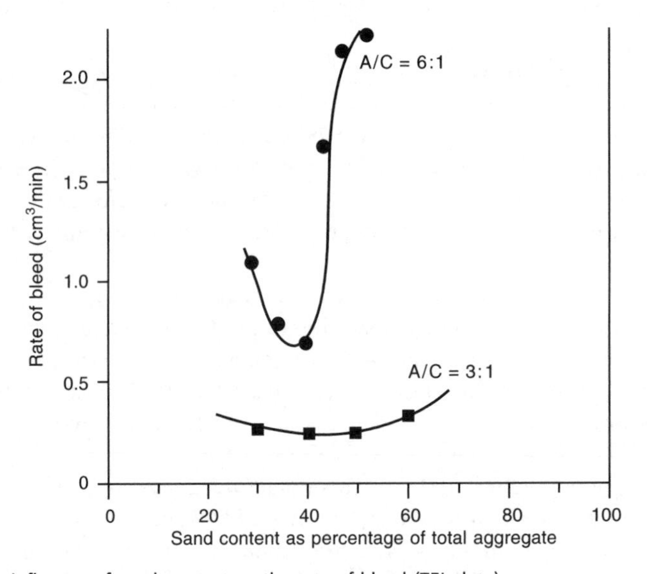


**Figure 13.4** The influence of sand content on the rate of bleed (TEL data).

A reduction in the water content of the mix will also reduce bleed, but where this is achieved with admixtures which also retard the set this may partially offset the benefit by increasing the time over which bleed can occur (Bamforth, 1986) as shown in Figure 13.5.

Another effective remedy for reducing bleed is air entrainment. The entrained air bubbles act in a similar way to very fine aggregate particles thus decreasing bleed, but without a deleterious effect on workability. However, air entrainment is not always justifiable, either economically or practically as it causes a reduction in strength, which must be offset by reducing the w/c ratio, and this may require an increase in cement content. It can be valuable, however, if the available materials are such that excessive bleed is unavoidable. The relationship between bleed, air content and w/c ratio is shown in Figure 13.6.

In view of the numerous factors which influence bleed and settlement, trial mixes should be undertaken to produce the optimum mix proportions for minimizing bleed. Full-scale placing trials prior to construction are also recommended for identifying any potential problems.

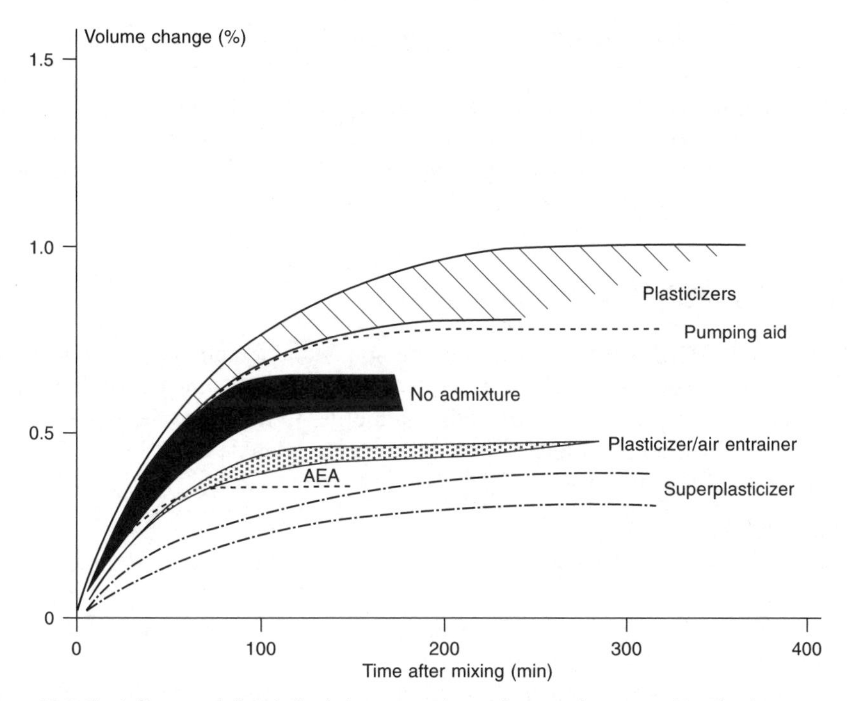


**Figure 13.5** The influence of chemical admixtures on the settlement of concrete (TEL data).

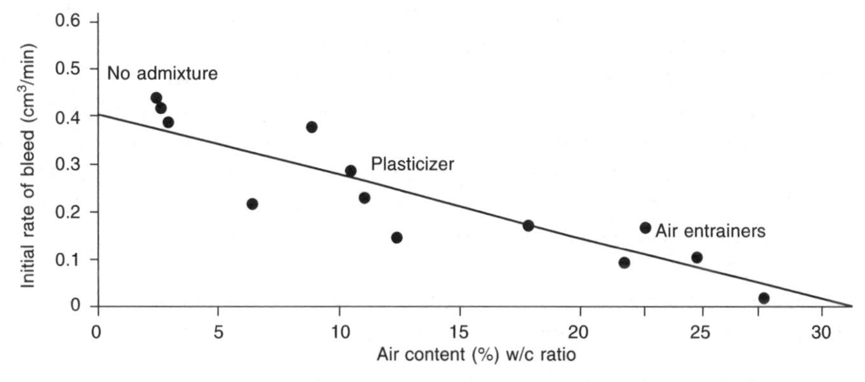


**Figure 13.6** A typical relationship between the initial rate of bleed and air content and w/c ratio of the concrete (TEL data).

## *Stiffening time*

When concrete is placed in large volumes it is advisable that rapid stiffening is avoided. This enables the upper layers to be compacted without damaging lower layers by transmitted vibration. The risk of cold joints is also reduced. However, excessive retardation of set can lead to increased bleed and increased formwork pressures. Concretes containing pfa will exhibit retarded set of roughly half an hour for each 10 per cent of pfa by weight of the composite cement. Ggbs has very little effect at replacement levels below 50 per cent

(Concrete Society, 1991) but will exhibit a retardation of about $1^1/_2$ hours at the 70 per cent level.

However, admixtures are often the most influential factors affecting concrete stiffening time. Even if retarders are not used, plasticizers based on lignosulphonates often cause retardation and superplasticizers at high levels of addition may act in a similar manner (Rixom and Mailvaganam, 1986).

## 13.4.3 Early age thermal cracking

In addition to the normal requirements for strength and workability, and the additional requirements described above, concrete for placing in a large volume has certain specific requirements regarding thermal characteristics in order to minimize the risk of thermal cracking. The following additional factors must be considered in relation to the risk of cracking due to early age thermal effects;

- Heat evolution
- Thermal expansion coefficient
- Tensile strain capacity

### *Heat evolution*

Hydration of cement is an exothermic reaction and in large-volume pours where heat dissipation is low, the temperature within the pour can rise significantly. In the centre of sections greater than 2 m thick, the temperature rise will be nearly adiabatic and be proportional to the cement content of the concrete mix (Bamforth, 1980). Values of maximum temperature of the order of 60–70°C are common.

In smaller pours where heat is more readily lost to the environment, the temperature rise is also affected by the rate at which heat is developed (Harrison, 1992). The rate and amount of heat generated by Portland cements depends on the fineness of grinding and the chemical composition. For example, there are data which indicate that cements with a high $C_3A$ content will develop the highest temperature rises (Bamforth; Lerch and Bogue 1934). Conversely cement with a low $C_3A$ content, e.g. sulphate-resisting cement, will tend to develop less heat.

With regard to concrete mix design, there are several factors which will influence both the rate of hydration of the cement, and the ultimate heat generated:

- Total content of cementitious material
- Type and source of Portland cement
- The type and proportions of composite cements which utilize either pfa or ggbs.

Other factors, such as the type of formwork, the geometry of the pour and the concrete mix temperature are also influential. These are beyond the scope of this section on mix design, and are discussed later in sections 13.5 and 13.6 appropriate to the nature of the pour being cast.

### *Cement content*

In the centre of a large-volume pour, which is close to adiabatic conditions, the temperature rise is approximately proportional to the cement content. Hence a reduction in cement content will effect a proportional reduction in temperature rise (Figure 13.7).

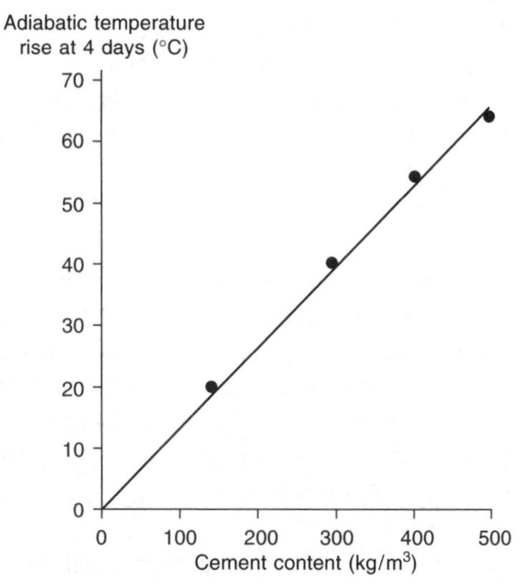


**Figure 13.7** Adiabatic temperature rise as affected by cement content (TEL data).

Water-reducing admixtures are beneficial in enabling a reduction in the water content required for a given workability and hence a reduction in cement content at a given level of w/c ratio (i.e. strength) and workability. Plasticizers will permit a reduction in water and cement of the order of 6–8 per cent while reductions of up to 20 per cent may be achieved with superplasticizers (Rixom and Mailvaganam, 1986).

Increasing the maximum aggregate size compatible with the section dimensions and reinforcement spacing will also contribute to a reduction in total cement content. However, there is evidence that this will also cause a reduction in the strain capacity of the concrete, and that this will more than offset the benefit of reduced temperature rise in relation to risk of cracking (Carlson *et al.*, 1979). This is illustrated in Figure 13.8.

A reduction in cement content can also be achieved by the use of microsilica, which is particularly effective in lower-grade concretes. For strength, the cementing efficiency

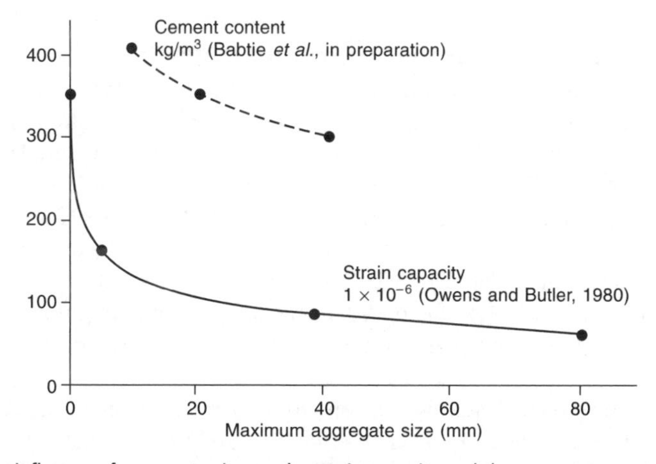


**Figure 13.8** The influence of aggregate size on the strain capacity and the cement content of concrete (Carlson *et al.*, 1979).

of microsilica is about 4, being part chemical (reactive silica) and part physical (pore blocking of fine particles), while its contribution to heat of hydration is only 1–3 times that of Portland cement (PC) (Concrete Society, 1993). Hence lower temperature rise can be achieved by the use of microsilica concrete, compared with PC concrete of the same strength grade.

### *Composite cements*

Composite cements based on combinations of Portland cement with either pfa (pulverized-fuel ash to BS 3892 Part 1) or ggbs (ground granulated blastfurnace slag to BS 6699) generally hydrate at a slower rate than Portland cements to BS 12 and produce a lower temperature rise. This applies to both manufactured blends, and blends produced in the mixer. The level of pfa in such blends is usually up to 40 per cent, whereas ggbs is used in larger amounts, often up to 75 per cent of the blend. The relative effects of different composite cements on the typical temperature rise in sections of different thickness is shown in Figure 13.9. See also Figure 13.10.

It must be appreciated, however, that not all ggbss exhibit the same performance. Some current sources are more reactive than those commonly used over the last 20 years, and provide less benefit in terms of reduced rate of heat generation (Coole, 1988) as shown in Figure 13.11. Similarly, blends of the various sources of pfa and PC may result in different hydration characteristics. It is, therefore, advisable to seek the manufacturers advice, or to obtain test results for the proposed materials and mix designs.

Numerous methods are available for measuring the heat generation or temperature rise in concrete. For thick sections adiabatic or near-adiabatic testing is preferred (Bamforth, 1978; Lerch and Bogue, 1934) as this represents most closely the conditions at the centre of the element. For practical site purposes, it is common to cast an insulated 1 $m^3$ block and to measure the temperature rise at the centre. A simple plywood form lined with 50 mm thick expanded polystyrene is adequate to provide comparative results.

It must also be appreciated that, while the weight for weight temperature rise of composite cements is reduced, the benefit may be partially offset by the need to use higher total cement contents if the concrete is required to meet the same 28-day standard cured strength.

### *Influence of early age temperature on strength*

While all concretes are adversely affected by high early age temperatures, there is now substantial evidence that in thick sections, the *in-situ* (heat cycled) strength of composite cement mixes will be higher than PC concrete, designed to the same grade. Results from six studies (Bamforth, 1980, 1984b; Owens and Buttler 1980; Dhir *et al.*, 1984; China Light and Power, 1991; Coole and Harris) are shown in Figure 13.12 for pfa concretes. In the most extreme cases, the difference in *in-situ* strength was in excess of 20 MPa, with more typical values of the order of 10 MPa. Additional benefit could be achieved, therefore, by reducing the materials safety factor, $\gamma_m$, for concretes containing pfa according to the predicted temperature rise in the element, hence enabling a lower grade of concrete to be used.

Table 13.4 shows typical mix proportions for concretes designed in accordance with the DOE mix design method (Teychenne *et al.*, 1988). PC mixes have been designed to achieve mean strengths of 30, 40 and 50 MPa and PC/pfa mixes (with 30 per cent pfa) have then been designed with the same total cement content. The standard 28-day strengths

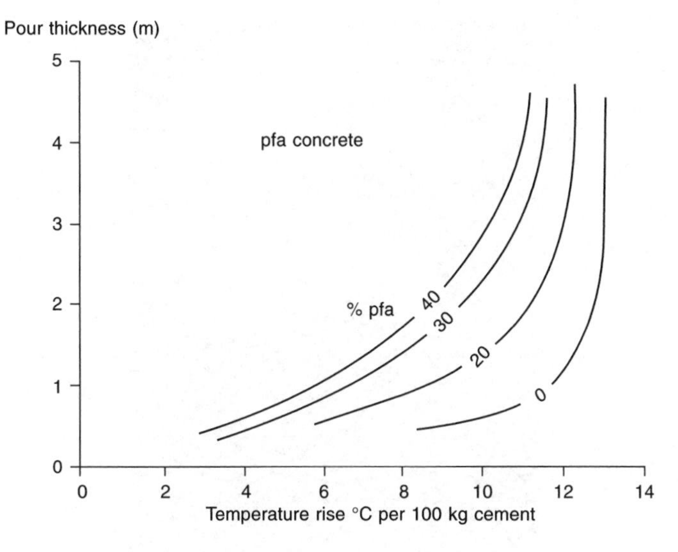


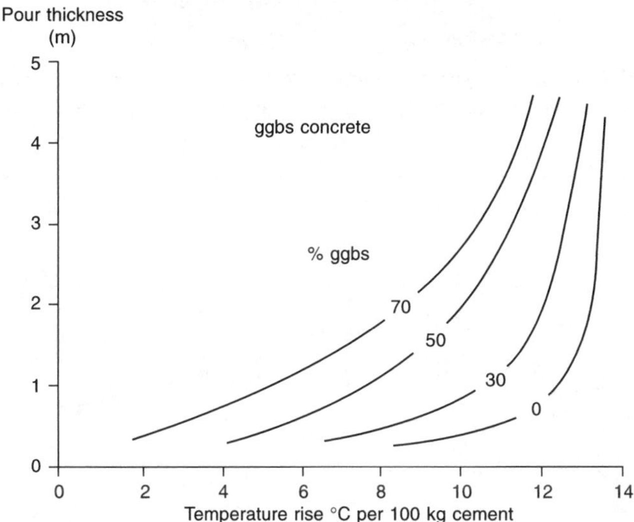


**Figure 13.9** The influence of pfa and ggbs on the temperature rise in concrete elements of different thickness (Bamforth, 1980).

of the latter were 5–8 MPa lower. This is of the same order as the benefit achieved as a result of the *in-situ* heat cycle for a temperature rise of about 30°C (i.e. consistent with 300 kg of PC/pfa in a thick section). This indicates that mixes with the same total cement content using either PC or composite cement with PC/pfa, will achieve the same *in-situ* strength at 28 days.

Thus in sections which will be subjected to a significant rise in temperature due to hydration of cement, it is suggested that the materials safety factor, $\gamma_m$, can be reduced to 1.25 when pfa is used in the mix or alternatively the strength grade may be specified at a later age, say 56 or 90 days.

Similar, but less substantial benefits may be achieved using ggbs (Bamforth, 1980;

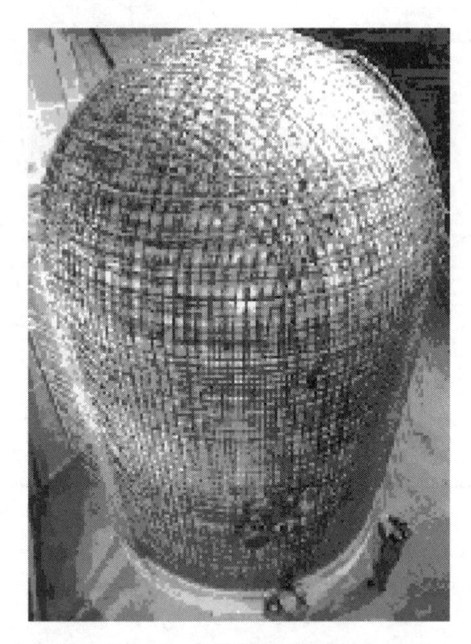

**Figure 13.10** One-tenth scale model of the Sizewell B Nuclear Power Station containment building. PFA was used to replace up to 40 per cent of the PC to minimize the risk of early thermal cracking.

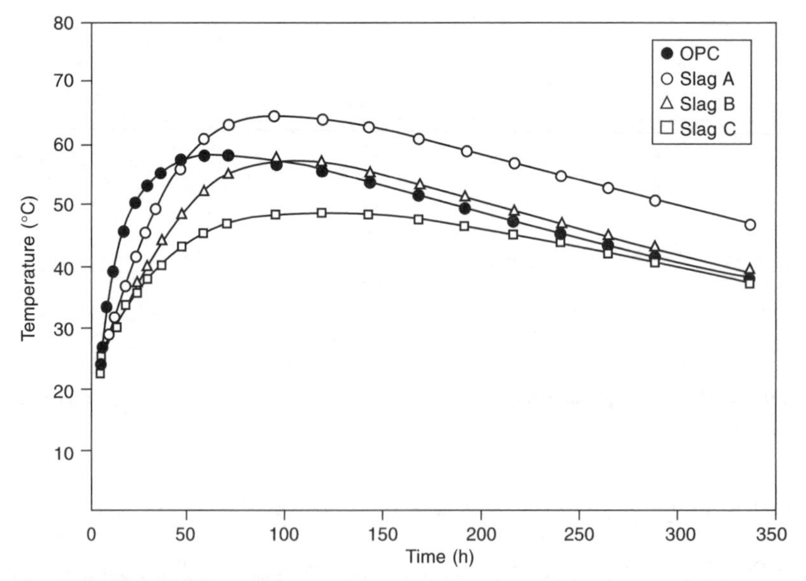


**Figure 13.11** The effect of different sources of slag on the temperature development in laboratory simulated 3 m deep pours (Coole, 1988).

Coole, 1988), although at this time insufficient data are available to enable specific recommendations regarding the change in $\gamma_m$.

The long-term strength gain of the pfa concrete is greater. Measurements obtained by the Building Research Establishment on cores taken from structures up to 33 years old (Thomas and Matthews, 1991) demonstrated that 'the ratio of core strength to 28-day

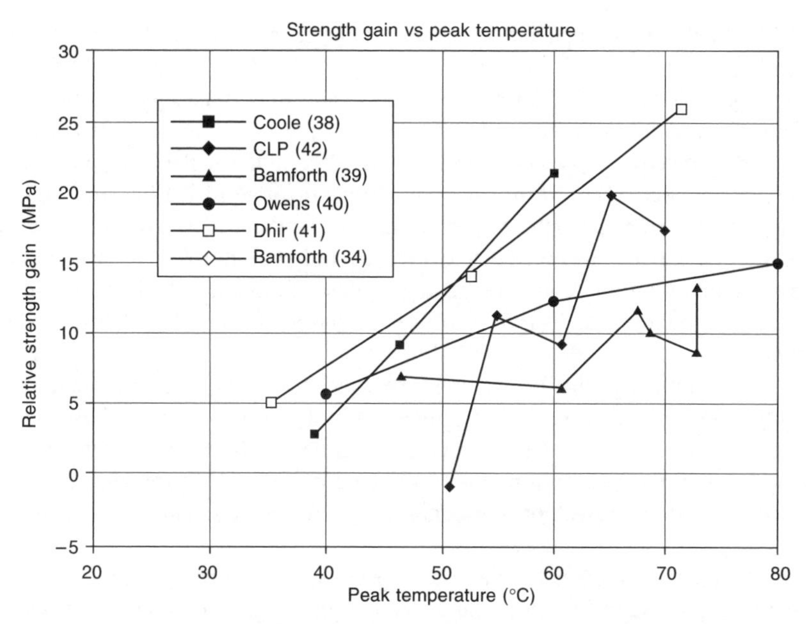


**Figure 13.12** The relative effect of pfa on the heat cycled strength of concrete, compared with Portland cement concretes.

**Table 13.4** Typical effect of composite cement (70% PC/30% pfa) on 28-day strength at a fixed total cement content

| | | | |
|---|---|---|---|
| PC content (kg/m$^3$) | 290 | 345 | 410 |
| W/C | 0.62 | 0.52 | 0.44 |
| Mean cube strength (MPa) | 30 | 40 | 50 |
| PC/pfa content (kg/m$^3$) | 290 | 345 | 410 |
| W/(C+F) | 0.57 | 0.48 | 0.40 |
| Mean cube strength (MPa) | 25 | 32 | 43 |
| Reduction in standard cube strength (MPa) | 5 | 8 | 7 |
| Enhancement under thermal cycling (MPa) | 6 | 8 | 10 |
| Net benefit of pfa on heat cycled strength (MPa) | 1 | 0 | 3 |

cube strength ranged from 146% to 240% (mean 193%) for the pfa concrete, and from 116% to 150% (mean 138%) for the PC concrete'.

The lower temperature rise and the enhanced resistance of composite cements to sulphate attack (Concrete Society, 1991) may also reduce the risk of delayed ettringite formation and its disruptive effects.

The concrete mix design should, therefore, contain the minimum cement content compatible with the specified strength and durability requirements, and preferably use a low-heat composite cement. In addition, the designer should consider compliance based on a later age strength, to take benefit from the continuing strength gain of concretes using these cement types.

## *Thermal expansion coefficient and strain capacity*

Thermal contraction cracking is discussed in greater detail in section 13.5. However, a reduced thermal expansion (and contraction) coefficient of concrete, when combined with a high tensile strain capacity, is of great benefit in reducing the potential risk of cracking.

The aggregate fraction of concrete (both coarse and fine) makes up the majority of its volume and thus exerts a major influence over the thermal expansion of the concrete. Different aggregate types produce concretes with different thermal expansion coefficients and some typical values (Bamforth, 1984a) are given in Table 13.2. Limestone aggregates may be particularly beneficial in this respect. Concretes containing some limestone aggregates can have thermal expansion coefficients which are only two thirds of the value exhibited by concrete containing siliceous gravel aggregates. Although not often used in large-volume pours, lightweight aggregate concrete also has a relatively low thermal expansion coefficient.

The tensile strain capacity $\varepsilon_{tsc}$ of concrete (and hence its crack resistance) is also significantly affected by the choice of aggregate (Bamforth, 1984a), with those aggregates exhibiting low values of $\alpha$, also achieving concretes with the highest values of $\varepsilon_{tsc}$. Limestone is again the best option among the commonly used aggregates and siliceous gravel is the worst. Lightweight aggregate also offers significant benefits in relation to $\varepsilon_{tsc}$.

The combination of increased tensile strain capacity and low thermal expansion coefficient associated with the use of limestone (and lightweight) aggregate in concrete makes this aggregate type particularly suitable for large-volume pours where cracking is to be avoided. Table 13.2 gives estimated limits on temperature differentials for concretes with defined values of $\alpha_c$ and $\varepsilon_{tsc}$.

Under conditions of sustained loading, the tensile strain capacity is marginally reduced by the use of ggbs or pfa in concrete. This is believed to be due to the lower creep associated with these mix types (Concrete Society, 1991; Bamforth, 1978). In the absence of specific quantitative data, it is impossible to give revised values of tensile strain capacity. However, based on limited observations of large-volume pours, minimum levels of ggbs and pfa have been proposed (Bamforth, 1980) in relation to the pour thickness, to ensure that the reduction in temperature rise more than offsets the reduction in strain capacity (see Table 13.5)

**Table 13.5** Recommended minimum levels of ggbs and pfa

| Pour thickness (m) | Minimum percentage level of: ggbs | pfa |
|---|---|---|
| Up to 1.0 | 40 | 20 |
| 1.0 to 1.5 | 50 | 25 |
| 1.5 to 2.0 | 60 | 30 |
| 2.0 to 2.5 | 70 | 35 |

These figures are for guidance only, and it is recommended that further research be carried out in this area.

## *Summary*

A successful concrete mix design for large-volume pours should have the following characteristics:

- Appropriate workability and strength

- Cohesiveness and low bleed
- Extended stiffening time to minimize cold joints
- Low-expansion aggregate such as limestone
- Minimum cement content compatible with strength and durability and the requirements for placing
- Low-heat cement, probably based on blends of Portland cement with either pfa or ggbs.

## 13.5 Construction of large-volume pours

This section provides guidance on the practical aspects of construction of large volume pours and means for the control of early age thermal cracking.

### 13.5.1 Reinforcement design and detailing

When designing reinforcement for large elements which have the potential to be cast as single-pours, in addition to the normal structural requirements, the designer must also consider the following aspects:

1 Control of early age thermal cracking (section 13.2)
2 Avoidance of steel congestion to enable access for vibrators through the top mat steel and to enable proper compaction around the steel and at faces and corners. Pokers vary in size from 25 mm to 75 mm diameter, with the larger-sizes being most appropriate in large-volume pours, and should be lowered vertically at about 0.6 m centres.

### 13.5.2 Formwork

As large-volume pours are generally relatively low height (a few metres at the most) there are no special requirements for the formwork, other than to provide vents through which rainwater can be expelled. The vents must be located having first defined the sequence of concreting.

When adjacent pours are to be cast, it is common to use expanded metal formwork at joints. This is supported during casting, but the supports can be removed within a few hours due to the accelerated strength development in the large pour.

### 13.5.3 Construction procedure to minimize restraints

When an element is being cast as a number of individual large-volume pours, the sequence of construction should aim to minimize restraints. Infill pours should be avoided as far as is practically possible. End-on-end construction provides a free end to accommodate movement, minimizing restraint to bulk thermal contraction.

With regard to the delay between adjacent casts or lifts, the following general rules apply;

1 The time between adjacent strips or lifts should be minimized, as the principal restraint acts along the direction of the joint (Figure 13.13(a)). If the previous pour is still warm, temperature (and hence strain) differentials between new and old concrete will be minimized as the two pours contract together.
2 The time between end-on elements should be maximized, as the principal restraint acts perpendicular to the joint (Figure 13.13(b)). This eliminates contractions in the old pour prior to casting the new element.

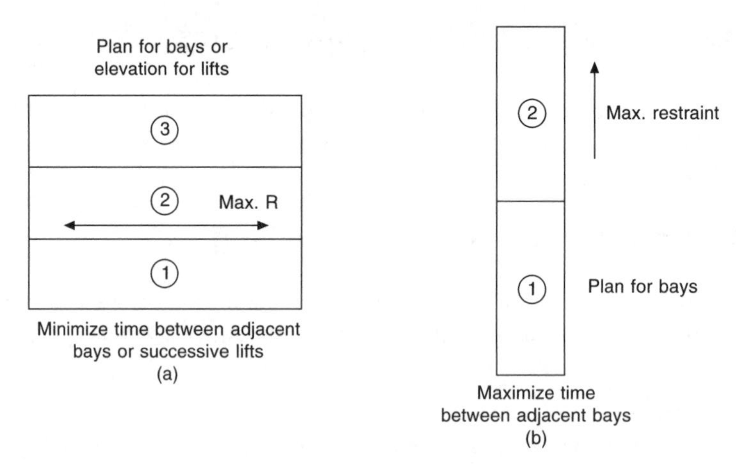


**Figure 13.13** Pour configuration as it affects restraint and delay between casting adjacent bays.

3 To avoid excessive restraint in either direction the preferred shape for a given volume of concrete is as close to square as possible.
4 If a source of local high restraint falls within the area of the proposed pour, it may be prudent to introduce a construction joint close to it. The gap, to be filled once the large pour has fully contracted, should be as small as practically possible.

In relation to restraint, it is also important to recognize the following:

- In very thick sections, external restraint is likely to be much less significant, as large forces can be generated even though the concrete may be relatively young. In such case, cracking is more likely to result from internal restraints caused by temperature differentials.
- In sections which are relatively thin, and cast against existing mature concrete, external restraint will be predominant. The influence of internal restraints are diminished as the external restraint prevents differential internal strains developing.

In practice both internal and external restraint exist simultaneously, although in many cases one or the other is predominant. Where combined restraints exist, the effects are superimposed (Bamforth and Grace, 1988) as shown in Figure 13.14 for a thick wall cast onto a rigid foundation. This demonstrates the difficulty in predicting restraints, and the level of simplification which has necessarily been adopted by BS 8007 and CIRIA Report 91. The latter acknowledges that during heating, external restraint will cause a reduction in the tensile stresses in the surface zone as bulk expansion is resisted. It recommends that, where external restraint exists, precautions need not be taken to minimize differential temperature as such action may lead to increase the risk of cracking.

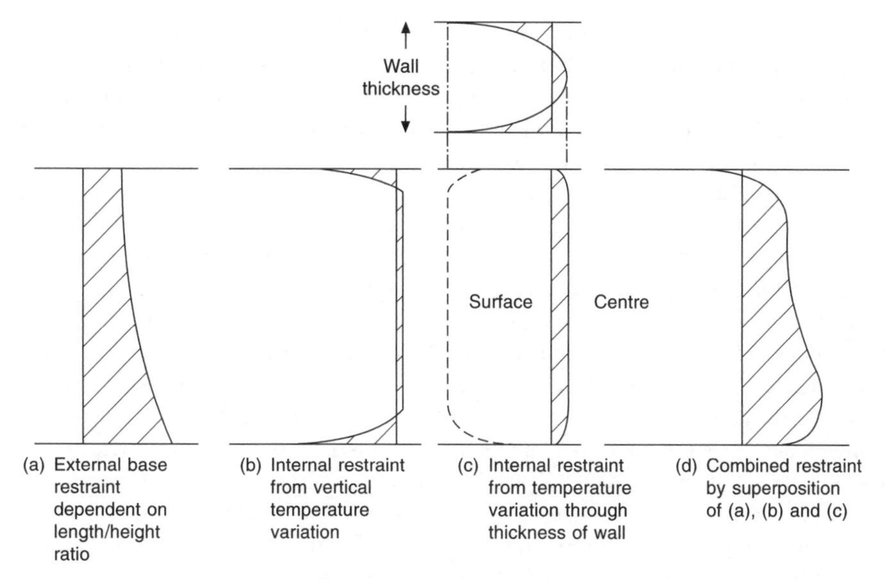


**Figure 13.14** The superposition of internal and external restraint (Bamforth and Grace, 1988).

Because of the inherent complexity of the problem, each case must be considered separately. However, the following broad guidelines may be followed:

1 For elements of large cross-section, where external restraint is low, steps should be taken to limit temperature differentials as defined in Table 13.2 for different mix types.
2 For thin elements (<500 mm) subject to high external restraint, temperature limits in excess of those given in Table 13.2 may be tolerated, subject to an analysis of the likelihood and/or extent of cracking.

Reasonable estimates of external restraint can be derived using the ACI method described in section 13.2.

## 13.5.4 Placing methods

The four principal methods of concrete delivery into large-volume pours are **skip**, **pump**, **chute** and **conveyor**. The method selected will depend on the required rate of placing, the ease of access and the plan area of the pour. When the plan area is relatively small, chutes provide an effective method of placing, but will only be successful for large-volume placements in which high-workability concrete is used. Using this system with superplasticized flowing or self-levelling concrete, placing rates of 80 $m^3$ per hour have been achieved, 400 $m^3$ in 5 hours. Conveyors provide access over greater distances, but are most successful for low-workability concrete. High delivery rates can be achieved (60 $m^3$/h), but this is dependent on the mix type, being higher for low-slump concretes.

However, for large-area, large-volume pours, mobile boom pumps provide the greatest degree of flexibility combined with very high placing rates, typically of the order of 40–50 $m^3$/h, but with reported throughputs approaching 100 $m^3$/h for individual pumps. This compares with typical rates of 8–12 $m^3$/h achieved using crane and skip.

The volume of concrete, the depth of the pour and its plan area all influence the sequence of placing and the number of placer units. Large-volume pours are best placed in successive layers. Each layer is normally up to about 600 mm thick. Shallower layers reduce the risk of poor compaction, but layers up to 600 mm thick can be properly compacted with sufficiently large vibrators.

The principal objective is to achieve a monolithic element with no planes of weakness, hence the construction sequence must be designed to minimize the period of exposure of the newly placed surfaces. This will lessen the effects of warming in hot weather, and reduce the effects of surface damage during rainfall.

For long, narrow pours, the step method is most effective in this respect and involves commencing successive layers after previous layers have been compacted sufficiently far ahead. Fitzgibbons (1977) proposed that the distance between the progressing faces could be ten times the layer thickness, i.e., up to about 5 m. The ACI (1984), however, recommends that this distance should not exceed about 1.5 m although this applies specifically to low-slump concrete. The actual distance used will depend on the thickness of each layer and the stiffness of the fresh concrete. For thin layers (300 mm) of low-slump concrete, 1.5 m is more appropriate. When thicker layers (500 mm) of higher-slump concrete are used 5 m may be necessary to avoid 'sagging' of the front.

To advance the 'stepped' front at a reasonable rate several placers are necessary, but if the pour is deep enough to complete in, say, three layers, the length of the section becomes unimportant.

When several placers are used, the options are numerous, and often involve dividing the pour into discrete segments, each of which is cast in layers. For circular elements, there is often the temptation to start at the centre and work towards the circumference. However, this makes it increasingly difficult to maintain a 'live' front, and working from side to side may offer better control.

## 13.5.5 Compaction

For large-volume pours, the only practical means of compaction is by the use of immersion (poker) vibrators. Recognized good practice involves placing the poker vertically into the concrete and at about 0.6 m centres. At each location, the time required to ensure expulsion of air will depend on the workability of the concrete.

A simple nomogram derived from data in Orchard (1973) is shown in Figure 13.15 which enables typical rates of compaction to be estimated. For a 75 mm slump concrete, an output of 20 $m^3/h$ per poker can be expected. This is consistent with ACI (1984d) recommendations which suggest rates of 6–20 $m^3/h$ for concretes with a slump less than 75 mm.

Where plastic settlement cracks are observed, it is acceptable to revibrate the surface layer. Concrete Society Report No. 22 (1992) confirms that this need not be damaging, provided that it is applied neither too soon, allowing further bleed, nor too late, causing disruption of partially 'set' concrete. Site trials are recommended to define the window within which to work.

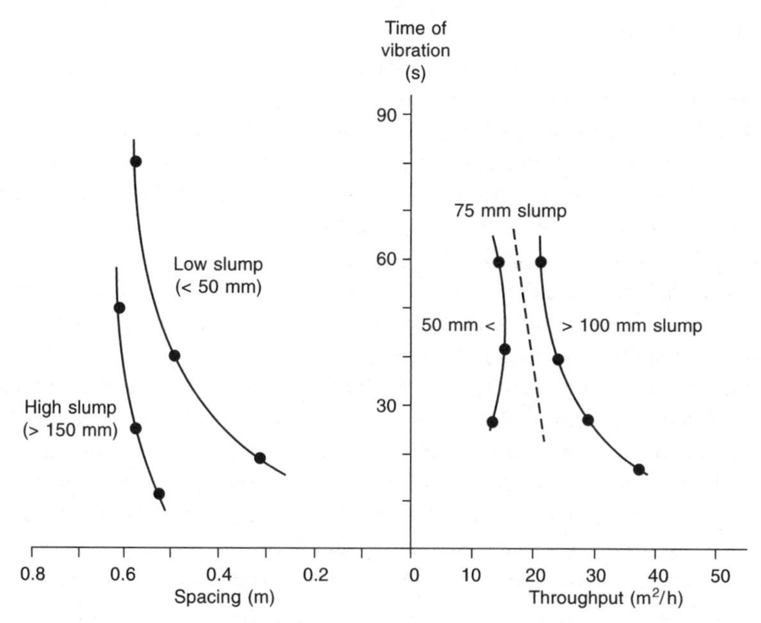


**Figure 13.15** Nomogram relating concrete workability, poker spacing and throughput.

## 13.5.6 Finishing

Large-volume pours will generally be either trowelled or floated. Normal practice will apply, but it must be recognized that the concrete mix may have been designed to have an extended stiffening time to avoid cold joints. For this reason, there may be a longer than usual delay before initial stiffening has taken place and the moisture film has disappeared from the surface.

## 13.5.7 Thermal control

The control of temperature rise and temperature gradients is often a key feature of large-volume concreting and various precautionary measures can be taken to minimize the risk of early age thermal cracking, or to limit its extent as follows:

- Select a low-heat generating mix by minimizing the cement content, within the constraints of the specification and requirements of the fresh concrete.
- Use composite cements (i.e. blends of PC with pfa or ggbs) to reduce the rate of heat evolution and the peak temperature rise.
- Select aggregates with a low coefficient of thermal expansion.
- Select aggregates which result in concrete with a high strain capacity.
- Reduce the concrete mix temperature by cooling of the component materials or the use of ice with the mix water.
- Use liquid nitrogen to cool the fresh concrete.
- Cool the concrete *in-situ* by the use of embedded cooling pipes.
- Use insulation to increase surface temperatures and to minimize thermal gradients.

## *Low-heat concretes*

Low heat means a low cement content mix or the use of a cement with low-heat generating characteristics. There is now considerable evidence supporting the beneficial use of composite cements (Portland cement blended with either pfa or ggbs) in large-volume pours (see section 13.4). However, as cements may vary between sources, tests are recommended to determine the temperature rise for the selected mix. Coole (1988) has reported that isothermal calorimetry cannot predict the effect of all cement types, being deficient for blends with ggbs in particular. Adiabatic, or near-adiabatic (heavily insulated block) tests provide data which are of more direct relevance. On site, a typical test block would involve a 1 metre cube insulated with 50 mm of expanded polystyrene, but may be of any dimension which simulates pours to be cast.

Reductions in cement content may be difficult to accommodate due to other requirements of the mix, although in section 13.4 it is proposed that the benefit of the enhanced *in-situ* strength of pfa concretes be used to permit a reduction in $\gamma_m$, the materials safety factor.

The use of microsilica (also known as silica fume) also enables the cement content to be reduced while maintaining the strength (Concrete Society, 1993). While the microsilica itself makes a contribution to heat evolution, the effect on strength is greater with an overall benefit in terms of reduced temperature rise (Fédération Internationale de la Précontrainte, 1988).

Increasing the maximum aggregate size will also enable a reduction in cement content and this is often proposed. However, there is also a change in strain capacity resulting from the use of larger aggregate/lower cement content mixes (Figure 13.8) and the detriment appears to be greater than the benefit of reduced heat.

## *Aggregate type*

As discussed in section 13.4, those aggregates which have a low coefficient of thermal expansion also tend to exhibit higher strain capacity, and considerable benefits can be achieved in terms of relaxation of allowable temperature differentials by judicial aggregate selection. Although, for most contracts, the locally available aggregates can be accommodated in the design process, it may be appropriate for critical structures, e.g. where either radiation shielding or containment is a principal performance criterion, to use aggregates which will minimize the risk of cracking. As discussed in section 13.4, of the naturally occurring aggregates limestone will produce concrete with the greatest resistance to cracking, while siliceous gravel is least resistant.

Lightweight aggregate concrete outperforms all naturally occurring aggregates in terms of low $\alpha_c$ and high $\boldsymbol{\varepsilon}_{tsc}$, but also requires a higher cement content for the same grade, which partially offsets some of this benefit. Acceptable limiting temperature changes and differentials are given in Table 13.2 which illustrate the significance of aggregate type.

## *Reducing the mix temperature*

Reducing the mix temperature has a number of benefits:

- It reduces the rate of workability loss
- It delays the stiffening time
- It slows down the rate of heat generation
- It reduces the peak temperature, and thus the temperature drop from peak to ambient

The most common method for cooling the mix involves cooling one or all of the individual mix constituents. As the aggregate comprises the largest single component of the mix, cooling the aggregate will have the greatest effect on the concrete mix (except where ice is used). This can be achieved practically by:

- Shading aggregate stockpiles, to prevent solar gain
- Sprinkling the stockpiles with water (preferably chilled)

Other more extreme measures include immersion in tanks of chilled water, spraying chilled water on aggregate on a slow-moving belt, or blowing chilled air through the stockpiles (American Concrete Institute, 1984e). When the aggregates are cooled with water, this water must be taken into account during batching, by adjustment to the added mix water.

An alternative method is to cool the aggregate using liquid nitrogen ($LN_2$). This process has been used in Japan (Kurita *et al.*, 1990) to cool sand to –140°C and achieve a reduction in mix temperature of about 10°C.

While the aggregate constitutes the greatest mass in the mix, the water has the greatest heat capacity, and hence cooling efficiency. The specific heat of water is about five times that of the aggregate and cement (4.18 kJ/kgK compared with 0.75 kJ/kgK respectively). In addition, water is much easier to cool and the temperature can be controlled more accurately. As it is practical to cool water to about 2°C this is a very effective method of cooling the mix.

For very effective cooling, ice can be used. The latent heat of ice is 334 kJ/kg, and the heat absorbed by 1 kg of melting ice is equivalent to cooling 1 kg of water through about 80°C or 1 kg of aggregate through 445°C. Hence, a relatively small volume of ice can have a significant cooling effect. Ice is usually added to the mix in the form of crushed or shaved ice as it is important to avoid incorporating larger fragments of ice that melt slowly leading to the formation of voids in the hardened concrete. To obtain an indication of the requirements for cooling the individual mix constituents, the following general rules may be applied (Nambiar and Krishnamurthy, 1984). To cool the concrete by 1°C requires that:

(a) the aggregate is cooled by 3°C
(b) the mixing water is cooled by 7°C
(c) 7 kg of mixing water is replaced by ice

A nomogram illustrating the effect of cooling the various mix constituents is given in Figure 13.16. This has been developed for a specific mix to provide a rapid means for identifying what steps are needed in specific cases. The nomogram is based on the simple method of mixture as follows:

$$\text{Concrete temperature} = \frac{0.75(T_c M_c + T_a M_a) + 4.18 T_w M_w - 334 M_i}{0.75(M_c + M_a) + 4.18(M_w + M_i)} \qquad (13.6)$$

where $T$ is temperature, °C, and $M$ is mass, kg/m$^3$, and the subscripts c, a, w and i represent cement, aggregate, water and ice.

Where greater accuracy is required for a mix of different proportions, this equation can be used.

Recent developments with the use of liquid nitrogen ($LN_2$) now enable the mixed concrete to be cooled on site. The method involves spraying a mist of $LN_2$ (which has a

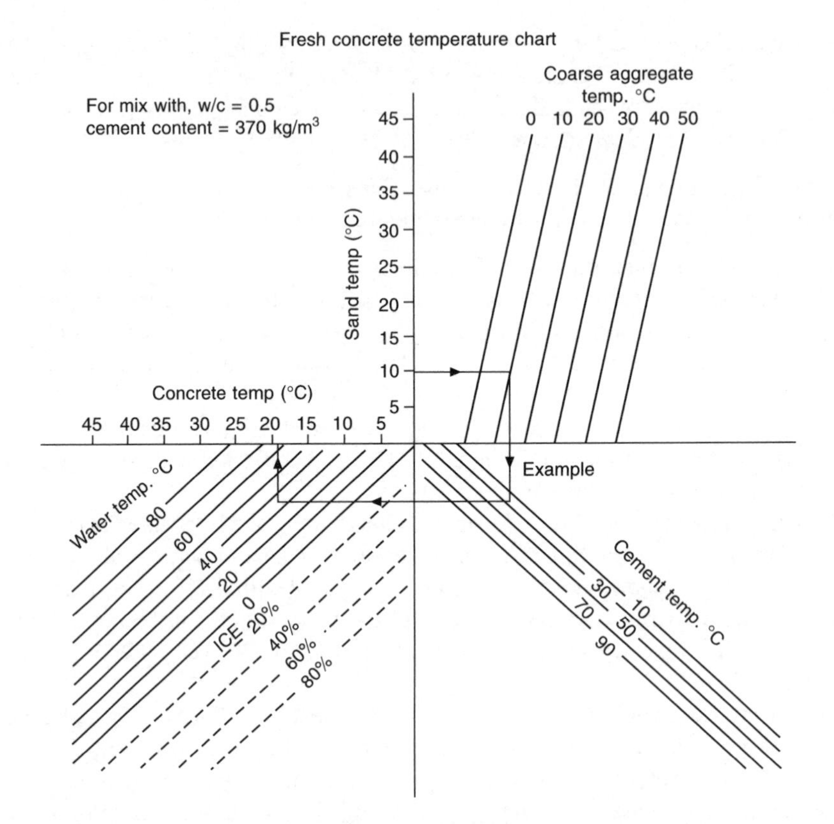


**Figure 13.16** Nomogram for estimating concrete mix temperatures.

boiling point of 77 K (–196°C) into the mixer at a controlled rate. This is achieved with a customized lance which is inserted into a mixer truck.

The latent heat of boiling of the $LN_2$ is 199 kJ/kg, about 60 per cent of that of melting ice. With 100 per cent efficiency, about 12 kg of $LN_2$ is, therefore, needed to cool 1 $m^3$ of concrete by 1°C. In practice, the efficiency is not much less than 100 per cent. During construction of an X-ray facility for Maidstone Hospital (Bobbins, 1991) about 1 tonne of $LN_2$ was used to cool a 6 $m^3$ mixer load through about 13°C. This is equivalent to about 13 kg of $LN_2$ per °C change in temperature per $m^3$ of concrete (or 16 litres/°C/$m^3$). A similar rate of consumption of 15 litres/°C/$m^3$ was recorded during construction of the Faro Bridges in Denmark (Henriksen, 1983).

## *Embedded cooling pipes*

When the specification prevents the use of concrete with low-heat generating characteristics an alternative means for reducing temperature is the incorporation of an embedded cooling system. This has the advantage that the system can be designed to accommodate any mix type. The cooling system must be designed to remove heat at the required rate without inducing excessive internal temperature differentials. For this reason, plastic pipes may be preferred to metal pipes as the heat flow into the coolant is limited by the conductivity of the pipe itself.

A method for designing a cooling system is given in ACI Report 207.IR-35 (1984e).

Typical pipe spacings are likely to be of the order of 1 m in large-volume pours with relatively low heat-generating capacity. In elements cast using high-grade structural concrete, closer spacing, of the order of 400–500 mm may be necessary.

The location of internal cooling pipes is unlikely to coincide with the reinforcement, as the latter is generally concentrated near the surface. However, there may need to be some collaboration between the designer and the contractor when this approach is adopted. This technique is often used to construct water retaining structures. For example, in Germany (Anon., 1982), a rainwater reservoir was cast as a single element using this technique to control the thermal stresses. The 2 m thick base (plan area 27.4 × 28.2 m) and the 1.2 m thick 12 m high walls were continuously cast (1100 $m^3$ of concrete) over a 32-hour period, to avoid joints which were prohibited by the specification.

## *Insulation*

Insulation is used to minimize temperature differentials by enabling the surface temperature to increase. It should be restated, however, that where the external restraint is predominant, the temperature differentials required to cause cracking are increased and insulation may be unnecessary. Furthermore, the use of insulation, by increasing surface temperatures, in such cases will increase the risk of surface cracking. However, when appropriate, insulation can be achieved by various methods including:

- Maintaining the formwork in place if it has insulating properties, e.g. plywood or steel backed with expanded polystyrene
- Foam mats, blankets and quilts
- Soft board
- Sand on polythene
- Tenting
- Ponding

The method adopted will depend on whether the surface is vertical or horizontal, the period for which the insulation must be maintained, and the access required to the surface.

Guidance on the minimum period for insulation in relation to the minimum pour dimension is given in (Bamforth 1984a) and is reproduced in Table 13.6. For very thick elements, the insulation may need to remain in place for up to 3 weeks.

**Table 13.6** Minimum periods of insulation to avoid excessive temperature differentials

| Minimum pour dimension (m) | Minimum period of insulation |
|---|---|
| 0.5 | 3 |
| 1.0 | 5 |
| 1.5 | 7 |
| 2.0 | 9 |
| 2.5 | 11 |
| 4.5 | 21 |

When using insulation care must be taken to avoid its early age removal, as this can generate thermal differentials which are much worse than might have occurred if no insulation had been used (Figure 13.17).

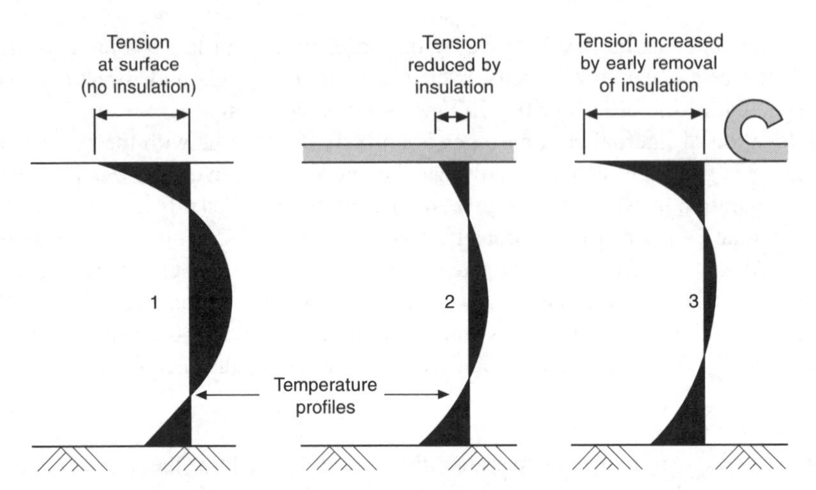


**Figure 13.17** Gradients developed in thick sections if insulation is removed too early.

Once the commitment to insulate has been made, it must be maintained until the temperature within the pour has cooled to a level which can accommodate the change which will occur when the insulation is removed. In this situation, the surface zone is heavily restrained by the bulk of the concrete and the expected temperature drop in the near surface zone must not exceed the limiting values given in Table 13.2 for a restraint factor of 1.

### *Curing*

The curing requirements for large-volume pours are no different from those of concrete in general, with the exception of the thermal curing requirements to avoid excessive temperature differentials. However, the sealing of the surface associated with thermal curing, which generally extends over several days, is likely to be more than adequate. In most cases, additional curing beyond the end of the thermal curing period will be unnecessary.

## 13.5.8 Monitoring and actions in the event of non-compliance

Temperature monitoring is generally required in order to demonstrate compliance with specified temperature limits. The simplest and most cost-effective method, particularly in very deep or thick pours, is to use embedded thermocouples. These are simple to install and to monitor. The output can be monitored with simple manual devices. On large projects, automatic logging systems can be set up.

In its simplest form, the monitoring will involve thermocouples at the centre of a section and close to the surface (about 10 mm). The maximum value at the centre and the difference between the centre and the surface will then be assessed against specified limits.

When the specification is more complex, for example by limiting differences between a mean value and a surface value, the designer must specify how the mean is to be determined. This involves defining how many locations through a section must be monitored, where the thermocouples are to be located, and how the results are to be analysed.

Where contracts specify the need to meet temperature limits, the contractor should obtain specific details from the designer on how compliance is to be demonstrated in relation to the location and number of monitoring points and the interpretation of the results obtained.

Full-scale mock-ups are very useful for providing data on temperature profiles and hence for identifying the precautions needed in terms of insulation, and increasing use is being made of computer models.

If the pour is a one-off, or if it is the first of many, pairs of thermocouples should be installed at critical locations to provide back up in the event of a failure. The mix selection and adopted site procedures should be such that the temperature limits will be met. However, contingency plans should be defined which can be implemented when the results exceed 'action levels'. For example, if the allowable temperature differential is being approached, and is likely to be exceeded, additional insulation may be applied.

If the maximum allowable temperature is approaching its limit, no actions can be taken in thick sections unless embedded cooling pipes are being used. However, this value is much easier to predict based either on the mix constituents and the mix temperature or if available, on laboratory test results for adiabatic temperature rise (Coole, 1988).

If allowable limits on temperature differential are exceeded the following action can be taken:

- Inspect the surface for cracks after the pour has cooled back to ambient.
- If the cracks are excessive, initiate remedial measures.
- Consider the consequences of internal cracking – it is possible that internal cracking may occur even if there is no evidence of surface cracking. In this case, it is difficult to establish the extent of cracking without extensive coring and even this is unlikely to indicate crack widths.

The value in carrying out a detailed investigation to find cracks which do not propagate to the surface is, therefore, still doubtful, except in circumstances where such cracks are clearly detrimental to the performance and safety of the structure.

If cracks are to be repaired, the method used will depend on whether the cracks are 'dead' or 'alive', and on the serviceability requirements of the structure. In a massive raft for example, surface cracks are unlikely to increase after the element has cooled and sealing will be primarily for durability. In this case injection grouting is most common, using an epoxy resin when the cracks are less than about 1 mm width. Procedures are defined in Ed Allen and Edwards (1987).

For cracks which are likely to continue moving, it is usual to make provision for further movement to continue after repair.

The allowable maximum temperature is applied to limit the detrimental effect of temperature on the strength of concrete and sometimes to prevent delayed ettringite formation. If this value is exceeded, core testing will provide evidence of whether the deterioration in strength is excessive. BS 6089 (1981) provides guidance on the interpretation on *in-situ* strength, and it is advisable that the designer and the contractor agree the acceptance level of core strength before results are obtained.

## 13.6 Review

**Advantages** – The principal advantages of large-volume pours are as follows:

- Avoidance of construction joints, with their consequential implications on cost and timescale.
- The construction of monolithic units.

**Achieving successful large-volume pours** – The sizes of pours now being cast internationally demonstrate that large-volume pours are feasible and cost-effective. However, their successful completion relies upon:

- Recognition of the principal technical requirements at the design stage
- Cooperation between the designer, the contractor and the concrete supplier
- Adequate notice of start of casting given by contractor to all other parties
- Consideration of technical factors relating to the concrete including:
  - stiffening time
  - plastic settlement
  - heat of hydration
  - selection of materials and mix design
- Planning to ensure that the concrete can be continuously delivered, transported, placed and compacted within the allocated timescale and in such a way as to avoid cold joints and other construction defects.
- Contingency plans to take account of breakdown at any of the stages in the construction process, and the possibility of having to stop concreting, or to extend beyond the allocated timescale.

Numerous large pours have been cast worldwide and a limited review of some of the larger pours is given in Table 13.3. Some specific examples are given below:

- **Thames Barrier cofferdams** (Grace, and Hepplewhite, 1993) A total of 6600 $m^3$ of concrete was cast over a 3-day period. The mix used a 50:50 blend of PC and pfa to reduce heat of hydration, and a retarding admixture to achieve a delay in stiffening time of up to 36 hours. The 150 mm slump concrete was placed under water using 300 mm diameter tremie pipes on a 7 m grid fed by mobile boom pumps.
- **The 'Messeturm' Mat foundation** (Anon., 1998) The largest continuous concrete pour reported to date is the $5^1/_2$ m thick 17 000 $m^3$ mat foundation for a 260 m high office tower in Frankfurt, Germany. The pour took 78 hours to complete using four mobile boom pumps and 90 readymix trucks from six plants were used for concrete delivery. Workers operated on a 12-hour shift basis, with 120 workers per shift.
- **LINAC facilities** (Robbins, 1991) Although of relatively modest volume, liquid nitrogen was used to reduce the mix temperature to below 10°C for the construction of the joint free walls and roof of a radiotherapy centre at Maidstone, Kent. The 375 $m^3$ of concrete for the 1–2.5 m thick walls was cast over a 13-hour period using a single mobile boom pump, using a low-heat grade C30 mix comprising 70 per cent ggbs, 30 per cent PC and a 20 mm limestone aggregate. The formwork was removed after 4 to 5 days and no cracking was reported.
- **Foundation mat, California, USA** (Laning, 1990) The most rapid reported rate of continuous output from a pump over a prolonged period was achieved during construction

of a 5700 m$^3$ foundation mat in California. Five pumps achieved an average throughput of 496 m$^3$/hr (99 m$^3$/hr per pump) to complete the pour in only 11$^1/_2$ hours using a grade C20 mix with pfa.

- **LLRF facility, Devonport** (Cullen, 2001) An 11 000 m$^3$, 10 m thick foundation for a low-level refuelling facility was cast, under water, in layers up tp 3.75 m thick. The foundation was unreinforced but required to exhibit early thermal cracking no more severe than had it been designed as a reinforced element. A combination of cement comprising 75 per cent ggbs and the use of limestone aggregate (coarse and fine) achieved the characteristics of low heat, low thermal expansion and crack resistance needed to achieve this demanding requirement. High-flow, self-compacting concrete was placed by tremie in volume of up to 3600 m$^3$, over periods of up to 90 hours. No compaction was possible. Temperatures were monitored and cores cut to record *in-situ* performance. The pours were all completed satisfactorily and within the specification requirements (see Figure 13.18).

**Figure 13.18** LLRF facility at Devonport.

## References


American Concrete Institute (1984a) Mass concrete for dams and other massive structures. *Manual of Concrete Practice*, 207, IR-70.

American Concrete Institute (1984b) The effect of restraint, volume change and reinforcement on cracking of massive concrete. *ACI Manual of Concrete Practice*, 207.2R-73.

American Concrete Institute (1884c) Standard practice for selecting proportions for normal, heavyweight and mass concrete. ACI *Manual of Concrete Practice*, 211.1-81.

American Concrete Institute (1984d) Standard practice for consolidation of concrete. *Manual of Concrete Practice,* 309-72 (revised 1982), ACI.

American Concrete Institute (1984e) Cooling and insulating systems for mass concrete. *ACI Manual of Concrete Practice*, 207.4R-80.

Anon. (1982) West Germany – Sub surface rainwater reservoir. *Construction Industry International*, August, 17–18.

Anon. (1988) Pour sets world record. *ENR*, 1 December 14.

Anon. (1990) Largest concrete pour. *ICI Bulletin*, No. 32, September.

Bamforth, P.B. (1980) *In-situ* measurement of the effect of partial Portland cement replacement using either fly ash or ground granulated blastfurnace slag on the performance of mass concrete. *Proc. of Instn. of Civ Engrs*, Part 2, **69**, September, 777–800.

Bamforth, P.B. (1982) *Early age thermal cracking in concrete*. Institute of Concrete Technology, Technical Note TN/2.

Bamforth, P.B. (1984a) *Mass Concrete*. Concrete Society, Digest No. 2.

Bamforth, P.B. (1984b) The effect of heat of hydration of pfa concrete and its effect on strength. In *Ashtech 84, Second Int. Conf. on Ash Technology and Marketing*, London, September pp 287–294.

Bamforth, P.B. (1986) Admixture – A contractor's view. In *World of Concrete, Europe 86*, London.

Bamforth, P.B. (1978) An investigation into the influence of partial Portland cement replacement using either fly ash or ground granulated blastfurnace slag on the early age and long term behaviour of concrete. Taywood Engineering Limited Research Report No. 914J/78/2067.

Bamforth, P.B. and Grace, W.R. (1988) Early age thermal cracking in large sections – towards a design approach, In *Proceedings of Asia Pacific Conference in Roads, Highways and Bridges*, Institute for International Research, Hong Kong.

British Concrete Pumping Association (1988) *The manual and advisory safety code of practice for concrete pumping*. BCPA, revised edition.

British Standards Institution (1981) Guide to assessment of concrete strength in existing structures. BS 6089.

British Standards Institution (1982) Pulverised Fuel Ash, Part 1, Specification for pulverised-fuel ash for use as a cementitious component in concrete. BS 3892: Part 1.

British Standards Institution (1985) Structural Use of Concrete: Part 1, Code of Practice for Design and Construction, BS 8110.

British Standards Institution (1986) Specification for ground granulated blastfurnace slag for use with Portland cement. BS 6699.

British Standards Institution (1987) Design of concrete structures for retaining aqueous liquids, BS 8007.

British Standards Institution (1991) Concrete: Methods for specifying concrete mixes. BS 5328, Part 2.

Carlson, R.W., Houghton, D.L. and Polivka, M. (1979) Causes and control of cracking in unreinforced mass concrete. *ACI Journal*, July.

China Light and Power (1991) *PFA Concrete Studies, 1988–1998*. Prepared by CLP, Taywood Engineering Limited and L.G. Mouchell and Ptnrs (Asia), Vol, 9, Supplementary Report at Year 1 and Appendix, CLAP.

Clear, C.A. and Harrison, T.A. (1985). *Concrete pressure on formwork*. CIRIA Report 26 No. 108.

Coole, M.J. and Harris, A.M. The effect of simulated large pour curing on the temperature rise and strength growth of pfa containing concrete. Blue Circle Industries Plc, Research Division, Greenhithe, Kent.

Concrete Society (1991) *The use of ggbs and pfa in concrete*. Concrete Society, Technical Report, 40.

Concrete Society (1992) *Non-structural cracks in concrete*. Concrete Society, Technical Report 22, 3rd edn.

Concrete Society (1993) The use of microsilica in concrete. Report of a Working Party, Technical Report No. 41.

Cooke, T.H. (1990) *Concrete Pumping and Spraying – a practical guide*. Thomas Telford, London.

Coole, M.J. (1988) Heat release characteristics of concrete containing ground granulated blastfurnace slag in simulated large pours. *Mag. of Concr. Res.*, **40,** No. 144, 152–158.

Cullen, D. (2001) Construction beneath the waves. *Concrete Engineering*, Summer.

Danish Concrete and Structural Research Institute (1987) CIMS – Computer integrated test and monitoring systems for construction sites. CSRI, Dr Neergaards Vej 13, Postboks 82, DK-2970, Horsholm, February.

Dhir, R.K., Munday, J.G.L. and Ong, L.T. (1984) Investigations of the engineering properties of OPC/pulverised fuel ash concrete: Strength development and maturity. *Proc. Inst. Civ. Engrs*, Part 2, Vol. 77, No. 6, June, pp 239–254.

Ed Allen, R.T.L. and Edwards, S.C. (1987) *Repair of Concrete Structures*. Blackie and Sons Ltd, London.

Emborg, M. (1989) *Thermal stresses in concrete structures at early ages*. Doctoral Thesis, Lulea University of Technology.

Fédération Internationale de la Précontrainte (1988) Condensed silica fume in concrete. FIP State-of-the-Art Report, Thomas Telford, London.

Fitzgibbon, M.E. (1971) Large pours – 3, continuous casting. *Concrete*, February 35–36.

Grace, J.R. and Hepplewhite, (1993) Design and construction of the Thames Barrier cofferdams. *Proc. Instn. Civ. Engrs*, Part 1, **74**, May, 191–224.

Harrison, T. (1992) *Early age thermal crack control in concrete – revised edition*. CIRIA Report 91.

Henriksen, K.R. (1983) Avoidance of cracking at construction joints and between solid sections. *Nordisk Betong*, 1, 17–27.

Kurita, M., Goto, S., Minehishi, K., Negami, Y. and Kuwahara, T. (1990) Pre-cooling concrete using frozen sand. *Concrete International*, June, 60–65.

Laning, A. (1990) Ocean-side foundation mat pour presents challenges for concrete. *Concrete Construction*, September, 669–670.

Lerch, W. and Bogue, R.H. (1934) Heat of hydration of Portland cement pastes. *Journal of Res. Nat. Bur. Stand*, **12**, No. 5, May, 645–664.

Masterton, G.T. and Wilson, R.A. The planning and design of concrete mixes for transportation, placing and finishing, CIRIA, London, Report 165.

Nambiar, O.N.N. and Krishnamurthy, V. (1984) Control of temperature in mass concrete pours. *Indian Concrete Journal*, March, 67–73.

Orchard, D.F. (1973) *Concrete Technology, Vol, 2 – Practice*. Applied Science Publishers Ltd, London.

Owens, P.L. and Buttler, F.G. (1980) The reactions of fly ash and Portland cement with relation to strength of concrete as a function of time and temperature. *Proc. of 7th Int. Conf. on Chemistry of Cement, 3*, IV-60, Paris.

Randall, F.A. (1989) Concrete pumps complete massive foundation pour in $13^1/_2$ hours. *Concrete Construction*, February 158–160.

Rixom, M.R. and Mailvaganam, M.P. (1986) *Chemical Admixtures for Concrete* E&FN Spon, London.

Robbins, J. (1991) Cool customer. *New Civil Engineer*, 12 September.

Teychenne, D.C. *et al.* (1988) Design of normal mixes. Dept of Environment, London.

Thomas, M.D.A. and Matthews, J.D. (1991) Durability studies of pfa concrete structures. BRE Information Paper 11/91, June.

# 14

# Slipform

*Reg Horne*

Slipform refers to a method of constructing vertical concrete structures using a self-climbing formwork system that had its origins in the USA over one hundred years ago and is capable of climbing at a rate of 30–40 m per week. However, in the early 1970s an automated paving system was introduced from the USA (by Gomaco) which is often referred to as 'slipform'. To differentiate, the latter is classified as horizontal paving or horizontal slipforming.

Very little has been written about slipforming, and there are no British or US Standards, Codes of Practice, or similar documents to refer to for guidance. The following is a brief outline of the history, development, method of operations, and concrete practice for vertical slipform. It is not intended in any way to be a method statement for the contractor.

## 14.1 Vertical slipforming

### 14.1.1 Types of structure suitable for slipform construction

After a hundred years of operations, slipform is still the fastest method of construction for designated vertical structures, whether the system is operated on a 24-hour basis or on a dayshift-only basis, yet there are many in the construction industry today who have never had the opportunity to become involved with this specialized system.

The technique has been used to create some of the world's tallest and largest structures ever built in reinforced concrete, including:

- Straight and tapering chimneys
- Observation towers

- Concrete gravity structures (oil platforms)
- Bridge pylons and piers
- Mine-shaft headgear towers

It has also been used to build innumerable basic structures, such as:

- Silos
- Linings to shafts below ground
- Surge shafts
- Liquid containment vessels
- Service cores for commercial buildings
- Lift and stair shafts

Typical structures, suitable for slipform construction, are usually over 25 m in height, for economical reasons, stable in the free-standing condition, or are able to be stabilized with temporary or permanent bracing, and benefit from a very short construction period. The economical qualifying height for each structure may be less if repetition is involved, which may be the case for bridge piers or caissons.

Slipform construction is an essential method for constructing concrete silos and chimneys over, say, 4 m in diameter, quickly and economically, with other structures needing to be assessed for plan configuration and height. Tapering structures with wall reductions, either gradual, over a short height, or stepped, can be accommodated. However, it is advisable to seek the opinion of specialist contractors at an early stage in the proceedings to establish the suitability of structures for slipforming.

## 14.1.2 Standard equipment

The equipment generally used for slipform is similar in format to that used for normal wall formwork, consisting of face panels backed by walings and supported by strongbacks, for either single- or double-face 'shutters'. However, the height of the face panels is usually limited to 1.2 m, and the strong backs are formed into yokes straddling the wall which are mechanically lifted by the jacks. These jacks are attached to the crossbars of the 'H'-shaped yokes which are in turn supported by rods or tubes positioned within the wall or structure. The tubes are supported from the foundation and are laterally 'braced' by the formed concrete during the slipform operations, by the frames in openings or by temporary columns formed specifically for the purpose.

The early designs consisted of simple two-deck systems with timber face panels and walings. The yokes were also constructed from timber and required low crossbars to cater for the induced stresses. The slipforms were raised manually by screw jacks, all necessitating labour-intensive operations. These systems prevailed until after the Second World War, when the lifting systems were replaced with synchronized hydraulic circuits and hydraulic jacks. The later development of stronger steel yoke frames allowed the critical distances under the cross-bars to be increased substantially, making reinforcement fixing and general operations far easier.

Two main types of system have evolved over the years,

- A lighter two-deck system, generally utilizing 3 tonne capacity hydraulic jacks and climbing on 25 mm diameter rods (Figure 14.1)

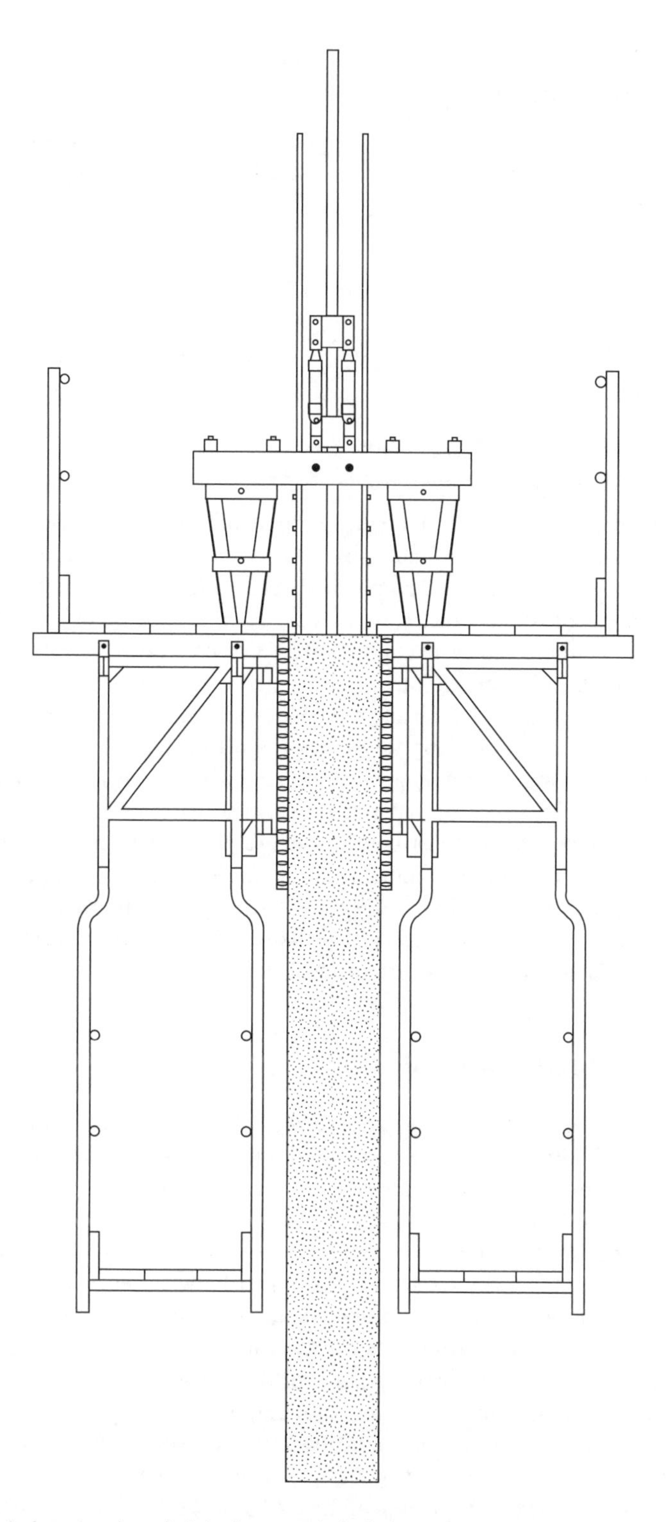

**Figure 14.1** Typical section through light-duty two-deck slipform.

- A heavy-duty three-deck system, using 6 tonne capacity jacks and climbing on 48 mm diameter tubes (Figure 14.2).

The former lightweight system is only suitable for the construction of small simple structures, whereas the heavy-duty system is suitable for more complex structures and provides numerous advantages, including:

- Greater safety.
- Additional storage
- Weather protection
- Greater accuracy
- Good access
- Easier distribution
- Wider spacing between lifting positions
- Better and easier support for ancillary services

The systems are composed of standard components, resembling Meccano sets, that can be built to various plan shapes and re-used many times for maximum economy. Steel has replaced timber for all the major components providing better strength and durability. Only the decks are made up in either softwood or plywood for ease of shaping and fixing, although in certain circumstances even these areas can be arranged with standard panels, composed of metal decking or similar.

Complex plan shapes can be built using modular panels and standard slipform equipment, only occasionally requiring specially fabricated items to complete the arrangements, but at the same time allowing economical re-use for the majority of the component parts.

These systems, employed on non-tapering projects, are assembled at the commencement level, usually a base or foundation structure, and are plan-braced before commencement to maintain the correct profile throughout the operations. Minor changes, such as wall reductions, can easily be accomplished during slipforming. The introduction of additional walls at higher levels or the termination of walls before completion of the structure can also be accommodated, albeit with certain restrictions.

## 14.1.3 Tapering equipment

The heavy-duty systems referred to earlier have been adapted to cater for tapering structures and are significantly different from the standard arrangements for uniform sections. The lifting unit of the slipform operates in a similar manner for both systems. The formwork sections between the lifting units are rigid for non-tapering work, whereas they need to cater for reductions in wall length, as the tapering takes effect. This is achieved with overlapping panel arrangements, usually positioned between each lifting point, which take up the variation in each section. Large reductions may necessitate the removal of formwork panels as the slide progresses or even the removal of certain lifting units. A stop to the sliding process may be necessary for this purpose.

For circular tapering structures, a system of radial beams or a special truss arrangement allow the lifting positions to be traversed towards the centre, effectively reducing the diameter. Wall reductions are brought about by 'squeezing' either the inside shutter, or the outside shutter, or both towards the centre of the wall. A force is applied, either mechanically or hydraulically, to move the shutter, on either face, away from the yoke leg, whilst still

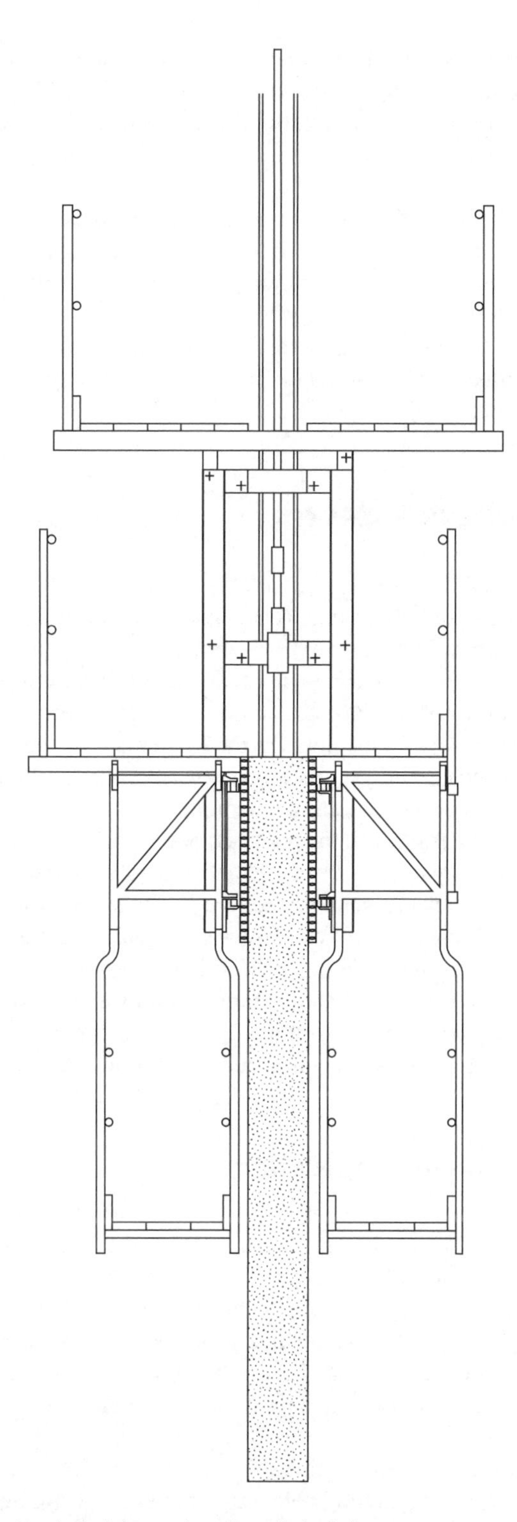

**Figure 14.2** Typical section through heavy-duty three-deck slipform.

supporting the main sections of walings. Alternatively, the positions of the vertical members of the yoke frame are moved towards the centre of each wall achieving the same effect. All incremental adjustments are absolutely critical, need to be fully synchronized and need to maintain the correct degree of taper on the shutter, as referred to in the next section.

H-shaped, rectangular, or hollow structures may also be tapered, but in every case the tapering operations must maintain equilibrium, or gain independent support to remain on line. The reductions must be carefully monitored and this is normally achieved by making the incremental reductions against fixed, accurate scales.

Tapering structures are usually in excess of 80 m high and consist of observation towers, chimneys, bridge pylons and piers. They reduce for aesthetic reasons as well as for design consideration.

## 14.2 Design of the slipform

It is not proposed to detail design calculations for slipform rigs in this chapter. The design of slipform shutters has mainly been empirical. Operating the slipform produces a tide of changing forces, but only when the operations are poorly carried out or the concrete mix is poorly designed do the forces become unacceptable. Filling the forms with concrete is a slow process and does not produce the normal pattern of pressures associated with standard concrete pressure diagrams. Initially, the shutter is propped at the base, changing the yoke legs into propped cantilevers. As soon as the slipform is lifted off the ground, the yoke legs revert to their normal operating situation.

The main concern is the friction at the interface of the concrete and the shutter and that any induced forces that may occur do not become excessive. Other main considerations are the adequacy of the decks to support the live loadings, and the combination of the loads, plus any frictional loads that may overpower the jacks. This is one more reason for using a heavy-duty system that is better able to cope with a combination of loadings. However, it is reassuring to note that slipforming is a failsafe method of operations. It is impossible for the slipform to fall as it would only attach itself to the structure, irrevocably as far as the operation of the jacks is concerned, should this ever happen.

### 14.2.1 Standard system operations

The jack capacities have evolved from the strength of the rods or tubes used with the systems – 3 tonne for 25 mm rods or 6 tonne for 48 mm tubes – and the style of slipform rig is related to the inherent carrying capacity. Jacks are generally composed of lower and upper clamp heads supported by twin or annular rams. The rams open and close hydraulically to climb the jack up the support and lift the slipform. The lifting units are usually controlled by one hydraulic circuit, enabling the system to lift in increments according to the 'set' of the rams.

The tubes or rods are generally withdrawn and re-used for economical reasons. To achieve this, sleeves extending slightly below the shutter and hanging from the crossbars are drawn through the concrete as the system climbs, leaving circular voids for the tubes to sit in. This allows them to be withdrawn upon completion or at intervals for high structures, without adhesion to the concrete structure.

Positioning of the lifting units is critical for the correct operation of the slipform system. Whereas the lighter systems are limited and the jack positions are usually placed at 1.8 m centres maximum along a wall, the heavy-duty system allows the centres of the jacking units to be safely increased to 2.5 m, making steelfixing and other activities so much easier. Many factors influence the design of the jack layout and retribution would be swift for an ill-conceived arrangement.

The shutters are assembled to the plan layout, preferably without the aid, or otherwise, of kickers. The yoke frames are installed, followed by the working decks, before the hydraulic circuits and jacking positions are completed. The electrical supply and the circuits for the lighting, poker vibrators and power tools, but not least, the power units for lifting the slipform also have to be installed.

When the system is complete and ready to operate, the shutter is filled with concrete, in layers, over several hours and in accordance with the setting time of the concrete, so that once the climbing begins, it becomes a continuous operation. As the climbing commences the inspection platforms can be completed.

The placing of the concrete is a systematic layering system based on maintaining the initial set of the concrete at the mid-position of the shutter. A critical operation before commencing the concreting is to set the 'taper' on the shutter. This involves accurately adjusting the verticality of the shutter, relative to the line of the wall, so that the form panels lean in at the top and lean out at the base position to provide a 'batter'. This has the effect of reducing the wall width at the top and increasing the width at the bottom of the shutter by a few millimetres and the wall is formed halfway down at the true thickness.

The success of a slipform relies upon several factors including a well-assembled rig, good planning to match plant and material requirements, good discipline throughout the execution of the operations, but more importantly, the design and provision of the correct concrete mix.

## 14.3 Concrete mix

### 14.3.1 Performance and development

The taper on the shutter, referred to in section 14.1.3, is the key to the slipform technique. Concrete is systematically placed in the forms in layers of approximately 150–200 mm deep and immediately vibrated. The concrete slumps and starts to form the shape of the structure, half-way down the shutter, as the initial stiffening commences. The concrete forms as it separates from the face panels due to the taper on the shutter. It is therefore essential that the initial stiffening occurs in the centre of the shutter, to form the wall at the correct thickness.

The concrete emerging at the base of the shutter, for a regular slipform, is often only 4 or 5 hours old and in a green condition.

Traditional construction requires that the concrete is sufficiently workable to ensure full compaction and that it meets the minimum strength requirement at a given age. Slipform construction, however, not only demands that the concrete remains workable during distribution and placement to ensure full compaction, but it must be cohesive to avoid segregation or bleeding. It must remain plastic in the top half of the shutter with a capability of internal mobility plus external deformation without suffering planes of weakness or actual fracture. Furthermore, it must also provide a laitence for the lubrication of the forms at the interface to prevent anything other than nominal friction.

Until the advent of pumping, there was always an air of mystique surrounding slipform concrete design, but today a standard 'pump mix' often requires very little adjustment to provide a suitable slipform mix, as they both have to meet similar criteria, with the exception of the setting times.

A typical slipform mix, therefore, requires a higher volume of fine material than would normally be required for maximum cement economy with the effect of providing a matrix in which coarse aggregate can rotate, whilst also providing a zone of low shear strength at the shutter face. It must also retain the correct amount of water for the hydration process to be completed, without any tendency to bleed. This is particularly important considering the method of vibration that is adopted for slipforming.

The speeds of the slipforms in the early years were purely related to the performance of the cements, aided and abetted by the ambient temperatures and the amount of water added to the mix. OPC in the UK, for instance, provided a concrete mix that would normally attain slipforming casting rates of between 150 mm/h and 300 mm/h. The water content was the sole 'accelerator' or 'retarder' for the mix concerned. The casting rate and the minor control that was available fortunately matched the outputs of the workforce and the available plant.

The 1960s and 1970s provided a more adventurous era, when larger, more complex structures were undertaken by slipform construction. Typical examples that were constructed by slipform were the Toronto CN Tower, a tapering structure 350 m high, the 'NatWest' building in London, a complex plan arrangement with a height of 190 m, and oil platforms of large diametric proportions. The oil platforms were slipformed in shallow coastal waters in one operation, to provide flotation, whilst maintaining an even keel.

This period also saw an increase in the use of admixtures. Retarders were mainly used for slipforming, allowing the speeds to be adjusted to match the distribution and placement of materials. With large floating structures the speed of sliding was reduced, in some instances, to as low as 75 mm per hour, which necessitated careful monitoring of the concrete for successful results.

New materials, particularly cement substitutes, were also developed and later incorporated into slipforming operations, providing long-term advantages for the concrete and some disadvantages for the slipforming operations. PFA, with its low heat generation and slow gain in strength, produced varying results with the initial set of the concrete and proved even more unpredictable when combined with plasticizers to meet the lower w/c ratios demanded during the 1980s and 1990s. Although ggbs is a little more predictable, it is now usual to limit the cement substitution for slipforming, whenever possible, to 30 per cent of total cementitious content in order to achieve some consistency in the setting times.

PFA and ggbs produce good finishes due to the migration of the fines to the surface during vibration. Any other fillers, such as kaolins, if approved for use, have the same effect.

Microsilica is an extremely fine pozzolan which is used to improve the properties of concrete. Unfortunately, it has a very resinous effect and does not assist the slipform method of operations. Its use, therefore, must be carefully monitored, if specified, and it is advisable to limit the quantity used to 5 per cent of the cementitious weight. Like all pozzolans it reduces the heat of hydration and slows down the strength development in the early stages.

## 14.3.2 Mix design

The concrete mix for slipform should be designed to comply with the contract specification and the relevant standards, that would be expected of any normal mix. As referred to earlier, an easy reference would be akin to a 'pump mix' based on a 75 mm or 100 mm slump, depending on the likely ambient temperatures to be encountered.

The initial stiffening of the concrete should occur at a given time after batching, to allow for delivery, distribution, and the time taken to slide to the half-way position in the forms – for a single silo, say 18 m diameter, with normal site conditions, 3.5 hours would be anticipated.

## 14.3.3 Cement

The cement and aggregate selection may be determined by regions, but must comply with the job specifications. The cement used will also have a bearing on the possible use and type of admixtures.

It is usual practice to use solely OPC, or OPC combined with either pfa or ggbs. Only occasionally is sulphate-resisting cement specified as this has been largely superseded by the use of pozzalans. The setting time of the cement, given on the certificate, is a guide as to how the mix may react, but the standard paste is very different from the cement when combined with the rest of the materials. Other important factors affecting the performance of the mix are:

- The cement temperature
- The 24-hour strength
- Ambient temperatures.

## 14.3.4 Aggregates

### *Coarse aggregate*

Coarse aggregates should be considered for the overall design of the concrete mix and not for the slipform requirements. Rounded aggregates are certainly not a prerequisite, as many successful slides have been carried out using sea-dredged materials or crushed rock. The normal size for the aggregate used is 5–20 mm. Occasionally, the size has been limited to 10 mm for areas that are very congested with reinforcement, and in special circumstances lightweight aggregate has been used, although this application requires the addition of sand to ensure compatibility with slipforming. A structure that is suitable for slipforming is unlikely to require aggregate in excess of 20 mm.

### *Fine aggregate*

The success of the slipform mix revolves around the combination of the fine aggregates and the cement. Of the total aggregate (fine + coarse) a good slipform mix should require between 38 per cent and 48 per cent of fine material depending on the other parameters. The fine aggregate grading should meet the medium category, as too coarse a grading would permit bleeding, whilst too fine a grading would retain too much water for too long a period and cause slumping. For the fine aggregate grading the sieve analysis should indicate around 50 per cent passing the 600 micron sieve with a good distribution throughout the lower banding, to provide the correct mass for retaining the water, to ensure good workability, a good finish and complete hydration. Fine aggregate achieved from crushed rock would probably be too coarse and would have to be complemented with a dune sand or similar fine material. If cement substitutes are chosen, the higher proportion of cementitious material will often compensate for a dearth of fines.

### 14.3.5 Admixtures

The development of high-strength concretes has brought about the prolific use of plasticizing admixtures to control the w/c ratio, increase workability and maintain high strengths. The strength of concrete used in slipforms has increased considerably in recent years to meet the demands of the designers. Normal core design now generally requires concrete strengths of C35 to C40 and civil engineering projects often call for strengths ranging up to C80 or greater.

The plasticizer often has a retarding action particularly when overdosed, and when used in conjunction with pozzolans. However, pozzolans are often required as partial cement replacement for large projects to lower the heat of hydration and improve durability. In some of these cases a plasticizer still needs to be used in conjunction with a retarding agent. The recent use of plasticizers formed from polymer systems has improved the performance of concrete with cement replacement and reduced the variability with regard to setting times. As in all cases with admixtures used under such stringent conditions, trial mixes are necessary to establish addition rates and setting times. Other admixtures, such as air entrainment are rarely used or required for vertical slipforming.

### 14.3.6 Distribution on the slipform rig

The concrete is placed in layers, systematically around the slipform, and distribution can be achieved in many ways. The plant chosen for servicing the slipform combined with the layout of the structure will often dictate the methods to be employed. Logistics will often affect the thinking as well as the labour rates prevailing in the area of operations.

For medium-sized cores or circular structures requiring between 4 and 8 cubic metres per hour, concrete could be delivered towards the walls, by chute, directly from the crane-skip with a small clearing-up operation required afterwards. Alternatively, the concrete could be deposited on the working deck, by skip or pump, with the labour force being responsible for final distribution, by shovel to the allotted areas (Figure 14.3). This system would cater for in excess of 6 cubic metres per hour. A rate of over 6 metres per hour would allow distribution to the walls to be carried out directly by pump. This method is often ideally suited for circular structures, providing the pump boom is able to reach.

For silos or cores serviced by mobile crane, the provision of one or two wet hoppers,

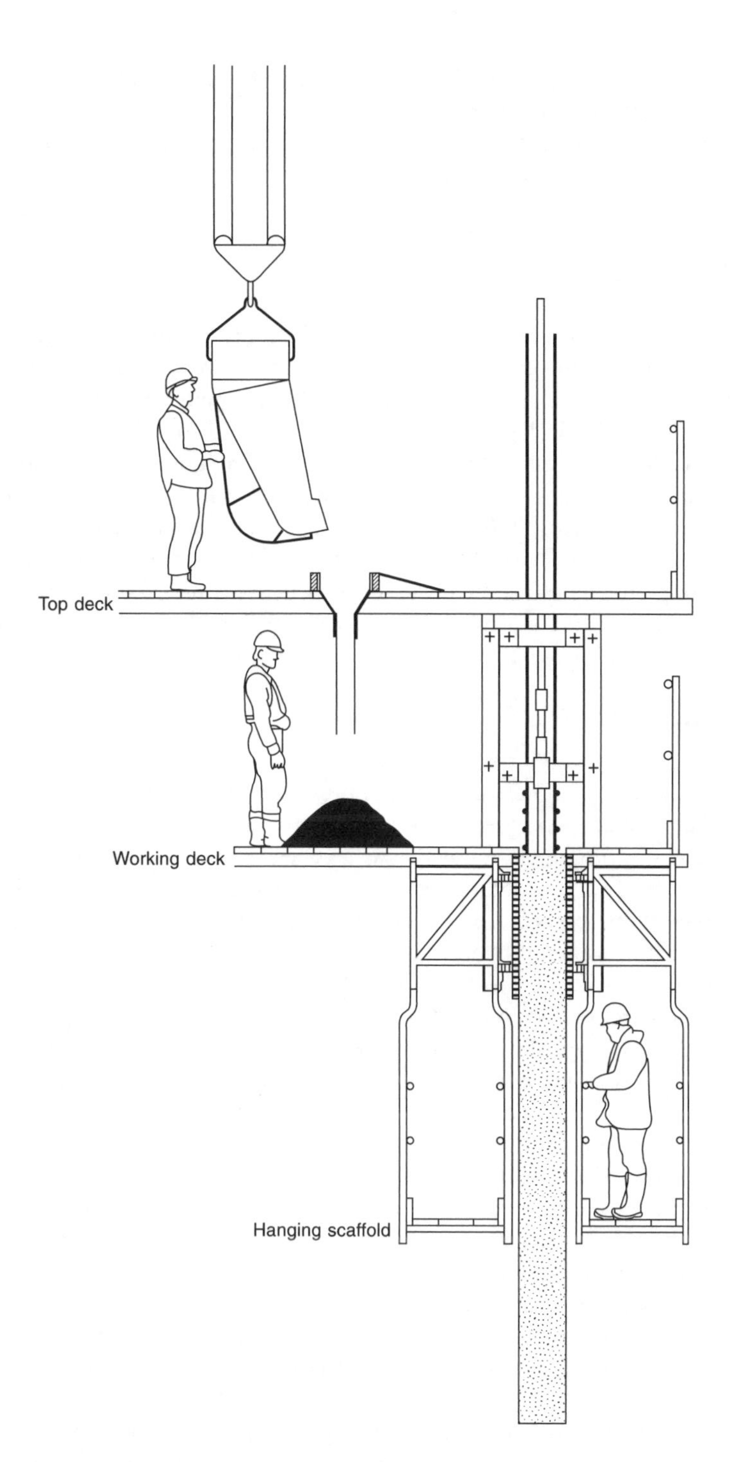


**Figure 14.3** Typical section through heavy-duty three-deck slipform.

strategically placed, would allow final distribution to be achieved by wheelbarrow. This is an old-style method of distribution that is still proving useful.

For any method chosen, it is still necessary to clean up after each operation to avoid old concrete being placed with the next fresh layer.

## 14.3.7 Vibration

Vibration is a major contribution to the success of a slipformed structure, as it affects the finish, the strength and the curing. The placement of concrete for a slipformed structure is very different from the method associated with normal pours. To reiterate, the concrete is placed systematically in layers, with each layer formed from a similarly aged batch. Each layer should be placed and immediately vibrated by immersing the poker vibrator at approximately 600 mm centres for several seconds, purely to compact the concrete. The vibrator must not be used for placing the concrete, or penetrating too far beyond the surface of the proceeding layer. Prolonged vibration will cause the concrete to act as a liquid and will gradually affect the lower levels, where the process of setting has started. A well-designed mix should prove easy to place and should compact relatively easy.

## 14.3.8 Curing

Curing of concrete is a process designed to provide an environment that will allow the hydration of the cement to be completed satisfactorily, and is affected by the local climates and seasons. A slipform operation demands a cohesive concrete mix and thus a greater degree of fine material than that required for normal design. It therefore has a better ability to retain the water required for the total period of hydration.

The specification generally demands a free w/c ratio of between 0.4 and 0.6, which satisfies the strength requirement and also provides sufficient water for the hydration process under controlled conditions. Due to the nature of the vertical slipform process, it is not practicable to 'flood' the structure or to cover it in wet hessian, polythene or similar, so as the concrete emerges from the forms it is, therefore, necessary to protect the outer face to prevent water loss for a prescribed period until the ambient humidity is sufficient to provide the moisture required.

Shrouding the slipform rig, from the top deck to the underside of the hanging scaffold, will provide a natural curing zone immediately around the shutter and the exposed element, maintaining a high humidity, and preventing shrinkage cracks. It will protect the surface from the direct sunlight and wind at a critical period, or in cold climates it will also provide an insulation barrier to maintain a reasonable temperature gradient. It has a distinct advantage over traditional construction, as the protection is in place beforehand and there is no time lag, allowing the concrete to be exposed to the elements at a vulnerable time.

Extremes of climate must be given careful consideration and examples of such detailed planning have allowed slipforms to be carried out successfully in temperatures ranging from –14°C in the UK to 48°C in the Middle East, whilst in areas such as Northern Europe and Scandinavia, slipforms have been completed in far lower temperatures.

Spray-on curing membranes are often used in warmer climates in addition to shade netting, but they require careful consideration, not only to monitor application but also because following trades may be affected. White dyes are often added to sprayed curing compounds, to ease the application visually and reflect the sunlight when the surface is exposed. Curing membranes work on the principle that the hydration process requires the amount of water retained to provide maximum efficiency, as no additional water is used in the curing process.

Another method of curing, which is often specified in hot climates, is a mist of sprayed water. This can be achieved by suspending perforated pipes from the inspection platform with a gravity-fed or pumped water supply, that sprays the outside surface after a suitable time. This method requires considerable organization and maintenance throughout the slipform operation.

For average ambient conditions experience has shown that shrouding the slipform has provided adequate curing conditions.

## 14.3.9 Methods of concrete distribution

### Tower crane

*Advantages*

This is generally the best and simplest solution for delivery and initial distribution. The crane driver usually has total vision and hence saves time on delivery. A one-cubic-metre skip would be used for most operations as the weight of a full skip is within the normal minimum capacity of the crane. The 'hook-height' would need to be 7 metres above the finished height of the slipform to allow for delivery and dismantling.

*Disadvantages*

It is a very costly item of plant with expensive set-up costs. Generally it is underutilized for servicing small one-off structures. For very tall structures, it is often necessary to raise and 'tie in' the crane (possibly above 50 m in height).

### Mobile crane

*Advantages*

It is easier to gain access to various parts of the site and generally far cheaper to use than a tower crane. Hire periods relate to the actual requirements on-site.

*Disadvantages*

The driver is often working blind and relies on instructions from the banksmen, therefore it is slower to operate than a tower crane. It is limited in height and, more importantly, in reach. This method often requires a different method of distribution for the concrete on the platform, i.e. wet hopper and barrows.

### Hoist

*Advantages*

For small structures on plan, requiring a maximum of 3 $m^3$ of concrete per hour a 'self-erecting' rack and pinion hoist can be used quite successfully. Concrete can be transported in dobbin barrows in the goods/passenger cage or by a special skip that travels on one side of the mast with the cage operating on the opposite side. This method is ideal for servicing small to medium chimneys and other small towers.

*Disadvantages*
This system is limited for quantities. Ties are required between the mast and the structures at approximately 6 m centres (half the normal distance required) because of the height required above the last tie to discharge onto the slipform shutter. Rolling ties attached to the slipform assist the reduction of the spacing for the static ties but can present problems on small structures due to large forces introduced by a cantilever action.

*Note*: Rope-guided hoists can be used supported from 'cat heads' on the slipform shutter. However, this type of plant is not readily available and faces stringent regulations with regard to set-up and operation. Each application would have to be designed to meet individual needs.

### *Pump*

*Advantages*
Static and mobile pumps can be used successfully because of the similarity of the concrete mixes. It is very useful for supplying concrete to slipforms carried out in difficult locations, and for slipforms requiring relatively large volumes of concrete. Height of pumping is not usually a governing factor for static pumps.

*Disadvantages*
The system is not suitable for small shafts where the volume of concrete required is low on an hourly basis. The lower limit should be 6 $m^3/h$ and ambient conditions must be carefully considered. Mobile pumps have a limited range, mainly due to height versus reach, posing similar problems to mobile cranage.

In selecting systems for distribution of concrete it is vitally important to choose the most direct method to avoid:

- Time delays
- Loss or gain of heat from the concrete
- General loss of moisture.

Obviously the cost of setting up a system must be taken into consideration. This would be based on availability of plant and general running costs and utilization during the period on-site.

## 14.3.10 Problems that may arise with slipforming

The slipform method of construction is a highly skilled operation and is more akin to the process industry than the construction industry. It is a combination of mechanical operations, material supply and human resources that may give rise to occasional problems.

However, as slipform is an unusual method of construction for most teams, there is an inducement for greater research, planning and preparation which tends to reduce complacency and thus help to eliminate major defects. As with all operations, monies allocated to this effect would be substantially less than costs attributable to remedial operations.

Problems can emanate from the main operations listed below:

- Steering of the forms
- Mechanical or hydraulic operations

- Concrete placement
- Fixing of reinforcement
- Positioning of inserts

Quality assurance, method statements, risk assessments and well-trained labour should eliminate the majority of problems.

The provision, operation and steering of the equipment is usually entrusted to specialist contractors, as only the experience of fully trained personnel is able to ensure that satisfactory structures are constructed.

The poor performance of the equipment, poor quality control and distribution of the concrete, poorly fixed reinforcement, and badly distributed loads can affect steering of the forms.

Embedments, particularly of a heavy nature, fixed proud of the correct position and not taking account of the taper of the forms, can also affect the steering and correct operation of the forms.

The design of a suitable concrete mix should allow the slipform to progress within a range of speeds that take account of:

- The complexity of the structure
- Volume of concrete and reinforcement
- Amount of embedments

As the laitence on the surface of the formed wall provides the lubrication for the steel forms, it is not surprising that problems evolve from the failure to produce the correct consistency with the concrete mix.

Minor defects in the concrete range from:

- Bleeding
- Surface honeycombing
- Limited slumping
- Surface tearing

to more serious defects such as:

- Loss of external arisses
- Larger areas of slumping
- Loss of sections of wall
- Severe surface tearing
- Lifting of sections of wall

The reinforcement fixing can affect the rate of climb and the steering of the shutter as well as producing problems that equally exist with traditional construction, such as misplaced or omitted starters.

Box-outs or inserts can also be wrongly positioned or omitted, but a misplaced heavy insert could impede the path of the shutter and severely affect the progress of the slide.

### 14.3.11 Remedial action

Good supervision, strict disciplines, correctly maintained plant, well-selected crews and contingency arrangements all combine to help eliminate remedial work. There is no substitution for good knowledge of the equipment and a wealth of experience. If major faults occur then the slipform operations can be halted in order to facilitate repairs or remedial work.

When assessing the mode of operation for remedial action for the concrete, there are two types of structure to take account of:

- Elements of buildings where access can be gained, at a later date, to most levels as building work progresses.

or

- Free-standing structures with completely exposed faces, where access would be difficult at a later date, namely towers, silos, chimneys, etc.

A well-designed mix and a steady rate of progress go a long way to producing a good structure with an excellent finish.

Adjustments to the mix should be made to cater for:

- Changes in ambient temperature
- Variations in material content including moisture
- Changes in the rate of progress

Admixtures, such as plasticizers/retarders, may be required to facilitate the above and provision for their inclusion should be made at the design stage. However, the easiest adjustment to make is by varying the slump within the allowable limits. The water content should be adjusted and used as a retarder or an accelerator. Bleeding is a function of the mix and should be eliminated, although some bleeding could be caused by excessive vibration.

The speed of the shutter should be balanced with the slump of the concrete distributed throughout the slipform and all the associated operations.

The concrete emerging from the shutter after a few hours is in a 'green' condition and as it is easily accessible from the hanging scaffold or inspection platform, provides the perfect opportunity to make good minor surface blemishes. A mixture of similar fine materials can be dressed onto the surface to cater for honeycombing or other imperfections. Bonding agents should not be required at this stage and if used will discolour the concrete.

Slumping below the shutter should not occur, as it is the result of incorrect speed of sliding, poor batching, or poor materials. However, should minor slumping occur, it is possible to rectify this from the hanging scaffold, by removing the offending section and rendering the void in similar material.

It is also possible to cut away poor areas of concrete and use localized static shutters to reform the affected parts whilst the concrete is still in a green condition. Thus a natural binding action would be induced.

To maintain progress it may be wiser to leave behind some repair work to be carried out at a later date. This would apply, in particular, to the repositioning of starter bars or inserts that would require concrete that had gained sufficient strength to cater for drilling and fixing. After slipforming, the structure would have gained sufficient strength for the works to be carried out in a similar manner to *in-situ* work.

The general philosophy for any remedial action is 'prevention is better than cure' and every effort should be made to uphold this theory.

## 14.3.12 Slipform mixes – examples

### *Slipform mix used in the south of England in winter conditions*

Specification. C40. Free w/c 0.5 OPC. Slump 75 m.

OPC 390 kg (Sp. Surface 365 units Initial Set 144 min.)
20 mm marine flint 664 kg
10 mm marine flint 287 kg
Sand marine 800 kg

| *Grading % passing BS sieve* | |
|---|---|
| 5 mm | 99 |
| 2.36 mm | 82 |
| 1.18 mm | 69 |
| 600 micron | 55 |
| 300 micron | 31 |
| 150 micron | 7 |
| 75 micron | 2 |
| Free w/c ratio | 0.47 (water heated for winter conditions) |
| No admixtures | |

### *Slipform mix London average autumn conditions*

Specification C35 Free w/c 0.56 OPC slump 100 mm

OPC 360 kg
20 mm land flint 765 kg
10 mm land flint 255 kg
Sand land flint 780 kg

| *Grading % passing BS sieve* | |
|---|---|
| 5 mm | 99 |
| 2.36 mm | 81 |
| 1.18 mm | 67 |
| 600 micron | 51 |
| 300 micron | 18 |
| 150 micron | 3 |
| Free w/c ratio | 0.56 |
| Water-reducing agent Cormix P108 16 ml | |

## 14.3.13 Operations in varying temperatures

Slipform operations are carried out in most areas of the world, presenting a vast range of climatic conditions. A slipform is able to cope with a far greater range of temperatures than conventional construction as the slipform rig can be shrouded for shade and therefore cooled or heated in tent-like conditions. Thus if the concrete is delivered to the forms within the specified limits, the operating or ambient temperatures can be controlled accordingly.

## 14.3.14 Hot weather concreting

Precautions with the concrete before and after batching:

- Protect cement storage areas from sun, induce cool air flow around silos.
- Cool aggregates by shading and inducing air flow, or spray coarse aggregate with water.
- Use refrigerated water in batching plant.
- Protect delivery vehicles from sunlight.
- Wrap readymix vehicle drum in hessian and spray regularly.
- Dampen distribution plant sufficiently to compensate for evaporation to avoid moisture loss from concrete.
- Vertical stand pipes for pumping should wrapped in hessian and dampened.
- Avoid time delays in delivery and placing.

All, or a combination of the above actions can be activated according to the severity of the heat and amount of humidity. Cooling by air flow or evaporation has a great effect in reducing temperatures with little cost.

Cooling the batching water by flaked ice can produce variable results due to time factors and size control and requires the utmost quality control.

The slipform decking can be sprayed with cool water to lower the ambient temperature and motorized plant should be avoided where possible to reduce heat build-up – or the effect of fumes in restricted areas. The shrouding should be created from shade netting and fans can used to induce air flows, if necessary. The use of low-heat cementitious materials is an advantage. Air circulation can be induced in vertical shafts by creating ducts or vents in the working decks.

## 14.3.15 Cold weather concreting

Precautions before and after batching:

- Protect aggregates from winds and frost.
- Heat batching water – insulate tank.
- Steam heat or use jet-air heaters on aggregates.
- Avoid delays in delivery and distribution.
- Use the minimum amount of plant for distribution.

The shrouding to the slipform rig should be complete to the underside of the hanging scaffold (inspection platform) and thoroughly tied or otherwise fixed to form a complete shelter avoiding all draughts. In extreme conditions, heaters can be deployed on the platforms providing they do not present hazards.

Generally in climates similar to the UK heated water for the batching plant is the most that is required to provide suitable concrete for the slipforming operations.

## 14.3.16 Catering for horizontal connections and openings

It is not the intention to provide detailed information for connections or void formers, but provision is required for constructing openings and horizontal slabs or steelwork on even the simplest of vertical structures. This can range from a few openings for pipework on civil engineering structures, with provision for accepting roof beams, to complex arrangements for providing slab and beam connections, door openings, service ducts and the like on large-core structures for high-rise office blocks.

## 14.3.17 Connections for concrete

### *Slabs*

This is usually achieved by the use of proprietary pull-out starters, which also provide a rebate, formed by expanded metal, to receive the slab. These are inserted into the slipformed wall ahead of the concrete. The maximum size of bar that can be successfully accommodated in this arrangement is 16 mm. For diameters above this, it would be necessary to use reinforcement couplers.

### *Beams*

Leave voids to receive the main reinforcement from the beams, or revert to couplers. Couplers are often attached to a former of plywood for more accurate positioning, and also to form a nominal rebate.

## 14.3.18 Connections for steelwork

### *Slabs*

If *in-situ* slabs are required, possibly supported by metal decking, it is usual to provide single pull-out starters as anti-crack reinforcement in the top half of the slab.

### *Beams*

Pockets formed in the walls or plates are the preferred methods to the receive steelwork. There are a variety of ways in which these can be achieved. For small beams, and edge supports, it is preferable to 'drill and fix' afterwards as the main beams or slabs are installed.

### *Box-outs or void formers*

All formers for openings are fixed into the reinforcement at the working deck level, with reinforcement required to pass through some of the smaller formers. Small formers can be made from polystyrene or similar, whilst the majority are manufactured in timber. Large openings may be slipformed. It should be noted that due to the taper on the shutter, the boxouts should be narrower than the width of the wall in order to pass through. By the same token, all plates or other face inserts will be positioned back from the surface by a few millimetres.

## 14.3.19 Example of slipformed project

### *The flour mill – Doha – Qatar*

*Main contractor – Taylor Woodrow International. Slipform subcontrator – RMD*

This was a fine example of slipforming, demonstrating economical construction over a short period – programme time 9 months. Sixty-four grain silos, consisting of 4 no. blocks of sixteen were slipformed in eight sections and a further 3 no. blocks of rectangular silos were slipformed in five sections, necessitating vertical slipformed joints on all structures. The arrangements allowed two sets of equipment to be used alternately over the construction period to provide good logistics. Each block was slipformed from the base level to the roof level to include the external walls and internal support columns. Temporary columns were used internally at the top to maintain continuity and support for the slipform. The two sets of equipment required 146 no. heavy-duty jacks, to complete the works. A climb rate of between 5 and 6 metres per day was achieved (See Figures 14.4 and 14.5).

The extremely smooth finish required for the flour silos was achieved despite temperatures reaching 48°C. Concrete was batched using air-cooled aggregates and refrigerated water from an automated batching unit set up on-site.

For vertical sliding joints, referred to above, the chosen positioning usually allows a stub section of wall, with starters projecting, on which the panels of the adjacent slide connect by approximately 100 mm. The panels are then guided by the existing section of wall.

The mix details were as follows:

Contents/m$^3$
Cement SRPC 400 kg/m$^3$
Coarse aggregate 1110 kg/m$^3$
Fine aggregate 650 kg/m$^3$
Water 180 litres/m$^3$
Retarding plasticizer 1.6 litres/m$^3$
Slump 75 mm

## 14.3.20 Dayshift-only sliding

This system, often referred to as 'intermittent sliding' was introduced thirty years ago by Slipform International Ltd for use in environmentally sensitive areas, or where the supply of concrete proved difficult or costly on a 24-hour basis. It has been used for the construction of many service cores in cities throughout the world from the UK to Australia.

The method of construction allows for concreting with the slipform throughout the normal daily working hours, avoiding additional costs, and for the slipform shutters to be pressure washed in the early evening. This creates a 'dayswork' joint and, hence would not be suitable for liquid-retaining structures, or exposed surfaces such as chimneys, although linings for storm water overflow tanks have been constructed by this method.

Allowing for the intermittent nature of the work, this method is still faster than other systems such as climbing forms, as 'Dayshift Only' still achieves 15 m in height per week.

**Figure 14.4** Second group of circular silos slipformed.

## 14.3.21 Conclusions

The versitility and economy provided by concrete will ensure a long future for the material in the construction industry, whether utilizing the present ingredients, or by-products from future manufacturing. Slipform construction cannot be ignored as it is the cheapest and fastest method of construction for many types of structure. It is an operation that places demands on everyone associated with the work, but the results are there to see. Structures built forty to fifty years ago by the slipform process remain in better condition than similar structures constructed conventionally. Like every building, poor workmanship or materials will manifest themselves, but the slipformed monuments are there as living proof – oil platforms, telecommunication towers, large cores, bridge pylons. The systems will develop with the benefits provided by the electronic age.

# 14.4 Horizontal slipforming

Developed in the USA in the early 1970s and born of a need to rebuild mile upon mile of highway, horizontal slipforming in two modes – paving and kerbing – is now carried out in many countries throughout the world. This concrete construction is an instantaneous extrusion that requires the moulded shape to be self-supporting and thus has limitations in shape. The two operations require different machines to carry out the construction work, but both are propelled forward very accurately leaving in place the formed concrete.

**Figure 14.5** Second half of rectangular block slipformed.

## 14.4.1 Paving

Used for constructing concrete aircraft landing strips, roadways, railtracks, aprons and other large areas, paving machines are generally capable of laying up to 90 linear m/h. Panel widths are usually from 3 m to 10.5 m wide although 4 m to 6 m are more typical.

The paving machine is constructed on a transverse beam design with drive units at each side, providing the support, height control and guidance. Depending on the application, the paver may have two, three or four drive units, usually of the tracked variety. For large contracts 'paving trains' are set up, consisting of three separate items of plant – a spreader, a placer and a finisher. Concrete is fed directly to the fore of the spreader by readymix truck(s). This concrete is distributed by a transverse worm (auger) on the spreader, with the placer following a few metres behind completing the placing and vibrating. The basic finished surface is achieved with a hydraulically controlled skid. The finisher provides any surface texture that may be required and puts in place either a sprayed or plastic curing membrane. For smaller contracts all the actions are carried out using one machine.

Each machine is driven by its own diesel engine which is also responsible for supplying the power to the hydraulic systems, and supplying the electrical power. Electronic sensors steer the paving machines, guided by string or laser lines set up ahead of the concreting operations. Hydraulic vibrators are set up in a bank and are controlled to provide the correct level of vibration, which is critical for the durability of the concrete.

## 14.4.2 Kerbing

More compact versions of the paving machines are used to form edge kerbs, barriers, drainage channels, irrigation ditches or similar. A special mould is designed to suit the required shape of the kerb or barrier. This is attached to the rear of the machine and extrudes the concrete to the given shape. Guidance of the slipformer is again achieved by electronic sensors to lines. The height and pressure to control the mould is governed by hydraulic rams. The concrete is fed from the readymix truck to a hopper via a chute or elevator, and then by auger to the mould. Controlled vibrators are again used in the placement of the concrete. A vast variety of shapes can be formed providing that they are self-supporting immediately upon exposure.

The present limitation on height is approximately 1.85 m, and rates of pour have reached 250 linear m/h. The choice of a smaller slipforming machine allows the formation of kerbs to very tight radii – 600 mm has been achieved using wheeled machines, and 1.2 m using tracked machines.

## 14.4.3 Concrete mix

Within an hour of batching, the concrete supplied for horizontal slipforming may be required to be freestanding in its final location. It therefore requires to be a cohesive mix that is able to provide a suitable finish with one pass. As with a pump mix and a vertical slipform mix, a higher fines content is required to provide these characteristics. However, whereas vertical slipforming can be achieved with high slumps of 50–150 mm, dependent on ambient temperatures and operating speeds, it is essential for horizontal kerbing to maintain very tight slumps of 20–40 mm to cater for this immediate exposure, free of any external side supports. The slump for paved areas can be increased accordingly, and may depend on edge restraints.

To expedite the placing of the concrete by this extrusion method, an air-entraining admixture is used which improves workability and cohesion, whilst in the long term also improves durability through increased resistance to de-icing salt and frost attack.

### 14.4.4 Vibration

Vibration is critical for vertical and horizontal slipforming. Experience has shown that over-vibration with horizontal slipforming has reduced the air containment to less than 5 per cent and has caused premature deterioration of the concrete. This has led to the introduction of special individual controls to vibrators which are fixed in banks, in order to monitor the operation of each unit.

# 15

# Pumped concrete

*Tony Binns*

The flow of fluids and suspensions through pipelines has been the subject of study by scientists including Newton (1687), Bernoulli (1724, 1738), Stokes (1842, 1846), Poiseuille (1844) (after whom the unit of viscosity – the Poise – was named), and Reynolds (1883) (whose Reynolds Number is used in the modelling of fluid flow).

Newton discovered that the flow of fluids depends on pressure overcoming friction which is minimal at the centre of the pipe, progressively increasing towards the fluid/pipe interface. The following principles have been found to apply:

- Material sliding under axial pressure through a cylinder is subject to the shearing effect of frictional resistance between the material and the cylinder lining.
- Unsaturated materials pass stress by inter-particle contact and give rise on commencement of movement to resistance which, in a non-tapering duct, is exponential with distance moved.
- Frictional resistance to the flow of liquids and saturated suspensions through non-tapering straight ducts increases in direct proportion to the length of the duct and is constant for equal length/diameter ratios.
- Increase in resistance causes a corresponding reduction in pressure to the point at which movement eventually ceases altogether.

## 15.1 Liquids

Pure liquids such as water, simple solutions and suspensions that are sufficiently dilute for there to be no forces between the suspended particles are known as Newtonian fluids and the relationship between the rate of flow (or shear) of the liquid and the stress resisting that shear is constant at constant temperature. The reciprocal of the slope, i.e. stress/shear rate, is a measure of its viscosity (Figure 15.1).

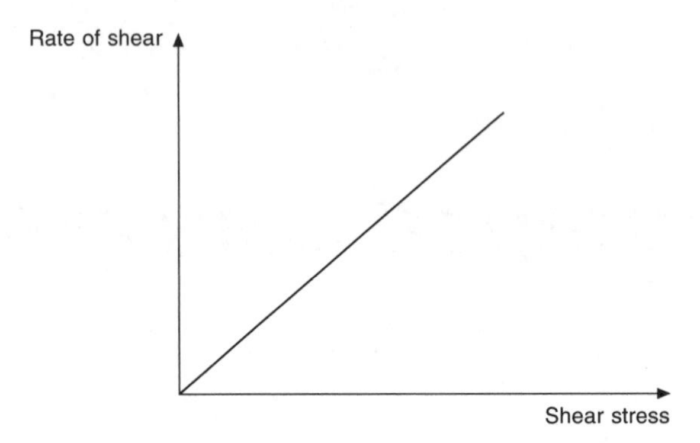


**Figure 15.1** Viscous relationship for a Newtonian fluid.

Because the relationship between stress and shear is linear, passing through the origin, a single-point test is adequate for determining the viscosity or flow rate of Newtonian fluids.

When liquid flows like water along a pipe the layer of fluid adjacent to the wall of the pipe remains stationary and the rate of flow increases in parabolic fashion to a maximum at the centre (Figure 15.2). An important fact derived from this relationship is that a Newtonian fluid will always move, albeit slowly, under an infinitesimally small force.

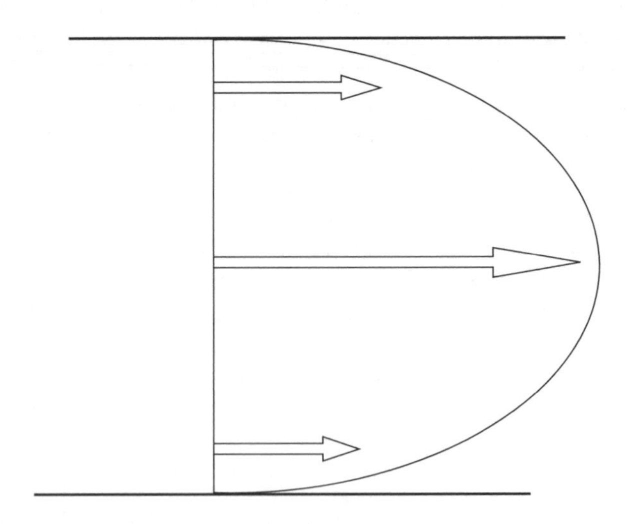

**Figure 15.2** Parabolic velocity profile for liquid flow in a pipe.

## 15.2 Suspensions

Concrete is not a Newtonian fluid and does not obey the rule of liquids; its flow behaviour is complex due to the wide range of its constituent materials and the changing nature of concrete as it stiffens and cement sets. Concrete closely resembles a Bingham body, which is a suspension that can be made to flow only after a critical force, known as the yield point, is exceeded (Figure 15.3).

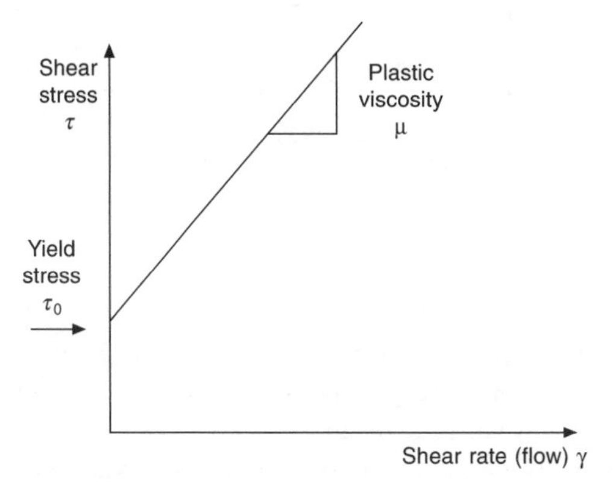


**Figure 15.3** Plastic viscosity of a Bingham fluid.

Shear stress $\tau$ is the pressure applied to the suspension; shear rate $\gamma$ is the rate of change of velocity at which one layer of fluid passes over an adjacent layer. The slope of the line (stress/strain) is described as the plastic viscosity ($\mu$). The flow properties of the fluid are characterized by the values of $\tau_0$ and $\mu$ while the pressure applied is derived from the equation $\tau = \tau_0 + \mu\gamma$.

Concrete may be considered to have an at-rest structure with a characteristic breaking force. The yield stress $\tau_0$ is the pressure at which movement starts. It is seen, therefore, that the flow of a Bingham fluid is quite different in character from that of a Newtonian liquid. The shearing process is now limited to the interface with the pipe walls, so that the central body moves as a plug (Figures 15.4 and 15.5).

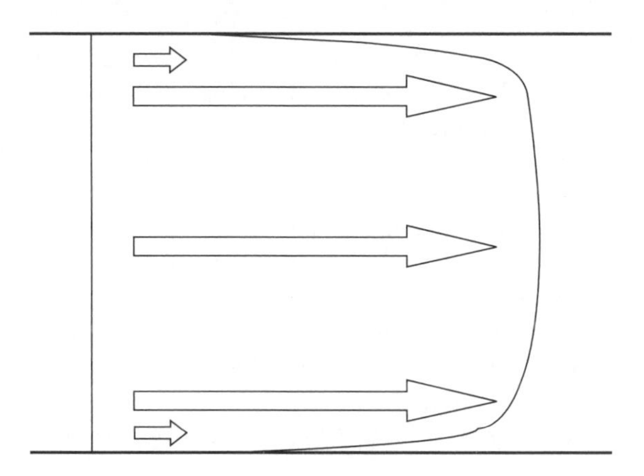

**Figure 15.4** Velocity profile for a Bingham fluid in a circular pipe.

In plug flow solid particles undergo only small internal movements within the body while friction at the concrete/pipe interface creates a shearing effect, forming a thin zone of fine particles that act as a lubricating layer as they slide relative both to the progressing body and to the wall of the pipe. Energy used in the shearing process causes pressure to

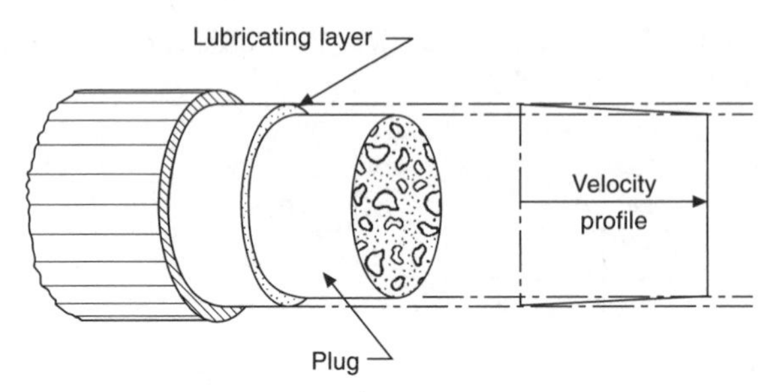


**Figure 15.5** Plug flow.

drop and movement eventually to cease but, provided the Bingham body maintains its lubricating layer, low plastic viscosity suspensions such as concrete may be pumped over considerable horizontal and vertical distances.

The thickness of the lubricating layer and the plug size vary with the rate of shear and the properties of the suspension. The lubricating layer in concete comprises a grout of cement and fine sand particles in suspension in water. Concrete being pumped at the rate of 30 $m^3/h$ through a 100 mm diameter pipe has been found to produce a 3 mm thick lubricating layer.

Concretes that are deficient in mortar cannot achieve adequate migration of grout to the pipe/concrete interface while the mortar phase of other concretes is of a glutinous nature that prevents the migration of grout. In both cases friction along the pipe wall is increased and pumping becomes difficult or impossible over any useful length, and is often associated with unacceptably high rates of wear on the pipes, particularly at bends.

## 15.3 Rheology

The movement of concrete during pumping, its response to stress and its behaviour on shearing depend on its rheological properties. Rheology is the name applied to the study of the flow of currents and, in the case of concrete, is concerned with behaviour of the suspension that goes beyond the simple concept of consistence (workability). The process of shearing and sliding of one layer relative to another within the suspension requires a degree of saturation and in concrete this usually relates to the quantity of water.

Every concrete has its saturation point at which the unsaturated materials pass through a transition phase to become oversaturated and axial pressure required to overcome resistance drops abruptly and pumping becomes possible. This is illustrated in Figure 15.6, but it is emphasized that the saturation point depends upon materials types and mix design for each individual concrete.

In a well-designed pumpable concrete the grout is drawn towards the pipe wall by two processes:

1. The compactive effect of the pump pressure causes the coarser particles of aggregates (including the largest sand particles) to become more consolidated, forcing some of the grout out of the body of the concrete.
2. The dragging effect of friction against the pipe wall pulls grout out of the concrete in a process similar to that of trowelling.

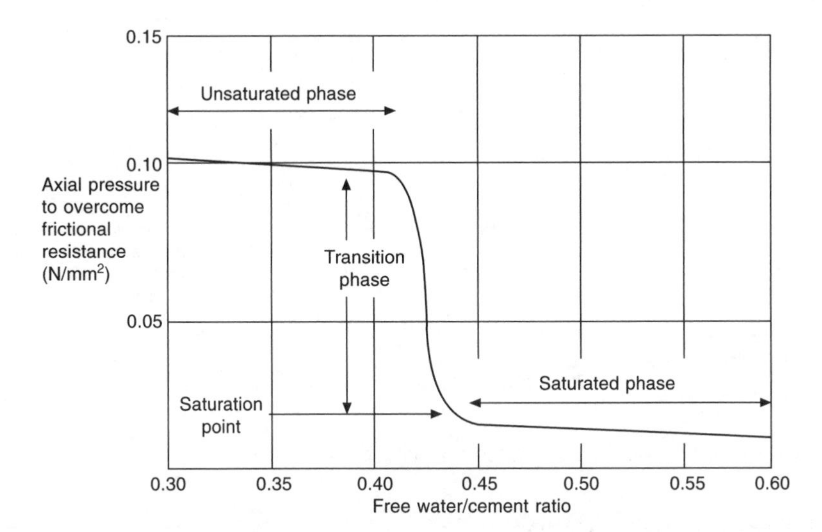


**Figure 15.6** Example of the relationship between pressure, free water/cement ratio and saturation point of concrete.

Because the pipe wall is smooth, it requires only a small amount of grout to be used in the lubricating process but, even so, the depletion of grout from the leading edge of the concrete plug would, in a dry pipeline, rapidly put a stop to the formation of a lubricating layer and the excessive friction would cause the pump to become blocked. For this reason it is a normal requirement that dry pipelines are primed immediately before the commencement of pumping operations with either a workable cement–water grout or, for longer pipe runs, a flowing cement–sand mortar. The quantity of cement required for priming the pipeline is approximately 0.5 kg per metre run, although, for the short booms of some mobile pumps, priming with grout may not be necessary at all, provided the concrete is well designed.

The ability of concrete to maintain its homogeneity and resist segregation when under pressure is one of the more important requirements for pumping. Unlike liquids, suspensions such as concrete have a segregation pressure at which the liquid phase separates out from the solids and becomes unpumpable. It is essential that the operating pressure of the pump does not exceed the segregation pressure of the concrete. The yield stress is largely determined by the aggregate characteristics, the plastic viscosity is more affected by the quantity of water and the nature of the grout. High values of yield stress and plastic viscosity give rise to higher operating pressures.

If grout separates out too easily from the body of concrete, the concrete becomes an unsaturated suspension and the coarser aggregate particles become locked together. This may favour the formation of plug flow in a straight non-tapering pipe but, when concrete is being pumped, it becomes deformed when passing through tapers or travelling around bends, for example, and a degree of internal lubrication is essential if aggregate interlock is to be avoided, (Figure 15.7).

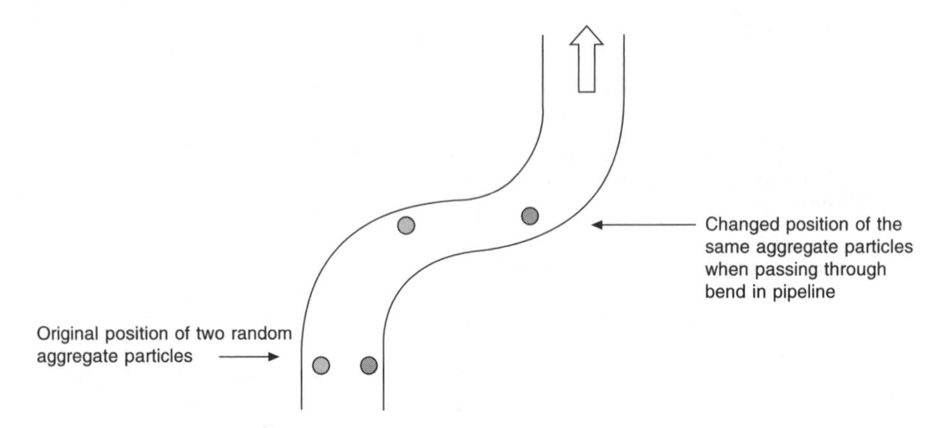


**Figure 15.7** Deformation and shear of concrete through a bend.

## 15.4 Pressure gradients

Pumps create pressure which is transferred to the material and progressively reduces with distance pumped. Once movement has been achieved, the distance and height to which the pump pressure can convey the material depend upon pump design and the rheology of the concrete. Changes in shear stress commence immediately as concrete is drawn into the first cylinder of the concrete pump.

Concrete must first be capable of flowing satisfactorily through the grille and withstand the reversals of high pressure imposed in the cylinders. Approximately half the pressure developed by a pump is normally expended in passing the concrete from the hopper into the pipeline.

A strong negative pressure is exerted as concrete is drawn into the pump. The piston then changes direction and a sudden reversal of pressure is applied. This sequence is one of the earliest stages in the pumping process where segregation can start to occur but any change in the rheology of the concrete might not become apparent until a blockage occurs at a later stage in the system.

Provided the segregation pressure of the concrete is greater than the pump pressure, the yield point is reached and the concrete begins to move, immediately passing through a taper where the concrete accelerates. If the cylinder diameter were 200 mm and the pipeline diameter 100 mm, the concrete would be accelerated to four times its original velocity (Figure 15.8).

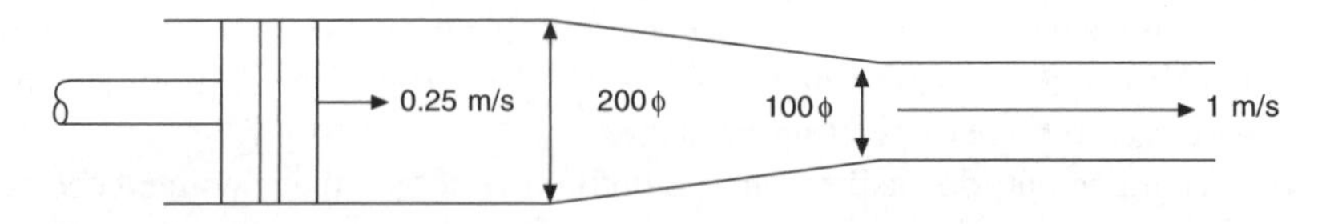


**Figure 15.8** Linear acceleration through a taper.

If concrete passes through the pump without change it is likely to continue successfully through the pipeline and, therefore, failures along the pipeline can often be attributed to changes in the mix that occur in the pump.

The pressure gradient gradually drops as frictional resistance to flow increases in direct proportion to increased length of pipeline. In straight horizontal pipes a linear relationship exists between pressure and distance but upwards vertical movement causes the pressure gradient to drop rapidly due to hydrostatic head and it is conventionally accepted that every 1 m of vertical pipeline is equivalent to 10 m horizontal. Frictional resistance also builds up at bends where, depending on their radius, 1 m of bend can be equivament to 3 m of straight pipe.

The most severe bends are in the swivel joints of the placing boom on mobile pumps. In addition to the abrupt changes in direction that occur at the joints, a vertical component also normally exists, imposing some hydrostatic head. When such rapid changes in pressure gradient occur, there is a risk that frictional resistance to flow exceeds the residual pumping pressure.

In the case of pumping concrete in a downward direction, gravitational effects encourage the more massive aggregate particles to separate out. Unless precautions are taken to ensure that the pipeline remains permanently filled, the concrete can become completely segregated.

## 15.5 Types of concrete pump

### 15.5.1 General description

The first patent for a concrete pump was registered in 1913 in Germany but it was not until 1931 that the German company Torkret produced the first commercial model. Early ball valves were superseded by sliding gate valves and concrete pumps were manufactured under licence in France, the Netherlands, the UK and the USA until the outbreak of the Second World War. Post-war rebuilding gave an impetus to the pumping of concrete and to the performance of pumps, particularly in Germany where the two dominant manufacturers, Putzmeister and Schwing, compete to develop more efficient and robust equipment up to the present time.

Concrete pumps comprise a machine capable of applying pressure to fresh concrete and transmitting it along a pipeline. The pipeline may be a boom, a static pipe run or a combination of both. A holding hopper is required to regulate the delivery of concrete and it is preferable to position the pump so that two delivery vehicles (such as truck mixers) may stand alongside each other; when discharge is complete from the first vehicle, the second can then take over seamlessly. The hopper normally includes rotating blades or paddles that not only keep the concrete free from segregation but also direct it towards the inlet of the pump. The rotation of the blades makes the hopper the most dangerous part of any pump. It is protected by a grille which must not be removed while the blades are turning. The grille minimizes the risk of accidents and prevents unwanted foreign objects entering the pump. Concrete is forced along metal pipelines by the pumping action, usually starting with a taper to reduce the size down from the diameter of the cylinder to the pipe diameter of 100 mm or 125 mm.

The final length of pipe is normally made from flexible rubber so that the workforce handling it can make small changes to the point of discharge. Sometimes, however, it is possible to replace the flexible pipe by a metal taper fixed to the end of a placing boom for very precise and deep insertion into inaccessible locations. Tapered pipes terminating in a diameter of 60 mm or less have been used in cases of congested reinforcement and narrow sections to ensure that concrete is correctly placed.

## 15.5.2 Reciprocating piston pumps

The majority of concrete pumps are twin-cylinder hydraulically operated machines with twin reciprocating pistons (Figure 15.9). The piston in one cylinder discharges concrete along the pipeline at the same time as the other piston draws concrete down and into its cylinder from a hopper in which concrete is kept agitated.Valves are synchronized to open and close at the beginning and end of each stroke: one opens the inlet to the empty cylinder at the same time as closing the inlet to the other cylinder which is now full; the valve opens the outlet from the full cylinder at the same time as closing the outlet from the empty cylinder. In this way the induction stroke is engaged exclusively in drawing concrete from the holding hopper and the driving stroke in pushing it along the pipeline.

A simpler design is the flapper valve, where the valve gear comprises a single moving component located inside a box. The opening of the flapper is synchronized with the cylinder exerting positive pressure on the concrete, at the same time preventing concrete from being drawn into the cylinder with negative pressure.

A supply of water permanently surrounds the push rods of the pistons to clean the cylinder bores. The flapper valve has proved to be less reliable than the gate valve mechanism, due to mechanical failure in overcoming resistance to rotation caused by the concrete and the valve-gear on most current concrete pumps is currently S-type, elephant trunk or rock valves (Figure 15.10). At the end of each stroke with all twin reciprocating pumps there is a momentary pause when the pistons change direction. If necessary, the pumping action can be reversed, drawing unused concrete back into the holding hopper and, in the event of a delay, concrete in the pipeline can be kept continuously mobile by pulsating.

The ratio of the size of hydraulic pistons to that of working pistons is generally about 0.25 so that the hydraulic pressure of about 16 bar is reduced to 4 bar acting on the concrete. The hydraulic pump is driven on mobile units by the diesel engine of the vehicle; static pumps may be either diesel- or electric-powered.

## 15.5.3 Peristaltic pumps

Sometimes called the Squeezecrete pump, peristaltic concrete pumps are enlargements of the type of hospital pump used for supplying fluids at a smooth, controlled rate. Concrete is delivered into a holding hopper where rotating agitator blades not only maintain the concrete in a fresh, cohesive condition but are designed to spin the concrete and accelerate it towards the inlet tube, forcing it into the pump. The tube is usually 75 mm in diameter, constructed from flexible nylon rubber which is successively flattened by a linked pair of rollers rotating in planetary action against the walls of the pump chamber. Each advancing roller squeezes the tube flat, driving the concrete forwards along the tube and out into the pipeline which may be 75 mm or 100 mm in diameter (Figure 15.11).

The technique is described as a 'live seal' and no significant amount of solids are trapped under the rollers. A vacuum of up to 1 bar is maintained in the pumping chamber,

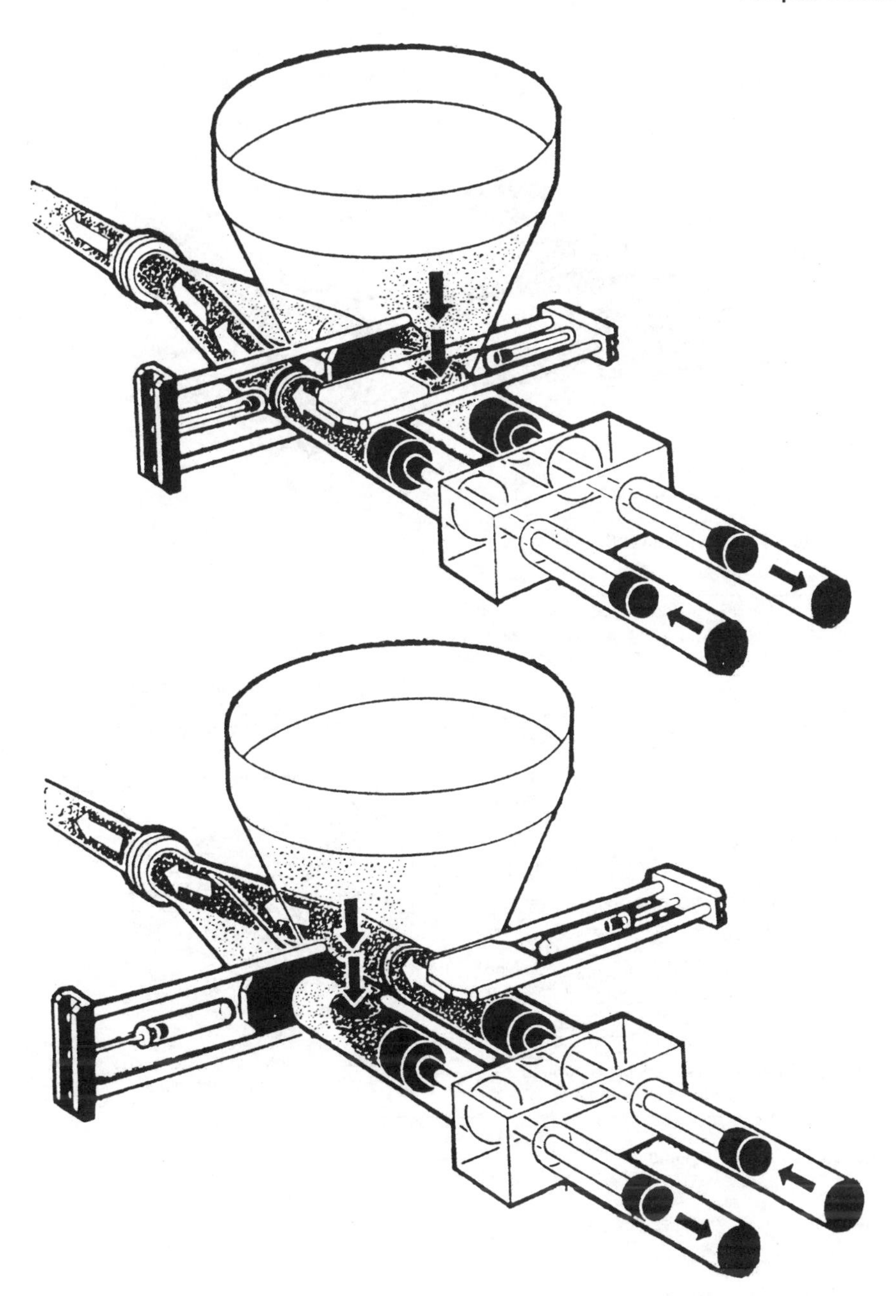

**Figure 15.9** Pumping action: reciprocating piston pump.

allowing the tube to be restored to its normal shape as atmospheric pressure assists in drawing more concrete from the hopper into the tube. Truck-mounted peristaltic pumps have conveyed concrete up to 100 m horizontally and 30 m vertically.

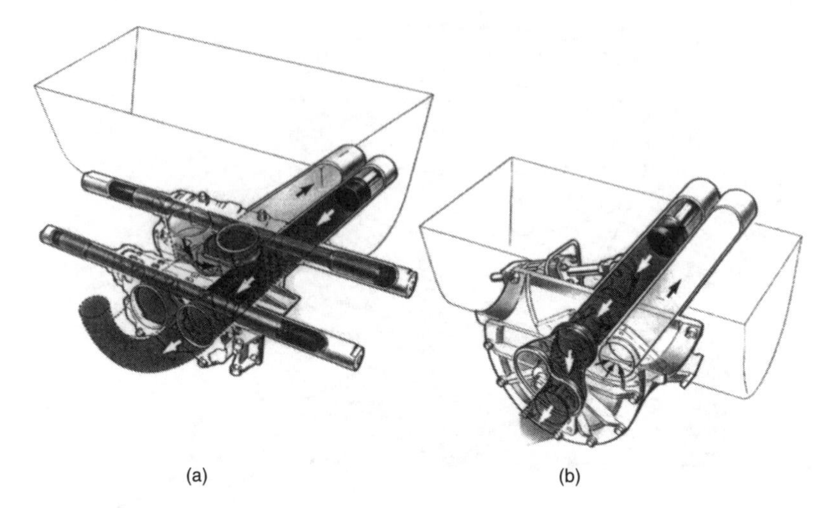


**Figure 15.10** (a) Flat gate valve; (b) rock valve.

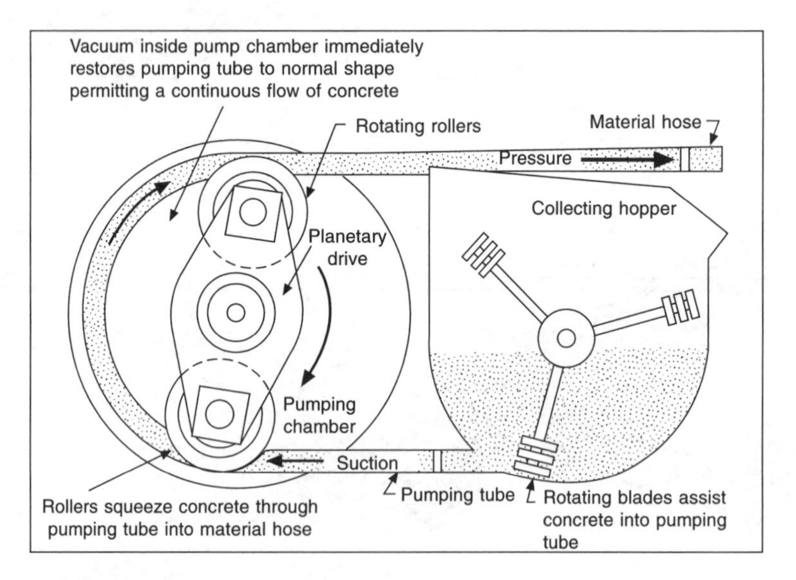


**Figure 15.11** Peristaltic pump.

## 15.5.4 Static pumps

Static pumps may be any of the types described above but, by virtue of not having to be fitted to the chassis of a vehicle, can be heavy-duty and of large capacity. The most common type is the twin reciprocating piston model. It is often convenient to position them directly beneath concrete batching plants for large civil engineering projects or, alternatively, they are placed alongside the project and supplied with concrete by conventional transport in order to minimize the length of pipeline.

## 15.5.5 Mobile pumps

### *Mobile boom pumps*

Most concrete pumps are mobile boom pumps, mounted on conventional truck chassis (Figure 15.12). The larger pumps and longer booms require the largest chassis and must conform to motor vehicle construction and use regulations, including maximum permitted gross weight. The smallest size in regular usage is known as the 'city' pump with a two-section 16 m boom that is quickly manoeuvrable for short-distance deliveries. The most common sizes are 24 m, 32 m, 42 m and 52 m booms. The larger sizes currently include 58 m booms with four folding sections.

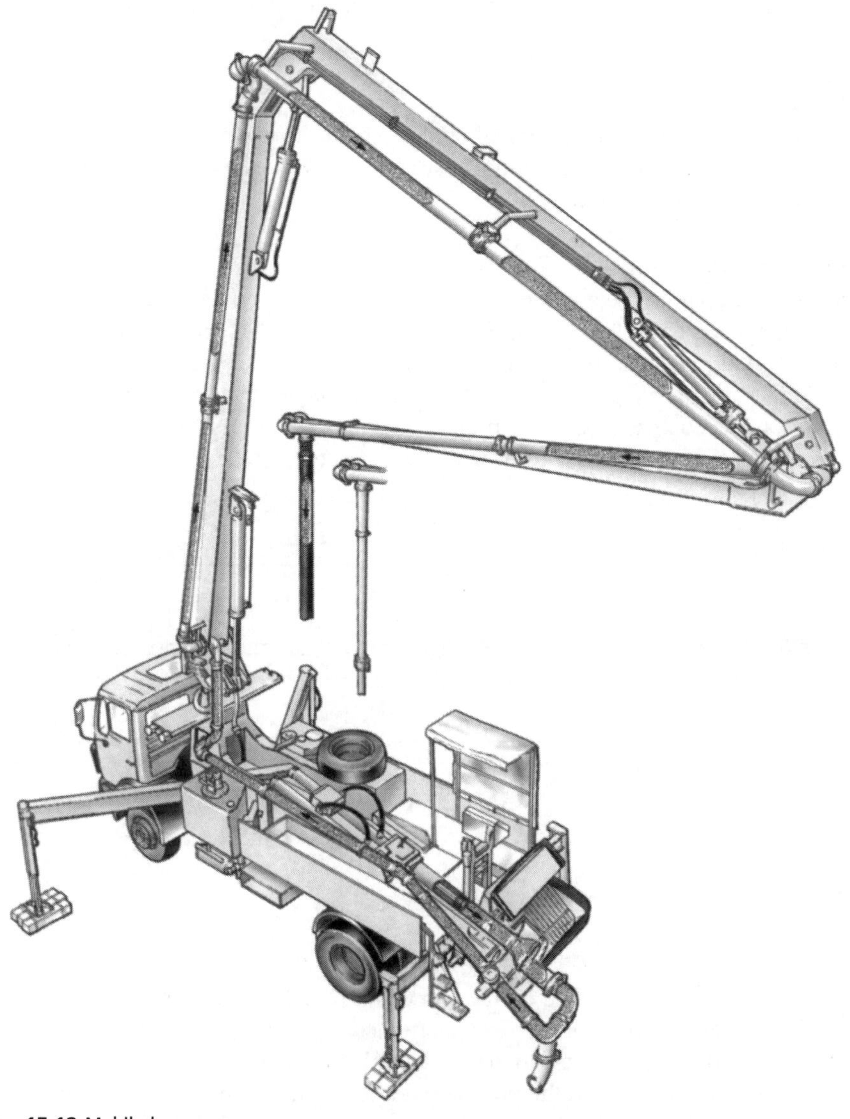

**Figure 15.12** Mobile boom pump.

Setting up a mobile boom pump calls for careful planning. Soft ground conditions of a typical construction site can fail to support the vehicle; its outriggers must be firmly placed on strong plates or timbers. The vehicle should be level if possible. Where a slope is unavoidable, the vehicle should face down the slope to allow easier discharge from truck mixers. Positioning the mobile pump sideways across a slope is most unstable, particularly when the boom is being manoeuvred, and should be avoided if possible. The boom must be folded away before the vehicle is driven to a new position. When positioning a mobile pump on a road or street, not only must permission be obtained from statutory authorities but enough space needs to be provided for the very long 'footprint' of pump plus truckmixer that may extend to a total of 20 to 30 m.

On large construction sites the pump operator will be positioned strategically, often some distance away from the pump, the remote control box linked by an umbilical cable or radio to slave controls on the machine.

### *Line pumps*

Line pumps are mobile, truck-mounted pumps which carry instead of a placing boom, a large number of pipes. The unit is more cost-effective where concrete needs to be moved some distance along the ground.

### *Trailer pumps*

Trailer-mounted pumps are not normally equipped with any boom, discharge being along a static pipeline usually laid out on the ground. The mechanism is twin reciprocating piston with either elephant trunk valves or rock valves. Some single-cylinder pumps are available in trailer form and their weight is light enough for them to be towed by very small trucks. The compactness of the trailer pump gives it the advantage of having a small 'footprint' in a busy road or on a congested construction site (Figure 15.13).

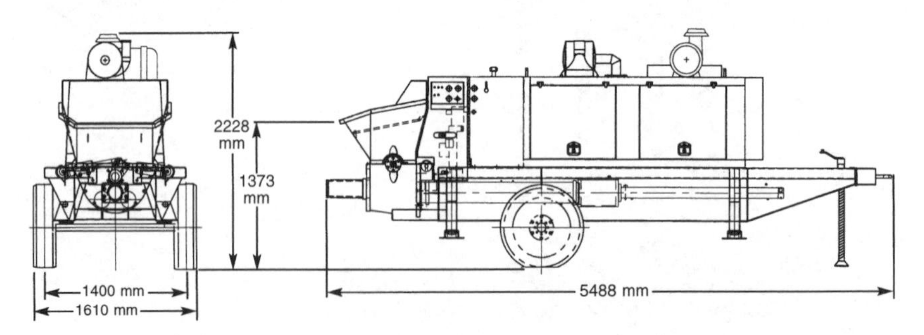


**Figure 15.13** Trailer pump.

## 15.5.6 Satellite booms

There are situations when it is not satisfactory or safe to have the flexible end of a pipeline being manually handled by the workforce. Satellite placing booms are fixed to plinths or pedestals whose location is predetermined at planning stage. Concrete is pumped from either mobile or static pump to the satellite placing boom where an operator controls both the speed of delivery and the precise point of discharge over a distance up to 25 m through a full 360° arc. The height of the pedestal can be extended as construction levels rise.

### 15.5.7 Pneumatic placers

Pneumatic placers are not true pumps for concrete is moved along a pipeline not by mechanical pumping action but by compressed air. Concrete is supplied into a large vessel capable of holding up to 0.5 cubic metres. A counterbalanced inlet door closes automatically when the vessel is full. A valve then allows compressed air into the vessel and the concrete is forced out through a taper into a pipeline which tends to be wider than that of conventional pumps: 150 mm or 200 mm in diameter.

It is important not to allow too much concrete to leave the vessel; it should be topped up before becoming empty, otherwise there is a risk of a pocket of compressed air travelling along the pipeline between successive batches of concrete, with violent and dangerous consequences at the discharge end of the pipeline.

The equipment is relatively inexpensive and occupies a small amount of space on site. But the rate of delivery from pneumatic placers is quite slow compared to that of reciprocating pumps and the large pipes are difficult to handle when setting up or changing position.

### 15.5.8 Condition of pumping equipment

A programme of planned maintenance, refurbishment, repair and replacement of vital parts is an essential requirement for successful pumping. Old equipment can cause blockages by uneven application of pressure and by losing grout through leaking joints. Grout provides the essential lubrication without which pumping rapidly becomes impossible. This also applies to old pipes (including flexibles) and their fastening clips. Breakdowns can be costly if a concrete pour has to be abandoned.

Regardless of the type of pump in use, modern well-maintained pumps work safely and efficiently with the ability to pump difficult concretes that would have blocked the pumps of an earlier generation.

## 15.6 Requirements of a concrete for pumping

British attempts to understand the physics of concrete pumping included work by Dawson (1949) who demonstrated that power demand increased in a linear relationship with distance pumped, while Joisel (1952) in France recognized the existence of plug flow.

Ede (1957) at the University of Cambridge applied to fresh concrete the principles of soil mechanics in triaxial compression and found that Joisel's hypothesis linking radial stress to resistance was flawed. Ede discovered that material sliding under axial pressure through a cylinder is subjected to the shearing effect of friction between the material and the cylinder wall. He called this frictional resistance 'adhesive resistance' and the main conclusions of his work are:

- Resistance is linear and, therefore, directly proportional to distance pumped.
- Resistance is constant for equal length/diameter ratios.
- Resistance is independent of pressure.

Ede recognized the importance of aggregate properties and, in his quest to determine optimum gradings, developed apparatus in which the ability of different sands to resist the passage of water could be timed. This approach to evaluating the pumpability of concrete is known as the 'blocked filter' technique (Figure 15.14).

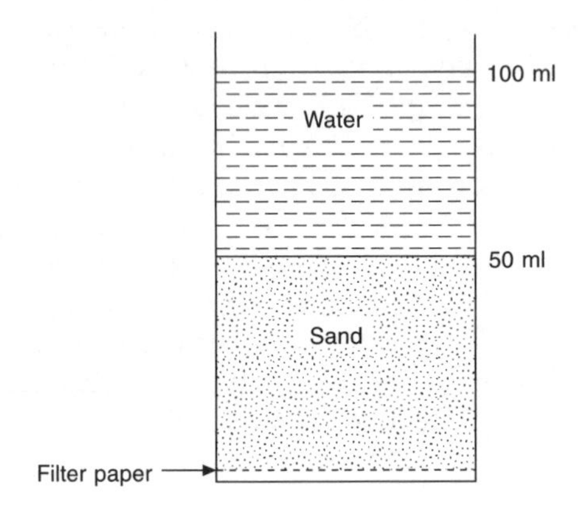


**Figure 15.14** Blocked filter apparatus (Ede).

## 15.6.1 Segregation pressure

Any saturated suspension such as concrete has a segregation pressure. This is a stress at which the flow reaches a maximum. Increased pressure beyond this stage is not accompanied by faster rate of shear but the liquid phase can separate out from the solids, thus transferring pressure directly to the aggregate particles because the concrete is now in an unsaturated state. The frictional resistance exerted by the packed aggregates renders the mix unpumpable.

The main requirement of a concrete for pumping, therefore, is for a combination of liquid and solids with a segregation pressure greater than the pressure required to pass it through the pump and pipeline.

## 15.6.2 Consistence class

The appropriate level of workability or consistence for pumping concrete is frequently misunderstood. While it is essential for the concrete to have some ability to move, the precise consistence class should normally be determined by the requirements for placing and compaction on-site and it would be a mistake to assume that concrete needs to be highly fluid in order merely to pass through a pump: a high consistence class could be totally inappropriate for some end uses – the construction of ramps, for example. Concrete has been successfully pumped at 50 mm slump and Consistence Class S2 (50 to 90 mm slump) is normally recommended for pumping where no special construction demands exist. If high levels of consistence class are required for ease of placing and/or compaction on-site, then special care needs to be taken in the design of the concrete if segregation and excessive bleeding are to be avoided.

## 15.6.3 Controlled bleeding

It is essential that concrete under pressure does not easily lose its mixing water. Nevertheless it has been noted that an adequate quantity of grout with a low resistance to shear must be allowed to migrate to the pipe/concrete interface. Concrete suitable for pumping will, therefore, permit an amount of bleeding to occur. Excessive bleeding is undesirable because it would be accompanied by segregation where the result of applying shear stress to concrete in the form of pump pressure could be to move some of the mixing water through a packed blockage of unsaturated and interlocked solids. Conversely, a satisfactory lubricating layer cannot form in a concrete that does not bleed sufficiently, or one that creates a cement-rich grout of high viscosity.

A controlled amount of bleeding is required to bring a constantly renewed supply of low-viscosity grout into the shearing layer without depriving the concrete of an excessive quantity of water and fine material.

## 15.6.4 Factors that control bleeding

### *Consistence class*

Bleeding becomes progressively more difficult to control at higher consistence classes. Bleeding and segregation increase as free water content increases. Not all concretes behave in the same way, however, depending on their cohesion (Figure 15.15). In general, high consistence classes of concrete usually have lower plastic viscosity with low yield stress, provided they do not exhibit poor cohesion, excessive bleeding and segregation (Figure 15.16). There are exceptions, however, and it is possible for different concretes to have the same plastic viscosity with the different yield points or, conversely, different plastic viscosities but the same yield point (Figure 15.17).

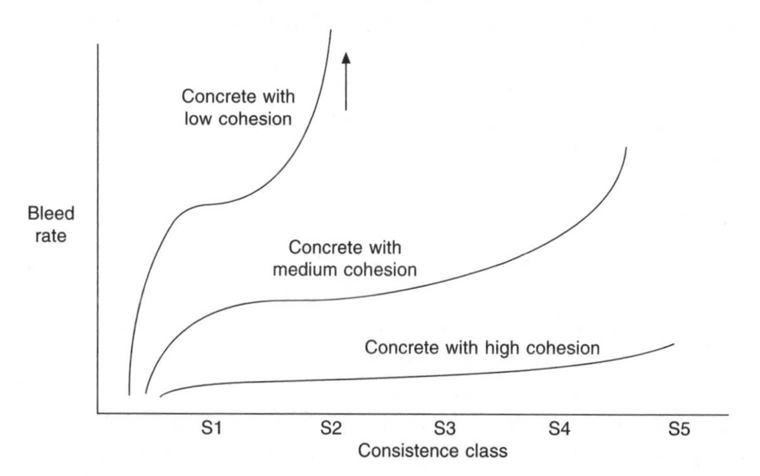


**Figure 15.15** Relationship between bleed rate and consistence for concretes with different cohesions.

It can be seen that the classification of concretes into standard consistence classes gives little indication of their ability to produce a lubricating layer through controlled

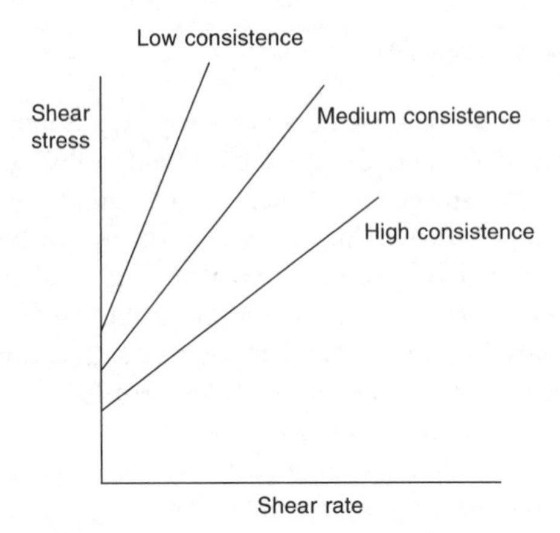


**Figure 15.16** Normal relationship between consistence and plastic viscosity.

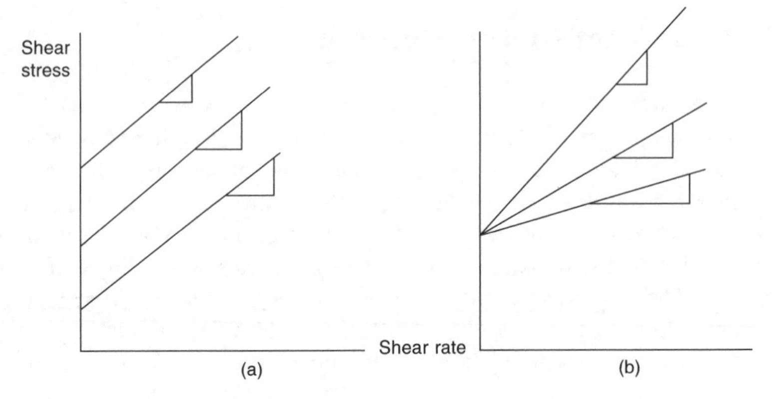


**Figure 15.17** (a) Concretes with same plastic viscosity but different yield points; (b) concretes with same yield point but different plastic viscosities.

bleeding other than a simplistic rule that concretes in the lowest consistence class may not bleed sufficiently and those in the highest classes may bleed excessively.

## *Cohesion*

The property that describes the ability of concrete to resist segregation is conventionally known as cohesion. The definition of cohesion is somewhat elusive other than in the negative way of stating that it is the opposite of segregation and, currently, no standard test exists for quantifying it. Cohesion cannot, therefore, be precisely specified and it is not possible to apply any conformity criteria to it. In certain counties including The Netherlands cohesion classes are allocated to different rankings of concrete displaying specific characteristics of smoothness, harshness, bleeding and segregation, but the assessment, nevertheless, remains largely a subjective one based on how the concrete 'feels'.

There is frequent reference in the following sections to two distinct types of concrete that are difficult to pump. Both of them are unable to produce the essential lubrication layer of grout: the first due to insufficient cohesion, the second due to an excess of it. The correct degree of cohesion for a pumpable concrete depends both on correct proportioning of the constituents through mix design and the characteristics of the constituents themselves.

An additional requirement for all concrete of consistently high quality is that is thoroughly mixed. Pumpability is particularly sensitive to within-batch variability created by inadequate mixing; greater care in weigh-batching, longer mixing times than usual and strict control of consistence are recommended when producing concrete for pumping.

Concrete pumps may be regarded as being their own quality control tools: it should be possible to pump existing well-proportioned general-purpose concretes over reasonable distances. For greater distances, for complex pipelines involving numerous bends and vertical lift, or for concretes that cannot be pumped at all, specific requirements relating to the rheology of the concrete need to be addressed.

### *Materials*

It will be seen in the following section that cementitious materials, aggregates and admixture can greatly influence the rate of bleeding and, therefore the pumpability of concrete.

## 15.7 Effects of aggregates, cement and admixtures on the pumpability of concrete

### 15.7.1 Cement types

In general, finer materials reduce bleed and this is true for cements. The standardization of Portland cements has resulted in a more uniform specific surface following many years of increase. Controlled-fineness cement was specially developed to assist the rate of bleed but is normally limited to the manufacture of spun precast concrete products and is not directly relevant to the subject of placing concrete by pump.

Pulverized-fuel ash is finer than cement; its spherical particle shape and glassy texture are acknowledged factors in reducing the water demand and increasing the cohesion of concrete. Accordingly, pfa makes a significant contribution to the pumpability of concrete with low cohesion, particularly when combined with a low cementitious content. Pfa has also been used specifically to improve the grading of some coarse sands. Conversely, pfa may hinder the formation of a suitable low-viscosity grout due to an excessive degree of cohesion in high-strength concrete of low consistence class, particularly where the aggregates themselves contribute to cohesion.

Condensed silica fume can make a considerable increase in the cohesion of concrete. It is beneficial in concrete with low cohesion but, being a colloidal material, its use in high-performance concrete requires care in the selection of aggregates and admixtures if excessive cohesion is to be avoided.

Ground granulated blastfurnace slag has little effect on pumpability as the particle shape and size resemble those of most Portland cements. However, it should be remembered that total cementitious contents tend to be higher for a given strength class when using ggbs or pfa and this could give rise to excessive cohesion. Conversely, if the total cementitious content is low, some bleeding can occur when high proportions of ggbs are used in combination with difficult aggregates and specific admixtures but in most structural pumping grades of concrete the only influences that can be detected by the presence of ggbs in fresh concrete are delayed stiffening (a distinct advantage in the event of a delay) and its lighter colour.

Metakaolin has little effect on the pumpability of concrete but it has been beneficial in the case of concretes with low cohesion.

## 15.7.2 Aggregates

Virtually all commercially produced aggregates in the UK can be used to make pumpable concrete. Some are more favourable than others but it is a mistake to believe that certain types and shapes are completely unsuitable.

### *Aggregate size*

A conventionally accepted rule is that the maximum nominal size of aggregate should be not more than one quarter of the pump diameter: i.e. minimum pipe diameter = $4D_{max}$. It is recognized that the risk of segregation increases with greater values of $D_{max}$ and extra care is required at mix design stage to maintain cohesion.

### *Aggregate grading*

A suitable aggregate for pumping will have a continuous grading that allows controlled migration of water and fine material towards the pipe/concrete interface. Excessive bleeding is caused by gap-graded aggregates or by coarse sands that present discontinuity between the coarser cement grains and the finest aggregate particles. Well-graded coarse aggregates combined with a sand that contains an adequate proportion of 'fine fines' will produce a continuous and smooth grading curve that goes some way to meeting the requirements of particle packing theory and pumpable concrete. Trials are essential to determine the best particle packing, the minmum voids volume and optimum compacted bulk density.

It is difficult to make a pumpable concrete with excessively coarse sand unless a finer sand or mineral addition such as pfa is included; merely increasing the sand content in an attempt to improve pumpability can make matters worse because of the absence of 'fine fines' between the gradings of the sand and cement.

Sand with a broad grading curve that connects both with the 10 mm coarse aggregate and the cement at 150 μm particle size contributes to the control of bleeding. This may require the importation of a special 'pumping grade' sand or the blending of two different sands to achieve a satisfactory grading.

One rule of thumb relates to the total fine fines: that is, all the cement plus the fraction of sand passing 300 μm. The recommended quantity increases with reducing aggregate size, thus:

| $D_{max}$ | 40 mm | 20 mm | 10 mm |
|---|---|---|---|
| Minimum mass of particles finer than 300 μm | 380 kg | 440 kg | 520 kg |

This approach, however, does not take account of aggregate shape or voids ratio and is, accordingly, simplistic as a guide to pumpability.

Standard techniques for combining aggregates to give a required grading are suitable for optimizing available materials in a pump mix. Some computer programs enable more accurate estimates to be made for trial mixes by entering aggregate data in respect of mean particle size, loose bulk density and particle density (relative density). The computer models eliminate much of the trial-and-error component of investigations by optimizing the theories of particle packing, voids ratios and particle interference to produce pumpable concrete with economic cement contents.

Having determined the optimum aggregate grading it is essential that uniformity is maintained. Dialogue with aggregate producers coupled with extra diligence and increased frequency of aggregate inspection, sampling and testing contribute towards uniformity.

Single-sized coarse aggregates are preferred to graded materials; significant changes can be made to the rate of bleeding or tendency towards segregation by adjusting the 10 mm coarse aggregate size in concrete with a 20 mm $D_{max}$ material. The current trend for concrete producers to use 20–5 mm graded coarse aggregate creates difficulties in making these adjustments to the coarse aggegate grading.

## Aggregate shape

Most shapes of aggregate can make pumpable concrete provided the volume of voids is kept to a minimum. Rounded gravels are popular because the high rates of shear that occur, particularly at tapers and bends, favour rounded particles as they are more easily rearranged in moving relative to each other. It is a mistake, however, to disregard an aggregate if it has an angular or irregular shape; it is possible for a well-graded crushed rock aggregate to have a lower voids ratio than a rounded gravel material and, consequently, produce a pumpable concrete.

High indices of flakiness and elongation can create difficulties in pumping due to aggregate interlock and bridging at bends in the pipeline. The sand properties and proportions need extra consideration when using any badly shaped coarse aggregate.

## Aggregate density

Any problems with cohesion and segregation can be aggravated by differences in density between one size of aggegate and another (for example, the sand and the coarse aggregate) or between the combined aggregate and the cement paste. In a badly proportioned lightweight concrete the coarse aggregate tends to float. This can be seen in the pump hopper and, once drawn into the pump, the concentration of aggregates may cause a blockage. Flotation may also be seen after discharge.

Lightweight aggregates are normally porous with a cellular structure. A significant quantity of mixing water may be forced into the aggregate under the pressure of pumping and the concrete can become unsaturated, putting an end to further progress. Unless special measures are taken, such as pre-soaking the aggregate and/or incorporating a special pore-blocking admixture, lightweight concrete can be pumped only over short distances and up low heights. When the concrete leaves the pipeline and is subjected again to normal atmospheric pressure, much of the water contained in the aggregate particles is emitted, creating difficulties with excessive bleeding, particularly in slabs.

High-density aggregates can sink to the invert of horizontal lengths of pipeline. As in the case of lightweight aggregates, the concentration of large particles represents an unsaturated suspension which is likely to cause a blockage. After completing a high density pumping operation, the unsaturated concrete in the invert is difficult to remove by normal pipe-cleaning methods. If concrete is allowed to set in the invert the effective diameter of the pipe is reduced, possibly to the critical dimension of approximately $4D_{max}$

or less. On subsequent pumping, long sections of this hardened concrete can become loosened and cause immediate blockage.

Gravitational factors associated with pumping downward (see section 15.4) are normally magnified by high density aggregates and the need to maintain a full pipeline at all times becomes paramount. In extreme cases of long pipe runs and deep descents it is advisable to avoid an abrupt change from priming grout to concrete by gradually introducing coarse aggregate in stages through transitional or 'weaning' batches.

### 15.7.3 Admixtures

#### *Pumping aids*

Thickening agents, originally based on cellulose ethers, have been marketed as admixtures to enhance general-purpose concrete to a pumpable mix without any further modification. It is now considered that an unpumpable concrete has a serious mix design deficiency and its pumpability is more likely to be achieved by addressing that deficiency than by introducing an exotic chemical.

Two useful contributions that may be made by thickening agents are when attempting to pump concrete with (a) low cement content combined with high consistence class over a long distance or up a great height and (b) lightweight aggregate where loss of water inside the porous particles needs to be avoided. The thickener is often combined with a dispersant in order both to limit the activity of the admixture specifically to blocking the pores in the aggregate and to act as a water reducer, maintaining strength for a given consistence class. A viscous residue may remain on the lining of all machinery involved in mixing and placing concrete incorporating a thickening agent, including mixers and pumps, which need to be thoroughly cleaned after use.

#### *Water-reducing admixtures, plasticizers, high-range water reducers and superplasticizers*

Water-reducing admixtures are recommended both for avoiding excessive cement content for a given free water/cement ratio at the required consistence class and to reduce any tendency towards excessive bleeding by minimizing the free water content. The selection of an appropriate water-reducing admixture is essential, depending on the cohesion of the fresh concrete: those that are formulated to enhance cohesion, sometimes entraining a small quantity of air, should be used on mixes that exhibit excessive bleeding and/or segregation whilst cement-rich concretes with a glutinous mortar phase benefit from water-reducing admixtures that act by lowering the viscosity of water and dispersing cement particles by deflocculation.

#### *Air-entraining agents*

The small air bubbles in air-entrained concrete behave like fine fines and can, accordingly, compensate for the lack of this size of material in coarse sand. The ability of air-entrainment to increase cohesion is well-established. Concrete that would normally be unpumpable due to low cohesion may be improved by the addition of only a small quantity of air-entraining agent.

Air entrainment is of no benefit to the pumpability of cement-rich concretes, particularly with a fine sand, and can give rise to a glutinous mortar that does not bleed, creating very high friction against the pipe wall. Difficulties of pumping high-strength air-entrained concrete may be partly overcome by adopting a high consistence class and/or using a high-performance plasticizer that encourages bleeding. The use of single-product, combined, air-entraining water-reducing admixtures is not recommended because it is essential that consistence, cohesion and air content can all be independently controlled.

One effect of pump pressure on air entrainment is to compress the bubbles. The energy used in compressing the air results in a rapid drop in pressure gradient, limiting the lengths and heights to which air-entrained concrete may be pumped. Air-entrained concretes with high air content – for example, those with a small $D_{max}$ – are particularly prone to air bubble compression which, in vertical pumping, can cause blockage within the boom of a mobile pump.

It can be seen that air entrainment places constraints on pumpability. Where air-entrained concrete must be pumped over long distances or through a great height, stage pumping may be used whereby the concrete is pumped from one machine to another.

Air-entrained concrete can lose a significant volume of air while being pumped. Current standards (BS 5328, BS 206-1) require that full-scale trials be established to estimate this loss and to make an appropriate correction so that the concrete has the specified entrained air content as it leaves the pipeline and enters the works. Sampling for conformity testing should technically take place at this location, although contractual arrangements may dictate that concrete is sampled as it leaves the delivery vehicle and enters the pump.

### *Retarders*

Retarding plasticizing admixtures are advisable in hot climates where pump pipelines are exposed to solar gain and, in all environments, they serve to extend the available time before concrete starts to stiffen in the pump and pipelines in the event of unexpected delays.

Commercially available retarders are, in fact, retarding water-reducers and, therefore, the points made earlier regarding the effects of plasticizers on cohesion, bleeding and pumpability should be considered whenever the use of retarders is proposed.

Where excessive bleeding is a persistent problem, incorrect selection of retarder can make matters worse by extending the time during which bleeding can occur.

### *Accelerators*

Accelerating water-reducing admixtures are now little used in the UK for general-purpose concrete construction. If used, their significance to concrete pumping would be associated to loss of workability, reducing the time available in the event of an unexpected delay, particularly in hot weather. Alternative ways of accelerating strength development, by reducing the free water/cement ratio or raising the concrete temperature may be more appropriate. Sprayed concrete, however, usually incorporates an accelerator and speed of use is of paramount importance in this specialist industry.

### *Other admixtures*

Integral waterproofers, corrosion inhibitors, bonding agents etc. have little if any effect on the pumpability of a well-designed concrete.

### 15.7.4 Batching and mixing of materials

All high-quality concrete should be accurately batched and thoroughly mixed. Pumped concrete is sensitive to batching errors, cross-contamination of aggregate sizes and incorrect measurement of mixing water. Forced-action mixers should be operated to their manufacturers' recommended mixing cycle while freefall mixers including truck-mixers need the benefit of a greater number of mixing revolutions at a higher speed of rotation than normal. Readymixed concrete should always be remixed immediately before discharge into a pump.

## 15.8 Modifications to the concrete mix design to ensure pumpability

It should be noted that the pumpability of concrete cannot be considered in isolation: the specification must be met in full including consistence and strength classes, durability clauses including maximum free water/cement ratio, minimum/maximum cement contents and any other specific requirements.

Sometimes the demands of the specification make it difficult to design a concrete that is also pumpable. No-fines concrete and some gap-graded concretes designed for exposed aggregate architectural finishes do not contain sufficient grout to be pumped and most concretes in consistence class S1 are too stiff to allow their grout to migrate to the pipe wall. Accordingly, consistence class S2 should be assumed as suitable for pumped concrete in the majority of normal types of construction. Where higher consistence classes are needed due to on-site complexity (congested reinforcement, narrow formwork etc.), modifications are required to the mix design to control cohesion, bleeding and segregation.

In principle, concrete is most likely to be pumpable provided a continuous combined aggregate grading curve can be achieved and the cement content is sufficient to fill voids within the combined aggregate. As with mix design for normal concretes, angular coarse aggregates, coarse sand grading and high consistence classes require higher sand contents.

It is important that concrete to be pumped is capable of replenishing the lubricating layer of low-viscosity grout in a continuous process of controlled bleeding. Accordingly, concretes for pumping contain approximately 5 per cent more sand than normal.

Where concrete lacks cohesion, an increase in sand content and a corresponding reduction in coarse aggregate content of 50 kg/m$^3$ to 90 kg/m$^3$ may be appropriate. The greater water demand that is the result of this cohesion correction can require 10 kg/m$^3$ additional cement which, in turn, futher improves cohesion. In the case of over-rich concretes that exhibit low rates of bleeding, a negative cohesion correction is appropriate.

If single-sized coarse aggregates are available, adjustment to the proportion of 10 mm aggregate can have a large effect on the rate of bleeding. Where aggregates are rounded gravels, bleeding can be reduced by increasing the quantity of 10 mm material but this has the opposite effect with crushed angular aggregates if it results in an increase in voids content.

Laboratory trials should be conducted to find the optimum particle packing by combining aggregates in different proportions until minimal voids exist. It is generally true that, provided the cement content is sufficient to fill the remaining voids, the concrete will be pumpable. Taking the typical bulk density of cement to be 1440 kg/m$^3$, the minimum cement content for pumpable concrete can be calculated from the equation:

$$\text{Minimum cement content} = \text{voids \%} \times 1440 \text{ kg/m}^3$$

The consequences of modifying the cement content of a concrete could affect free water/cement ratio, free water demand, sand:coarse aggregate ratio and plastic density, calling for a fundamental recalculation of the mix design.

Sometimes a change of sand source, or the blending of two or more sands, is required to obtain a suitable 'pumping grade' sand that assists in achieving a continuous combined aggregate grading with minimum voids content. Approximate guides to suitable aggregate grading curves are shown in Figure 15.18, from research carried out in the UK by the Concrete Pumping Association (which no longer exists) and the Building Research Establishment (BRE) (Kempster, 1969a). The BRE also published work by Ted Kempster (Kempster, 1969b) on apparatus for making a direct measurement of voids in aggregates.

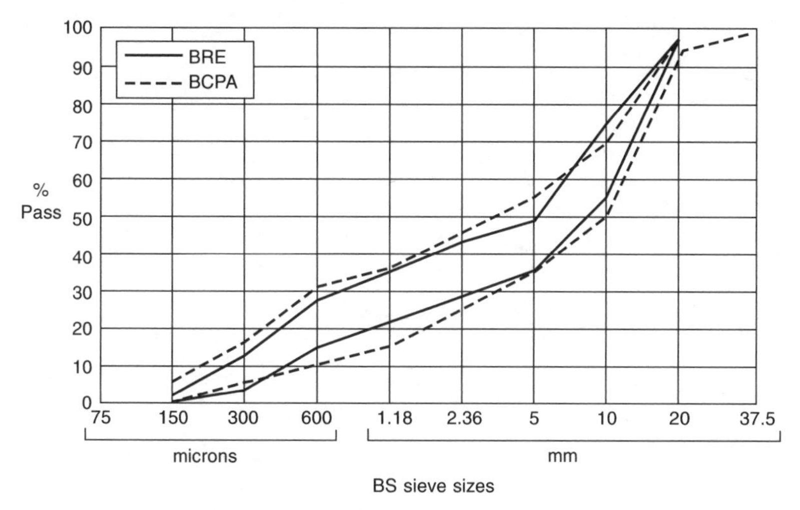


**Figure 15.18** Grading curves for pumpable concrete.

## 15.9 Void content of aggregate and procedures for measuring void content in combined aggregate grading

The voids that occupy the interstices between aggregate particles demand a large proportion of grout to fill them. If they are not filled, uncontrolled bleeding, segregation and aggregate interlock are promoted by the presence of voids. Voids need to be kept to a minimum if the segregation pressure of concrete is to be high (Figure 15.19).

Overfilling of voids with an excessive quantity of cementitious material can produce the high friction condition referred to earlier, which can be as unfavourable to pumping as a deficiency of fine material (Figure 15.20).

Optimum particle packing is required if void content is to be minimal and this state may be confirmed by finding the maximum rodded bulk density of the combined aggregate. The voids volume can be calculated from the expression:

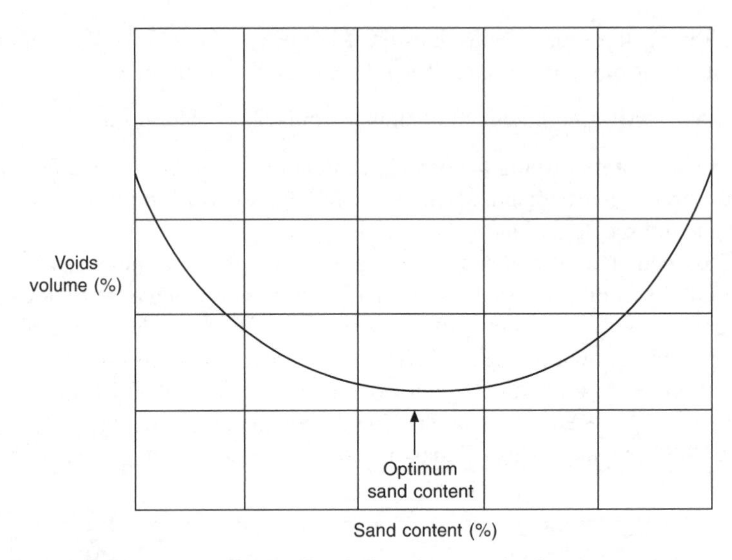


**Figure 15.19** Relationship between voids content and sand content for determining optimum aggregate grading.

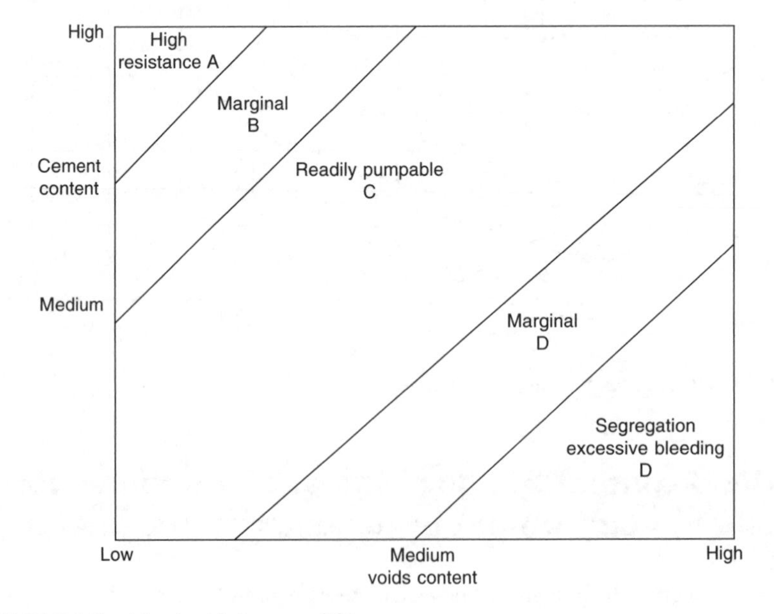


**Figure 15.20** Relationship of voids to pumpability.

$$\text{Voids} = 100 - \frac{\text{Bulk density} \times 100\%}{\text{Particle density}}$$

Alternatively a direct measure of the voids content can be made by use of the void meter apparatus (Kempster, 1969b) developed by Ted Kempster of the then Building Research Station (now Building Research Establishment). Details of the test procedure are given in Box 15.1. The void meter gives a clear indication of air voids content (Figure 15.21).

**Box 15.1** Void meter test procedure

The void meter apparatus comprises a graduated 3.5 litre glass cylinder with an airtight lid. The lid is connected through a tap to a graduated water column which can apply a constant partial vacuum to the contents of the cylinder.

Calibration involves partially filling the cylinder with water in 10 per cent increments and checking the accuracy of the corresponding readings on the graduated water column.

The test is made by preparing a sample of aggregate that has been combined in the proportions expected to be used. The sample needs to be at least air dry and throughly mixed.

The cylinder is filled in 10 equal increments, each layer being thoroughly compacted in a standard procedure by tamping with a piston weighing 3.15 kg. The piston is allowed to fall through 25 mm ten times per layer.

After compacting the final layer and ensuring that it is level with the top graduation mark, the system is made airtight by closing the bleed valve and creating a partial vacuum by allowing the water column to drop. The water column falls until equilibrium is reached and it comes to rest in the sight glass.

A direct reading is taken which, provided the apparatus is calibrated, represents the volume of air voids in the aggregate sample.

Some skill is required when filling the cylinder with aggregate to avoid creating unwanted voids through segregation within the narrow confines of the glass cylinder.

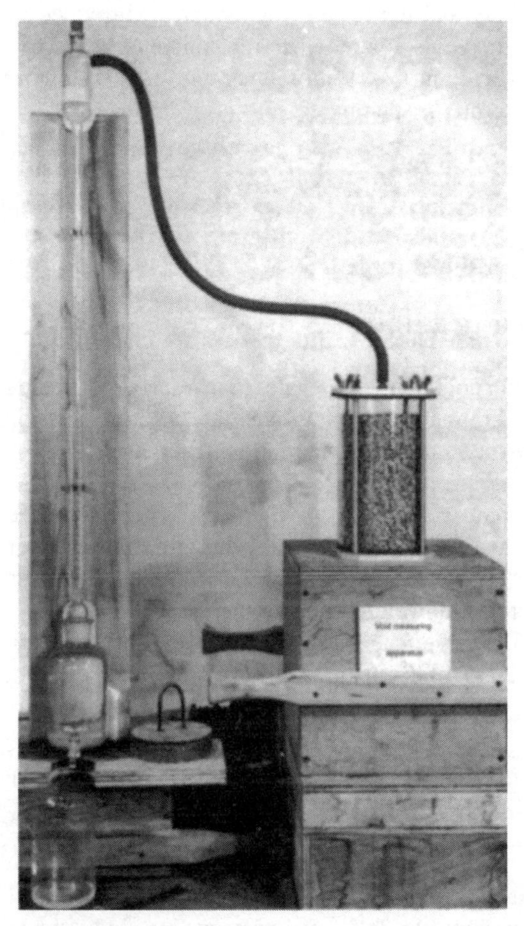

**Figure 15.21** Void measurement apparatus.

It was demonstrated that concrete was generally pumpable provided the volume of cement was at least sufficient to fill the volume of voids. It is convenient to calculate the minimum cement content by mass from the expression:

$$C_{\text{min}} = \frac{\text{Bulk density of cement} \times \text{voids volume}}{100} \ (\text{kg/m}^3)$$

### *Example*

If the measured voids content was indicated to be 25 per cent, the minimum cement content per cubic metre for pumpability would be 25 per cent of the bulk density of cement (1440 kg/m$^3$), i.e. 360 kg/m$^3$. Kempster's work also confirmed that a large excess of cement presents difficulties because of high plastic viscosity.

Voids content may alternatively be calculated from measured values for compacted bulk density and particle density (relative density × 1000) of the aggregate:

$$\text{Voids (\%)} = \frac{\text{Paticle density} - \text{bulk density}}{\text{Particle density}} \times 100$$

A series of laboratory trials are required to find the aggregate proportions that produce the highest bulk density and the lowest voids content. Provided the cement volume is slightly greater than the voids volume, then the concrete should be cohesive, containing sufficient grout to provide the lubrication layer and to allow deformation at tapers and bends. It should, therefore, be pumpable.

## 15.10 Suitable combinations of aggregates for pumpable concrete

### 15.10.1 Suitable gradings

Established methods for combining aggregates to fit a specified grading curve are applicable to pumped concrete. Computer programs exist that determine the optimum combination of aggregates to produce minimal voids ratios. Alternatively, the following traditional 'best fit' technique may be used.

A good starting point would be the middle of the grading curves proposed by BRE (Kempster, 1969a). A slight upward bias above the mid-point, particularly in the 'fine fines' (300 μm and 150 μm size material) should be applied in the case of high free water/cement ratio concretes: a downward bias in the case of concretes with a very low free water/cement ratio and those made with very fine cement, pfa or silica fume.

Referring to Figure 15.18 the midpoints of the BCPA recommended grading curve are seen to be:

| Sieve size | 150 μm | 300 μm | 600 μm | 1.18 mm | 2.36 mm | 5 mm | 10 mm | 20 mm | 37.5 mm |
|---|---|---|---|---|---|---|---|---|---|
| % passing | 4 | 11 | 21 | 26 | 36 | 45 | 60 | 97 | 100 |

The target grading values are superimposed on a straight diagonal line (Figure 15.22). The sieve analyses in Figure 15.23 are examples of three individual aggregate sizes to be

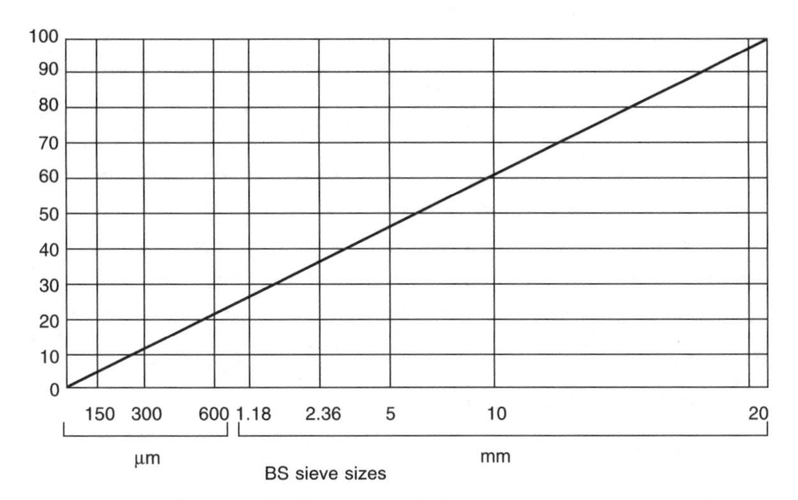


**Figure 15.22** Target aggregate grading.

| | A | B | C | D |
|---|---|---|---|---|
| BS sieve size | 20 mm | 10 mm | Sand | Required grading |
| 37.5 mm | 100 | 100 | 100 | 100 |
| 20 mm | 95 | 100 | 100 | 97 |
| 10 mm | 25 | 98 | 100 | 60 |
| 5 mm | 3 | 40 | 98 | 45 |
| 2.36 mm | 1 | 12 | 80 | 36 |
| 1.18 mm | 0 | 2 | 65 | 26 |
| 600 µm | | 0 | 55 | 21 |
| 300 µm | | | 24 | 11 |
| 150 µm | | | 10 | 4 |
| 75 µm | | | 2 | |

**Figure 15.23** Typical aggregate gradings for: (A) 20 mm single-sized coarse aggregate; (B) 10 mm coarse aggregate; sand (C); (D) target grading.

combined to a required grading. Gradings of the individual aggregates are now superimposed on the sieve sizes and best-fit straight lines are drawn to represent each aggregate (Figure 15.24). The ends of the aggregate lines are joined to each other and the points at which they interesect the diagonal are marked (x, y) (Figure 15.25). The proportions of aggregates are derived from:

Aggregate A = $100 - y$ = 45%
Aggregate B = $y - x$ = 19%
Aggregate C = $x - O$ = 36%

An arithmetic check should be made to assess the accuracy with which the proposed combination fits the required gradings shown in Figure 15.26.

Minor adjustments can be made if the grading is required to be closer. In this example a small reduction (say 1 per cent) to the proportion of 10 mm aggregate and a corresponding increase in the sand would bring the achieved grading closer to the required grading.

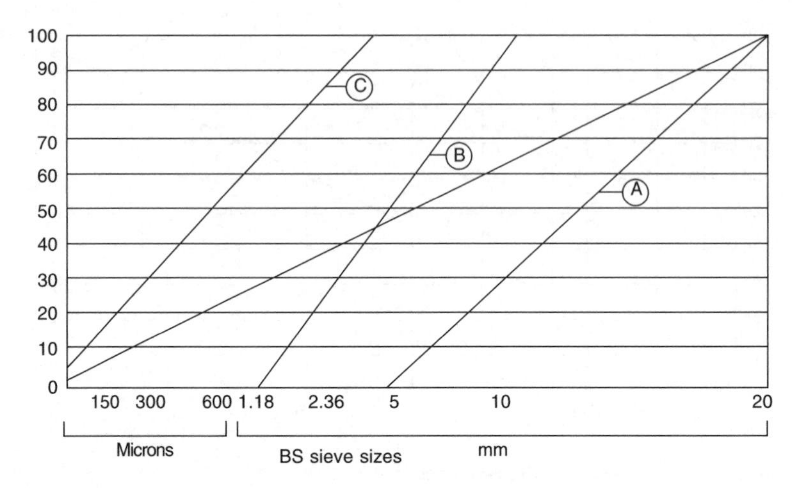


**Figure 15.24** Superimposition of gradings of individual aggregates in sieve sizes.

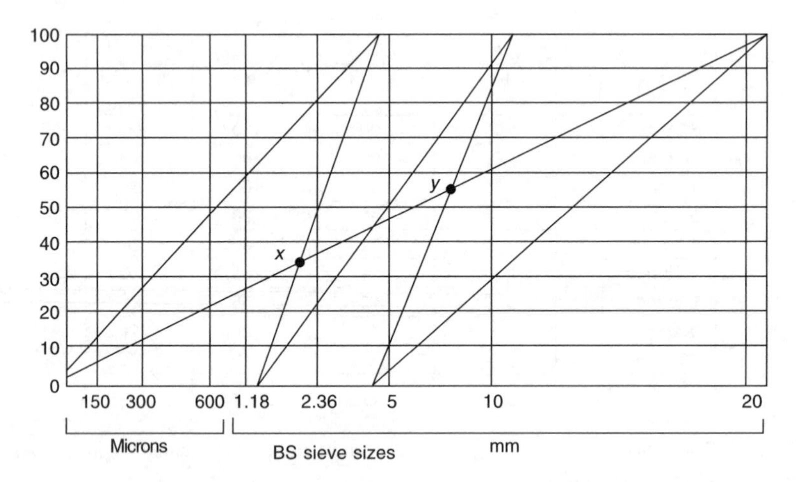


**Figure 15.25** Joining of the ends of aggregate lines.

| | A (45%) | B (19%) | C (36%) | Sum of A + B + C | D |
|---|---|---|---|---|---|
| BS sieve size | 20 mm | 10 mm | Sand | Combined grading | Required grading |
| 37.5 mm | 45 | 19 | 36 | 100 | 100 |
| 20 mm | 43 | 19 | 36 | 98 | 97 |
| 10 mm | 10 | 17 | 36 | 63 | 60 |
| 5 mm | 1 | 8 | 35 | 44 | 45 |
| 2.36 mm | 1 | 3 | 29 | 33 | 36 |
| 1.18 mm | 0 | 2 | 24 | 26 | 26 |
| 600 μm | | 1 | 20 | 21 | 21 |
| 300 μm | | 0 | 9 | 9 | 11 |
| 150 μm | | | 4 | 4 | 4 |
| 75 μm | | | 2 | 1 | 0 |

**Figure 15.26** Arithmetic check on combined aggregate grading.

## 15.11 Recognizing pumpable concrete

The need to measure the ease of placing concrete has led to the development of tests for workability/consistence as defined in national standards. The tests measure a fluid property of concrete and it is understandable that attempts would be made to apply them in the quest for determining the suitability of a particular concrete for placing by pump. The tests have proved unsuitable for this purpose because they are single-point tests.

The slump test to BS 1881: Part 102/BS EN 12350-2 and the flow table test to BS 1881: Part 105/BS EN 12350-5 involve measurement of the concrete when deformation under a reducing stress has ceased. They are thereby an inverse measure of the yield value of concrete – i.e. the higher the consistence class, the lower the yield value.

The vibration applied in the VeBe test to BS 1881: Part 104/BS EN 12350-3 imposes an arbitrary and fairly high rate of shear on the concrete and the time taken to achieve a given amount of deformation is measured. The test therefore measures flow at a high rate of shear.

The compacting factor test to BS 1881: Part 103 (now withdrawn) and the compactability test to BS EN 12350-4 measure the extent by which loose uncompacted concrete becomes consolidated. Concrete in a pump pipeline has already been compacted, first by the remix paddles and again when it is compressed by the pump. A measure of the degree of compaction is not related to the behaviour of concrete flowing along a pipe. Current consistence tests are single-point tests, all of which fail to determine rate of shear relative to the applied stress.

In order to measure viscosity of concrete the values of shear rate must be measured at two different shear stresses, requiring a test technique based on the principles applied by Tattersall (1991) at the University of Sheffield and developed into a two-point workability test apparatus. Figure 15.29 shows that a single-point test can give misleading information regarding the rheological properties of concrete.

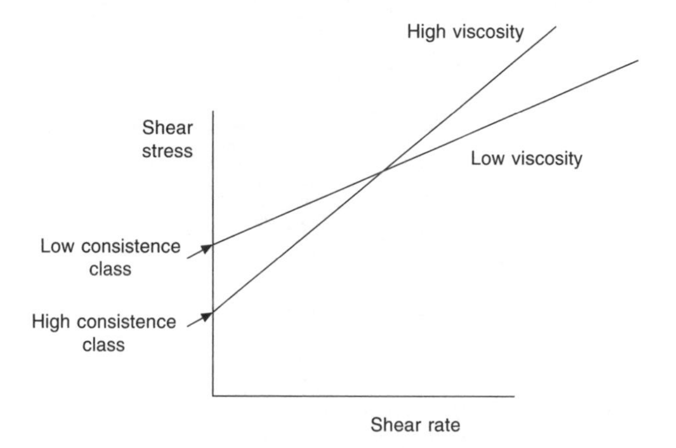


**Figure 15.27** Two-point testing is required for assessing rheological characteristics.

Ferraris and de Larrard (1998) developed a new test method based on a modification of the slump test after attempting to adapt rheometers used in soils engineering. The method is claimed to be a two-point test with a promising relationship to the rheological nature of concrete.

ASTM test C 232 may be used for determination of bleeding and bleed rate, but no test currently exists to quantify cohesion. The slump test should always be checked for cohesion, as advised in the British Standard; the flow table is also an effective indicator of excessive bleeding and segregation.

Monitoring of pump pressures and observation of segregation and/or excessive bleeding can give warning that a concrete may present difficulties when being pumped. Thorough mixing is essential and truck mixers must remix concrete before discharge into pumps.

## 15.12 Pressure bleed test apparatus

The suitability of fresh concrete can be confirmed by subjecting trial mixes, or fresh concrete sampled from full-scale production, to a pressure bleed test developed by Brown of Taylor Woodrow Materials Research Laboratory (Brown, 1975) for the Construction Industry Research and Information Association (Loadwick and Brown) (Figure 15.28).

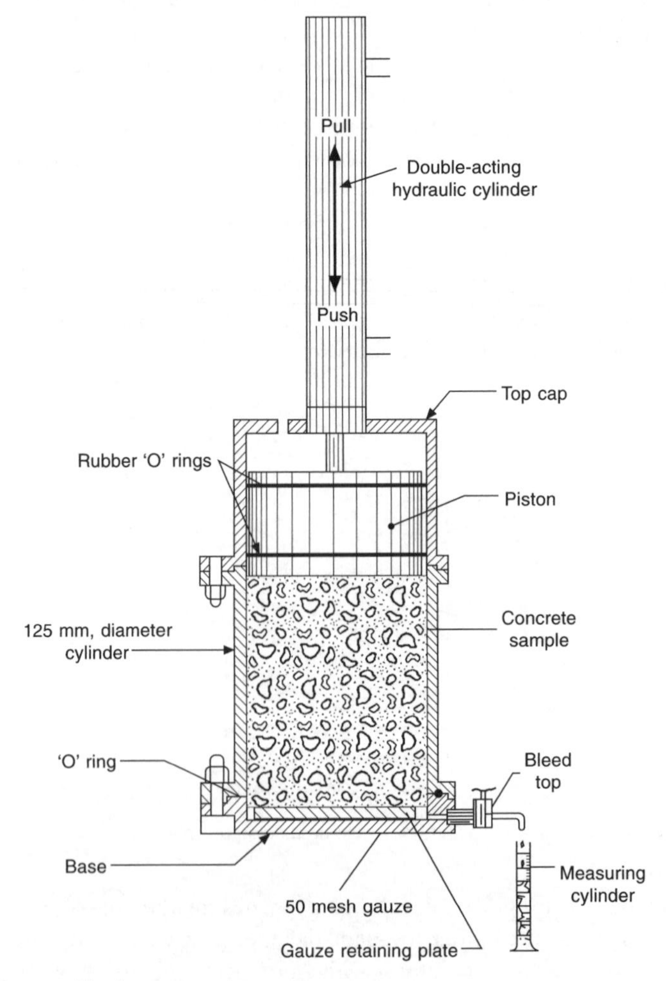


**Figure 15.28** Pressure bleed test apparatus.

In the pressure bleed test a sample of fresh concrete is placed inside a cylinder fitted with a fine mesh to allow grout to escape. Pressure is applied by a hydraulic pump to a piston acting directly on the concrete until an operating pressure of 3.5 $N/mm^2$ is achieved, replicating typical concrete pump hydraulic pressure of 390 $kg/cm^2$. A tap is opened and the volume of grout emitted under pressure is measured. Two readings are required: the first after 10 seconds and the other after 140 seconds. Most concretes give up all their bleedwater in well under 140 seconds but more important than the total volume of bleedwater is the difference between the volume emitted after 10 seconds and total volume.

In principle, if concrete under pressure loses a large volume of water initially and very little thereafter, it exhibits uncontrolled bleeding and is unlikely to be pumpable. Concrete is more likely to be pumpable if it emits water at a more steady, controlled rate.

In the research programme a number of different concretes were tested whose pumpability was well known; those extreme concretes which bled very rapidly and those which bled hardly at all were known to be unpumpable and all gave very low values of $V_{140} - V_{10}$, while the concretes displaying a controlled rate of bleed and satisfactory pumpability gave higher values of $V_{140} - V_{10}$ (Figure 15.29). Operating instructions for the pressure bleed test apparatus are give in Box 15.2.

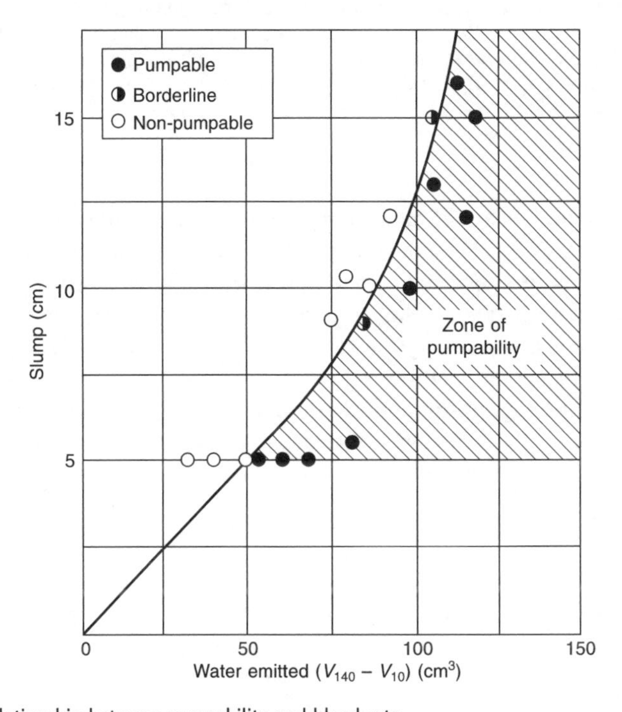


**Figure 15.29** Relationship between pumpability and bleed rate.

## 15.13 Conclusion

The study of pumped concrete is closely related to the discipline of rheology in which the flow of suspensions is seen to depend upon the properties of the materials in suspension.

**Box 15.2** Pressure bleed test procedure

1. Clean apparatus thoroughly before use, particularly at the interconnecting surfaces between the cylinder, base and top cap. If during the test a significant amount of water is lost other than through the bleed tap, then the interpretation of the test result will be rendered meaningless.
2 Grease each of the 'O' rings on the cylinder and base plate.
3 Bolt the base plate securely to the cylinder and close the bleed tap.
4 Pour 30 ml of water into the cylinder and run off 15 ml through the bleed tap to remove any air from the system. The remaining 15 ml will be sufficient to fill the channel around the inside of the base and prevent air from being trapped beneath the gauze.
5 Place the fresh concrete sample into the cylinder in two layers up to a level approximately 15 mm from the top of the cylinder. Each layer should be tamped sufficiently to remove any pockets of air but maintained in a relatively uncompacted state, as it would be in a pump.
6 Bolt the top can to the top of the cylinder, having ensured once again that the surfaces are clean.
7 Using the hand pump, apply a pressure of 3.5 N/mm$^2$ to the concrete. This is equivalent to an oil pressure of 389 kg/cm$^2$.
8 Open the bleed tap when the pressure has been maintained for 10 seconds and collect the water emitted in measuring cylinders at time intervals of 10, 20, 30, 60 and 140 seconds, while maintaining the pressure at 3.5 N/mm$^2$.
9 Close the bleed tap and release the pressure.
10 Release the nuts on the base plate and push out the compressed concrete plug.
11 Return the piston in the top cap housing by reversing the valve on the hand pump.
12 Separate the top cap and cylinder and clean the apparatus thoroughly for the next test.
13 The test should be carried out twice on each concrete and an average taken. If the results are significantly different a third test should be carried out.

Consistence and cohesiveness of concete must be considered together if the concrete is to be pumped successfully. The concrete needs to be in a saturated phase, it requires a high segregation pressure, low yield stress and low plastic viscosity.

The cohesion at a particular level of consistence is required to be such that a controlled amount of bleeding is promoted by the sliding action of concrete through a pipeline.

Prior assurance of the pumpability of a particular concrete remains elusive. Because they are single-point tests, existing standard concrete test procedures have only limited capabilities for assessing the behaviour of fresh concrete and a test is required for determining the concrete's yield stress and plastic viscosity and thereby quantifying the suitability of a mix design for pumped concrete.

## References

ASTM. ASTM test C 232 Method for Bleeding of Concrete. ASTM, Conshohocken, Pennsylvania, USA.

Bernoulli, *Mathematical Exercises* (Part 2) (1724), *Hydrodynamica* (1738).

BS 1881: Part 102. Testing concrete: Method for determination of slump. British Standards Institution, London.

BS 1881: Part 103 (now withdrawn) testing concrete: Method for determination of compacting factor. British Standards Institution, London.

BS 1881: Part 104 Testing concrete: Method for detemination of Vebe time. British Standards Institution, London.

BS 1881: Part 105 Testing concrete: Method for determination of flow. British Standards Institution, London.
BS EN 12350-2 Testing fresh concrete: Slump test. British Standards Institution, London.
BS EN 12350-3 Testing concrete: Vebe test. British Standards Institution, London.
BS EN 12350-4 Testing concrete: Degree of compactability test. British Standards Institution, London.
BS EN 12350-5. Testing concrete: Flow table test. British Standards Institution, London.
BS 5328 Concrete: Part 4. Specification for the procedures to be used in sampling, testing and assessing compliance of concrete. British Standards Institution, London.
BS EN 206-1. Concrete – specification, performance, production and conformity. British Standards Institution, London.
Brown, N.E. (1975) What bleeding concrete do you need for pumping? Taylor Woodrow Materials Research Laboratory, Southall, Middlesex, UK.
Dawson, O. (1949) Pumping concrete – friction between concrete and pipeline. *Magazine of Concrete Research*, **1**, No. 3 December.
Ede, A.N. (1957) The resistance of concrete pumped through pipelines. *Magazine of Concrete Research*, November.
Ferraris, C.F. and de Larrard, F. (1998) Modified slump test to measure rheological parameters of fresh concrete. *Journal of Cement, Concrete and Aggregates*, **20**, No. 2 ASTM, Conshohocken, Pennsylvania, USA.
Joisel, A. (1952) *Composition des betons hydrauliques.*
Kempster, E. (1969a) Pumpable concrete, Current Paper 29/69, Building Research Station, Garston.
Kempster, E. (1969b). Measuring void content: new apparatus for aggregates, sands and fillers. Current Paper 19/69, Building Research Station, Garston.
Loadwick, F. and Browne, R.D. The mechanics of concrete pumping. CIRIA Report No. 27. Construction Industry Research and Information Association. London.
Newton, Sir Isaac (1687) *Principia*. Trinity College, Cambridge.
Poiseuille, J.-L.M. (1844) Movement of liquids in small-diameter pipes.
Reynolds, O. (1883) An experimental investigation of the circumstances which determine whether the motion of water in parallel channels shall be direct or sinuous and the law of resistance in parallel channels.
Stokes, G.G. (1846) On the steady motion of incompressible fluids, 1842. Report on recent researches in hydrodynamics. British Association for the Advancement of Science.
Tattersall, G.H. (1991) *Workability and Quality Control of Concrete*. E&FN Spon, London.

# 16

# Concrete construction for liquid-retaining structures

*Tony Threlfall*

This chapter sets out the special requirements for reinforced concrete liquid-retaining structures. Particular attention is given to the following requirements: durability and watertightness; specification of concrete and constituent materials to British and European Standards; cracking and autogenous healing; control of cracking due to temperature and moisture change; joints and jointing materials.

## 16.1 Introduction

Reinforced concrete construction is commonly used, without an internal lining or external tanking, for a wide variety of liquid-containing and liquid-excluding structures. Typical examples are reservoirs and water towers, settlement tanks for sewage treatment, storage tanks for agricultural and industrial effluents, swimming pools and utility-grade basements. Prestressed concrete is also used, particularly for cylindrical tanks.

Such structures are required to be watertight but not vapour proof. Containment vessels are normally tested for liquid retention during construction and some initial short-term seepage may be tolerated at this stage. Failure to achieve an adequate standard of watertightness may result from factors such as porous concrete, uncontrolled cracking, defective joints, continuous leakage paths caused by formwork ties and other embedded items.

The design and construction of liquid-retaining structures in the UK are generally carried out in accordance with the recommendations of BS 8007. Further guidance is given in BS 8102 and CIRIA Report 139 for liquid-excluding structures. As an alternative to BS 8007, DD ENV 1992-4 (hereafter referred to as EC2: Part 3) may be used in conjunction with DD ENV 206 (effectively replaced by BS EN 206-1).

## 16.2 Permeability and durability

The concrete should have low permeability. This is important not only for its direct bearing on watertightness but also because of its influence on durability and the protection from corrosion of embedded steel. In BS 8007 the recommendations for concrete mixes and materials are intended to ensure that a water-impermeable concrete is obtained, provided complete compaction without segregation is obtained on-site. Both internal and external surfaces are taken to be in at least the severe exposure category, as defined in BS 8110. The severe condition would apply typically to a covered reservoir containing potable water. Surfaces in contact with more aggressive liquids, such as sewage or silage effluent, or exposed to corrosive external environments are taken to be in the very severe category.

In BS EN 206-1 concrete is considered to be water impermeable if, when tested according to BS EN 12390-8, the maximum water penetration is less than 50 mm and mean average water penetration is less than 20 mm. The exposure class for carbonation-induced corrosion would generally be taken as XC 4.

### 16.2.1 Concrete grades

In BS 8007 for severe exposure, the minimum cement content is required to be 325 kg/m$^3$ and the maximum water/cement ratio 0.55 for Portland cement and Portland-slag cement and 0.50 for Portland-fly ash cement. The 28-day characteristic cube strength of the concrete is to be not less than 35 N/mm$^2$ with the concrete classed as grade C35A. This is a non-standard classification and a grade C40 would be recommended in BS 8110. The lower strength grade was considered appropriate in BS 8007 in order to avoid cases where considerably more cement would be needed to achieve the strength of 40 N/mm$^2$, with a corresponding increase in heat of hydration effects and associated cracking problems. Blinding concrete is required to be not less than 75 mm thick and not weaker than grade C20 normally or grade C25 for aggressive soil conditions.

In BS EN 206-1 for exposure class XC 4, the minimum cement content is required to be 300 kg/m$^3$ and the maximum water/cement ratio 0.50. The minimum strength class of the concrete is given as C30/37.

### 16.2.2 Maximum cement content

Maximum cement contents are restricted in order to minimize cracking due to thermal contraction and drying shrinkage. In BS 8007 the limits for reinforced concrete are 400 kg/m$^3$ for Portland cement and Portland-slag cement and 450 kg/m$^3$ for Portland-fly ash cement. Values of 500 kg/m$^3$ and 550 kg/m$^3$ respectively are allowed for concrete that is to be pre stressed, on the basis that any cracking would be counteracted by the pre stress.

### 16.2.3 Concrete cover

In BS 8007 for severe exposure, the nominal cover (i.e. cover including an allowance for tolerance) to ordinary reinforcement and other embedded steel is required to be 40 mm. The combination of grade C 35A concrete and 40 mm nominal cover is considered appropriate for structures with an intended design life of 40 to 60 years. When a longer design life is required, additional precautions are recommended such as increasing the cement content, increasing the cover, and using special reinforcement (galvanized, epoxy-coated or stainless steel).

## 16.3 Constituent materials

The water absorption of aggregates and its effect on drying shrinkage has been an on-going point of contention for many years in British codes. Although there appears to be no particular relationship between absorption and durability, which relates more to overall concrete porosity, an absorption limit of 3 per cent is imposed in BS 8007.

Cements as specified in EN 206–1/BS 8500 are used, depending on the nature of the contained liquid and, in the case of buried structures, the composition of the soil and groundwater. Portland-slag and Portland-fly ash cements are recommended, not only for their beneficial effect on sulphate-resistance but also for their lower heat properties particularly in thicker sections.

Water reducers and superplasticizers are used when necessary to obtain workability with reduced water and cement contents, but the use of waterproofing admixtures is rare as reported by de Vries and Polder (1995). For concrete in exposed situations, air-entraining admixtures are used where freeze/thaw resistance is required.

### 16.3.1 Use of ground granulated blastfurnace slag

In BS 8007 it is stated that for normal use the target mean proportion of ggbs should not exceed 50 per cent. At first sight this value seems to be unreasonably low, since the full benefits of enhanced durability and resistance to both chloride ingress and sulphate attack cannot be achieved at proportions below 70 per cent. In practice, the restriction is taken to apply to concretes for *normal* use (i.e. exposed to carbonation-induced corrosion) and not to apply to concretes for *special* use (i.e. exposed to chloride-induced corrosion and/or aggressive chemical environments).

The Water Authorities Association Specification (1998) defines the ggbs contents for concrete that is designed to retain aqueous liquids as *normal* 0–50 per cent and *special* 70–90 per cent.

The use of high ggbs contents is also particularly beneficial in reducing the temperature rise of the concrete in thick sections, as reported by Henderson and Clear (2000). Typical values for the temperature rise of concretes containing different proportions of ggbs cast in sections with a minimum dimension of 1 m are shown in Table 16.1. It can be seen that the BS 8007 limit of 400 $kg/m^3$ for the maximum cement content could be raised to a value exceeding 500 $kg/m^3$ for concretes containing high proportions of ggbs.

**Table 16.1** Temperature data for concrete sections with a minimum dimension of 1 m

| Binder combination %PC/%ggbs | Temperature rise (°C) per 100 kg of binder | Temperature rise for various binder contents (°C) | | | |
|---|---|---|---|---|---|
| | | 350 $kg/m^3$ | 400 $kg/m^3$ | 450 $kg/m^3$ | 500 $kg/m^3$ |
| 100PC | 12 | 42 | 48 | – | – |
| 50PC/50ggbs | 7 | 25 | 28 | 32 | 35 |
| 30PC/70ggbs | 5 | 18 | 20 | 23 | 25 |

## 16.3.2 Use of limestone aggregates

Though not specifically excluded in BS 8007, limestone aggregates are sometimes prohibited in practice. Some softer oolitic limestone materials may have high water absorptions but many limestone aggregates exhibit low absorption and are of a consistently good quality. Also, limestone aggregates are advantageous in reducing both initial and longer-term thermal movements because of their lower coefficients of linear expansion.

Concerns over the possibility of attack by aggressive water have been raised. Although limestone aggregates may be vulnerable to attack by acidic water, their use has been shown to be generally beneficial in prolonging the life of the structure as the attack is more uniform. By contrast, inert aggregates may eventually stand proud of the surface, become detached and produce more detritus.

## 16.3.3 Aqueous leaching of metals

Concrete has traditionally been used in contact with water for public supply with no apparent risk to health. All water undertakers, however, under regulations issued by the Drinking Water Inspectorate in 2000, are required to use only approved products as a safeguard against any potential hazard with regard to the leaching of metals.

BS 6920 specifies methods of test and maximum admissible concentrations (MAC) for the aqueous leaching of metals, and tests on readymixed concrete have been carried out at the Water Research Centre. Six different mix compositions each with two types of aggregate, gravel and granite, were investigated. Aluminium was detected in all the aqueous leachates, with the highest concentrations (generally above the MAC of 200 μg/litre) being in the samples containing slag and/or gravel aggregate. Small quantities of barium, chromium and iron, less than 10 per cent of the MAC values, were detected in some aqueous leachates.

It was concluded that the concentrations of metals found in the tests were not a cause for concern, and that the actual testing of admixture-free concretes to obtain an authorization for use should not be necessary. Where admixtures are used, the chemical identity of the components must comply with the substances specified in a List of Authorized Cement Admixture Components, and the admixture must be used at a concentration that does not exceed the manufacturer's recommended dosage. The situation may well change significantly in the future, depending on the final form of the European Acceptance Scheme. In the currently proposed form, there is no provision for gaining generic approval that would exempt concrete from further testing.

## 16.4 Cracking and autogenous healing

All cracks are a potential source of concrete deterioration and steel corrosion by allowing ingress to deleterious substances. Even small cracks that pass through the full thickness of a section will also allow water percolation resulting in damp patches and surface staining. However, investigations by Clear (1985) and Edvardsen (1999) have shown that a process of autogenous healing can occur that causes the water flow to gradually reduce and even cease altogether. The healing process was found to be a combination of mechanical blocking and the chemical precipitation of calcium carbonate (Figures 16.1 and 16.2). Self-healing has been shown to occur in active cracks if they are narrower than 0.3 mm and depart less than 30 per cent from the minimum width.

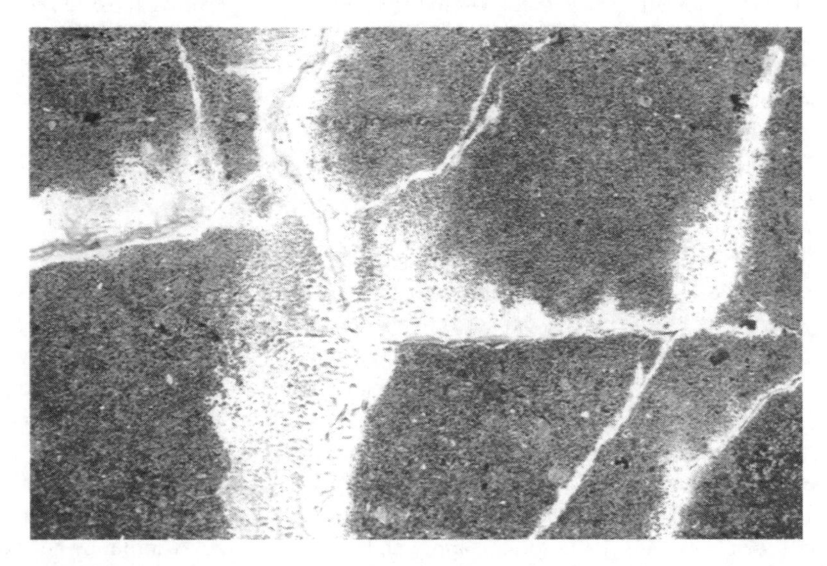

**Figure 16.1** White calcium carbonate at the surface of a concrete structure indicating self-healing of cracks (*Concrete*, April 1999).

The model used to analyse the initial water flow before self-healing is derived from the theory of laminar flow of an incompressible fluid between smooth parallel-sided plates, attributed to Poiseuille. The model shows the crack dimension to be the dominant factor because the water flow is proportional to the cube of the crack width. However, in practice the water flow will be much less than Poiseuille's formula predicts because of factors such as the roughness of the crack surfaces, reductions in crack width at the reinforcement, crack branching, adhesion and cohesion effects. The model therefore needs to be modified by a reduction factor derived from tests. From a regression analysis of test results, a reduction factor of about 0.25 appears to be reasonable.

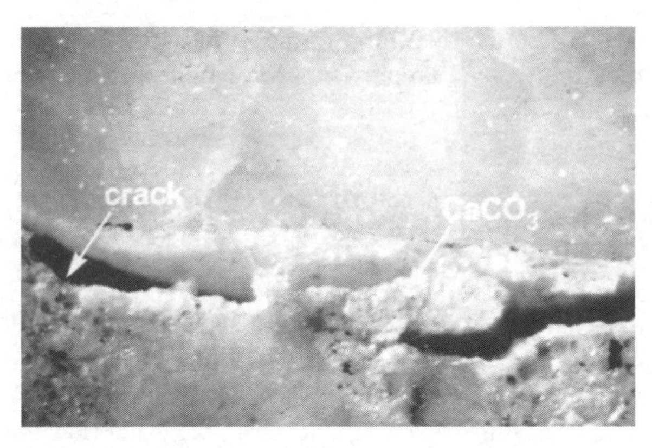


**Figure 16.2** Blocking of the crack path (crack width 0.2 mm) by $CaCO_3$ crystals (*Concrete*, April 1999).

## 16.4.1 Design crack width

In BS 8007, the design surface crack width for severe or very severe exposure is 0.2 mm. For structures whose appearance is considered to be aesthetically critical, the limit is 0.1 mm. These values apply to cracks that pass through the full thickness of a section and to those that form on one face only. When a containment structure is first filled the liquid level is maintained steady for a stabilizing period, while absorption and autogenous healing take place, before the structure is tested for water retention. The recommended stabilizing periods are 7 days for a design crack width of 0.1 mm and 21 days for a design crack width of 0.2 mm.

In EC2: Part 3, design crack widths are related to the requirements for leakage according to the classification of the element or structure as shown in Table 16.2. For class 1, the widths of any cracks that may be expected to pass through the full thickness of the section are limited to 0.2 mm where healing can be expected and 0.1 mm where healing is not expected. For class 2, cracks that may be expected to pass through the full thickness of the section are to be avoided unless measures such as liners are provided to ensure that leakage does not occur. Cracks may be expected to heal in members where the annual range of strain at a section is less than $150 \times 10^{-6}$. Where cracks do not pass through the full thickness of the section and a depth of at least 50 mm remains in compression, the requirements of EC2: Part 3 do not apply and the design crack width may be taken as 0.3 mm.

**Table 16.2** Classification of liquid-retaining structures in EC2: Part 3

| Class | Leakage requirements |
|---|---|
| 1 | Global tightness: leakage limited to a minimal amount, surface staining or damp patches acceptable |
| 2 | Local tightness: leakage generally not permitted, appearance not to be impaired by surface staining |

The following relationship is given for the prediction of leakage through a crack with no self-healing:

$$Q = (K/\eta)(\Delta p/h) L_c w_{eff}^3 \, m^3/s$$

where: $K$ = coefficient depending on crack surface characteristics (taken as 1/50)
$\eta$ = dynamic viscosity of the liquid (kg/ms)
$\Delta p$ = pressure difference across the element (Pa)
$h$ = thickness of the element (m)
$L_c$ = crack length (m)
$w_{eff}$ = effective crack width (m)

## 16.5 Cracking due to temperature and moisture effects

When water is added to a concrete mix, heat is evolved due to the hydration of the cement and the temperature of the concrete increases. During this period the concrete has a low elastic modulus and high creep characteristic so that any compressive stresses due to restrained expansion are easily relieved. The concrete matures during the subsequent cooling period and any tensile stresses generated by restrained contraction are less easily relieved.

Sections up to about 1 m thick achieve a temperature peak within a day of the concrete being placed and then cool rapidly over the next two days. Cracking often occurs at this time due to restraint to the contraction of the concrete while it is still weak. Subsequent falls in temperature and loss of moisture in the concrete during the construction period and later in normal service may cause the original cracks to widen.

### 16.5.1 Early thermal cracking

Figure 16.3 shows early temperature changes in °C for a 450 mm thick wall section using plywood formwork. The concrete mix contained 344 kg/m$^3$ of Portland cement and the formwork was struck after 24 hours. Removal of the formwork increased the temperature gradient during the cooling-down period but had no effect on the total temperature change since the peak temperature had developed within 12 hours. The fall in temperature from the peak to the mean ambient value is about 35°C in this example. For construction during the summer, the use of steel formwork is generally beneficial in allowing a more rapid dissipation of heat.

External restraint to the contraction of the concrete as it cools occurs mainly as a result of placing the concrete in separate stages rather than as a continuous pour. Wherever construction joints are introduced, restraint occurs to the thermal contraction of the concrete that is placed at a later stage. Figure 16.4 shows some typical examples of cracking due to restraint at construction joints in cases where insufficient reinforcement has been provided.

The first example shows an infill bay where restraint may result in an opening-up of the construction joints and the formation of one or more transverse cracks depending on the length of the bay. The second example shows a wall panel on a previously cast base, with the other edges of the panel unrestrained. The warping of the panel in this case may result in an opening-up of the construction joint at each end of the panel and the formation of one or more vertical cracks according to the length and aspect ratio of the panel. The third example shows a wall panel restrained along two adjacent edges where the restraint at the vertical edge may result in a horizontal crack near the mid-height.

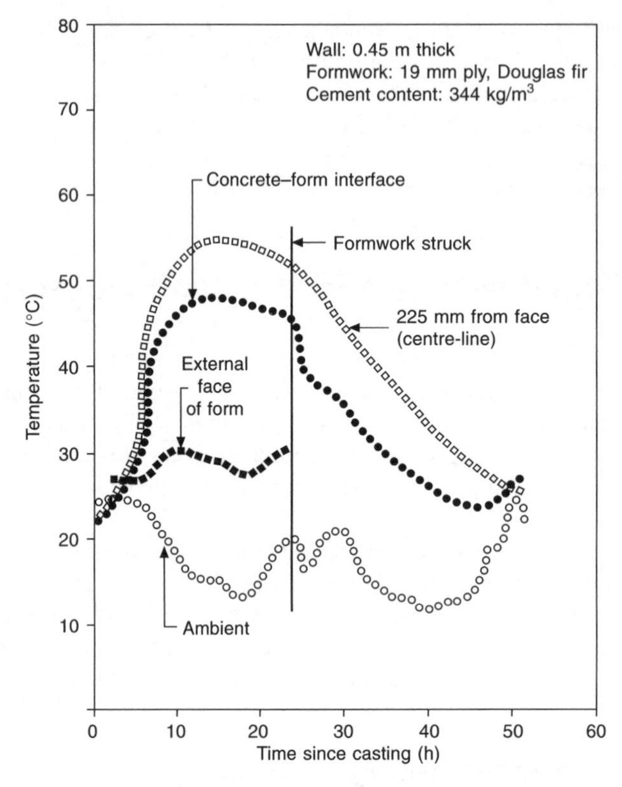


**Figure 16.3** Early temperature changes in concrete wall section showing effect of striking timber formwork (*Watertight concrete construction*, C&CA, 1978).

Internal restraint occurs in thick sections where large temperature gradients can develop. Surface zones, which are in direct contact with the environment, lose heat more rapidly than the inner core and undergo a smaller rise in temperature. In very thick sections it may take several days for the core temperature to reach a peak value and the continuing expansion of the hot core can stretch the cooler surface zones to the extent that cracking occurs. Such cracks may sometimes close during subsequent cooling of the core, if there are no other external restraints. Otherwise, the surface cracks may gradually extend into the core and link-up to form continuous cracks through the full thickness of the section. In this case, it is important to control the initial cracking by providing sufficient reinforcement within the surface zones.

## 16.5.2 External restraint factors and critical temperature changes

The magnitude of the tensile strain depends in part on the degree of restraint provided to the thermal contraction. For most situations there is always some restraint but complete restraint is rare. BS 8110: Part 2 provides typical values of external restraint factors recorded for a range of pour configurations, as given in Table 16.3.

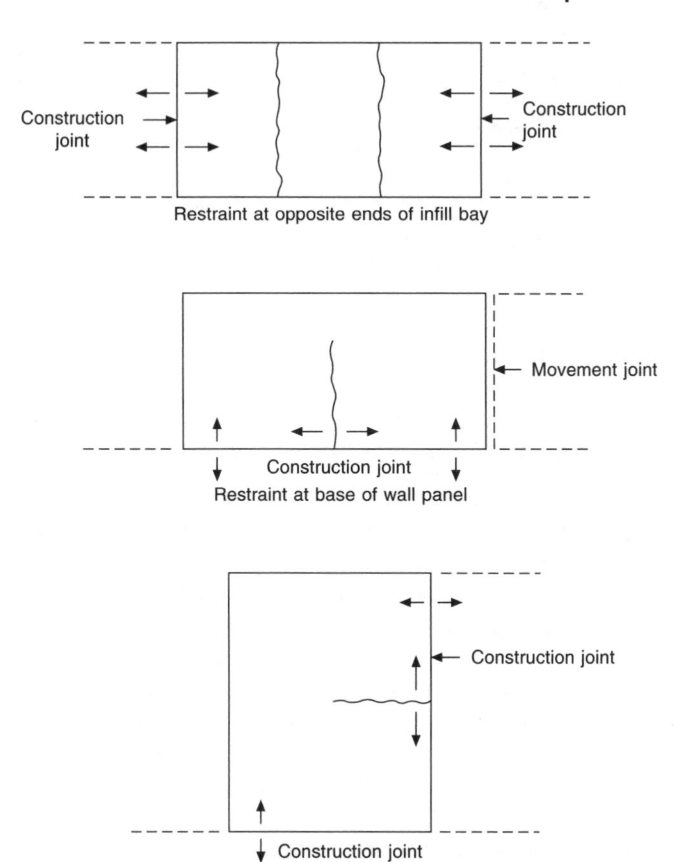


**Figure 16.4** Examples of cracking due to external restraint at construction joints.

**Table 16.3** Values of external restraint recorded in various structures

| Pour configuration | Restraint factor $R$ |
|---|---|
| Thin wall cast on to massive concrete base | 0.6 to 0.8 at base, 0.1 to 0.2 at top |
| Massive pour cast on to blinding | 0.1 to 0.2 |
| Massive pour cast on to existing mass concrete | 0.3 to 0.4 at base, 0.1 to 0.2 at top |
| Suspended slabs | 0.2 to 0.4 |
| Infill bays, i.e. rigid restraint | 0.8 to 1.0 |

In principle, cracking should not occur in well-defined forms of thermal restraint if the restrained component of the thermal strain does not exceed the tensile strain capacity of the concrete. BS 8110: Part 2 provides estimated values of maximum acceptable temperature changes to avoid cracking, as given in Table 16.4. The tabulated values have been derived from the relationship $\Delta t = \varepsilon_{ct}/0.8\alpha R$, where 0.8 is a creep related coefficient.

**Table 16.4** Estimated limiting temperature changes to avoid cracking

| Aggregate type | Thermal expansion coefficient α $10^{-6}$/°C | Tensile strain capacity $\varepsilon_{ct}$ $10^{-6}$ | Limiting temperature fall Δ*t* for varying restraint factor *R* | | | |
|---|---|---|---|---|---|---|
| | | | 1.00 °C | 0.75 °C | 0.50 °C | 0.25 °C |
| Gravel | 12 | 70 | 7 | 10 | 14 | 28 |
| Granite | 10 | 80 | 10 | 13 | 20 | 40 |
| Limestone | 8 | 90 | 14 | 19 | 28 | 56 |

## 16.5.3 Methods of control

BS 8007 recommends that in order to control cracking that may result from temperature and moisture changes in the concrete it is desirable to limit the following factors:

(a) The maximum temperature and moisture changes during construction by:
  (1) using aggregates having low or medium coefficients of thermal expansion and avoiding the use of shrinkable aggregates
  (2) using the minimum cement content consistent with the requirements for durability and when necessary sulphate resistance
  (3) using cements with lower rates of heat evolution
  (4) keeping concrete from drying out until the structure is filled or enclosed
  (5) avoiding thermal shock or over-rapid cooling of a concrete surface
(b) Restraints to expansion and contraction by the provision of movement joints
(c) Restraints from adjacent sections of the work by using a planned sequence of construction
(d) Localized cracking within a member between movement joints by using reinforcement or pre stress
(e) Rate of first filling with liquid
(f) Thermal shock caused by filling a cold structure with a warm liquid or vice versa.

## 16.5.4 Movement joints

Movement joints are intended to accommodate relative movement between adjoining parts of a structure with special provision made to maintain the watertightness of the joint. Freedom of movement may be required for structural reasons, as in the case of hinged and sliding joints, or to relieve restraints to temperature and moisture effects. Examples of expansion and contraction joints in walls and floors are shown in Figure 16.5.

Expansion joints are intended to accommodate either expansion or contraction of the concrete. An initial gap is formed between adjoining parts of the structure and the use of a joint filler, water-stop and sealant is essential.

Contraction joints may be complete, with all of the reinforcement discontinued at the joint, or partial, with only 50 per cent of the reinforcement discontinued at the joint. The recommended maximum spacing for complete joints is twice that given for partial joints. Contraction joints may be either formed or induced. In the first case, stop-ends are used and concreting is discontinued at each joint. In the second case, the thickness of the

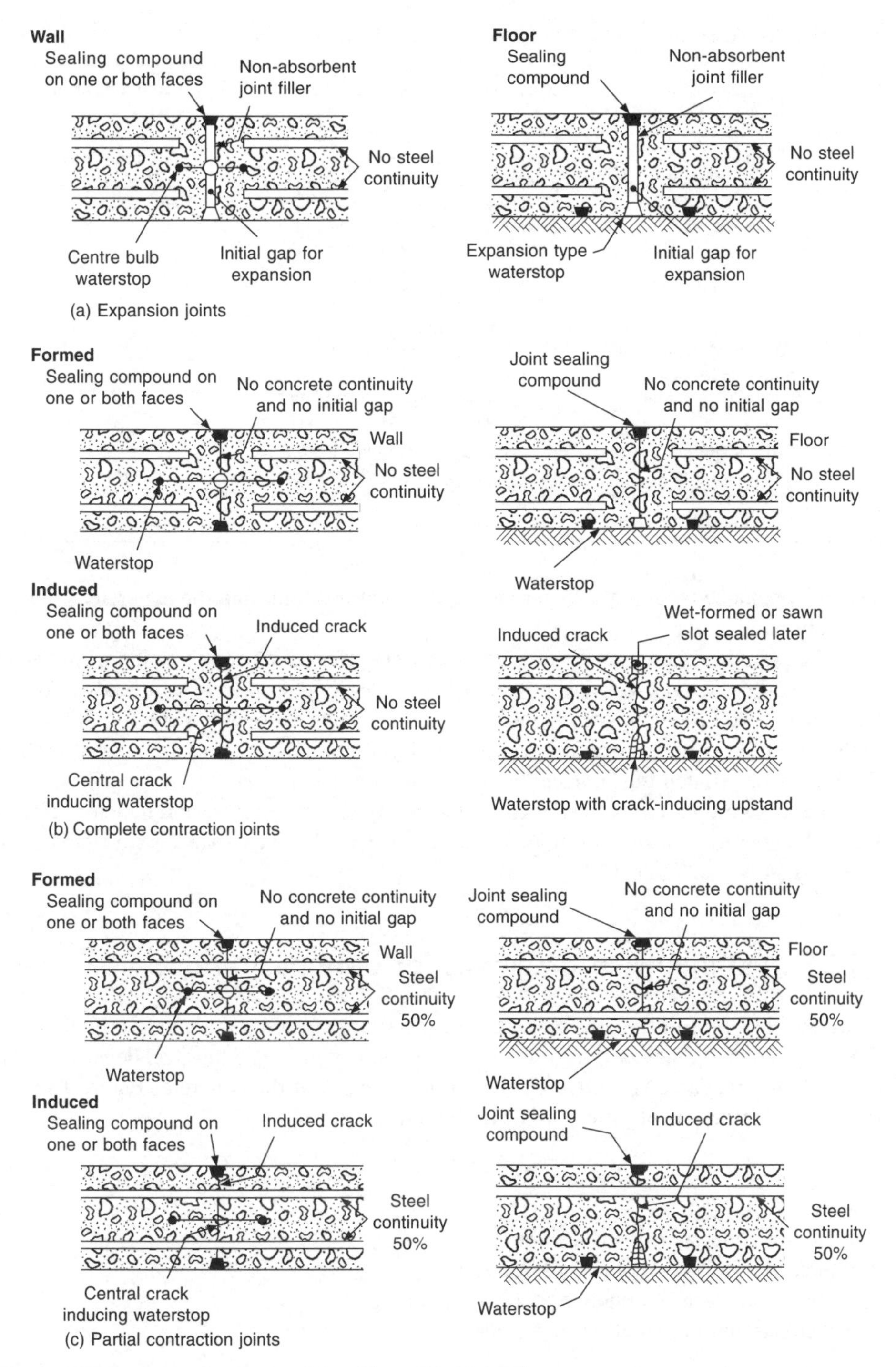


**Figure 16.5** Examples of movement joints (Figure 5.1, BS 8007).

**Table 16.5** Design options for control of temperature and moisture effects in BS 8007

| Option | Type of construction and method of control | Movement joints at maximum spacing (m) | Reinforcement ratio |
|---|---|---|---|
| 1 | Continuous.<br>Restrained movement (at all stages) and controlled cracking | No joints, except for expansion joints at wide spacing, which may be desirable in walls and roofs that are not protected from solar heat gain or where a contained liquid is subjected to a substantial temperature range. | Determine by calculation but not less than $\rho_{crit}$ |
| 2 | Semi-continuous.<br>Restrained movement (early thermal) and controlled cracking | (a) Complete joints [15]<br>(b) Alternate partial and complete joints [11.25]<br>(given spacing by interpolation)<br>(c) Partial joints [7.5] | Ditto |
| 3 | Closely spaced movement joints.<br>Freedom of movement and minimal cracking | (a) Complete joints [$4.8 + w/\varepsilon$]<br>(b) Alternate partial and complete joints [$2.4 + 0.5s_{max} + w/\varepsilon$]<br>(c) Partial joints [$s_{max} + w/\varepsilon$] | $(2/3)\rho_{crit}$ |

section is reduced by at least 25 per cent at the joint positions and the concrete is placed in a continuous pour.

In the examples shown, crack-inducing water-stops are used in conjunction with sealant slots to reduce the thickness of the section. The centrally placed water-stop shown for the wall section would be replaced by an outer surface water-stop with a crack-inducing rib in the case of a basement wall. For thicker walls, additional measures are needed such as rigid pipes or inflated tubes placed vertically within the section.

Three design options are described for the control of temperature and moisture effects where the reinforcement requirements are related to the incidence of any movement joints, as shown in Table 16.5.

## 16.5.5 Minimum reinforcement for options 1 and 2

If the tensile strain in the concrete exceeds the strain capacity, the concrete will crack. The initial cracking may be controlled by reinforcement provided the steel does not yield. Since the tensile force is limited by the tensile strength of the concrete (Figure 16.6), a minimum reinforcement ratio may be derived in the form:

$$\rho_{crit} = (A_s/A_c)_{crit} = f_{ct}/f_y$$

where:
$A_c$ = area of concrete (in surface zone)
$A_s$ = area of reinforcement (in surface zone)
$f_{ct}$ = tensile strength of concrete at age of cracking
$f_y$ = yield strength of steel reinforcement

The value of $f_{ct}$ is usually taken at the age of 3 days to be given by the relationship $f_{ct} = 0.12 f_{cu}^{0.7}$ μ 1.6 N/mm$^2$. For a C35A concrete and grade 460 reinforcement, $f_{ct}$ = 1.6 N/mm$^2$ and $\rho_{crit} = 0.0035$.

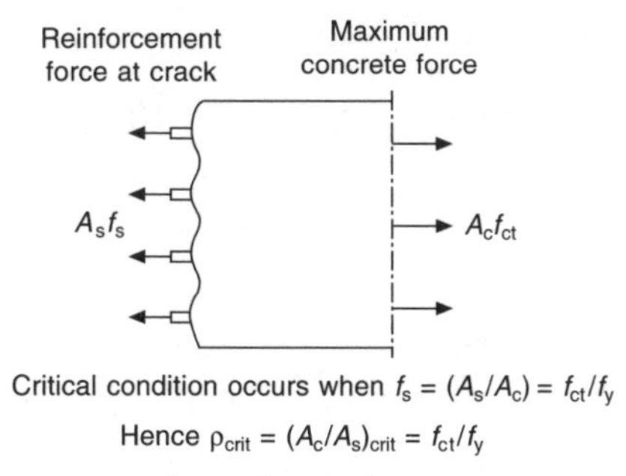


**Figure 16.6** Minimum reinforcement ratio for crack control.

## 16.5.6 Crack spacing and crack width

The approach in BS 8007 to the calculation of crack spacing and crack width in immature concrete subject to pure tension derives from work by Evans and Hughes (1968). When a crack forms in the concrete, bond-slip occurs and the force is resisted entirely by the reinforcement. With increasing distance from the crack there is a transfer of force from the reinforcement to the concrete by bond, until a position is reached where the concrete stress has again reached the tensile strength of the concrete (Figure 16.7).

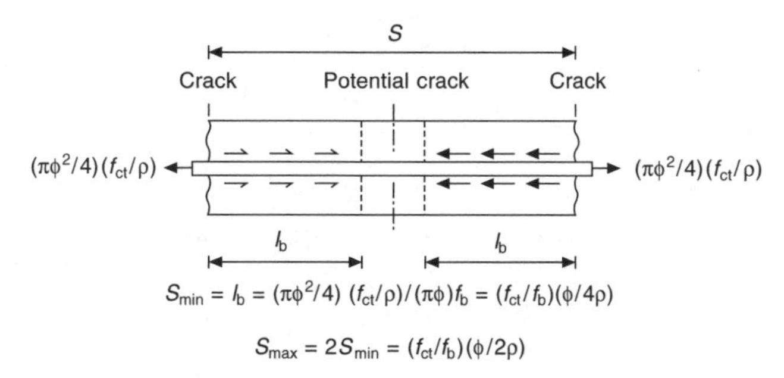


**Figure 16.7** Crack spacing limits according to bond-slip theory.

This concept permits a relationship to be derived for the minimum distance between two cracks by equating bond length to tensile force divided by bar perimeter times bond strength in the form:

$$s_{min} = (\pi\phi^2/4)(f_{ct}/\rho)/(\pi\phi)f_b = (f_{ct}/f_b)(\phi/4\rho)$$

where:
$f_b$ = bond strength at age of cracking
$\phi$ = bar size
$f_{ct}$ = tensile strength of concrete at age of cracking
$\rho$ = reinforcement ratio ($\geq \rho_{crit}$)

In a fully developed crack system, the maximum crack spacing, $s_{max} = 2s_{min}$ and the maximum crack width due to strain $\varepsilon$ is given by:

$$w_{max} = s_{max}\ \varepsilon = (f_{ct}/f_b)(\phi/2\rho)\varepsilon$$

The effective strain ε to be used in the calculations is determined by the relationship:

$$\varepsilon = (\varepsilon_{cs} + \varepsilon_{te} - 100 \times 10^{-6})$$

where:
$\varepsilon_{cs}$ = estimated drying shrinkage
$\varepsilon_{te}$ = estimated thermal contraction
For concrete made with good quality dense aggregate in normal UK climatic conditions, $\varepsilon_{cs} \leq$ [100 × $10^{-6}$ and ε may be taken as $\varepsilon_{te}$ given by the relationship:

$$\varepsilon_{te} = R\alpha(T_1 + T_2)$$

where:
$R$ = effective restraint factor usually taken as 0.5
$T_1$ = initial temperature fall from hydration peak value to mean ambient value at time of construction
$T_2$ = subsequent temperature fall due to temperature variations during service life
α = coefficient of thermal expansion of concrete (according to aggregate type)

Values of α (per °C) are given in BS 8110: Part 2 as 12 × $10^{-6}$ (flint or quartzite), 10 × $10^{-6}$ (granite or basalt) and 8 × $10^{-6}$ (limestone). Typical values of $T_1$, taken from CIRIA Report 91, are given in BS 8007 and Table 16.6.

**Table 16.6** Typical values of $T_1$ (°C) for Portland cement concretes

| Section thickness mm | Walls: Steel formwork Cement content kg/m³ | | | Walls: 18 mm plywood formwork Cement content kg/m³ | | | Ground slabs Cement content kg/m³ | | |
|---|---|---|---|---|---|---|---|---|---|
| | 325 | 360 | 400 | 325 | 360 | 400 | 325 | 360 | 400 |
| 300 | 11 | 13 | 15 | 23 | 25 | 31 | 15 | 17 | 21 |
| 500 | 20 | 22 | 27 | 32 | 35 | 43 | 25 | 28 | 34 |
| 700 | 28 | 32 | 39 | 38 | 42 | 49 | – | – | – |
| 1000 | 38 | 42 | 49 | 42 | 47 | 56 | – | – | – |

Designs are to be based on the specified maximum cement content unless the actual cement content is known. The tabulated values of $T_1$ are based on a mean ambient temperature of 15°C and a concrete placing temperature of 20°C. Designers are required to consider whether to allow for a higher concrete placing temperature to cater for hot weather conditions and long haul distances.

## 16.5.7 Summary of design options and reinforcement requirements in BS 8007

For option 1 where no contraction joints are provided, reinforcement is required to control cracking due to $T_1$ and $T_2$. For option 2 where contraction joints are provided at a spacing that is considered sufficiently close to accommodate seasonal thermal movements, reinforcement is required to control cracking due to $T_1$ only. For option 3 where contraction joints are provided at a spacing that is considered close enough to accommodate all movements, only a nominal amount of reinforcement given by $(2/3)\rho_{crit}$ is required.

The reinforcement required for options 1 and 2 may be obtained by rearranging the expression for $w_{max}$ to give:

$$\rho = (f_{ct}/f_b)(\varepsilon\phi/2w_{lim}) \geq \rho_{crit} = f_{ct}/f_y$$

For a C35A concrete and grade 460 deformed type 2 reinforcement, this becomes:

$$\rho = \varepsilon\phi/3w_{lim} \geq \rho_{crit} = 0.0035$$

Reinforcement is required in surface zones. For walls and suspended slabs, the thickness of each surface zone is taken as $0.5 \times$ thickness of section $\leq 250$ mm. Values of maximum effective strain $\varepsilon \times (10^{-6})$ for a 0.2 mm design crack width, with given reinforcement details in specified surface zones, may be obtained as shown in Table 16.7.

**Table 16.7** Values of maximum effective strain $\varepsilon$ ($\times 10^{-6}$) for 0.2 mm design crack width

| Zone thickness mm | Bar size mm | Bar spacing mm | | | | | | | |
|---|---|---|---|---|---|---|---|---|---|
| | | 300 | 250 | 225 | 200 | 175 | 150 | 125 | 100 |
| 100 | T10 | | | | 235 | 269 | 314 | 377 | 471 |
| | T12 | | | | 282 | 323 | 377 | 452 | 565 |
| 125 | T12 | | | 201 | 226 | 258 | 301 | 362 | 452 |
| | T16 | | 241 | 268 | 301 | 344 | 402 | 482 | 602 |
| 150 | T12 | | | | 188 | 215 | 251 | 301 | 377 |
| | T16 | 167 | 201 | 223 | 251 | 287 | 335 | 402 | 502 |
| 175 | T12 | | | | | 184 | 215 | 258 | 323 |
| | T16 | 143 | 172 | 191 | 215 | 246 | 287 | 344 | 430 |
| 200 | T12 | | | | | | 188 | 226 | 282 |
| | T16 | | 150 | 167 | 188 | 215 | 251 | 301 | 377 |
| | T20 | 157 | 188 | 209 | 235 | 269 | 314 | 377 | 471 |
| 225 | T12 | | | | | | | 201 | 251 |
| | T16 | | 134 | 149 | 167 | 191 | 223 | 268 | 335 |
| | T20 | 139 | 167 | 186 | 209 | 239 | 279 | 335 | 418 |
| 250 | T12 | | | | | | | 181 | 226 |
| | T16 | | | 134 | 150 | 172 | 201 | 241 | 301 |
| | T20 | 125 | 150 | 167 | 188 | 215 | 251 | 301 | 377 |

*Example*

The perimeter wall of a large reservoir is designed as a free-standing vertical cantilever for hydrostatic loading. Determine suitable arrangements of complete contraction joints and horizontal reinforcement to limit the crack widths due to temperature and moisture effects to 0.2 mm, using each of the design options. The wall stem is 500 mm thick, the concrete is grade C35A containing flint aggregate and a maximum Portland cement content of 350 kg/m$^3$, and the reinforcement is grade 460 deformed type 2.

From Table 16.6, allowing for the use of plywood formwork, $T_1 = 35°C$. For design option 1, consider $T_2 = 15°C$. The effective strain will be taken as $\varepsilon_{te}$, with $R = 0.5$ and $\alpha = 12 \times 10^{-6}$ per °C, and the reinforcement may be selected for a surface zone thickness of 250 mm from Table 16.7. The results are shown in Table 16.8.

**Table 16.8** Typical design options (0.2 mm crack width) for a 500 mm thick wall

| Option | Complete contraction joints | Effective strain and reinforcement requirements (each face) | |
|---|---|---|---|
| 1 | None | $\varepsilon_{te} = 0.5\alpha\,(T_1 + T_2) = 300 \times 10^{-6}$ | T16 – 100 (EF) |
| 2 | Spacing [15 m] | $\varepsilon_{te} = 0.5\alpha\,T_1 = 210 \times 10^{-6}$ | T16 – 140 (EF) |
| 3 | Spacing $[4.8 + w/\varepsilon]$<br>$= 4.8 + 0.0002/210 \times 10^{-6}$<br>= 5.75 m | $\varepsilon_{te} = 0.5\alpha\,T_1 = 210 \times 10^{-6}$<br>$(2/3)\ \rho_{crit} = 0.0024$<br>$A_s = 600\ mm^2/m$ | T16 – 300(EF) |

## 16.5.8 Crack prediction models in BS 8007 and EC2: Part 3

The approach given in BS 8007 for cracking of immature concrete due to restrained contraction uses a bond-slip theory that gives simple relationships, in which crack spacing depends on reinforcement bond length, and crack width increases as the effective strain increases. The approach used in both BS 8007 and BS 8110 for cracking of mature concrete due to the effects of applied loads employs a no-slip theory, in which the crack width increases as the tensile force increases.

In EC2: Part 3, a no-slip theory is used in all cases. For cracking due to restrained contraction, crack spacing reduces as the effective strain increases (i.e. more cracks are assumed to form) and crack width is a function of the tensile strength of the concrete. The following approximate relationships may be derived from the equations given in EC2: Part 3.

$$w_{max} = (125 + 0.4\phi/\rho_{r.eff})(f_{ct.eff}/\rho_r) \times 10^{-6} \text{ for section thicknesses} \leq 300 \text{ mm}$$

$$w_{max} = (100 + 0.2\phi/\rho_{r.eff})(f_{ct.eff}/\rho_r) \times 10^{-6} \text{ for section thicknesses} \geq 800 \text{ mm}$$

where:

$\rho_r$ = reinforcement ratio (based on full section thickness)
$\rho_{r.eff}$ = reinforcement ratio (in surface zone), where surface zone thickness is taken as 2.5 × distance from surface to centroid of reinforcement ≤ 0.5 × section thickness
$f_{ct.eff}$ = tensile strength of concrete at time when cracks are first expected to occur

# 16.6 Workmanship

It is essential that the workability of the concrete and the methods of handling and compaction are such that the concrete can be placed without segregation, be fully compacted, surround all the reinforcement and completely fill the formwork. It is particularly important to ensure that full compaction is obtained in the vicinity of joints and around embedded items such as water-stops and pipes.

BS 8007 requires particular care to be taken when forming construction joints and specifies that vertical joints should be formed against a stop-end. The surface of the first pour should be roughened to increase bond strength and provide aggregate interlock. Care should be taken to ensure that the joint surface is clean immediately prior to the fresh concrete being placed against it. The surface may need to be dampened to prevent excessive loss of mix water by absorption. Particular care should be taken to ensure that the new concrete close to the joint has an adequate fines content and is fully compacted.

Although water-stops should not be necessary in properly made construction joints, it is common practice to incorporate an impermeable hydrophilic material for this purpose. The material expands in the presence of water and creates a watertight seal when used in a confined zone. The Concrete Society Current Practice Sheet No. 119 (2000) provides further details of the materials and their use.

## 16.7 Summary

The requirements of reinforced concrete liquid-retaining structures have been examined with particular regard to the recommendations in BS 8007. Attention has been drawn to the use of ggbs where departure from the code restrictions is necessary for practical reasons. The relationships between crack width, leakage and autogenous healing have been examined and the watertightness requirements of BS 8007 and EC2: Part 3 have been given.

Considerable attention has been given to the control of cracking due to temperature and moisture effects and the design approach used in BS 8007 has been explored in detail. This method is also widely used in the UK for basements, bridges and earth-retaining structures. Reference has been made also to the crack prediction model incorporated in Eurocode 2, which is fundamentally different from the approach in BS 8007.

## References

BS 6920: 1996 *Suitability of non-metallic products for use in contact with water intended for human consumption with regard to their effect on the quality of water*. BSI, London.

BS 8007: 1987 *Design of concrete structures for retaining aqueous liquids*. BSI, London.

BS 8102: 1990 *Protection of structures against water from the ground*. BSI, London.

BS EN 206-1: 2000 *Concrete – Part 1: Specification, performance, production and conformity*. BSI, London.

BS EN 12390-8: 2000 *Testing hardened concrete – Part 8: Depth of penetration of water under pressure*. BSI, London.

DD ENV 1992-4: 2000 *Design of concrete structures: Liquid-retaining and containment structures*. BSI, London.

CIRIA Report 91: *Early-age thermal crack control in concrete*. CIRIA 1992.

CIRIA Report 139: *Water-resisting basement construction*. CIRIA 1995.

Clear, C.A. (1985) *The effects of autogenous healing upon the leakage of water through cracks in concrete*. C&CA Technical Report 559. British Cement Association.

Concrete Society Current Practice Sheet No. 119 (2000) Use of expanding (hydrophilic) water-stops in concrete construction. *Concrete*, February, 37–38.

De Vries, J. and Polder, R.B. (1995) Hydrophobic treatment of concrete. *Construction and Repair*, September/October, 42–47.

Edvardsen, C. (1999) Self-healing of concrete cracks. *Concrete*, April, 36–37.

Evans, E.P. and Hughes, B.P. (1968) Shrinkage and thermal cracking in a reinforced concrete retaining wall. *Proc. Institution Civil Engineers*, vol. 39, January, 111–125.

Henderson, N. and Clear, C.A. (2000) BS 8007 & the use of ggbs concrete. *Concrete*, February, 32–33.

Water Authorities Association (1998) Civil Engineering Specification for the Water Industry: 5th edition.

# 17

# Coatings

*Shaun A. Hurley*

## 17.1 Introduction

This chapter considers:

- Film-forming surface coatings
- Related treatments that are absorbed by, and/or that interact chemically with, typical concrete surfaces.

These materials, normally in the form of proprietary products, are commonly used on mass or reinforced concrete to:

- Prevent premature deterioration
- Limit or control ingress
- Enhance appearance
- Modify other surface properties

Products are available for internal, external, underwater and trafficked surfaces.

The following related products are outside the scope of this chapter:

- Thick renders, floor toppings and sprayed concrete
- Overlays for roads and bridge-decks
- Preformed sheet membranes for waterproofing
- Protective tapes
- Curing membranes
- Fire protection systems
- Coatings associated with electrochemical techniques that halt or prevent reinforcement corrosion (e.g. cathodic protection, desalination or realkalization)
- Rebar coatings and active corrosion inhibitors

# 17.2 Reasons for use

## 17.2.1 Summary of uses

Uncoated concrete provides a long service life in many environments. In overly aggressive conditions requiring additional surface protection, concrete can still remain an attractive construction material due to its versatility and relatively low cost. For some applications, protection against various forms of deterioration and/or ingress/transmission may be essential; for others, it may be optional, giving increased assurance of satisfactory durability. However, coatings should not be viewed as a basis for reducing cover or for inadequate mix design, placement and curing. The need for enhancement/change of appearance and/or the modification of other surface properties may also vary from being essential to optional.

Wherever surface coating/treatment is optional, increased initial costs, and the inevitable maintenance costs, must be balanced against the projected in-service benefits.

A summary of the common reasons for applying coatings/surface treatments to concrete is given in Table 17.1. Although these applications are well proven, specific local conditions can affect performance significantly. Consequently, particular requirements should always be discussed with suppliers.

More specifically in relation to the protection and repair of concrete structures, general principles for the use of products and systems – including coatings and related treatments – are set out in the European Prestandard DD ENV 1504-9: 1997.

These principles, and the related methods that are relevant here, are summarized in Table 17.2. As shown, there are eleven principles, six of which are concerned primarily with defects caused by mechanical, chemical or physical actions on the concrete itself, and five of which are specifically concerned with reinforcement corrosion. A related European Standard (prEN 1504-2) will deal in detail with the performance properties of these products and systems.

## 17.2.2 The specification of surface coatings/treatments

The use of protective surface barriers can be related to the age and condition of the concrete, as discussed below.

### *New concrete, satisfactory quality: 'normal specification' of surface treatment*

Surface barriers would normally be specified at the design stage where there is an obvious incompatibility between the performance of concrete and a particular service environment or demand. Hence, the predominant reasons for use are likely to be the prevention of direct deterioration and the control of ingress/contact.

In exceptionally aggressive environments, they may be used to enhance the resistance to indirect deterioration due to reinforcement corrosion. However, as surface treatments inevitably require maintenance, other options may be preferable (e.g. improved concrete mix design, increased cover, alternative forms of reinforcement).

Coatings may also be specified initially where both chlorides and sulphates are present

**Table 17.1** Common reasons for using surface coatings/treatments

| Reasons for surface coating/treatment | | Examples/comments |
|---|---|---|
| To prevent direct deterioration | Chemical attack | Attack by aggressive chemicals such as acids, sulphates, sugars and fertilizers |
| | Physical effects | Deterioration due to erosion/abrasion, salt crystallization and freeze–thaw action (surface scaling or deeper disintegration) |
| To prevent indirect deterioration due to reinforcement corrosion | Loss of concrete alkalinity and steel passivation due to the ingress of acidic gases | Carbonation |
| | Premature initiation of corrosion due to ionic ingress | Ingress of chlorides in coastal environments or from de-icing salts |
| To limit or control ingress/contact | Waterproofing | Barriers to liquid water that can vary widely in their resistance to moisture vapour transmission; some systems are approved for contact with potable water |
| | Vapour/gas barriers | Barriers to moisture vapour, methane, radon and acidic gases, e.g. $CO_2$, $SO_2$, $(NO)_x$ |
| | Ease of cleaning and decontamination | Floors, walls in food-processing areas, hospitals and nuclear installations |
| To enhance/maintain appearance | Colour and texture | Building facades |
| | Reflectance | Road tunnels and car parks |
| | Prevention of mould growth and dirt staining | Walls and floors |
| | Anti-graffiti coating/treatment | Assisting removal of graffiti |
| | Uniformity after repair | Overcoating following patch repairs |
| To enhance safety | Anti-slip/skid | Used with a scatter of fine aggregate on floors/roads |
| | Anti-static/electrically conductive systems | Floor coatings in manufacturing areas |
| | Road/floor markings | Defining specific areas by colour |

(even for the lower classes of sulphate exposure) as sulphate-resisting Portland cement has a relatively low resistance to chloride ingress.

## *New concrete, unsatisfactory quality: 'remedial specification' of a surface treatment*

Here, surface barriers may be used to alleviate potential deterioration due to reinforcement corrosion arising from an inadequate mix design, insufficient cover or the poor compaction/curing of concrete. It is well established that proprietary products can provide effective chloride and carbonation barriers over lengthy periods of service (although they are not a substitute for correct initial specification and placement of the concrete).

## *Concrete undergoing deterioration: 'repair specification' of a surface treatment*

Where direct deterioration due to chemical/physical effects has occurred, reinstatement by the simple application of a surface treatment may be feasible at an early stage, particularly

**Table 17.2** Principles and methods according to ENV 1504-9: 1997

| Principle no. and definition | Methods based on the principle |
|---|---|
| 1 Protection against ingress<br>Reducing/preventing the ingress of adverse agents | Pore blocking impregnation*<br>Surface coating with or without the ability to accommodate crack formation and movement |
| 2 Moisture control<br>Adjusting and maintaining the moisture content in the concrete within a specified range of values | Hydrophobic impregnation*<br>Surface coating |
| 3 Concrete restoration | – |
| 4 Structural strengthening | – |
| 5 Physical resistance<br>Increasing resistance to physical/mechanical attack | Overlays or coatings<br>Pore blocking impregnation |
| 6 Resistance to chemicals<br>Increasing resistance of the concrete surface to deterioration by chemical attack | Overlays or coatings<br>Pore blocking impregnation |
| 7 Preserving or restoring passivity | – |
| 8 Increasing resistivity<br>Increasing the electrical resistivity of the concrete | Limiting moisture content by surface treatments or coatings |
| 9 Cathodic control<br>Creating conditions in which potentially cathodic areas of reinforcement are unable to drive an anodic reaction | Limiting oxygen content (at the cathode) by saturation or surface coating |
| 10 Cathodic protection | – |
| 11 Control of anodic areas | – |

*The distinction between pore blocking and hydrophobic impregnation is discussed in section 17.3

if a thick lining or render is used. It is more likely, however, that significant preparation will be required before coating (i.e. cutting back, cleaning and profile restoration).

For structures that have been in service for a significant period, a distinction must be made between the potential effectiveness of surface treatments applied before/after the initiation of reinforcement corrosion; also between chloride ingress and carbonation:

1 *Prior to initiation* Appropriate surface treatments can effectively inhibit the progress of carbonation, reducing its rate to a negligible level. Hence, the service life of a structure can be extended considerably where carbonated concrete in the cover zone has not reached the reinforcement. The value of surface treatments is more debatable when chloride ingress has occurred. The rate of chloride migration to the reinforcement may be reduced by preventing additional penetration and by limiting the internal moisture state of the concrete. However, it is unlikely that corrosion will be prevented in the longer term – or even the short/medium term if chloride ingress is at a high level prior to coating.
2 *After initiation (following carbonation or chloride ingress)* In this case, surface barriers can only be effective by limiting the ingress of oxygen and/or moisture. As only very small amounts of oxygen are required to fuel the corrosion process, even the

most (oxygen) impermeable surface treatment is unlikely to be effective. However, attempting to maintain the concrete in either a very dry or a water-saturated condition may assist in reducing the corrosion rate. After initiation, therefore, surface treatments are likely to be cost-effective only where a relatively short extension of service life is required.

Prior to carrying out any repair, it is essential that a proper assessment is made of the defects in the concrete and their causes, and of the ability of the structure to perform its function.

## 17.2.3 Additional comments

Although a detailed discussion of the applications summarized in Tables 17.1 and 17.2 cannot be given here, concise information, in most cases, can be obtained readily from the publications listed in the Further reading section. Some brief comments regarding several particular uses are given below, however, as published information in these areas is more scattered. These examples also serve to illustrate the more general point that specialist knowledge is often essential, even when considering seemingly simple applications.

### *Alkali–silica reaction (ASR)*

The hypothesis has often been advanced that above-ground structures subject to ASR can be protected by surface waterproofing, thus maintaining the concrete in a sufficiently dry state to inhibit or arrest deterioration. The results of various laboratory studies seemingly support this hypothesis. However, there appears to be little, if any, documented evidence for the effectiveness of this approach on real structures. Field trials, and more general considerations, have also led to the conclusion that there are serious practical difficulties in attaining the ideal conditions within the concrete for the inhibition of ASR.

These difficulties may be influenced by a variable microclimate in the vicinity of the affected concrete and by moisture ingress from other parts of the structure. Furthermore, many coatings and related treatments that provide excellent barriers against liquid water ingress have a relatively low resistance to moisture vapour transmission. Consequently, for this form of deterioration in particular, the simple extrapolation of laboratory results to in-service conditions is likely to be very speculative.

### *Waterproofing and carbonation*

Concerns have arisen over the use of pore-lining penetrants, such as silanes (see section 17.3), that eliminate periods of high internal moisture content while offering little, if any, protection against carbonation. However, the view that carbonation may be encouraged by the maintenance of an optimum internal moisture state (50–70 per cent RH) does not appear to be supported by experience from real structures.

### *Control of reinforcement corrosion by the limitation of oxygen ingress*

Oxygen is required both to sustain the electrochemical corrosion reaction at steel reinforcement and, also, for the formation of the expansive rust that causes cracking and spalling of the concrete cover zone. The potential use of surface coatings to control reinforcement corrosion by limiting oxygen ingress is noted in Table 17.2 (Principle 9 of DD ENV 1504-9: 1997). However, in general, this approach is most unlikely to be effective because, as noted earlier, only very small levels of oxygen are required to fuel the corrosion process, while even the most impermeable coatings are not completely resistant to oxygen transmission. Furthermore, without total encapsulation, oxygen is likely to reach the reinforcement via other pathways.

## 17.3 Materials for surface coating/treatment

Many proprietary products are available for the surface treatment of concrete. They can be classified conveniently according to the main generic component, as shown in Figure 17.1.

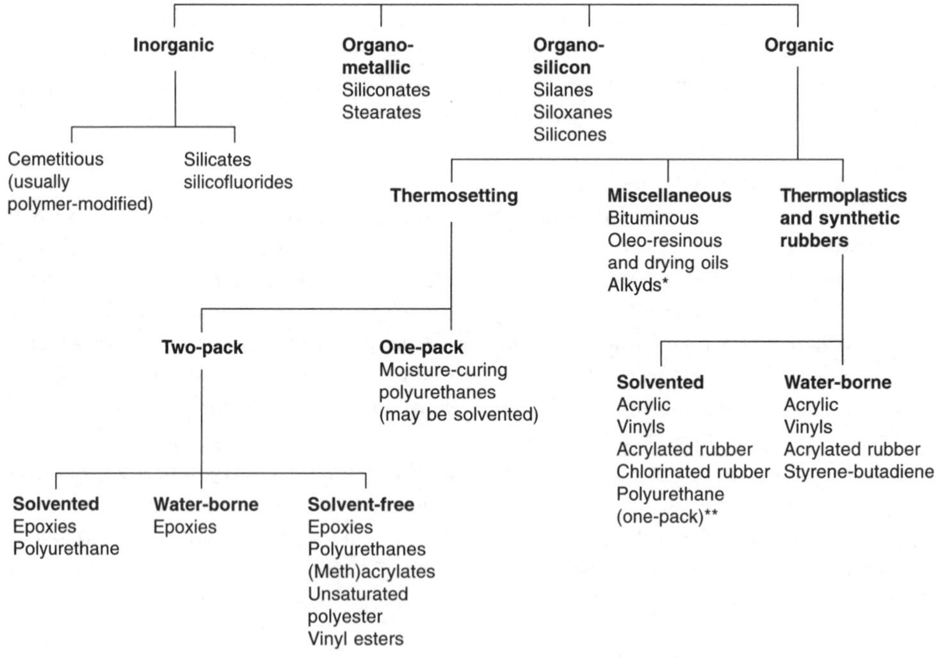


**Figure 17.1** A classification of surface treatments

An alternative classification scheme, based on the following categories, is also useful as it can be related more easily to particular applications. This approach has been adopted in European Standards DD ENV 1504-9: 1997; prEN 1504-2:

1 Materials that impregnate the concrete and that do not depend on the formation of a significant surface film. These may function in two ways:
   - Penetrants that line the pores with a water-repellent/hydrophobic layer, relying on surface tension effects to prevent the ingress of liquids by absorption (capillary action).

- Sealers that fill and block the pores, thus offering some resistance to the ingress of liquids under a pressure gradient. A distinct surface film can be formed if the application rate is sufficient (usually >2 or 3 coats) or if the concrete is particularly dense and impermeable.

2 Systems that depend on the formation of a continuous layer, thus shielding the concrete surface completely. Here, the following sub-division is useful:
   - Coatings, including high-build systems, that have a dry film thickness up to approximately 1–2 mm.
   - Heavy-duty linings/renders with a greater thickness.

This scheme is used in Tables 17.3, 17.4 and 17.5 where brief comments are given on the main performance characteristics of the more common systems (thick linings and renders have been excluded, to avoid repetition, as they are often based on similar generic components to those used in coatings).

**Table 17.3** Classification and performance characteristics of hydrophobic penetrants/pore-liners (water-repellents)

| Common generic types | Main performance characteristics |
|---|---|
| Silanes<br>Siloxanes<br>Silicones | • Protect against water/chloride ingress in the absence of hydrostatic pressure (questionable on ponded surfaces)<br>• Some systems are approved for use on UK highway structures |
| Numerous variations in chemical type, content of active material, and solvent system are available | • Negligible effect on ingress of moisture vapour, carbon dioxide or oxygen<br>• No benefit for resistance to chemicals, abrasion for impact<br>• Negligible effect on concrete appearance |
| May be 100% active | • Durability generally expected to be good, more so for silanes/siloxanes, but claims vary |
| Silanes/siloxanes react *in-situ* with the concrete; silicones generally undergo drying only. In both cases, the pores/capillaries are lined with a hydrophobic layer | • Relatively easy to apply/reapply with minimal surface preparation<br>• Substantially dry concrete required<br>• Low temperatures acceptable but hot/windy conditions unfavourable (use of a 'paste' form preferred) |

**Table 17.4** Classification and performance characteristics of pore-blocking sealers

| Common generic types | Main performance characteristics |
|---|---|
| Generally solvented | • Enhanced resistance to ingress of liquids, gases, vapours, chemical attack and abrasion |
| Thermoplastic<br>(acrylics, pus)<br>Physically drying only | • Effectiveness depends on severity of service conditions, specific requirements and porosity of concrete/no. of applications (2/3 is usual)<br>• Some slight darkening of the concrete is usual and the formation of a surface glaze may occur |
| Thermosetting<br>(epoxy, pus)<br>Chemically reactive and physically drying<br>May be one- or two-pack systems<br>Wide variations in solids content, solvent/polymer type are possible | • Generally have reasonably good long-term effectiveness as there is no dependence on surface adhesion<br>• Relatively easy to apply/reapply with minimal surface preparation<br>• Can also be used effectively as a primer/sealer prior to coating on porous/friable surfaces |

**Table 17.5** Classification and performance characteristics of surface coatings

| Common generic types | Main performance characteristics |
|---|---|
| **Cementitious, polymer-modified**<br>Generally ~ 2 mm thick (2 coats)<br>Supplied as two-pack systems (blended cement/fine fillers and aqueous polymer dispersion) or as one-pack (blended cement/filler/redispersible powdered polymer that is mixed with a specified quantity of water) | • Generally provide excellent barrier properties (water, chlorides, carbon dioxide, oxygen)<br>• Relatively low resistance to moisture vapour transmission<br>• Generally unsuitable for exposure to strong acids but can provide some resistance to sulphate attack<br>• Can possess a useful degree of flexibility<br>• Good durability under a wide range of service conditions (including immersed and below ground)<br>• Reasonably tolerant of various surface/application conditions; temperature extremes may require special precautions or limit use<br>• Acceptable maintenance requirements |
| **Silicates**<br>Filled, pigmented systems (much simpler products are used as low-coast dust-proofers)<br>Reaction with the concrete surface assists adhesion | • Excellent durability as breathable and weatherproof decorative coatings<br>• Some anti-carbonation properties<br>• Resistant to some chemicals<br>• Reasonably tolerant of various surface and application conditions<br>• Easy to maintain |
| **Solvent and water-borne thermoplastics and elastomers**<br>Physically drying<br>Formulations and performance vary widely | • Can provide excellent barrier properties (water, chlorides, carbon dioxide) and good breathability<br>• Some systems suitable for exposure to aggressive chemicals<br>• Good durability and retention of appearance<br>• Generally not suitable for immersed/below ground applications or demanding abrasion resistance<br>• Can maintain a barrier when applied to some cracked concrete (fibre/fabric reinforcement may be required)<br>• Can be applied under widely varying conditions with relative ease (particularly when solvent-borne)<br>• Good surface preparation required – often need a fairing/levelling coat and/or primer<br>• Relatively easy to maintain |
| **Thermosetting systems**<br>Solvented, water-borne or solvent-free<br>Undergo chemical reaction during cure<br>Formulations and performance vary widely<br>Extremely versatile – can be 'tailored' to specific requirements<br>Usually two-pack<br>Solvent-free systems are particularly suited to very high-build applications (up to at least 250 µm/coat) | • Excellent barrier properties against liquids, gases and moisture vapour<br>• Acceptable appearance although chalking and colour instability prevalent with some products<br>• Excellent durability under all conditions including frequent contact with, or immersion in, aggressive chemicals<br>• Good impact/abrasion resistance<br>• May accommodate cracks in the concrete (likely to require fibre reinforcement)<br>• Generally require a high level of surface preparation for potential to be fully realized<br>• Systems available for application to wet surfaces (or, in some cases, under water)<br>• Use may be limited/not possible at low temperatures (generally below 5°C)<br>• Cure rate often slow, even at 'normal' temperatures<br>• Fairly demanding application requirements (specialist applicator usually advisable)<br>• Maintenance often difficult due to poor old/new intercoat adhesion |
| **Bituminous systems**<br>Hot-applied (generally with fabric reinforcement)<br>Cold-applied: water-borne emulsions or solvent-borne (± reinforcement) | • Used mainly for waterproofing applications below and above ground<br>• Reasonably durable but can tend to embrittle with age, particularly when exposed to sunlight and hot conditions<br>• Relatively easy to apply/reapply with good tolerance of surface and ambient conditions |

It should not be assumed that all proprietary products will conform to this overview, as many variations are possible; in addition, a combination of generic types may be used to provide a sealer/primer/intermediate and top-coat system. Generic characteristics (e.g. 'all epoxies bond well and have good chemical resistance') can also encompass significant variations, depending upon the detailed composition (formulation) of a product.

Properties are not discussed here in detail, but a summary of characteristics that may have to be considered for various applications is given in Table 17.6. Where corresponding information is not provided on data sheets, suppliers should be consulted and, occasionally, specific tests may need to be commissioned.

**Table 17.6** Properties of surface coatings/treatments required by the specifier/applicator

| Unmixed and freshly mixed | Transition to the dry/cured state | Fully cured | |
|---|---|---|---|
| | | Short term | Long term |
| Shelf life | Effects of ambient/substrate temperature | Dry film thickness | Change of short term properties on exposure to service conditions |
| Storage requirements, particularly temperature | Sensitivity to ambient/substrate moisture, % RH | Adhesion | Accommodation of crack formation and movement |
| Flash point | Usable (pot) life | Colour, texture, hiding power, gloss and reflectance | Abrasion resistance |
| Volatile components | Gel time | Barrier properties | Effects of thermal cycling/shock |
| Health, safety and environmental considerations | Reaction exotherm (for some thicker systems) | Mechanical properties | Resistance to water, chemicals, biological attack/mould growth, radiation |
| Taint, e.g. of nearby foodstuffs | Rate and extent of cure/drying vs time | Fire performance | Resistance to weathering |
| Density and coverage rate | Cure shrinkage (only for certain thick systems) | Electrical properties | Cleanability |
| Need for priming | Over-coating interval | Slip/skid resistance | Ease of maintenance |
| Mixing requirements | | Effect on potable water | |
| Application properties/methods | | Ease of cleaning and nuclear decontamination | |
| | | Resistance to graffiti and the ease of its removal | |

## 17.4 Surface preparation and application

Concrete is inherently suitable as a substrate for coatings although it can present problems that are not met with steel, for example:

- The (uncarbonated) surface has a high alkalinity.
- Concrete surfaces are invariably rougher and often have partly open air-voids ('blow-holes') and protrusions (e.g. fins/nibs or grout runs).
- Concrete is absorbent to a varying degree and frequently has a relatively high moisture content.
- Surfaces may be dusty and friable and often have a thin, relatively weak surface layer (laitance).
- Contamination due to mould release agents, curing membranes and fungal growth is common (note: curing membranes designed to degrade may persist in the absence of sunlight; some specific membrane types can be overcoated).

For a coating to perform satisfactorily, application to a clean, sound surface is normally essential, thus enabling the development and retention of the maximum bond strength. Certain demands may be relaxed somewhat for many penetrants and sealers as surface adhesion is less critical, although some preparatory work may still be necessary; contaminants that hinder penetration, for example, should be removed.

The need to remove an existing coating can vary, depending upon its condition, including adhesion, and its chemical/physical/adhesive compatibility with the new system. Identification of the existing coating will assist significantly in determining the best course of action; *in-situ* testing, e.g. for adhesion, can also be advisable. Specific performance properties required of the original system should be identified and maintained, where necessary, by the new coating.

Controlled permeability formwork can simplify surface preparation as it provides a sound surface with minimal blow-holes and, also, eliminates contamination by shutter release agents.

The more common methods of surface preparation and coating application are outlined in the following sections, where it is assumed that the surface is basically sound. Hence, the following processes are not discussed:

- patch repair/recasting to rectify reinforcement corrosion, spalling or mechanical damage;
- the treatment of joints and cracks; sealing against water ingress/seepage.

Health, safety and environmental issues are also not discussed here, but it must be noted that certain assessments and actions are covered by statutory regulations. Guidance can be obtained from published sources (see, for example, Concrete Society (1997) and references therein) and from product suppliers (Health & Safety Data Sheets with a regulated content must be made available).

## 17.4.1 Surface preparation

### *Cleaning*

Many techniques are used for preparatory cleaning and the most appropriate will depend upon individual circumstances. The following methods are given in approximate order of increasing severity:

- Wet scrubbing with emulsifying detergents and, where necessary, biocides
- Low-pressure water cleaning
- Steam cleaning
- Acid etching
- Hand/power wire brushing
- Grinding
- Wet or dry abrasive blasting (with vacuum recovery)
- Shotblasting
- Scarifying/planing (cutting teeth on rotating discs)
- Needle gunning
- Scabbling
- (Ultra) high-pressure water jetting (100–300 MPa)
- Flame blasting
- Milling (cutting teeth on a rotating drum)

Further advice on cleaning methods is given in several standards and other publications (Concrete Society, 1997; BS 8221-1: 2000; BS 6150: 1991; Bassi and Roy, 2002).

Additionally, a range of plaques has been produced that replicate the typical profiles given by various methods of preparation. These specimens are also linked to the use of coatings/treatments of different thickness (*Concrete Repair Manual*, 1999). However, it is recommended here that this correlation should be used for general guidance, rather than prescriptively, as suppliers' recommendations may differ and should take precedence.

Relatively simple washing techniques may not be effective where contaminants such as oil and grease have been absorbed by the concrete. They will also be ineffective if the contamination is merely spread further. Following wet cleaning, thorough washing with fresh water should be carried out, as some detergent and biocide residues can have an adverse effect on adhesion.

Acid etching is now used infrequently due, mainly, to the associated health and safety risks. Additionally, the absorption, via porous areas/cracks, or surface retention of certain ionic species, e.g. chlorides, can be detrimental.

Wet grit-blasting and vacuum dry-blasting are commonly favoured because they are effective techniques that allow good control of the cleaning process, while generally having no significantly detrimental effects on the surface. Health, safety and environmental risks are also reduced, compared to 'open dry-blasting', although the wet process may create containment and disposal problems. Many typical concrete surfaces require no more than very light blasting ('sweep-blasting') prior to coating.

High-pressure water jetting is an effective method for dealing with many contaminated or unsound surfaces. Equipment is available that allows good control of the process, giving an acceptable profile, and recycling of the jetting water.

Power-driven mechanical methods, such as scarifying, needle gunning and scabbling, are widely used and can be very effective for the removal of defective surface layers and firmly adhering or deeply absorbed contaminants. However, they can also be too aggressive, producing microcracking of the aggregate, which may be detrimental to bond strength, and an overly deep texture. Excessive power wire-brushing can lead to undesirable polishing of the surface.

Milling, which is also a very aggressive method, can remove an appreciable depth of concrete, but the large, heavy equipment is only suited to horizontal surfaces.

Flame blasting is used infrequently. It fractures and removes the concrete surface by superheating the pore water, thus generating expansive forces.

### *Void filling and levelling*

Durability (for other than permanently dry conditions) and aesthetics require that coatings form a continuous film, free of pinholes and with an adequate thickness over the entire surface. Consequently, any significant unevenness must be dealt with following initial preparation.

With some thicker products, this may be achieved by direct application to a prepared surface. In other cases, however, small voids and 'blow-holes' (which usually become more exposed and enlarged during initial preparation) must be filled and the surface must be levelled prior to coating. Filling and reprofiling is carried out by 'bagging-in', by use of a 'scrape/skim coat' or by application of a (typically) 1–2 mm thick 'fairing coat'.

Cementitious slurries and proprietary products, generally polymer-modified, are most commonly used, although pastes or fine mortars based on a reactive resin, e.g. an epoxy, can also be useful. Compatibility with the coating system and any other specific requirements should be checked with the supplier.

## 17.4.2 Application

It must always be ensured that the coating/treatment is suitable for use on an alkaline concrete surface – some products are readily degraded and debonded by alkalis, particularly in moist conditions.

Excessively absorbent or weak and friable/dusty surfaces may be brought to a satisfactory condition with a sealer/stabilizer. The over-application of some products must be avoided and, in general, the sealer and other components of the system should be obtained from the same supplier to ensure compatibility.

The required age of new concrete prior to coating varies, but a minimum of 21–28 days under reasonable ambient conditions is not uncommon. Normally, this allows for both early shrinkage effects and drying. Particular attention must be given to new floor slabs and screeds, as very long periods can be required to attain a moisture content that will not cause premature failure of a coating (*FeRFA Guide*). Specific products – surface damp-proof membranes – are available for circumventing this problem.

More generally, for mature concrete elements, the tolerable moisture level varies according to the particular coating system – its ability to displace water, the bonding characteristics and the permeability to moisture vapour are particularly relevant properties – and the supplier's recommendations should always be observed. Coatings are available for application to very wet surfaces, even under water. However, this capability is formulation dependent rather than being typical of a generic type. For the adequate absorption of penetrants and sealers, water-free pores/capillaries are obviously essential.

The most appropriate method of application depends on the product type/viscosity, the area to be treated and its continuity/accessibility. Common methods are:

- Spraying
- Rolling
- Brushing
- Flood coating

Airless spraying, which utilizes a high-pressure pump to atomize the material, is most suitable for the rapid application of relatively low viscosity products to unobstructed large areas. Air-spraying is suited to more viscous products that are difficult to atomize – in this case, masking of the surroundings is likely to be necessary, due to the overspray mist that is created. Both methods become difficult to use in windy conditions.

Air-assisted spraying, using plant similar to that employed for spraying concrete, can place very thick products containing significant amounts of sand, large filler particles or fibres. A finishing operation, typically trowelling, is required and wastage, due to rebound, is inevitable.

Application by roller is commonly used for areas of intermediate size, i.e. those that are too small/large for spraying/brushing, respectively. Brush application is most suited to relatively small and/or inaccessible areas, and to products, usually primers, that must be worked vigorously into the surface.

Flood coating is a technique that utilizes a low-pressure spray to saturate a surface uniformly with very low viscosity penetrants, particularly silanes/siloxanes. A free-running wet front is maintained, working from top to bottom on a vertical surface. A related technique, used on floors, employs a 'squeegee' to apply and spread flood coatings of penetrating sealers.

The range of ambient temperature and humidity that is acceptable during application and drying/curing varies, according to both generic type and formulation, and recommendations should always be available from the supplier. In general, it is good practice to store materials in a controlled temperature environment for at least 24 hours before use.

Particular care is required when temperatures are 'low and falling' as condensation can affect both adhesion and appearance. It can become essential to monitor ambient conditions and surface temperature to ensure that application is not carried out in proximity to the dew point.

For reactive coatings, epoxies in particular, the recommended maximum period between successive coats must be observed (with allowance being made for on-site temperatures that differ significantly from those given on the data sheet). An excessively long delay between coats can lead to poor intercoat adhesion because the surface of the initial coat becomes less receptive to bonding as cure progresses.

The period recommended for the attainment of full cure/drying, which can be as long as 10–14 days, must be allowed prior to imposing aggressive service conditions, for example: chemical contact, immersion in water or heavy trafficking. Where a data sheet refers only to a single cure temperature, such as 20°C, a significantly longer period is likely to be required under colder, more variable site conditions.

## 17.4.3 Quality control

Inspection and supervision should ensure that surfaces are well prepared and suitable for application of the coating/treatment. *In-situ* strength measurement (e.g. by pull-off testing) may be used to ensure that the substrate is mechanically sound. Determination of the moisture level in the concrete is often advisable; in the case of floor slabs/screeds, this measurement is usually required by specification (see, for example, *FeRFA Guide*).

Application of the correct coating thickness is normally essential for satisfactory appearance and durability. It can be monitored approximately by recording the consumption rate over known areas – with due allowance for the effects of surface roughness and wastage. Wet film gauges can be useful for carrying out rapid spot-checks during application, but their accuracy is limited unless the surface is reasonably smooth.

Dry film thickness is more difficult to check on concrete than on metallic substrates, although direct reading instruments based on ultrasonic methods are now available. Monitoring may be carried out via metallic coupons attached to the surface prior to coating, thus allowing use of the more common electronic thickness gauges. In the event of dispute, however, direct measurement on cored specimens may be necessary.

Most coating systems require the application of at least two coats and, in some cases, different colours can assist in achieving the correct build-up of successive layers.

Once the coating has cured, adhesion can be checked by pull-off tests using one of the many instruments available commercially. Procedural details, the extent of testing and performance criteria should be agreed beforehand.

Trial areas can be useful, particularly on large contracts and where maintenance of a high level of performance is critical. By this means, standards of workmanship can be agreed – and any problems resolved – at an early stage. Where necessary, extensive *in-situ* testing, or core removal for laboratory assessment, can also be carried out conveniently at this stage.

## References


Bassi, R. and Roy, S.K. (eds) (2002) *Handbook of Coatings for Concrete*. Whittles Publishing.

BS 6150: 1991. Code of Practice for Painting of Buildings.

BS 8221-1: 2000. Code of Practice for Cleaning and Surface Repair of Buildings – Part 1: Cleaning of Natural Stones, Brick, Terracotta and Concrete.

*Concrete Repair Manual* (1999) International Concrete Repair Institute/ACI International, pp. 619–661.

Concrete Society (1997) *Guide to Surface Treatments for Protection and Enhancement of Concrete*. Technical Report No. 50. The Concrete Society, UK.

DD ENV 1504-9 (1997) Products and Systems for the Protection and Repair of Concrete Structures. Definitions, Requirements, Quality Control and Evaluation of Conformity. Part 9: General Principles for the Use of Products and Systems.

*FeRFA Guide to the Specification and Application of Synthetic Resin Flooring*. The Resin Flooring Association, Aldershot, Hampshire. (The current version of this document can be downloaded from www.ferfa.org.uk)

prEN 1504-2. Products and Systems for the Protection and Repair of Concrete Structures. Definitions, Requirements, Quality Control and Evaluation of Conformity. Part 2: Surface Protection Systems (presently in draft form only).


## Further reading


Allen, R.T.L., Edwards, S.C. and Shaw, J.D.N. (1993) *The Repair of Concrete Structures* (2nd edn). Blackie Academic and Professional, London.

American Concrete Institute (1998) A Guide for the Use of Waterproofing, Dampproofing, Protective and Decorative Barrier Systems for Concrete, ACI 515. IR-79 (revised 1985), ACI Manual of Concrete Practice.

Concrete Repair Association (1992) *The Application and Measurement of Protective Coatings for Concrete – Guidance Note*.

Department of Transport/Highways Agency (1986). *Materials for the Repair of Concrete Highway Structures*. Departmental Standard BD 27/86.

Department of Transport/Highways Agency (1990) *Criteria and Materials for the Impregnation of Concrete Highway Structures*, Departmental Standard BD 43/90.

Department of Transport/Highways Agency (1990) *Impregnation of Concrete Highway Structures*. Departmental Advice Note BA 33/90.

Dhir, R.K. and Green, J.W. (eds) (1990). Protection of concrete. Proceedings of the International Conference held at the University of Dundee, Scotland, UK on 11–13 September 1990. E&FN Spon, London.

Doran, D.K. (ed.) (1992) *Construction Materials Reference Book*. Butterworth-Heinemann, Oxford.

Mailvaganam, N.P. (ed.) (1991) *Repair and Protection of Concrete Structures*. CRC Press, Boca Raton, FL.

Mays, G. (ed.) (1992) *Durability of Concrete Structures – Investigation, Repair, Protection*. E&FN Spon, London.

Schmid, E.V. (1988) *Exterior Durability of Organic Coatings*. FMJ International Publications Ltd.

# Part 4

# Readymixed concrete

# 18

# Production of readymixed concrete

*Steve Crompton*

## 18.1 Development of readymixed concrete

The use of readymixed concrete was debated as far back as the 1870s. A well-known British civil engineer, George Deacon, commented in 1872 'If concrete were supplied to the site as a ready usable product this would without doubt, provide considerable advantages'. Towards the end of the nineteenth century some contractors were mixing their own supplies and hauling them for use on-site but the concept had not been adopted widely.

The breakthrough first came in the USA where the scale of construction created a demand for large volumes of building materials. The first recorded delivery of sold readymixed concrete was made in Baltimore in 1913. The next step in the development of readymixed concrete was in 1916 when Stephen Stepanian applied for a truck mixer patent but was unsuccessful. It was not until much later in 1926 that the transit-mixer was born and the first concrete delivered in a truck mixer. Rapid development in the USA led to some 52 per cent of cement production being used by the readymixed industry in 1959.

The UK readymixed concrete industry started in the early 1930s with the construction of the first plant at Bedfont in Middlesex by the Danish engineer Kjeld Ammentorp. This plant was capable of producing 40 cubic yards of concrete per hour. Aggregates and bags of cement were taken to overhead storage areas by a chain and bucket arrangement. They were mixed in a 2 cubic yard mixer and the 'readymixed' concrete discharged into $1^2/_3$ cubic yard agitators. These agitators were shaped like milk churns and tilted upright for filling, a lid was fitted and the drum lowered.

In its first year of operation the plant produced 8636 $yd^3$ (approximately 6600 $m^3$) of concrete and had a turnover of £10000. Over the next twenty years growth was slow but the period from late 1950s to the mid-1970s saw a rapid growth in the industry. In the mid-1970s annual production in the UK peaked at 32 million cubic metres and has broadly followed economic cycles ever since with annual production between 26 million and 30 million cubic metres (Figure 18.1).

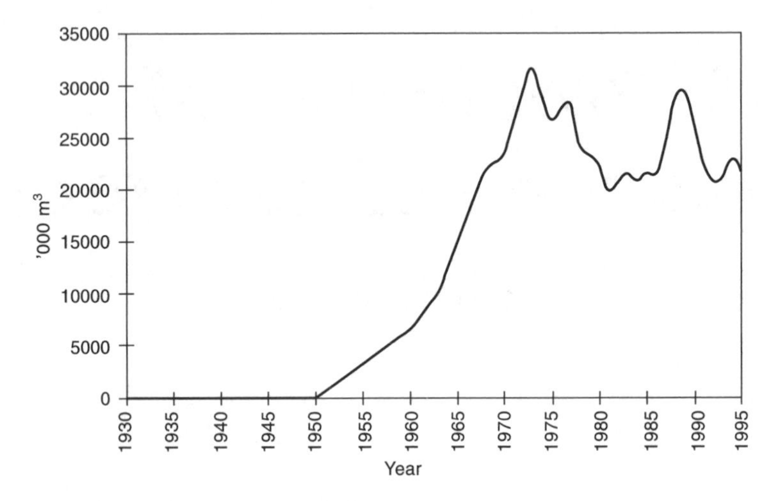


**Figure 18.1** Growth of readymixed concrete in the UK.

The industry has grown into a £1 billion a year business in the UK with approximately 90 per cent of all concrete now supplied readymixed; almost half of all cement produced in the UK is used in readymixed concrete.

To meet this demand a network of 1200 readymixed concrete plants has been established across the country. There is a wide variety of plants in operation and this chapter outlines the practical issues which determine the type and location of concrete production units and identifies the different types of plant currently in use. Differences between readymixed concrete and site-batched concrete are also discussed.

## 18.2 Effects of transportation on concrete properties

The principal difference between readymixed and site- or laboratory-mixed concrete is the time and method of transportation from batching to the point of placing. There are four major influences which arise during delivery and/or agitation of concrete:

- Evaporation
- Hydration
- Absorption
- Abrasion

Evaporation and hydration effects are the most significant while absorption is only significant with dry or highly absorptive aggregates. Abrasion is only an issue where abradable aggregates are used leading to increased fineness. It has been suggested some limited grinding of cement may also occur due to abrasion processes although this is less well documented.

The effects of these influences are:

- Evaporation and absorption leads to a lower effective water/cement ratio in the paste thus reducing workability and potentially enhancing strength.
- Hydration reduces water available for workability but has no real effect on the effective water/cement ratio.
- Abrasion increases fineness and reduces workability.

The rate of workability loss is influenced by a number of factors:

- **Cement content** Dewar (1973) showed that mixes of lower cement content lose workability at a lower rate because a smaller proportion of water is utilized in hydration (see Figure 18.2).

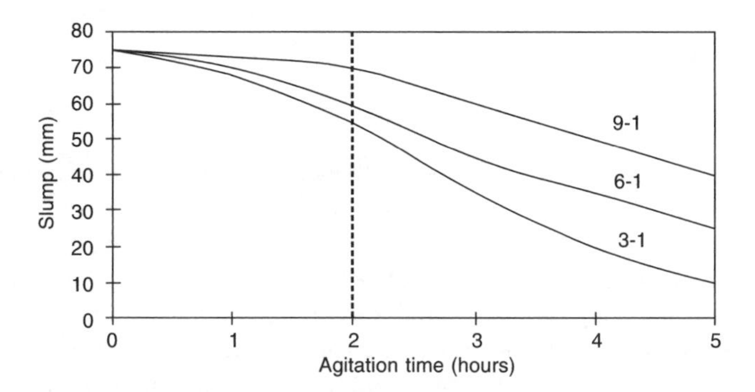


**Figure 18.2** Workability loss versus cement content.

- **Water content** Mixes of higher water content or higher initial workability lose workability at a lower rate because the effect of a given water loss due to evaporation or hydration is diluted (see Figure 18.3).

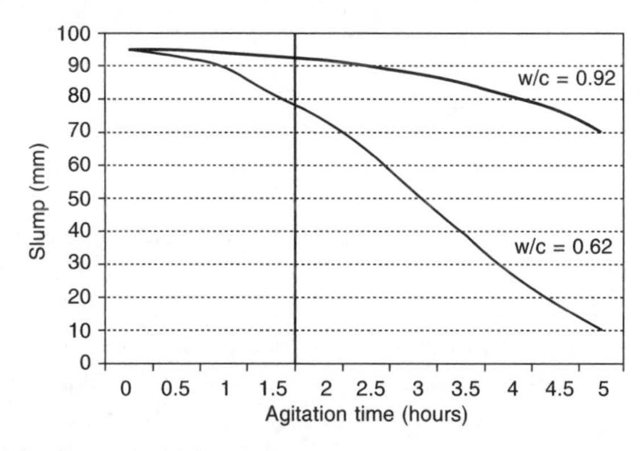


**Figure 18.3** Workability loss and initial water content.

- **Admixtures** Admixtures can influence the water content, hydration characteristics, rheology and air content of the concrete and thus the rate of workability loss. Mixers incorporating water-reducing admixtures (wra) can lose workability faster than concrete without wra at the same initial workability. This is because the effects of water loss will be concentrated due to the lower initial water content.

- **Weather** Higher temperatures increase the rate of hydration and evaporation and thus increase the rate of workability loss.
- **Volume of concrete delivered** Larger volumes of concrete are less susceptible to workability loss because of a lower surface area/volume ratio which reduces the significance of the lost water. On the other hand, larger volumes retain more heat and may increase the rate of hydration although practical experience indicates that larger volumes lose workability at a slower rate than small loads.

Strength generally increases with time at about 5 per cent per hour provided concrete can still be compacted as the workability reduces, according to Dewar (1962). This value depends on a range of factors such as:

- cement content and type
- initial workability
- time of agitation
- ambient conditions

If concrete is agitated for extended periods and supplied at the same slump as concrete compacted soon after mixing then small strength reductions may occur as shown in Table 18.1.

The effects are small in the UK since delivery times are generally significantly less than 2 hours and the temperate climate reduces water loss through evaporation. However, since quality control samples are generally taken at the delivery site the effects, however small, are taken into account automatically. Table 18.1 indicates approximate agitation times for a reduction of 2 N/mm$^2$ for different initial workabilities.

**Table 18.1** Approximate time for same slump concrete to result in 2 N/mm$^2$ reduction in strength

| | Agitation time (h) | |
|---|---|---|
| A/C (by wt) | 25 mm slump | 125 mm slump |
| 3 | 1 | 2 |
| 4.5 | 2 | 3 |
| 6 | 3 | 4 |
| 9 | 4 | 5+ |

Water added some time after mixing (retempering water) can be used to restore concrete to its workability without loss of strength up to certain times. This applies where the retempering water is offsetting the effects of insufficient water batched initially or higher rates of evaporation and/or absorption than anticipated. However, once hydration has started a loss of strength can be expected. Dewar (Dewar and Anderson, 1987) suggests under average conditions this hydration effect only occurs after times shown in Table 18.2 for the indicated slumps.

Some confirmation of this data was produced by Beaufait (1975) (Figure 18.4) who retempered concrete for 2 hours to maintain slump before strength reduced below the initial level.

**Table 18.2** Approximate time after which relations between strength and workability change from that for fresh concrete

| | Agitation time (h) | |
|---|---|---|
| A/C (by wt) | 25 mm slump | 125 mm slump |
| 3 | 0.5 | 1 |
| 4.5 | 1 | 2 |
| 6 | 2 | 3 |
| 9 | 3 | 4 |

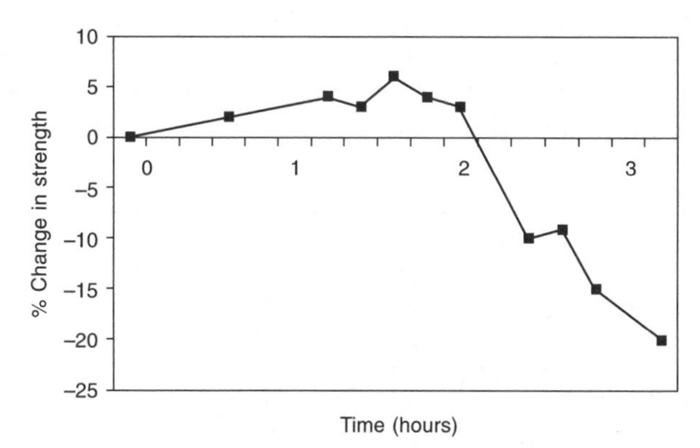


**Figure 18.4** Effects of prolonged agitation on strength (adapted from Beaufait, 1975).

Other properties may also be affected by prolonged agitation such as air content but this is a specialist subject. In general these effects are not excessive for normal periods of agitation unless high-workability concrete or mixes containing pfa are supplied.

## 18.3 Production methods

### 18.3.1 Constraints on plant design

The production of ready mixed concrete is a relatively straightforward process. In its simplest form it involves the integral mixing of cement, stone, sand and water. How and where this process is actually carried out is determined by a number of fundamental constraints which determine the final design of the plant:

- Setting time and workability loss
- Material availability
- Economics
- Market size and area
- Physical site restrictions
- Planning constraints
- Quality control requirements
- Health and Safety regulations
- Environmental restrictions

## *Setting time and loss of workability*

Readymixed concrete will begin to set in about four hours depending on the ambient conditions and mix characteristics. However, as we have seen in section 18.2 workability loss is more significant and the concrete must be placed before the concrete has too low a slump to be properly compacted.

In normal conditions the concrete will remain workable for over 2 hours and BS 8500–2 (BSI, 2002) requires that concrete is placed within 2 hours of cement coming into contact with water. In practice if 1 hour is allowed for unloading and placing on-site, 15 minutes to batch and mix the concrete at the plant, approximately 45 minutes remain to deliver the concrete. This will restrict the maximum distance that concrete can be transported and typically the average delivery distance is around 5 radial miles. In rural areas the figure will be higher and in larger urban area the average radial may be less than 3 miles. Figure 18.5 shows a typical distribution of deliveries.

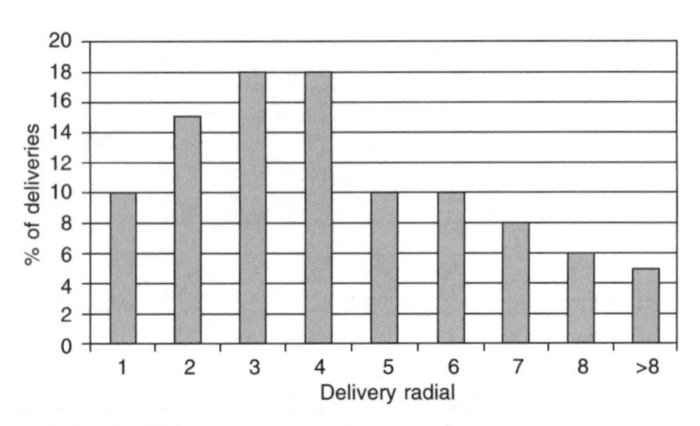


**Figure 18.5** Average delivery radials.

Admixtures can be used to extend the workable life of the concrete but these are rarely necessary in temperate climates.

## *Material type and availability*

*Cement type* The availability of different cement types will determine the number of silos required. Widespread availability of cement replacements such as ground granulated blastfurnace slag (ggbs) and pulverized-fuel ash (pfa) has led to almost all mainland plants having at least two silos and often three to allow the storage of a third cementitious material such as sulphate-resisting cement. It is critical to store different cementitious materials separately and to have robust procedures which ensure that the correct material is discharged into the appropriate silo. Such systems involve correct labelling of silos and 'lock-off' systems which prevent unauthorized discharge of cementitious materials without the supervision of the plant foreman.

The type of cement replacement will also influence the design of the silos. Pulverized-fuel ash requires enhanced aeration and filtration to ensure that it flows freely through the plant. It should also be noted that both pfa and ggbs occupy a larger volume per unit weight than cement and silos need to be larger to accommodate the same weight of material.

The amount of cement storage required will be determined by the distance from the source of cement, the delivery capacity of the supplier and the production rate of the concrete plant. Typically silo sizes range from 20 tonnes to 100 tonnes capacity.

*Aggregates* Aggregates are relatively inexpensive bulk materials which means that transport costs are a significant proportion of the final delivered cost. For this reason concrete plants are often situated very close to the source of aggregates and in many instances actually in the quarry itself. In this situation there is no need for large aggregate storage facilities. On the other hand, if the plant is situated some distance away from the aggregate source there may be a need for considerable storage facilities. Typically five to six times more aggregate than cement is used in a cubic metre of concrete and there may be a requirement to store up to a thousand tonnes of aggregate to balance the cement storage.

The type of products produced from the plant will also influence the nature and size of aggregate storage requirements. If mortar is produced from the same plant then additional storage will be required to allow for the finer grade sands used in mortar production. However, where mortar is produced it is critical to ensure that there is no cross-contamination of the mortar sand with even small amounts of coarse aggregates. This can be achieved by careful design and layout of the aggregate storage bins.

The type of aggregate will also influence the nature of the concrete plant. The use of single-sized aggregates, crushed rock fines and fine sands will increase the storage requirements beyond that required for graded aggregates and natural sands while the use of abrasive aggregates such as flint will significantly increase the wear rate compared to a plant using limestone aggregate.

*Admixtures* With the majority of concrete now incorporating at least one admixture frost-protected admixture storage areas are a prime requirement for all new plants.

## *Market size and area*

The size of the market and the geographical area served will have a significant influence on the size and type of cement plant. Large-volume city areas will require larger plants with a greater number of delivery trucks than a plant serving a predominantly rural area. Very large plants capable of producing more than 100 $m^3$/hour are the exception rather than the rule and typical plants are capable of producing between 30 $m^3$/hour and 50 $m^3$/hour.

## *Planning restrictions*

New plants are subject to strict planning regulations which often restrict the size of the plant and opening hours. In addition many local authorities and/or developers are not keen on locating concrete plants in light industrial areas because of perceived problems with noise and dust control. Increasingly plants are being located in less favourable locations subject to increased controls.

## *Environmental restrictions*

Environmental considerations covering dust emissions, waste control and noise also have to be considered when selecting an appropriate plant design. The Environmental Protection Act has imposed stringent requirements on concrete producers. In particular the control of run-off water, returned concrete and dust emissions has led to changes in the way that modern plants are designed and operated.

### Size of available site

The size of the site will often determine the type of plant selected. Small sites preclude the use of ground storage facilities and necessitate the use of overhead storage of all materials which increases the overall size and cost of the plant.

### Quality Assurance requirements

The Quality Assurance requirements also influence the way in which plants are designed particularly in relation to the design of storage capacity and weighing systems.

## 18.3.2 Types of plant

There are many types of plants with a significant number uniquely designed for particular locations. Standard plants such as those manufactured by Steelfields are available but even these are often modified for particular circumstances. A convenient classification for plants is the mixer type and these can be classified as:

- dry batch
- wet batch
- half-wet
- combination

### Dry batch systems

The majority of plants in the UK are dry batch units which rely on the mixing action of the truck mixer to produce a homogeneous concrete mix. The drum of the truck mixer consists of a cylindrical centre section with truncated cones at each end. The axis of the drum is inclined at an angle of between 12° and 20° and within the drum the two spiral blades are welded to the walls. The blades consist of solid steel sheets either plane in section or tipped with a T or L profile. The depths of the blades typically vary between 250 mm and 500 mm depending on their position in the drum.

The mixing action is essentially a twofold mechanism. Consider the complex movement of already well-mixed *plastic* concrete in the drum. Initially, ignore the action of the blades, then rotation of the drum and both the friction of the concrete against the inside surface of the drum and its cohesiveness (or internal friction) cause it to appear to climb up the right-hand wall (see Figure 18.6). The height to which the concrete climbs before gravity overcomes friction and causes it to fall back towards the 'bottom' of the drum is dependent upon two factors:

1. The linear velocity of the drum surface which also depends on its angular velocity and its internal diameter. The greater the linear velocity, the higher the break-away point of the concrete. At the far end of the drum where the diameter is larger, the concrete break-away point occurs further up the right-hand wall than at the mouth.
2. The workability of the concrete – the coefficients of friction both internal and between the concrete and the drum surface, increase with reducing workability. The lower the workability, the higher the break-away point.

An interesting point arises from these considerations. If the linear velocity is high enough, break-away is prevented and the concrete spins full circle in contact with the drum. This was observed at speeds of 22–27 rpm in the USA but this is far in excess of normal operating speeds of some 10–14 rpm.

As the concrete breaks away and rolls onto itself, highly efficient *local* mixing occurs. This motion on its own does not move the concrete in the axial direction to any great extent, although slight axial movement is caused by the inclination of the drum. The top surface of the concrete as it rolls tends to move away from the observer, and diagonally to the left. At any cross-section in the concrete, an axial movement in one area of the section must always be compensated for by reverse flow in another area, to maintain dynamic equilibrium.

Longitudinal motion is more positively effected by the spiral blades. As the blades rotate, they screw the concrete immediately in front of them towards the far end. As the concrete is fluid, a certain amount of spillage over the top of the blades occurs. Relative to the blade movement, spillage is towards the viewer at an angle to the left. In addition, as concrete near the interface with the drum surface is being forced to the far end of the drum, a compensating flow returns towards the observer, thus increasing the spillage over the blades.

The natural movement of the top surface of the concrete (due to gravity) is away from the observer while the action of the blades imparts top surface movement towards the observer. This leads to additional local mixing, although the latter movement so exceeds the former that, to the observer, the movement seems totally axial and towards him.

The interaction of rotational and longitudinal movement is very complex and what the observer sees is literally superficial.

Consider the locus of a point within concrete, initially in contact with the drum. At this time, the point is tending to move:

(a) up the right-hand wall and
(b) because it lies between the blades, towards the far end.

Half a revolution later, after break-away, it joins the surface flow of concrete towards the observer. After a complete revolution the point is 'picked up' again by the blades and moved towards the far end.

Summarizing in simple terms, rotation of the drum creates local mixing action, the spiral blades induce end-to-end mixing. As previously indicated, this initial appraisal is only a superficial introduction to mixing. Scales (1976) provides a much more informed consolidated appraisal in his project.

With low-slump mixes, both the internal friction and that between the concrete and the drum are increased. One result of this is that the break-away point occurs higher up the right-hand side of the drum.

## *Wet batch systems*

Wet batch systems are plant based mixers and there are a number of different types in use. They are normally higher output than dry batch plants and are therefore often seen in larger plants. Wet batch mixers are also normally required if mortar is also to be produced from the plant. Table 18.3 is a summary of the main types of wet batch mixers available.

The mixers are available in a range of sizes from 0.8 $m^3$ to 6.0 $m^3$ with the largest drum mixers capable of thoroughly mixing 6 $m^3$ of plastic concrete in 45 seconds. The final choice of which particular mixer unit will be used will depend on a number of factors such as required output, predominant type of mixes supplied, plant geometry, available space, initial and whole life costings.

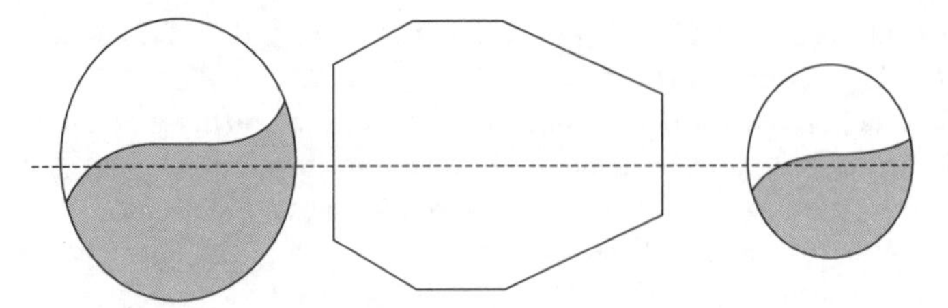

**Figure 18.6** Cross-section of truck mixing action.

**Table 18.3** Mixer types

| Basic mixer type | | Mixing action/variations |
|---|---|---|
| Rotating drum mixers | 1 | Rotating drum, freefall mixing action, non-tilting |
| | 2 | Rotating drum, freefall mixing action, tilting |
| Fixed trough mixers | 1 | Fixed mixing trough within which spiral blades revolve on horizontal shafts |
| | 2 | Fixed mixing trough within which paddles rotate on twin horizontal shafts |
| Pan mixers | 1 | Fixed horizontal pan in which mixing paddles travel around an annular channel |
| | 2 | Fixed horizontal pan in which mixing paddles travel around an annular channel while revolving about their own axis |
| | 3 | Fixed horizontal pan in which the mixing blades traverse the entire pan floor with a planetary motion |
| | 4 | Fixed horizontal pan in which two sets of mixing paddles travel around an annular channel in opposite directions |
| | 5 | Horizontal pan rotating beneath a stationary motor unit carrying paddles, the axes of rotation being non-coincident |
| Reversing drum mixers | | System of fixed blades and shovels within a non-tilting rotating drum giving a combined freefall and compulsory mixing action. |
| Continuous mixers | | Fixed trough mixer with twin rotating shafts and paddles angled at about 20° arranged to produce a continuous mixing action |

### *Half-wet system*

The half-wet system involves premixing sand, cement and water to form a slurry which is then added to the truck along with the aggregates. This type of system can significantly reduce batching times and reduce wear and tear on central mixer units.

### *Combined*

Many UK plants utilize a combined dry batch and wet batch system. The majority of concrete is batched through the dry leg but a small mixer, typically 0.8–2.0 $m^3$, is available to supply customers who wish to collect their own concrete or to batch very low workability concrete.

## 18.3.3 Weighing systems

Table 18.4 shows the options available for gauging the raw materials. Without exception cement and aggregate are batched by weight. Traditionally mechanical lever systems

**Table 18.4** Preferred methods of gauging materials

| | Weight | Volume | Flow |
|---|---|---|---|
| Cement | √ | × | × |
| Aggregates | √ | × | × |
| Water | (√) | √ | √ |
| Admixtures | (√) | √ | √ |

were used but these are now being phased out in favour of load cell type systems. These are more reliable and can easily link into computerized batching and recording systems. Water and admixtures can be either weighed or, more frequently, batched by volume or flow.

## 18.3.4 Computerization

All new plants covered by the Quality Scheme for Ready Mixed Concrete (QSRMC, 1997) are expected to be fully automated with autographic recording facilities. The benefits of computerized plants are enhanced retrospective quality control, accurate batching records, improved security and, above all, improved consistency. Despite the claims of the manufacturers of such systems computerization does not necessarily improve the speed of production and an experienced batcher can often match the speed of a computerized plant.

There are many computerized systems on the market and the final choice of a particular system is based on the following factors:

1. **Flexibility** The system must be capable of adapting to the numerous plant configurations in use.
2. **Ease of use** Readymix concrete is not a 'high-tech' industry and the IT systems used at plant level must be simple to use.
3. **Accuracy** There is little point in installing a system that cannot batch within accepted tolerance. Difficulties are often experienced with small loads and allowance for 'inflight' times for materials.
4. **Upgradeable** IT systems continue to improve and any system must be capable of being upgraded without the need to replace hardware.
5. **Reliability** Concrete plants are an aggressive environment and the system must be capable of performing reliably in such an environment.
6. **Speed** The system should not be slower than manual batching.
7. **Compatability** The system should have the capability of talking to other computers used in the company for quality control, order processing etc.

## 18.3.5 Environmental considerations

Environmental issues are a major concern of all industries and readymixed concrete is no exception. Emissions of cement dust are strictly controlled by the use of dust-extraction systems in loading areas, cement silo filter systems, restrictions on powder-blowing pressures and the provision of high-level silo alarm systems.

Many plants now have water-recycling systems which minimize the amount of water being discharged from the site. All wash-out and surface water is stored, any excess solids removed and the water re-used in concrete.

Some of the busiest plants also have complete concrete recycling units which separate returned concrete and/or wash-out water into individual reusable components.

### 18.3.6 Delivery

Deliveries are generally made in 6 $m^3$ or 8 $m^3$ purpose-made agitating/mixer units. Agitating speeds are generally low, around 1–2 rpm, although for some types of concrete non-agitating vehicles are used. This is particularly the case with low-workability mixes, such as cement bound granulated material (CBGM), which are often delivered in tippers. In such cases the tipper bodies should be sheeted to prevent moisture loss during transit. Very high workability concrete can restrict the carrying capacity of a truck particularly in hilly areas.

Although 6 $m^3$ trucks are the normal size some companies are now beginning to use larger vehicles and some 7 $m^3$ or 8 $m^3$ trucks are in use. In some continental European countries it is not unusual to see 12 $m^3$ articulated truck mixers. At the other end of the scale smaller 2–3 $m^3$ capacity trucks are often used to service smaller contracts.

A more recent development is the introduction of conveyor trucks which give added flexibility in placing concrete. These are standard trucks with an additional pneumatically controlled conveyor attached to the side of the truck. The conveyor extends the reach of the truck by up to 10 m and can assist in placing concrete in difficult to reach locations without the need for a concrete pump.

Traditionally their use has not been widespread due to the reduced payload and higher initial capital costs. However, the introduction of lighter alloy conveyors is changing the economics of the situation and a number of companies now operate such vehicles. Concrete trucks which incorporate integral concrete pumps are also available.

## References


Beaufait, F.W. (1975) Effects of improper handling of ready mixed concrete. In Dhir R.K. (ed.), *Proc. Int. Conf. on Advances in Ready Mixed Concrete Technology* Dundee, September/October, Pergamon Oxford, 359–366.

British Standards Institute (1996) BS 5328: Specifying concrete including ready mixed concrete.

Dewar, J.D. (1962) Some effects of prolonged agitation of concrete. *C & CA TRA/367*, December.

Dewar, J.D. (1973) The workability of ready mixed concrete. RILEM, Leeds.

Dewar, J.D. and Anderson, R. (1987) *Manual of Ready Mixed Concrete*. Blackie, Glasgow.

Quality Scheme for Ready Mixed Concrete (1997). Quality and Product Conformity Regulations.

Scales, R. (1970) Truck mixing. Individual ACT Project 1976 (confidential).

# Part 5

# Exposed concrete finishes

# 19

# Weathering of concrete

*Frank Hawes and the British Cement Association*

## 19.1 Introduction

For the purpose of this study weathering has been taken to mean the effects on well-made concrete of natural forces such as rain and sunlight and unnatural forces such as pollution. Such matters as rusting reinforcement and graffiti are not included, and neither are blemishes arising at the time of casting or placing of concrete nor damage occurring during the life of a building as a result of poor design or construction.

## 19.2 The inevitability of weathering

We are concerned principally with concrete but weathering affects all buildings and their materials to some degree. If men were to return to the moon, much of the delicate scientific equipment left by the Apollo astronauts would still be usable. On the other hand, while the Earth's atmosphere of air and water vapour permits us to live without space suits, it destroys delicate equipment if it is left unprotected. The natural environment imposes a regime of change on our buildings and building materials so that all buildings alter with time (Figure 19.1). Weathering is the natural effect of time on our architecture and it is unwise to ignore it or relegate it to a position of little importance in the design process.

It is not easy to plan for the succession of youth, maturity and old age through which buildings will inevitably pass but it seems that, for many buildings of the last 40 to 60 years, no such anticipation has been attempted. They have been designed with only their youth in mind and too many have proved quite incapable of gracefully accepting the imprint of the passing years.

**Figure 19.1** Weathering affects all buildings.

This tendency of many modern buildings to look spoiled by time is not restricted to concrete. After a few years of uncontrolled weathering, so-called self-cleansing materials will often look worse than a more robust material in a similar position, because the control of weathering involves more than just the choice of the building surface. Design and detailing must combine to control the flow of water on facades so that accumulations of dirt will not overwhelm the architect's design.

## 19.3 Alternative approaches

There could be said to be three basic ways for an architect to approach this aspect of design.

One approach, indicated by line 1 in (Figure 19.2), is to design for an eternal youth, defying the attempts of time and the elements to alter the appearance of the building. By the use of expensive materials and very careful design this might be achieved to a limited degree but it would never be cheap and would not necessarily be desirable. Would we want all our buildings to stand out from their environment for ever, always looking new?

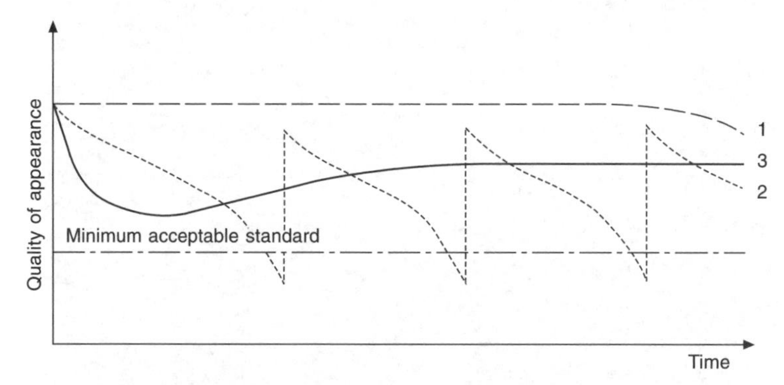


**Figure 19.2** Alternative approaches to design for weathering.

The second strategy, shown by line 2, is to design buildings that can be brought back to their original appearance at regular intervals by cleaning or painting or both. Repainting has been the traditional method for care of many handsome eighteenth and nineteenth century buildings in our towns and cities. It is a useful way of revitalizing certain buildings or locations but has two main drawbacks. It commits the owners to future maintenance expenditure, and it has to be accepted that the building will probably spend a substantial part of its life looking in need of maintenance.

The third option, indicated by line 3, is to attempt to design buildings that can grow old gracefully without expensive maintenance; buildings that will change with time but will not be spoiled. This is probably the most difficult strategy to follow but it is the one that this chapter attempts to describe.

If attempting strategy 3 proves less than 100% effective it will become as line 2 but with the horizontal (time) axis stretched and the vertical (cost) axis diminished to the benefit of everyone.

## 19.4 Weathering on old buildings

In their book *Salissures de façades,* Carrié and Morel (1975) illustrate the porch of the church of St Margaret at Westminster (Figure 19.3), and point out that while some people see it as an illustration of the objectionable grime that ravages our cities, others see only the noble patina of age.

Many buildings from former centuries have a visual strength and robustness which is able to carry considerable soiling without the architecture losing its character. Certainly,

**Figure 19.3** The church of St. Margaret, Westminster: porch.

modern structures in a similarly dirty state receive more criticism. Such dirt can, of course, harm some facade materials and must be removed, but where there is little likelihood of physical damage, it could be left. Soiling can sometimes give to a design an emphasis that it would lack for much of the year in the British climate due to the absence of sharp, well-defined shadows.

But the accumulations of dirt on these old buildings are not very precisely controlled (Figure 19.4). The design remains acceptable chiefly by providing places where dirt can build up without distracting the eye from the essential characteristics of the composition. By contrast, many of our more recent buildings are intended as clean compositions of straight lines and rectangles on which uncontrolled runs of dirt immediately look out of place.

**Figure 19.4** The church of St Margaret, Westminster: detail.

This is not to recommend the imitation of historic building forms nor the decoration of buildings with superfluous ornament. The ill-considered application of string courses and moulded architraves, without proper understanding of how they are to control the flow of water, may do little to help the appearance of a building. The function that a particular detail has to perform must first be decided. New details can then be devised or traditional ones copied provided that the chosen detail can be relied upon to perform the required task.

## 19.5 The weathering system

Whichever design approach is adopted, it is necessary to know something of the working of the complex system we have chosen to call weathering. In the simplest terms, surfaces get dirty and the dirt is then redistributed by rain under the influence of wind and gravity (Figure 19.5). But this interaction is complicated by variations in the type and amount of dirt, the shape, texture and materials of the surfaces and by the quantity, frequency, distribution and chemical characteristics of the local rainfall.

**Figure 19.5** An even layer of dirt on a north-facing stone wall being affected by concentrated flows of water.

To produce a building that will weather well, an architect should consider carefully three separate but closely interrelated aspects of his design:

- the overall massing, orientation and geometrical form of the building, and its relationship to other buildings and local topographical features;
- the materials and surface finishes of each external part of the building;
- the detailing required to control the flow of water over the facades.

These points must all be checked against the expected environmental conditions, which will vary both geographically and locally.

## 19.6 Characteristics of concrete

In the following sections of this chapter an attempt has been made to summarize the information that is available on these subjects, and to suggest how it may be used. In this section we will look first at the characteristics of concrete and concrete surfaces, then at the external influences which determine the conditions in which the concrete will exist. The final part of this chapter examines the visible effects of rain and dirt on buildings.

## 19.6.1 Properties of concrete surfaces

Concrete is a porous material. The pores are partly a result of the need to use more water to provide workability than is required for the hydration of the cement, and partly of the impossibility of removing absolutely all the air during compaction. The pores range in size from approximately one-tenth of a millimetre (100 μm) down to about one-millionth (0.001 μm). In spite of these pores, the surface of a good concrete, in comparison with other commonly used building materials, is dense and not very absorbent.

However, when considering the visual performance of concrete, we are concerned not with deep penetration but with water being taken up or released by the surface layer. The porosity of concrete is determined primarily by the water/cement ratio of the mix and the efficiency of placing and compaction, but the number, size and shape of the capillary pores at the surface are greatly affected by modifications of water/cement ratio in the surface layer. These modifications can arise from interaction with the formwork or during curing of the concrete. This layer, probably less than one millimetre thick, will have quite different characteristics from the general mass of the concrete, and its surface permeability is very difficult to control.

By the time concrete has dried after the formwork has been removed, a thin layer of calcite (calcium carbonate) crystals will normally have developed on the surface. Most concrete is intrinsically grey but this microscopic layer will make it look brighter or even white. The crystals form by reaction of the calcium hydroxide products of cement hydration with the carbon dioxide in the air. If this process is not controlled by careful curing, the white calcite crystals can be an embarrassment if the concrete is intended to be dark in colour. On most grey or white concrete, however, they represent the expected appearance of concrete, provided that there are no marked differences between adjacent areas.

Calcium carbonate crystals have a refractive index similar to that of water, so the colour of concrete appears to change when it is wetted (Figure 19.6). A film of water

**Figure 19.6** Concrete appears much darker when it is wet.

sufficient to cover the crystals prevents them from scattering light rays and allows the light to penetrate to the 'real' colour of the concrete which appears significantly darker than it would when dry.

If the as-cast skin is removed from the concrete by acid etching, washing, tooling or any other method, a surface is revealed for which it is usually easier to predict and control the performance (Figure 11.7). Additionally, the complexity of such a surface obscures some variations which would be noticed on an as-cast surface.

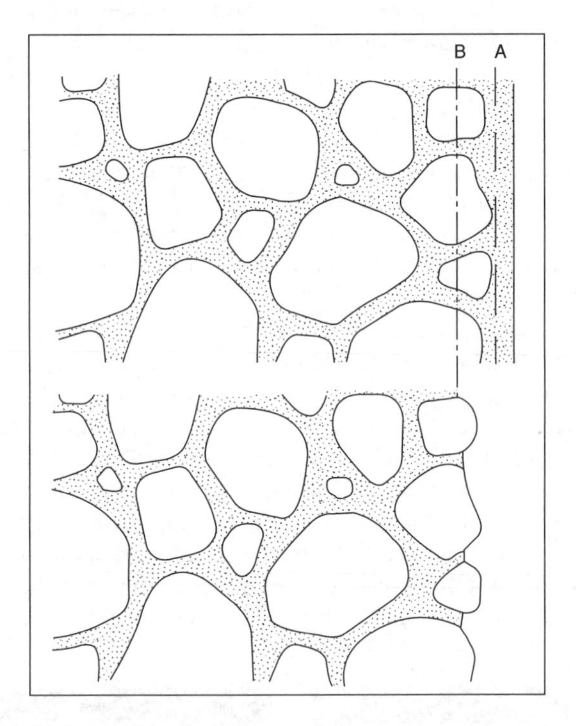


**Figure 19.7** The outer skin of as-cast concrete (to the right of line A) is not typical of the mass and may be more absorbent or more variable in its characteristics. The appearance of concrete can be improved by removing more or less of the matrix between lines A and B, but the depth of exposure should not exceed one-third of the size of the aggregate.

With the surface skin removed, the characteristics of the aggregate become as important as those of the concrete matrix. These characteristics – colour, size, shape and absorptivity – must be considered for their tendency to hold and show dirt or to facilitate biological colonization. Granites and similar dense materials predictably wash clean more easily than some of the more porous stones which may become visually unacceptable unless kept reasonably dry. Where surfaces are inevitably going to become soiled, the choice of darker aggregates may be sensible; though there is no reason why strong patterns of clean white and dirty black areas, such as we see on old limestone buildings, should not be allowed to develop if they are architecturally acceptable.

It is not reasonable, however, to expect the choice of a good exposed aggregate finish to deal with all problems of weathering. Water moves differently on such a surface but concentrated flows or their effects will still be visible and must be controlled.

## 19.6.2 Performance of concrete surfaces

Concrete, in common with other building materials, will change during its life. As the material matures colour changes will occur, either integrally or due to dirt accumulations, and growths can appear on the surface – either biological growths or inorganic ones, such as lime-bloom and efflorescence.

The basic colour of as-cast concrete is the colour of the finest particles in the mix which, unless a pigment has been used, normally means the cement in the surface skin. Exposing the aggregate makes control of the appearance much easier.

If the skin is left intact, colour changes must be expected during the life of the concrete due to the removal of the cement coating from the outermost granules by the action of rain water. This will lead to a change from the cement colour to the colour of the fine aggregate which, in exposed situations can be quite rapid.

Changes in appearance due to inorganic growths are a consequence of the interaction of the products of cement hydration with the atmosphere. They range from lime-bloom and efflorescence to the heavier deposits of lime arising from construction faults. These heavier deposits vary from a thin glaze to quite large stalactites (Figure 19.8), but all are the result of poor detailing or bad workmanship. If water is allowed to percolate through insufficiently compacted concrete from badly drained horizontal surfaces, or is permitted to leach through unprotected concrete retaining walls, these problems will inevitably occur.

**Figure 19.8** Stalacites formed as a result of poor detailing or bad workmanship.

The process by which calcite crystals are produced on concrete is highly dependent upon the surface permeability. Concrete can, as a result, become visually unacceptable if the surface permeability is uneven so that salts build up in patches on the surface.

Further changes in appearance occur in polluted atmospheres where sulphurous deposits, combining with the calcite on the face of the concrete, can form calcium sulphate or gypsum. This is much more soluble than calcium carbonate, and may be washed away taking dirt with it; but where there is insufficient water for washing, it can encapsulate the dirt and fix it.

## 19.6.3 Biological growths on concrete

Once its alkalinity has been reduced by carbonation, a concrete surface will provide suitable conditions for biological colonization. Many surfaces which appear to be dirty may be found on examination to have more biological contamination than mineral deposit. Areas of algae or decaying lichen on concrete can be ugly, but where attractive lichens occur on clean surfaces they often go unnoticed.

The first colonizers of a surface are micro-organisms: algae, fungi and associated bacteria; then visible growths of algae or lichens appear. Dirt sometimes collects on colonized surfaces and this, together with dead lichen, can provide a footing for mosses and later even for more developed plants.

Algae are free-living chlorophyll-using plants which derive their energy mainly from sunlight. There are thousands of species, and many can live on either acid or alkaline surfaces. They require moisture and some traces of mineral salts for survival but some organisms can adopt a resting state if water is not available for a period.

Fungi are parasitic or saprophytic non-chlorophyll-using organisms which get their energy from the starch-like products of decaying plant life. They need moisture and the appropriate nutrients but can grow without sunlight.

Lichens (Figure 19.9) are symbiotic combinations of algae and fungi which can readily reproduce on moist concrete. They do not usually develop until the alkalinity of the surface has reduced to about pH 7.5 but colonization can then be quite rapid.

Green or dark-coloured algae will grow on most concrete that remains damp. Proximity to vegetation or open soil encourages colonization by offering a vast source of propagating organisms and nutrients which can reach the concrete by dripping or splashing or on the wind.

## 19.6.4 Variability

Colonization by algae and lichen, in common with the other visible changes in the appearance of concrete described above, is dependent upon the surface permeability of the concrete. Microtexture, too, plays a part but the facility for water to pass into the surface is a most important factor. Poor quality concrete will, in general, become dirtier, grow more mould and suffer more efflorescence than dense concrete, and these phenomena will provoke and encourage each other. Dirt can accumulate on mould or be bound by salts; mould can initiate the process or grow on an accumulation of dirt.

Many structures have been built without sufficient attention to mix design, placing, compaction and curing of concrete.

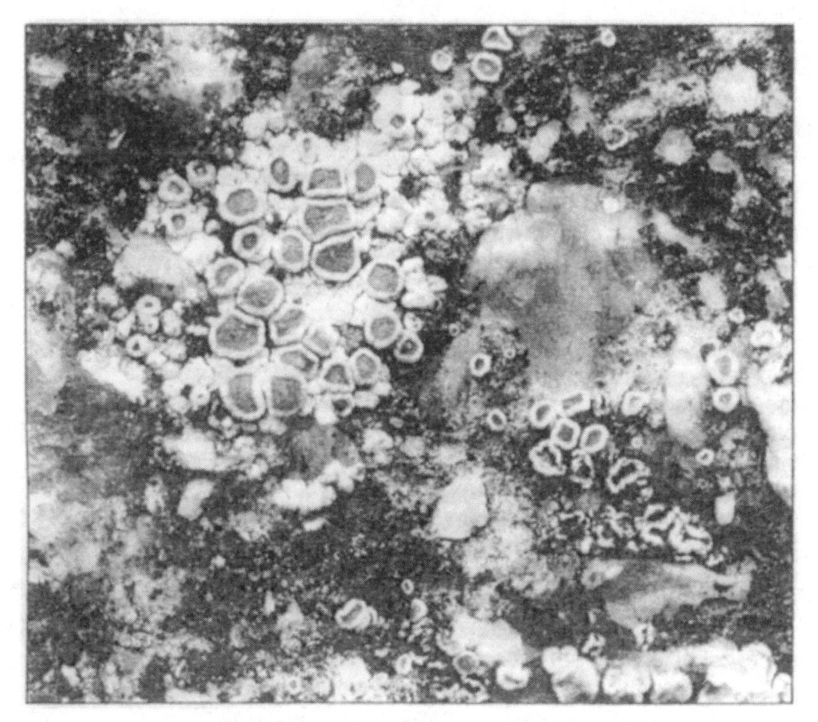

**Figure 19.9** Lichen *Lecanora dispersa* on slab of Rubislaw granite with granite sand and OPC (magnified ten times)

In practice, it is where mould and dirt are permitted to emphasize and exaggerate variations in the surface of poorly made concrete that the appearance is least acceptable.

## 19.7 External factors

### 19.7.1 Climate

An architect, if he is to control the way his building will weather, must add to an understanding of the material an appreciation of the environment in which the building will exist. Ideally, he should have information on the speed, direction and frequency of wind, quantities and frequency of rainfall, together with records of temperature and relative humidity, because all these will affect the way in which his building will get wet and will then dry out.

Meterologists recognize several scales of climate in their studies but in assessing the conditions around buildings the range has to be extended even further:

- *Macroclimate:* Country-wide, influenced by sea, mountains, etc.
- *Regional climate:* Covering a few square kilometres, and influenced by the elevation of the land, large forests and major valleys and lakes.
- *Local climate:* Covering smaller variations brought about by individual natural features, such as small hills, woods, rivers, etc.

- *Town climate:* The urban form of local climate due to the effects of built-up areas or groups of buildings.
- *Microclimate:* Variations at the scale of 10–100 m influenced by individual buildings or groups of trees – the climate of the building site.
- *Cryptoclimate or climatic sheath:* Describes the conditions close to the surfaces of individual buildings with a scale of one metre or even much less, and producing the climatic sheath in which the building's materials have to exist.

The terms micro and crypto refer only to the extent or areas over which changes are felt, not to the scale of those changes which can, paradoxically, be very large. There are often major climatic variations between one part of the climatic sheath of a building and another.

In general, when the weather is clear and winds are weak, variations will consist chiefly of large temperature differences. Cloudy skies or strong winds will almost eliminate these disparities of temperature but large variations in wind speeds can be expected to occur around the building, and the extent to which rain hits the surfaces of buildings will vary greatly.

These local climatic variations can often be seen when a building is being washed by driving rain or when trees or dust make wind patterns visible (Figure 19.10), while the effect of the variations can be observed in the way identical details or surfaces perform on various parts of the same building.

**Figure 19.10** Tree being blown away from a solid building by the vortex wind on the windward side

To predict likely weathering effects with accuracy, it would be necessary to be able to relate regional and local climate data to variations in the climatic sheath around a proposed building. We are a long way from being able to do that accurately. Until there have been more studies of the climatic sheath, designers will usually have to make assumptions based on macroscale or, at best, regional scale meteorological data.

Information on driving rain is available because there is a wide interest in preventing rain penetration. *The Driving-rain index* can give a useful comparison between the general exposure of one area and another provided that the necessary adjustments and allowances are made for altitude, shelter or other special conditions.

Because dampness facilitates the development of both organic and inorganic soiling, buildings in an oceanic climate like Britain's are more difficult to keep clean than those in a Continental or Mediterranean climate. Annual rainfall in many parts of Britain is lower than that for much of the Continent but their rain comes in heavy storms with long dry periods in between, whereas our buildings get few opportunities to dry out.

## 19.7.2 Wind and rain on buildings

The movement of wind around buildings has been studied in the field and in wind tunnels for many years to determine wind loads and dynamic behaviour of structures and to investigate environmental conditions in and around buildings.

The most important point to be learned about weathering is the very noticeable effect of the prevailing wind (Figure 19.11). During storms, driving rain can come from any direction, and sometimes local street patterns or topographical features override the geographical influences, but it seems to be the relationship of a facade to the prevailing wind and the intensity of rain from that direction which largely determine the effect of the weathering pattern. At the same time, Beijer's work in Stockholm has shown that vertically there are parts of a building which can be expected to be washed by rain and other areas which will never receive enough water to remove the accumulated dirt.

When the wind reaches a large building it forms a characteristic pattern of airflow (Figure 19.12), with high wind speeds round the edges and a horizontal vortex on the windward side. This phenomenon is strengthened if, as is most common in urban situations, there are lower buildings parallel with the windward facade. The lee sides of such buildings and, significantly, the lower part of the windward side also, receive very small amounts of rain on vertical surfaces compared with the amount striking the upper part of the windward face. Rain falls almost vertically close to these walls and will collect on any ledge, sill or moulding. Some of the most noticeable and least acceptable forms of uncontrolled weathering arise where this water is allowed to run onto the otherwise unwashed face (Figure 19.13).

If a building is intended to look the same on all faces, as many are, it may require either extraordinary means to get enough water on to the lee side faces to keep them clean or extremely bold details to protect the upper weather side faces from driving rain. The task of controlling weathering is greatly simplified when it is accepted that the rain, like the sun, does not visit all faces of buildings equally and, as a consequence, different solutions are often sensible.

## 19.7.3 Air pollution

The environment in which our buildings stand is not often clean. The air carries a certain amount of solid matter from natural sources – pollen, bacteria and dust swept up by the

**Figure 19.11** North, west and south-west facades with identical materials and details.

wind or emanating from volcanic eruptions. Mankind's activities have added to this natural pollution quantities of solid and gaseous products from our fires, industrial processes and traffic, Although air pollution has been measured only comparatively recently, it is not a new problem. Its effects on buildings and vegetation and on the light and weather of cities have been recorded in literature and paintings for centuries.

The atmosphere contains many pollutants, and they can be grouped usefully into three categories for the study of weathering.

- Particles larger than 1 μm in diameter, which will be referred to as grit and dust.
- Particles smaller than 1 μm, which we will call black smoke.
- The gaseous pollutants, of which sulphur dioxide is probably the most important.

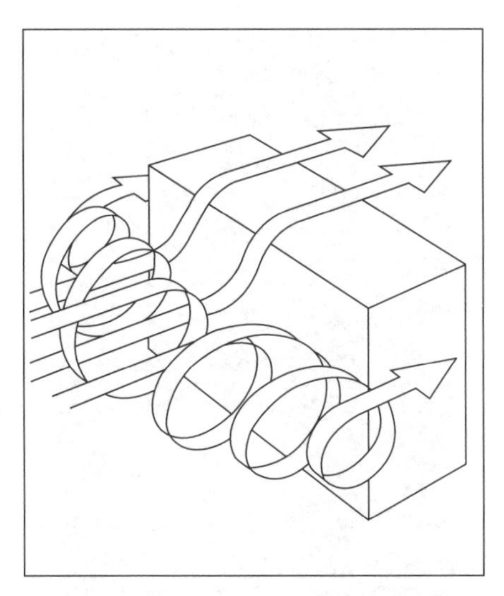

**Figure 19.12** Characteristic pattern of airflow on windward side of a large building.

Before discussing each group it is useful to explain this division of particulate matter. The larger particles, grit and dust, settle under the effects of gravity. In contrast, particles smaller than 1 μm stay suspended in the air for long periods because, when they move, the resultant friction forces are greater than the gravitational forces acting on them.

Grit and dust particles in the atmosphere consist largely of wind-borne dust from roads and industrial installations with some ash and unburnt solid fuel. Their concentration is difficult to measure accurately. Even when efficient equipment is available, it is not easy to relate readings to any other sites or even to the same site at different times. Atmospheric diffusion of these particles varies greatly and much of the fall-out occurs close to the source. Consequently, dust tends to be primarily a local nuisance, and deposition figures are valid for only a relatively small area around a gauge. Little published information is available that would be of use to designers. The seasonal fluctuations that are found in concentrations of smoke and sulphur dioxide do not seem to apply, suggesting that heating installations are not major sources of grit and dust. This is supported by the fact that there has been no noticeable fall in general grit and dust levels since the introduction in 1956 of the Clean Air Acts in Britain.

Because the fine particles in the atmosphere are efficiently distributed by wind, and because a proportion of them comes from natural sources anyway, black smoke pollution has never been confined to the towns. It is reasonable to say, however, that in the past smoke levels in towns were many times those in the country. Since the introduction of smoke-less zones in Britain, urban concentrations have fallen by 70 per cent. This is still about twice as much as is found in rural areas. Analysis of readings in city centres indicates that a high proportion of the particulate matter in the atmosphere (77 per cent in London) now comes from vehicles.

Sulphur dioxide is, for the most part, a product of the burning of fossil fuels, and has been a major contributor to the atmosphere of our towns and cities for hundreds of years. In sunlight, in the presence of hydrocarbons and nitrogen oxides, it can be oxidized to form sulphates and sulphuric acid.

Rainfall is naturally slightly acidic (pH 5–6) due to dissolution of atmospheric carbon

**Figure 19.13** Rain from a horizontal ledge producing uncontrolled effect on an otherwise unwashed area of wall.

dioxide but pollution increases the acidity (reduces the pH) to a level at which it can attack materials. Values below pH 4 are unusual but, in April 1974, Scotland experienced rain with a pH of 2.4. That rainfall was more acid than household vinegar!

There is some evidence to suggest that acid attack on building materials can be at least as bad in dry areas with heavy pollution as it is in wet climates. Sulfur dioxide is a gas but seems to be able to accumulate on surfaces in dry periods, possibly as a sulphite combining with solid pollutants or with materials in the surface itself. This may then produce quite strong acid attack, even though the acidity of the rainfall may be low (Figure 19.14).

**Figure 19.14** Effect of acid precipitation on Portland stone.

The creation of smokeless zones has not had as marked an effect on sulphur dioxide as it has on smoke. Concentrations in urban areas in Britain have decreased by 40 per cent, but the areas affected by sulphur dioxide pollution have spread. This is because the principal means of control adopted in industry has been to raise the heights of chimneys, and so spread the gases over a wider area.

Black smoke and sulphur dioxide levels are monitored throughout the British Isles, and the findings are collated by the Warren Springs Laboratory and published by the Department of the Environment.

## 19.8 The visible effects of weathering

### 19.8.1 Rain and dirt on buildings

The larger and smaller particles in the atmosphere affect buildings in different ways. In general, only the upward-facing surfaces of buildings – roofs, sills, ledges and mouldings – accumulate grit and dust, while all faces are soiled by the finer particles too small to be affected by gravity.

The way in which the fine particles can be attracted to and become attached to surfaces is explained by Carrié and Morel. Their adhesion to all surfaces, tops, sides and even soffits, is apparent from examination of any building. It seems, too, that the presence of moisture on the surface often just from condensation facilitates the bonding of particles to that surface.

Although the first few minutes of a rain shower after a dry period may carry dust from the atmosphere, most rainfall can be considered a cleansing agent provided there is enough of it to flow down the surface. A thin layer of water from drizzle or condensation, however, may encourage the attachment of dirt.

In conditions of driving rain, water will often start to flow down a large concrete surface but it will nearly always be absorbed before it reaches the lower levels. The factors that determine whether the surface will be cleaned or not include the quantity of rain and its cleanliness and acidity together with the texture, porosity and cleanliness of the surface. Beijer shows that the capacity of the wall material for absorbing water determines whether and how quickly rain run-off occurs. This capacity depends both on the material itself and on its moisture content. If water flowing down a face is absorbed before it reaches the bottom, a conspicuous wavy dirt line will be formed (Figure 19.15).

This phenomenon occurs at two scales on buildings. Beijer explained it in terms of large surfaces where driving rain on the upper part is absorbed before it reaches the lower storeys because of the major differences in quantities of water striking the building at different levels. The same wavy dirt lines are seen on a smaller, but no more acceptable, scale wherever water is permitted to run from a horizontal surface onto an otherwise unwashed vertical face (Figure 19.16). Water should, therefore, be kept off a concrete surface unless enough is available to flow across and clean the entire face.

An attempt was made by Carrié and Morel to calculate the angle of inclination necessary to achieve proper cleansing by rain. Recommended angles of inclination ranged from vertical on exposed faces to 40° on sheltered facades in zones with little rainfall. It has to be pointed out that these were regarded only as a guide or preferably as a starting point for experiment.

**Figure 19.15** Partial washing of large facade caused by variation in amount of rain reaching the facade.

The suggestions were made on the assumption that all faces should be clean and, if carried through logically, would mean that all buildings should be shaped like truncated asymmetrical Eiffel Towers! As this is clearly unreasonable it follows that some parts of buildings will remain dirty.

## 19.8.2 Movement of water on building facades

It has been demonstrated in the laboratory that, even on the smoothest materials, considerable quantities of water are required to ensure a complete film flowing on a vertical or inclined

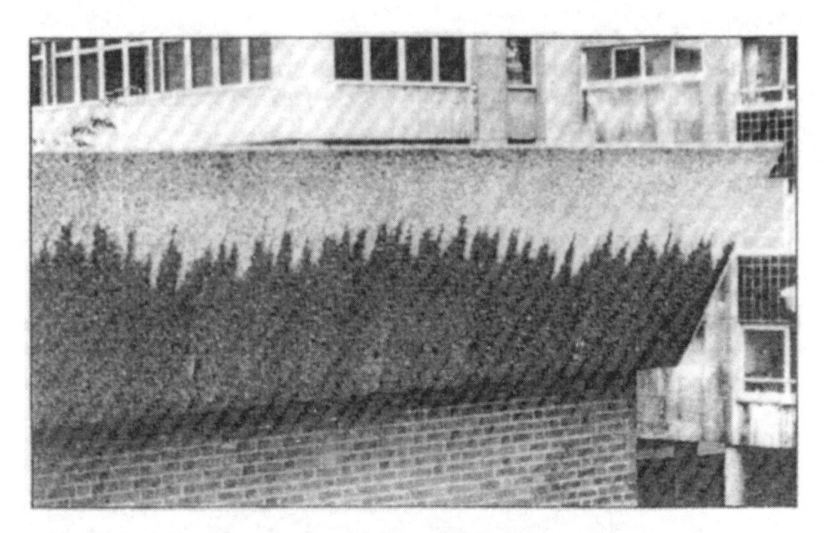

**Figure 19.16** Partial washing caused by water running off a horizontal surface.

surface. As soon as the film thickness decreases below a critical level, related to the viscosity of the water and the texture of the material, the flow divides into separate streams. The positions of the streams are determined by microscopic irregularities or differences in wettability of the surface. Experiments have shown that the features which cause the sub-division are often too small to be seen or measured but if the surface is dried and the experiment repeated the streams will re-form in the same places.

On building surfaces, dirt and salt deposits, and eventually even erosion, can reinforce such flow patterns and make them self-perpetuating. These random patterns can be avoided by controlling water flow so that its effects are related to the architectural form and detailing of the building.

On exposed facades, water tends to run in the sheltered internal angles of recesses, and observation shows that on vertically ribbed surfaces dirt seems to collect in these recesses, probably due to their dampness. There is reason to hope, therefore, that water can be controlled on facades but little formal research seems to have been attempted on this important subject.

Robinson and Baker, Carrié and Morel, Partridge and Beijer have all written usefully on the subject of the control of water but all of them have based their suggestions on observations of existing buildings. Controlled experiments have been very few.

Couper (1974) at CSIRO, Australia, measured the shedding characteristics of various detail shapes and showed that even small projections have the ability to shed up to 50 per cent of a fast-moving stream passing over them (Figure 19.17). In fact, Beijer's work indicates that water on facades rarely exceeds a speed of 0.5 m per minute so care must be exercised in relating Couper's work to real buildings.

Bielek (1977), while working at Trondheim studied the flow of water across and adjacent to joints between precast concrete panels. He experimented in the laboratory with a number of heights and shapes of ribs adjacent to joints, and recorded the way in which they controlled the flow of water under the influence of wind blowing parallel to the panel face (Figure 19.18).

Herbert (1974), at the Building Research Establishment, studied the effect of horizontal protective features on an exposure rig at Plymouth. Horizontal projecting fins towards the

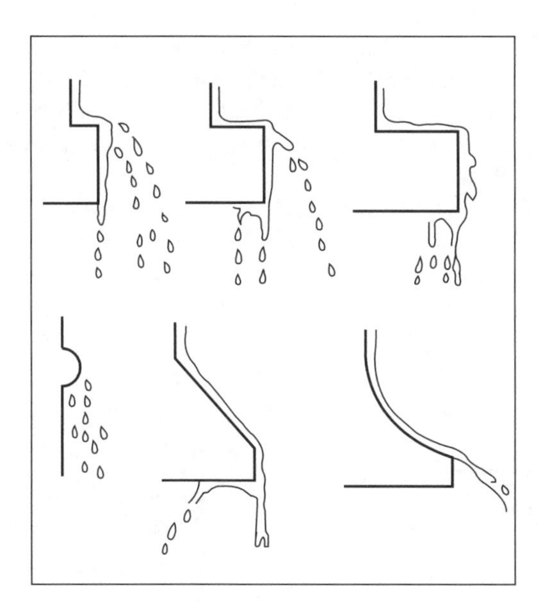

**Figure 19.17** Diagram from Couper's study which measured the ability of some not very typical details to shed a fast-flowing stream of water.

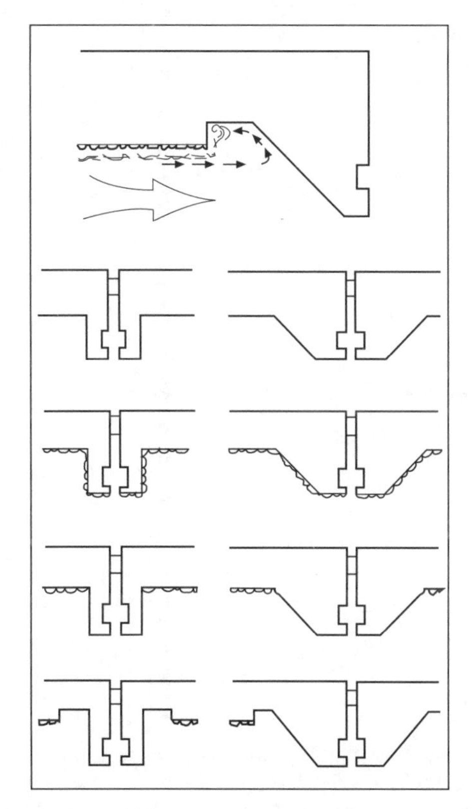

**Figure 19.18** Diagram from Bielek's study. In the most successful joint, water was induced to flow down the channel adjacent to the rib.

top of the rig were found to cause water to accumulate on the face just below. Projections of 50, 150, 200 and 300 mm were tested and the concentration was found to be related to wind speed but very strikingly to be worst for the smallest projection. A cushion of air appeared to build up below the larger projections which protected the wall face from the driving rain (Figure 19.19).

At lower levels, where the wind was no longer driving rain up the facade the fins had a normal protective action but, whilst small projections protected areas extending approximately twice their projection, the ratio was found to reduce as the projection increased so that larger fins projected areas of not much more than half the projection.

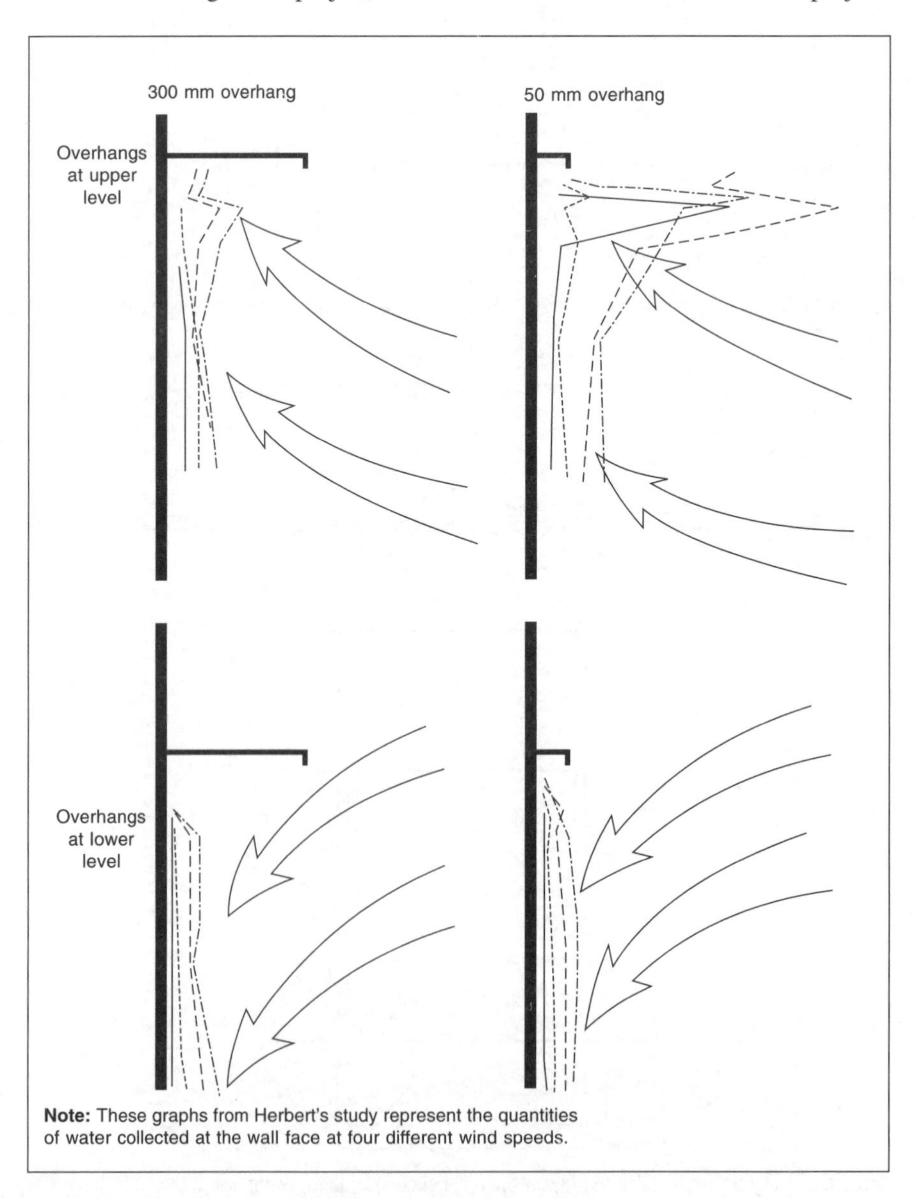


**Figure 19.19** Diagram from Hebert's study of the behaviour of weather-protective features.

This may, however, have been due to the exceptionally exposed site. On most facades, projecting features can be seen to give protection to considerable areas of wall below.

This is not a very impressive total of research into such an important subject. There is little data, apart from casual observation, related to the action of water on the more sheltered parts of buildings because all known work has been concerned principally with rain penetration.

## 19.9 Control of weathering

Buildings that weather well have details that control the flow of water on the facades, and materials and surfaces suitable for the conditions. These conditions will be determined by the location and form of the building and, locally, even by its details.

The design process starts, preferably, with the siting and massing of the building in such a way that it will not produce any special problems. Each facade or element of the design has then to be considered and broken down if necessary into zones in which conditions of exposure and dirt will be reasonably consistent.

It must then be decided which parts within these zones can be kept clean and which must be protected from partial washing. It is important to see that the details establish and maintain the differences between these areas, and that the finishes selected will perform well in the conditions that have been created.

Each of these steps will be discussed in this section but before the designer can make any of these decisions he must assess any difficulties and constraints imposed by the site and the environment.

### 19.9.1 Prediction

#### *Predicting local conditions*

There can be large differences between published meteorological data – which usually relates to countries or, at best, regions – and the conditions to be expected on a particular building site.

It would be useful if measurements could be taken on the site over a period prior to embarking on the design of the building. Obviously, it would rarely be possible to collect information for a long enough period for the figures to be used alone. However, it would be useful to compare site data with regional figures or with data collected, even over a short period, around existing buildings on which the weathering patterns were already established.

Valuable though such studies would be, they are most unlikely to be more than a rare procedure. On the other hand, every architect should learn to study and use the information that is available.

Nationally published information on climate and pollution should be obtained and considered in relation to local topography and adjacent buildings and trees which may affect the amount and position of driving rain hitting a building and the way in which water runs down the facades. In general terms, the way in which concrete and other materials respond to climatic conditions seems to be related less to their likelihood of getting saturated than to their opportunities for drying out. According to Beijer, this is strongly influenced by temperatures during the drying periods, so temperature, sunshine and wind information can be equally relevant.

The architect is likely to know the constraints of many sites from previous experience, and it is frequently possible to learn a great deal about the environmental conditions of a site by observation of existing buildings in the neighbourhood. They will often give a clear indication of the levels of pollution, the principal direction of driving rain and the likelihood of the rain being able to have a real effect on dirt accumulations.

### *The problems of an oceanic climate*

Because dampness facilitates the development of both organic and inorganic soiling, buildings of any material are more difficult to keep clean in an oceanic climate like Britain's. The Mediterranean region has warm damp winters and hot dry summers while most continental areas have long cold winters with low relative humidity and most of their rain coming in storms during their hot, predominantly dry summers. Unlike these, the British Isles have moderate rainfall spread over many raindays and only rare prolonged dry periods.

Relative humidity in Britain is usually at least 60 per cent and frequently it rises above 95 per cent for part of the day giving materials little opportunity to dry out. Lacy pointed out that even when we get long dry cold spells they are ended by the cold air mass being suddenly replaced by air which is warmer and more humid. In these conditions, all parts of a facade are likely to be wet from condensation; even those which are normally protected from rain.

In oceanic regions we must not only expect concrete to get wet more often from rain and condensation, but we must also expect it to remain wet longer because the rate of drying is related to ambient temperature and relative humidity.

## 19.9.2 Control

### *Choice of concrete surfaces*

On a surface with variable properties, patchy weathering will result even if the exposure is uniform. Therefore, whatever the choice of surface – as cast or exposed, precast or *in situ* – it is most important that the concrete should be as uniform in surface characteristics as possible.

Rough textures have the advantage of spreading water more evenly over the face and disguising any localized staining but will also hold more dirt during dry periods.

Surfaces might be tilted to collect more water and ensure thorough washing but the absorptivity of concrete in such positions becomes even more important. Obviously, the closer to the horizontal the slope, the more water will be collected; but it will flow more slowly, and the chance of its being absorbed and encouraging organic and inorganic growths on the surface will be greater. Here surface permeability becomes especially important. It is more difficult to achieve the necessary dense concrete surfaces when casting on site than under the controlled conditions of a precasting factory.

Some attempt at texturing or profiling or the exposure of interesting aggregates is more likely to be acceptable than smooth as-cast surfaces. Many dark aggregates are available, and they can be used to minimize the change in appearance as dirt is deposited.

Some flints and gravels have attractive mixtures of light and dark colours in the same material which can be useful in such situations.

Some surfaces look their best in exposed positions, and others look better if protected from rain but many are perfectly acceptable in either situation, provided that the detailing is good.

If the environment is clean and the weather pattern fairly uniform, the same finish may be acceptable on all parts of all facades. Where one finish is intended throughout the building, it must be suitable for protected positions and for the worst conditions of exposure to be expected. If the atmosphere is heavily polluted, different finishes for washed and unwashed parts or panels may be the best answer.

The greater the degree of control of weathering through detailing, the wider will be the choice of finish. If little control is attempted, only the densest surfaces with exposed non-absorbent aggregates can be expected to maintain their appearance in difficult conditions.

## *Maintaining uniform conditions*

Having attempted to predict the environmental factors affecting the building and having selected finishes suitable for the expected conditions, it is necessary by careful detailing to ensure that these conditions are maintained. We can learn a great deal by careful observation of successful and unsuccessful details on both new and old buildings. Concentrated streams of water on porous surfaces will promote mould and algae. Where surfaces are otherwise protected from washing, concentrated flows will produce clean streaks which can be equally unsightly.

The principle is to get an adequate, controlled flow of water on to the parts that are to be kept clean and to get it off again without allowing dirty water to affect any other clean areas or to disturb the dirt on areas which are to remain unwashed.

Site conditions must be studied in order to be able to define the expected quantity of water and the direction from which it will come. These will vary from one facade of a building to another and from top to bottom of each facade. It is not surprising, then, that details which work well in one position are not necessarily successful on another. However, if they are bold enough to cope with the worst conditions, their boldness is unlikely to be a problem where conditions are less severe.

In our present state of knowledge, complicated shapes are difficult to deal with except in very exposed situations. It is better to provide simple bold details that will maintain clear distinctions between washed and dirty areas. These differences can also, if required, be emphasized by the use of different finishes.

## *Getting water on to surfaces*

The principal influence on the appearance of most buildings is the prevailing wind and the intensity of rain from that direction but local street patterns or topographical features and height sometimes override the geographical influences and determine the local wind pattern. Weather protective details on the facade influence the quantity of water received while the texture and porosity of the materials involved determine how much of the water stays on the surface after impact to form a cleansing flow. These points must all be considered in assessing the possibility of getting enough water onto a surface to keep it clean.

It is possible to get more water on to particular parts of a building by giving them a batter or backward slope but this can cause problems if the surfaces are not very carefully chosen and if the effects are not fully considered. For instance, if a panel on a vertical facade is sloped to catch more rain it may prevent the panel below from being washed.

It is necessary to relate the surface characteristics – colour, texture and absorptivity – to the expected amounts of water and dirt and to the size of the panel or facade in question. More water will be required to wash a porous panel than an impervious one and to prevent the formation of the dirt line which will develop if water is absorbed before it reaches the bottom of the panel. A large area of concrete, extending perhaps over several storeys, would need an unusually impervious surface to be able to wash evenly from top to bottom. It is preferable to break up such facades with horizontal features which either collect or throw off the water at intermediate positions.

## *Throwing water clear of faces*

Copings, sills, string courses, cornices, etc., have always served a double purpose. Primarily intended to protect the top edges of facing materials, they also throw water clear of the face and limit the amount on the surface. Considering their importance, their effectiveness has had very little study. Observation of details on existing buildings supports the general conclusion of Couper that even small projections shed water. In the laboratory, he showed that eight or ten millimetres is the minimum effective dimension for the projection of such features but site conditions, coupled with the tolerances required for manufacture and assembly, dictate that these dimensions have to be considerably increased.

In our present state of knowledge subtlety is unwise and bold details are more trustworthy. On those very exposed parts of facades where rain is likely to be driven upwards by the wind, small horizontal projections may cause heavy concentrations of water, and they ought to be avoided in these positions.

It is difficult to make recommendations for the frequency with which water should be thrown clear of the face. This will vary according to the exposure of the facade and the type of detail but must also be considered in relation to the surface characteristics. What would be necessary on a rather porous white wall may be superfluous on a well-made dense panel with a dark non-absorbent exposed aggregate.

Lower down the facade, water which has been used to wash an area of concrete must be dealt with and must be removed from the face without permitting it to affect areas below. This is especially true on protected areas of facades where any water flow will leave the sort of clean streak so often seen at the ends of sills and hood mouldings. In general, it is easier to make such horizontal features continuous than to detail to avoid streaking where the projecting feature ends.

Every projection must have a drip groove on its underside to prevent water from flowing back onto the vertical face. Every book on detailing says this but, surprisingly, they are still omitted on many buildings.

Sills, copings, string courses and other details, such as changes in plane between one panel and another, consequently share an important role in throwing water off the building face. The shadows and the interest provided by these water-controlling details are the principal means by which the weathering patterns can be made subservient to the architecture.

## *Collecting water on facades*

Instead of throwing water clear of facades it is sometimes better to collect it and conduct it away in horizontal channels or gutters. Efforts have been made to calculate a minimum cross-sectional area of channel for every square metre of facade but the quantities of water involved are so small that in practice it is wiser to provide gutters which we know to be much larger than required.

The collected water can be discharged by pipes, gargoyles or by carefully detailed systems of vertical channels and grooves. If pipes are provided, regular inspection and maintenance must be ensured. Gargoyles are cheaper but probably suitable only for buildings of limited height.

It has been suggested that outlets from such gutters are unnecessary and that rainwater could be retained in the gutters and allowed to evaporate, whilst the occasional storm in which gutters overflowed would do less harm to the appearance of the building than the slow trickle of water from frequent light rain. This seems too extreme a solution but there may be an argument, when water is not to be discharged through pipes, for gutters to be detailed to retain somewater so that channels or gargoyles act only as overflows.

Removing water from elevations through concealed pipework is not cheap, and because of the expense, distances between outlets may become extended, with a consequent temptation to minimize the longitudinal fall on the gutters. It is wise, with such designs, to use properly formed and jointed gutters of generous cross-section, designed to ensure that if outlets should become blocked the water will overflow in a prominent position and has no opportunity of backing up into the building.

If distances between outlets can be kept short, generous longitudinal falls can be provided and much simpler systems of channels formed in the concrete can be used to steer the water down the face of the building in positions where it will cause no structural or visual problems.

## *Guiding water on facades*

Controlling water as it flows down building faces implies assessing the likely quantity of water and ensuring that it does not run in a random manner, producing visual changes disruptive of the architecture. Such haphazard staining is most noticeable on large flat surfaces. If plain flat areas cannot be avoided, they should at least have surfaces of exposed non-absorbent aggregate to help to diffuse the water courses.

Faces down which water is expected to flow are better modelled than flat but such modelling must be related to the water flow. Three-dimensional patterns based on historical forms or geometrical shapes will quickly be spoiled if this is ignored.

Water can be controlled by features of relatively small dimensions but, in our present state of knowledge, bold over-provision of guiding details is more sensible.

The incorporation of grooves, channel, ribs or other modelling into surfaces is only the first step. Great care must be taken to provide for the disposal of water right down the facade. One part must not be permitted to discharge water in such a way that it will spoil an area below.

Further research is needed into the best sizes and sections to suit specific applications but meanwhile we can learn from existing buildings to avoid the most visually disruptive effects of poor control.

## *Junctions*

Controlling the weathering of any part of a building entails predicting its exposure to the principal weathering agents – dirt and water – and then selecting suitable finishes and details. For that reason, junctions and edges, which are often the positions where such parameters change, are always going to present difficulties for the designer.

In still conditions, water seems to prefer to run down an arris, where it has reduced surface friction, rather than down an internal angle, where the surface contact and friction is greater. This variation in surface friction must apply to only a very narrow stream of water at the extreme edge of the detail but seems, in practice, to affect a wider stream and to become very noticeable.

Bielek has shown that in windy conditions water tends to run in the sheltered internal angles and it seems that these are the last areas to dry out, and for that reason perhaps acquire more dirt.

In addition, there are often concentrations of water at the edges of facades or panels due to rain being thrown there by the wind. Water, having greater mass, cannot change direction as quickly as air, so that as wind is deflected to squeeze round a corner, the rain continues on its original trajectory to hit the wall close to the corner.

For these reasons, concentrated flows often occur on external corners, while dirt tends to build up in internal angles. This effect can be observed at the scale of complete buildings as well as on individual arrises and on vertically ribbed surfaces.

In earlier times, these concentrations of water flow and abrupt changes in conditions would often have been disguised by a pilaster or by rusticated quoins. This century has produced one efficient alternative in the negative junction where two adjacent planes are made to float separately without actually appearing to meet and join. This detail sometimes works well to control the weathering differences on adjacent faces. Channels of this kind, formed either in flat planes or at internal or external corners, provide places for water to run without disfiguring the adjacent wall surfaces.

Rounded junctions, too, can help to avoid difficult arris effects but results are more difficult to predict.

Corners of openings in panels, and junctions between beams and columns, are sometimes rounded. This ought only to be done in situations of either full exposure or very sheltered conditions, and calls for careful selection of surface finish as curved shapes lead to diffuse but uncontrolled flow of water.

Any detail, whether borrowed from the past or newly developed, must be examined at both large and small scale to ensure that water will flow where it is intended. This may entail limiting the quantity of water to prevent it from spreading onto adjacent surfaces. It is best to avoid large concentrations of water building up at junctions by collecting it or rejecting it from the facade at frequent intervals in the height of the building.

## *Edges*

Vertical edges have been considered under junctions, so we are left to consider here what happens at top and bottom edges of facades, panels, beams or other horizontal features.

Top edges of facades are usually exposed to large amounts of rain. This applies also to edges of panels when they are not protected from above by overhanging features. Upward facing surfaces will normally accrue concentrations of airborne dust and dirt in dry periods. The most important function of any top edge detail, after ensuring that the edge of the facing material is protected, is to prevent that water and dirt from getting onto the face.

In most structures, such edge details can sensibly have overhangs of 40 or 50 mm to give protection to the wall below. Herbert's study, however, has made it clear that this can be unwise where rising currents of wind can cause water to accumulate under the overhang.

Any substantial horizontal surface, such as a flat roof, requires upstands at its perimeter to ensure that the wind cannot blow dirt and water over the edge and any such upstand must be given a steep crossfall back on to the roof.

Whatever the method of capping the top of wall or facade, special attention must be given to the jointing. Copings are exposed to extreme conditions of sun, wind and rain, so joints must be designed to accommodate both thermal and moisture movements without deterioration. Many buildings are visually ruined by streaks from badly made joints in copings and capping details. The difficulty of forming and maintaining efficient junctions in aluminium copings or trim may make it sensible to ensure that there is a vertical groove or feature under every joint to control and disguise the water that will discharge there.

At bottom edges water must be made to drip off without running back onto the soffit. Drip grooves must be provided in the undersides of every concrete element, and any other material which has a free bottom edge must similarly be detailed to ensure that water drips clear of any wall below. Drip grooves in concrete should be rounded or wedge shaped to facilitate casting, and should be at least 15 mm deep and 25 mm wide. The groove should be as close to the outer edge of the soffit as practical, bearing in mind that to have it too close may make the edge vulnerable to damage. As a general rule, any soffit in which there is room to form such a drip groove should have one and if there is not room the detail should be amended either to eliminate the soffit or to widen it!

Where soffits terminate against a wall or other vertical surface, the drip groove should be arranged to coincide with a channel or feature in the wall intended to collect and guide water. If this is not possible, it should stop about 50 mm short of the wall.

Any ribs, channels or other modelling on panel faces which are likely to concentrate water flow must extend to the bottom edge of the panel to permit discharge of the water. As water reaches the bottom edge of a vertical or inclined panel, surface tension effects cause it to slow down before dripping clear, and it tends to deposit any dirt it has been carrying. It is best, therefore, to attempt to limit the quantity of dirt carried by the water by reducing the distance it will have travelled before reaching the edge. We are back again to the dictum that water should be either collected or thrown clear of facades at frequent intervals.

## References

Ball, D.J. and Hume, R. (1977) The relative importance of vehicular and domestic emissions of dark smoke in Greater London in the mid-1970s. The significance of smoke shade measurements and an explanation of the relationship of smoke shade to gravimetric measurements of particulate. *Atmospheric Environment.* Vol. 11, London, Pergamon Press, pp. 1065–1078.

Beijer, O.F. (1976) *Water absorption in external wall surfaces of concrete.* Stockholm, Swedish Cement and Concrete Research Institute, 1976. 49 pp. CBI Report 6: 76.

Beijer, O.F. (1976) *Driving rain against external walls of concrete*, Stockholm, Swedish Cement and Concrete Research Institute, 92 pp. CIB Report 7: 76.

Beijer, O.F. (1980) *Weathering on external walls of concrete.* Stockholm, Swedish Cement and Concrete Research Institute, 70 pp. CBI Report 11:80.

Bielek, M. (1977) The main principles of water movement on the wall surfaces of buildings of various roughness. Paper based on research at the Norwegian Building Research Institute, Trondheim, presented at the RILEM/ASTM/CIB Symposium on *Evaluation of the Performance of External Vertical Surfaces of Buildings.* Otaniemi. 20 pp.

Brimblecome, P. (1976) *London air pollution 1200–1900*, Norwich University of East Anglia, 40 pp.

Carrie, C. and Morel, D. (1975) *Salissures de façades*, Paris, Editions Eyrolles, 140 pp.

Centre Scientifique et Technique de la Construction (CSTC). (1977) *Beton décoratif: Etude in-situ de la durabilité d'aspect des éléments de façade en beton décoratif.* Brussels Syndicat d'études interindustries – Construction, 43 pp. Final Report.

Couper, R.R. (1974) Factors affecting the production of surface run-off from wind-driven rain. Paper based on research in Australia presented at 2nd International CIB/RILEM Symposium on *Moisture Problems in Buildings*, Rotterdam, 10 pp.

Department of the Environment. *Winter mean concentrations of smoke and sulphur dioxide*. London, DoE, 1972–3 and 1975–6.

Department of the Environment. *Digest of environmental pollution statistics.* London, HMSO. No. 1: 1978; No. 2: 1979; No. 3: 1980 etc.

Herbert, M.R.M. (1974) *Some observations on the behaviour of weather protective features on external walls.* London, Department of the Environment. Building Research Establishment. 17 pp. BRE Current Paper CP 81/74.

Lacy, R.E. *The analysis of climatological data for the building industry*. Rotterdam, International Council for Building Research Studies and Documentation. 24 pp. CIB Report No. 39.

Lacy, R.E. (1976) *Driving-rain index*. London, Department of the Environment, Building Research Establishment, HMSO, 38 pp.

Lacy, R.E. (1977) *Climate and building in Britain*. London, HMSO. 185 pp.

Partridge, J.A. (1971) Architectural design and detailing. Paper to the Concrete Society Symposium: *The Weathering of Concrete*. London, Cement and Concrete Association. 12 pp.

Robinson, G. and Baker, M.C. (1975) *Wind-driven rain and buildings.* Ottawa, National Research Council of Canada, Division of Building Research, 34 pp. Technical Paper No. 445.

Ryd, H. (1983) Use of climatological data in building planning with respect to comfort. *Proceedings of Teaching the teachers on building climatology*. CIB Steering Group S4 Colloquium, Stockholm, September 1972 (Statens institut for byggnads-forskning) Stockholm, 306 pp.

Schwar, M.J.R. and Ball, D.J. (1983) *Thirty years on – a review of air pollution in London*. Greater London Council, 42 pp.

# Part 6

# Formwork

# 20

# Formwork and falsework

*P.F. Pallett*

## 20.1 Introduction

It is widely accepted in the building industry that formwork and falsework account for over 39 per cent of the cost of the concrete structure, and can be up to 55 per cent of the cost in civil engineering structures. Making the best use of the information and materials makes economic sense. This is a short review of formwork and falsework intended to give some practical guidance on the latest philosophies and economic use of the materials in use in the industry.

The two documents used in the industry on the subject are:

*Formwork – a guide to good practice* (2nd edn) (Concrete Society, 1995)
BS 5975: 1996 Code of Practice for Falsework

## 20.2 Definitions

**Formwork:** A structure, usually temporary, but in some cases wholly or partly permanent used to contain poured concrete to mould it to the required dimensions and support it until it is able to support itself. It consists primarily of the face contact material and the bearers that directly support the face contact material (See Figure 20.1).
**Falsework:** Any temporary structure used to support a permanent structure while it is not self-supporting.

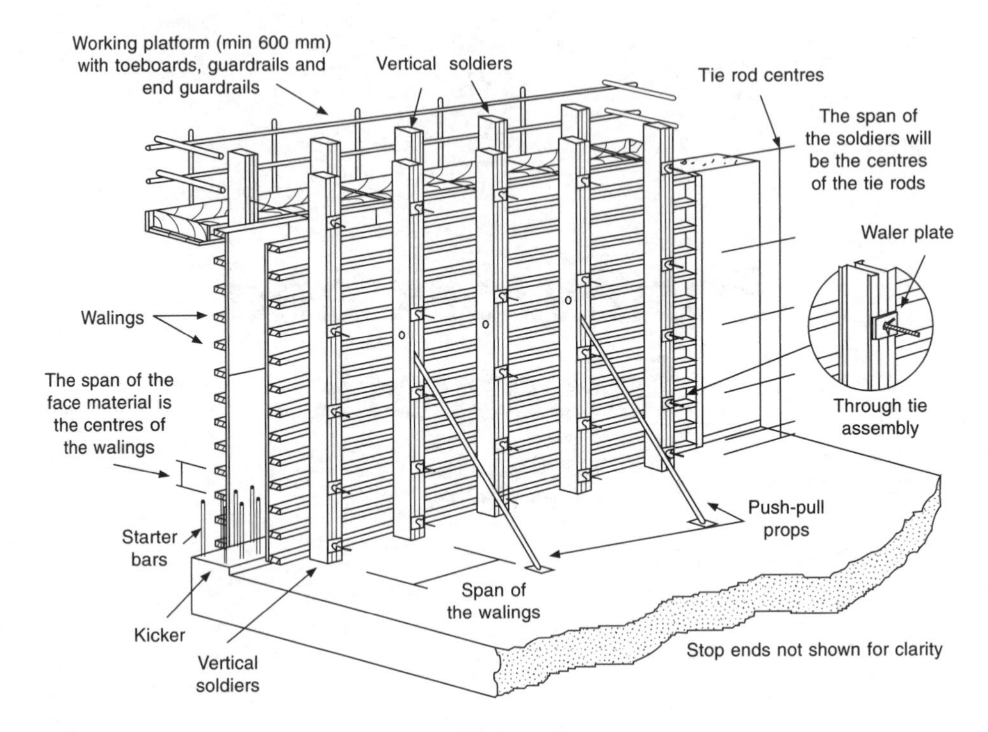


**Figure 20.1** A typical arrangement of double-faced wall formwork.

# 20.3 Formwork

## 20.3.1 Specifications and finishes

The formwork surface will be specified either by method or by performance. Most United Kingdom specifications are **performance specifications**, stating the standards and the tolerances required, but leaving the constructor to decide on the way to achieve it. By contrast, the **method specification** will tell the constructor exactly what to use – and in some ways it removes the onus from the constructor to think about the method adopted.

In building the two most common specifications are The National Structural Concrete Specification for building construction and the National Building Specification (NBS) Formwork for *in situ* concrete (Section E20). The more common performance specifications in civil works are the Water Services Association civil engineering specification for the water industry (Clause 4.28), the British Standard BS 8110 Structural use of concrete (Section 6.2.7) and the Highways Agency (DETR) Specification Volume 1 Clause 1708 which specifies five classes, namely:

Class F1 – basic, e.g. for pile caps etc.
Class F2 – plain, e.g. rear, unseen faces of retaining walls
(most UK panel systems will give Class F2)
Class F3 – high class but *no* ties, e.g. on visible parapets.
(the hardest to achieve!)

Class F4 – high class, e.g. quality surfaces to walls
Class F5 – high class, embedment of metal parts allowed
(intended for precast work)

The National Structural Concrete Specification uses Reference Panels for the finish, and seven sets of panels have been erected in locations throughout the UK (see BRE, 2000). There are also other special finishes often specific to one contract.

The final concrete surface together with the surface zone of the concrete will be affected by the choice of material used to create the face. It will also create a mirror image result in the structure – for example, the mark on the plywood from a hammer head caused when 'over-hitting' a nail will create a depression in the plywood, but leave a projection on the concrete. A useful guide to the understanding of blemishes in concrete surfaces is the BCA booklet *The control of blemishes in concrete* (Monks 1981). Other guides on finishes are Monks (1980) and Concrete Society (1999).

## 20.3.2 Tolerances

The three sources of deviations in the finished structure are:

**Inherent deviations**, such as the elastic movement under load (e.g. deflections)
**Induced deviations**, such as the lipping tolerance on two sheets of plywood (possibly both sheets within their manufacturing tolerance) and
**Errors**, such as inaccurate setting out.

The magnitude of the deviations will depend on whether the work is normal, high quality or special; and also on the distance from which the final surface is viewed. Typical deviations for normal work for verticality of a 3 m high wall form will be ± 20 mm from a grid line. See in particular Table 3 in the *Formwork Guide* (Concrete Society, 1991). The final position depends on the accuracy of the starting point, e.g. the kicker or base.

The deflection of the form is one of the inherent deviations. The rule in formwork is that the appearance and function is generally satisfied by limiting the deflection of formwork members to $^{1}/_{270\text{th}}$ of the span of the **individual** member. In certain cases, such as for direct decoration onto concrete walls in housing, then tighter deflections may need to be specified. Note that it is the deflection of the individual members and not the final concrete shape that is the limit.

## 20.3.3 Timber and wood-based products

The British Standard, BS 5268: Part 2: 2002 is for the permanent use of timber. We use timber in formwork as walings on wall formwork, and as bearers under the soffit in falsework. There are ten strength classes for softwood timber, so that it can be purchased related to its strength. In formwork and falsework the minimum class used is **C16** (used to be SC3). On-site several factors affect the strength of timber; it is exposed to the elements so the wet exposure service class three stresses are used (assumes moisture greater than 20 per cent): the duration of load, how the load is distribution, and the depth of the timber section used.

Timber is actually used in one of three ways on-site, and documents such as *Formwork – guide to good practice* (Concrete Society, 1995) and BS 5975: 1996 give safe working properties for:

1. **Wall formwork**, i.e. walings (where the load is applied for a short time and there are other timbers to share the load)
2. **General wall and soffit** members (where the duration of load is longer than on walls, and load sharing can be applied)
3. **Primary members** (where the duration of the loading is also longer than on walls, but the loads cannot be shared)

In falsework, there is an in-built stability requirement, with a limiting depth to breadth ratio of 2:1 when there is no lateral support, but ratio 3:1 when the ends are held in position (i.e. wedged).

### *Plywood*

Perhaps the most common sheet material in formwork is plywood – a layered material, made up in sheets (normally 1.22 × 2.44 m, i.e. imperial at 4′ × 8′), with properties related to the direction of the face grain of the outer ply. The stated shear stresses are low in BS 5268, and users should refer to Concrete Society (1995) which gives working properties for its use in either wall or soffit applications. Information is abstracted from BS 5268: Part 2: 2002 and, in Appendices G-W and G-S (Concrete Society, 1995), from trade associations and suppliers. For wall formwork applications G-W is reproduced in Table 20.1.

Users of wood-based sheets should be aware that their properties can vary depending on the orientation of the panel. Generally, panels are stronger when the face grain of the material is in the same direction as the span of the panel, referred to as 'face grain parallel'. The properties of 19 mm Douglas Fir Good One Side plywood used the strong way around are shown in Figure 20.2

## 20.3.4 Release agents

To obtain satisfactory finishes it is very important that the release agent on the formwork is correct. It should be stored, used and applied in accordance with the supplier's recommendations. Illustrated examples of good finishes using correct release agents are in Report TR52 (Concrete Society, 1999). Release agents are generally classified by numbers (see the *Formwork Guide*).

**Category 1: Neat oils and surfactants** – Not recommended for high-quality formwork
**Category 2: Neat oils** – Synthetic oils are now available which have good performance qualities, are totally non-harmful and non-flammable, with high rates of application. They have been developed for precast and automatic machinery use.
**Category 3: Mould cream emulsions** – Rarely used.
**Category 4: Water-soluble emulsions** – Quite successful in the precast industry, but can leave a dark porous skin on the concrete. Recent developments have improved the performance and usage rates, but rarely used on-site.

**Table 20.1** Wall formwork applications: Table G-W: working structural properties of sheet materials from trade associations and suppliers

| Permissible stresses wall formwork | | Plywoods (*Note 1*) | | | | | | | | | Oriented strand board (OSB) | | Expanded metal | |
|---|---|---|---|---|---|---|---|---|---|---|---|---|---|---|
| | | Canadian 17.5 mm | Canadian 19 mm | | APA 18 mm | | Finnish 18 mm | | | Indonesian 18 mm | Scottish | | English | |
| | SOURCE of information | Council of Forest Industries Canada (COFI) Engineering Information (Ref. 085-24/5M Rev. 1991) | | | American Plywood Association T.S.D., Tacoma USA | | Kymmene Schauman (UK) Ltd | | Koskisen Europe | PT Raja Garuda MAS | Norbord Highland Consultant | | Expamet Building Products | |
| | | | | | 6 Dec 1993 | 10 Jun 1994 | Tech Data Feb 1994 | Tech Data Dec 1994 | Data sheet Oct 1993 | Tech Data May 1992 | March 1994 | | Hy-Rib Designers' Guide Feb 1995 | |
| | Direction of face grain relative to span | COFI-FORM PLUS 17.5 mm 7 Play | COFI-FORM 19 mm 6 Ply or 7 Ply | Douglas Fir 19 mm 5, 6 or 7 Ply | B-C grade All Group I Exterior 5 Play 5 layer | 7 Ply 7 layer | Wisaform Special Birch through 13 Ply | Betofilm Mirror Constrn. Birch and spruce 9 Ply | Koskifilm Birch faced mixed cores 11 Ply | Garudaform Keruing face with hardwood veneers 7 Ply | Norbord Sterling Mainly Scots pine strands F2 Grade | F2 Grade | This product is only used in one direction, once Grade | Grade |
| | (*Note 2*) | overlaid | sanded | sanded | sanded | sanded | overlaid | overlaid | overlaid | overlaid | 18 mm | 22 mm | 2411 | 2611 |
| Bending Stiffness | Parallel | 3.93 | 3.72 | 3.26 | 4.29 | 3.50 | 3.28 | 2.34 | 3.23 | 3.77 | 1.44 | 2.63 | 3.94 | 2.78 |
| EI ($kNm^2/m$) | Perp'r | 2.08 | ns | ns | 1.35 | 2.13 | 2.42 | 2.44 | 1.87 | 4.00 | 0.52 | 0.96 | n/a | n/a |
| Moment of Resistance | Parallel | 0.645 | 0.574 | 0.489 | 0.508 | 0.400 | 0.782 | 0.563 | 0.823 | 0.904 | 0.340 | 0.457 | 0.431 | 0.339 |
| fZ (kNm/m) | Perp'r | 0.471 | ns | ns | 0.287 | 0.345 | 0.624 | 0.459 | 0.367 | 1.022 | 0.170 | 0.229 | – | – |
| Shear | Parallel | 8.70 | 7.18 | 6.79 | 10.19 | 8.39 | 16.55 | 10.53 | 7.89 | 5.24 | | | 9.97 | 7.68 |
| Load | | | | | | | | | | | 6.48 | 7.92 | | |
| qA (kN/m) | Perp'r | 7.09 | ns | ns | 5.58 | 8.18 | 16.55 | 16.55 | 6.92 | 5.91 | | | – | – |
| Minimum thickness (mm) | | 17.0 | 18.5 | 18.5 | 18.1 | 17.7 | 17.6 | 17.6 | 17.6 | 18.2 | 17.6 | 21.6 | 0.750 | 0.575 |
| Estimated weight ($kg/m^2$) | | 10.0 | 10.9 | 10.9 | 10.5 | 10.5 | 12.7 | 11.3 | 10.4 | 13.7 | 11.7 | 14.3 | 6.34 | 4.86 |
| Load duration factor used | | 1.5 | 1.5 | 1.5 | 1.5 | 1.5 | 1.5 | 1.5 | 1.5 | 1.5 | 1.5 | 1.5 | 1.0 | 1.0 |
| TRADE NAMES | | Pourform<br>Weldform+<br>Ultraform<br>Westform | | | | Intn. Paper<br>Gurdon-form<br>B-Matte<br>Stone F.I. | Wisaform Special | Betofilm | Koskifilm | Garudaform<br>Eagleform | Norbord Sterling | | Hy-Rid | |

*Notes:*

(1) The working properties for the wet (site) condition for plywood and OSB have been obtained from the source stated or from the supplier; where none stated marked 'ns'
(2) The face grain relative to its span indicates the disposition of the plywood face grain relative to its span.
(3) The duration of load factor used to determine the properties is shown for information.
(4) The values of bending stiffness are not modified by the duration of load factor.
(5) The shear load includes a modification factor of 1.1 for plywood.

*Source:* Concrete Society (1995).

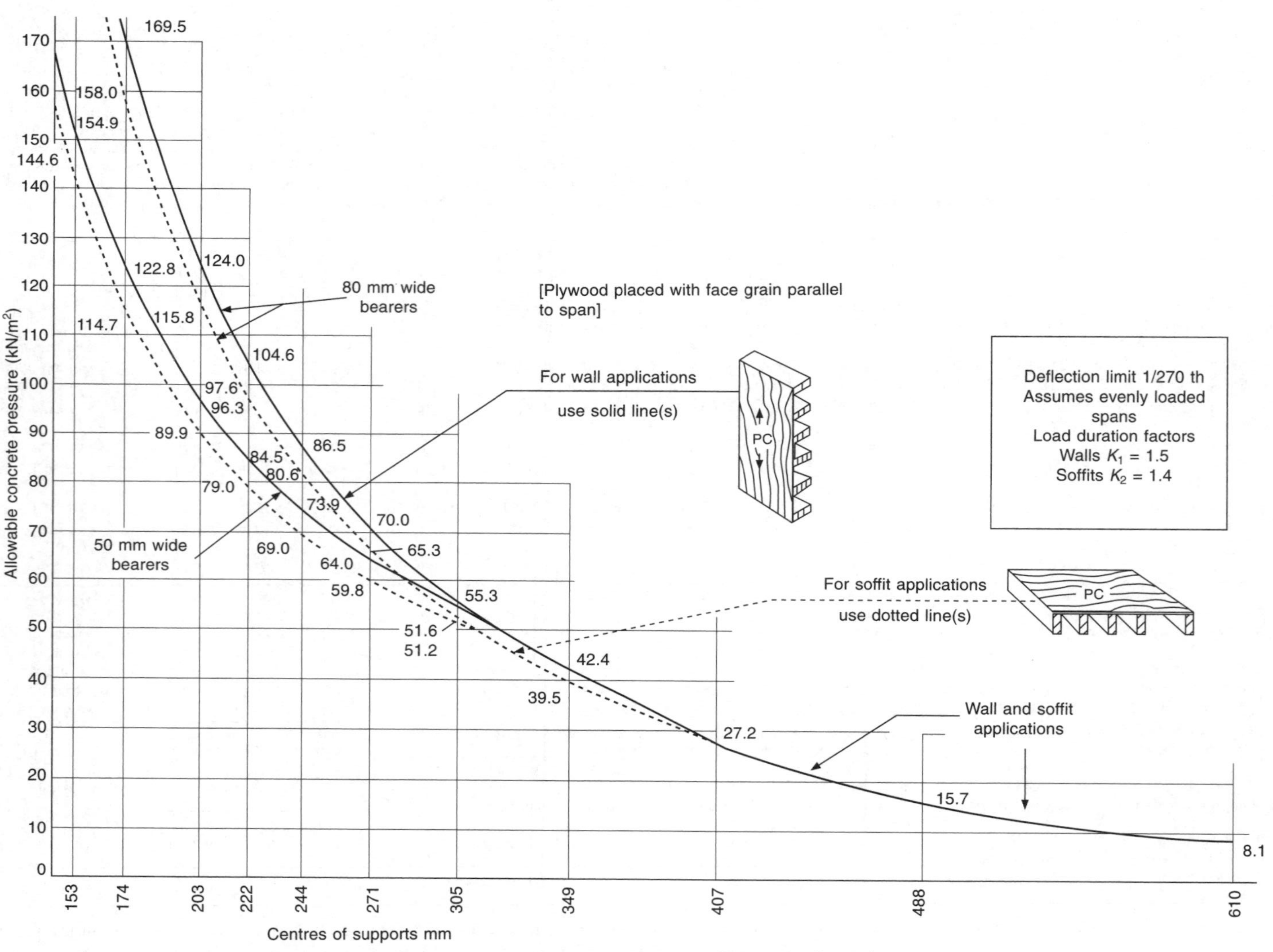


**Figure 20.2** Allowable concrete pressure for support centres for 19 mm Douglas Fir Good One Side Plywood.

**Category 5: Chemical release agents** – Remain the most popular. They act by having various blends of chemicals carried in light oil solvents. On spraying onto the formwork the light oil evaporates leaving a chemical to react with the cement to form the barrier. They are of a 'drying' nature, and suitable for plywood, timber, steel etc. and most forms. However, they do have the problem of over-application resulting in build-up on the forms, which leads to dusting of the resultant concrete surface. There are inherent health and safety implications as the chemical is dissolved in the oil.
**Category 6: Paints, lacquers and waxes** – Not strictly release agents, but the waxes are often very useful to provide the 'first' coat on very smooth impermeable surfaces, such as glass reinforced fibre moulds and brand new steel moulds. They can provide a surface for the normal release agent to 'stick'.
**Category 7: Other specialist release agents** – When using elastomeric (rubber) liners a specially formulated product is often required.
**Category 8: Vegetable Extract Release Agents (VERAs)** – The recent research from Europe into use of non-toxic biodegradable vegetable extract release agents, generally based on rape seed oil, provides a new and safe type of release agent, suitable for use in confined spaces, and for all formwork. Some are supplied in plastic cartons, and then diluted with water for use, but they do need more careful storage as they can freeze and separate out after a length of time. Use of VERAs on steel forms and moulds have shown improved finishes and rain resistance. Disposal after use is also improved as the products are non-toxic.

Site engineers should ensure that they have the product information and safety leaflets from the supplier, prior to use of the products. Subsequent treatments on the concrete surface can be affected, and the supplier's recommendations should always be sought.

## 20.3.5 Formwork ties

The most common type of tie rod is the **through tie** (Figure 20.3). It comprises a threaded bar, generally of 15 mm diameter, passed through an expendable plastic tube. Simple plastic cones are fitted at the ends of the tube. After use, the bar is removed and re-used. The load taken by the bar can be large and waler plates of sufficient size will be required to spread the load into the backing formwork members. A typical value of working stress in the bar is 700 $N/mm^2$. (Compare this to mild steel at only 180 $N/mm^2$.)

The *Formwork Guide* recommends that the **minimum factor of safety is 2** on these recoverable tie bars and assembly items. The safe tie load is based on the minimum guaranteed strength of the bars used. The plastic tube should be accurately cut to length to reduce grout loss on the surface of the concrete. The tube will leave a hole through the wall and will generally need to be plugged after striking the forms. When used on liquid retaining structures proprietary cementitious tie hole fillers are available to pressure seal the hole.

The **lost tie system** leaves the tie rod in the wall (Figure 20.4). A typical high tensile tie system is shown. The threaded ends of the tie rod are screwed into tapered ends of the 'she bolt'. This allows the entire assembly to be placed through the wall from one side during form erection, and permits the she bolts to be removed. The length of the lost tie rod is specified to suit the cover to the reinforcement. They are a quick and high-capacity

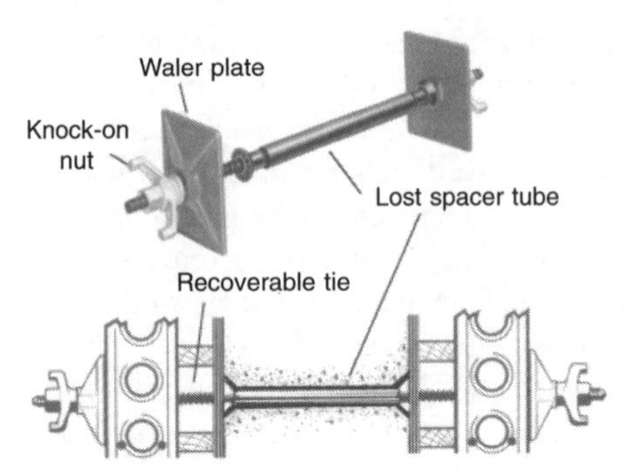


**Figure 20.3** Typical through-tie assembly.

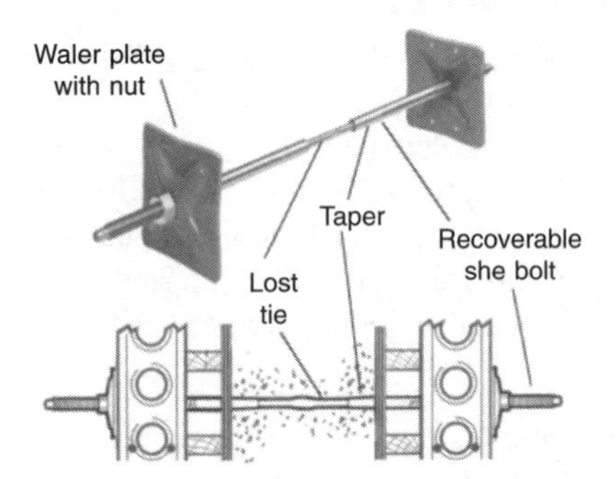


**Figure 20.4** Typical lost tie assembly.

tie system, but the tie rods have to be procured to the right length. As the tie is only loaded once, a typical factor of safety of 1.5 is used on these high tensile steel 'lost' ties.

A variation on the 'she bolt' system is the **taper tie** which is a long tapered bolt that is passed through the wall. It has different sized threaded nuts and waler plates to each end. After use, the entire taper tie is removed and re-used. The load capacity is governed by the diameter of the smaller size.

## 20.3.6 Proprietary formwork equipment

Many of the proprietary panel systems have plywood fitted within a steel frame. This gives a greater re-use of the plywood, and reduces damage to the sheets of plywood during handling. These can be up to panel sizes of 2700 × 2400 and often require a crane for handling. The significant advantage being the speed of initial use, with little 'make-up' time. It should be noted that such panel systems may not give high quality concrete surfaces direct from the form faces (see Figure 20.5).

There are a variety of **aluminium walings** available for hire or purchase from proprietary suppliers. These sections have high strength to weight ratios, allowing much stronger forms to be assembled for less weight than using steel. A benefit is also the increase in span, which when used as a waling means fewer soldiers. Designers might need to check the deflection of the walings at these larger centres as aluminium is five times more

**Figure 20.5** Formwork equipment.

flexible than steel. Special arrangements for access platforms for operatives might also be needed.

In use in UK are **proprietary timber beams**, generally from Europe, such as the Peri Vario beam. They are laminated and glued proprietary items, generally fitted vertically in formwork with stiff horizontal steel channels used as horizontal walings. The product standard is BS EN 13777: 2002.

**Proprietary soldiers** are the vertical stiff members associated with the wall formwork. They are available from different suppliers, and reference should be made to the supplier for the load characteristics of the soldiers. They should *not* be mixed on-site as they can have different properties and stiffness which would affect site safety.

**Permanent formwork** is increasingly being used on site. The CIRIA Report C558 (2001) gives detailed advice on applications with practical details; written specifically following problems in building and civil engineering, it gives detailed guidance on the use of many permanent formwork systems.

One of the common items being the **expanded metal** steel products, (e.g. Hy-Rib) which is shown in Figure 20.6 in use in a stop end. Hy-Rib has been shown by the British Cement Association to give significant reductions in concrete pressure. Trials on 5 m high walls using retarded concretes, poured at rates of rise in excess of 30 m/h, showed that the maximum likely concrete pressure was **only 38 kN/m$^2$**. This obviously reduces the number of tie rods/supports needed and can give significant economic advantages. It should be noted that such expanded metal products are left in place and are not struck off the stop end. Trials of the shear strength have shown the resulting joint stronger than a scabbled joint, with excellent compaction at the Hy-Rib.

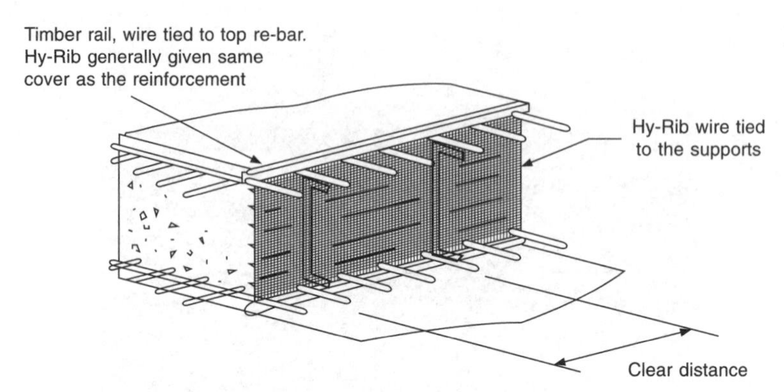


**Figure 20.6** Expanded metal sheet formwork.

## 20.3.7 The pressure of concrete – using a tabular approach

The pressure of the concrete affects the choice of face contact material, the waling centres, the soldier centres and the tie rod spacings. Correct calculation is very important in walls and columns where the pressures are high; typical values are 75 kN/m$^2$ on walls and 100 kN/m$^2$ on tall columns. Putting it in context; a pressure of 75 kN/m$^2$ represents *eight* cars parked on one square metre of formwork!

The maximum pressure (in kN/m$^2$) of a fluid at any position is the height in metres multiplied by the density in kN/m$^3$ such that:

$$P_{max} = h \times \text{density}$$

(usually density is taken as 25 kN/m$^3$).

The design pressure can be reduced because concrete is not a fluid and stiffens. CIRIA Report 108 (Harrison and Clear, 1985) includes a simple design pressure chart (see Table 20.2).

The factors that we need to know are:

- Is it a column or a wall/base? A dimension greater than 2 m is considered a wall.
- The tables are used for concrete temperatures of 5°C (winter), 10°C (spring and autumn) and 15°C (summer). Increases in design pressure will occur when the temperature drops.
- What is the height of the form (m) being used?
- The rate of rise (m/h) of the concrete uniformly vertically up the face of the forms measured in metres per hour. (*NOT the placing rate of the concrete in m$^3$/hour !*)
- The use of admixtures and normal superplasticizers will not generally alter the pressure of the concrete. But watch out for extended-life superplasticizers!
- The use of a retarder to delay the stiffening time of the concrete will increase the design pressure and separate classification of the concrete is needed.
- The use of additions to the mix, such as pulverized fuel ash (pfa) or ground granulated blast furnace slag (ggbs), as blended mixes *will* increase the concrete pressure. They cause the increase by retarding the stiffening of the concrete and this results in a greater fluid head of concrete. Different types of mix are categorized into 'groups'. The effects can be severe, so it is necessary to know the mix early on when designing or using the formwork.

The CIRIA R108 method gives the design pressure for both columns and wall forms. Part of the CIRIA table for walls only is reproduced in Table 20.2.

Similar information is presented as rate of rise tables in '*Formwork – a guide to good practice*. Thus knowing the designers' pressure for the formwork, the site can establish on a particular day the actual permitted rate of rise to keep the pressure to that as designed.

**Table 20.2** The maximum design pressure of concrete, $P_{max}$ (kN/m$^2$)

| | | | Walls and bases. A wall or base is a section where at least one of the plan dimensions is greater than 2 m | | | | | | |
|---|---|---|---|---|---|---|---|---|---|
| Concrete group | Conc. temp. (°C) | Form height (m) | Rate of rise (m/h) | | | | | | |
| | | | 0.5 | 1.0 | 1.5 | 2.0 | 3.0 | 5.0 | 10 |
| (1) PC 42.5, RHPC or SRPC without admixtures; (2) PC42.5, RHPC or SRPC with any admixture except a retarder | 5 | 2 | 40 | 45 | 50 | 50 | 50 | 50 | 50 |
| | | 3 | 50 | 55 | 60 | 65 | 70 | 75 | 75 |
| | | 4 | 60 | 65 | 65 | 70 | 75 | 85 | 100 |
| | | 6 | 70 | 75 | 80 | 80 | 90 | 100 | 115 |
| | | 10 | 85 | 90 | 95 | 100 | 105 | 115 | 135 |
| | 10 | 2 | 35 | 40 | 45 | 45 | 50 | 50 | 50 |
| | | 3 | 40 | 45 | 50 | 55 | 60 | 70 | 75 |
| | | 4 | 45 | 50 | 55 | 60 | 65 | 75 | 90 |
| | | 6 | 50 | 55 | 60 | 65 | 75 | 85 | 105 |
| | | 10 | 60 | 70 | 75 | 80 | 85 | 95 | 115 |
| | 15 | 2 | 30 | 35 | 40 | 45 | 50 | 50 | 50 |
| | | 3 | 35 | 40 | 45 | 50 | 55 | 65 | 75 |
| | | 4 | 35 | 45 | 50 | 50 | 60 | 70 | 90 |
| | | 6 | 40 | 50 | 55 | 60 | 65 | 75 | 95 |
| | | 10 | 50 | 55 | 60 | 65 | 75 | 85 | 105 |
| (3) PC42.5, RHPC or SRPC with a retarder; (4) LHPBFC, PBFC, PPFAC or a PC42.5 blend containing less than 70% ggbfs or 40% pfa without admixtures; (5) LHPBFC, PBFC, PPFAC or a PC42.5 blend containing less than 70% ggbfs or 40% pfa with any admixture except a retarder | 5 | 2 | 50 | 50 | 50 | 50 | 50 | 50 | 50 |
| | | 3 | 65 | 70 | 75 | 75 | 75 | 75 | 75 |
| | | 4 | 75 | 80 | 85 | 90 | 95 | 100 | 100 |
| | | 6 | 95 | 100 | 105 | 105 | 110 | 110 | 135 |
| | | 10 | 120 | 125 | 130 | 130 | 140 | 150 | 165 |
| | 10 | 2 | 40 | 45 | 50 | 50 | 50 | 50 | 50 |
| | | 3 | 50 | 55 | 60 | 65 | 70 | 75 | 75 |
| | | 4 | 60 | 60 | 65 | 70 | 75 | 85 | 100 |
| | | 6 | 70 | 75 | 80 | 80 | 90 | 100 | 115 |
| | | 10 | 85 | 90 | 95 | 100 | 105 | 115 | 135 |
| | 15 | 2 | 35 | 40 | 45 | 45 | 50 | 50 | 50 |
| | | 3 | 40 | 45 | 50 | 55 | 60 | 70 | 75 |
| | | 4 | 45 | 50 | 55 | 60 | 65 | 75 | 90 |
| | | 6 | 50 | 60 | 65 | 65 | 75 | 85 | 105 |
| | | 10 | 65 | 70 | 75 | 80 | 85 | 100 | 120 |

*Note:*
The extract of part of this table from CIRIA Report 108, reproduced by kind permission of CIRIA, does not include the CEM classifications of cement as CIRIA R108 was published in 1985.

## 20.3.8 Controlled permeability formwork

The use of a Controlled Permeability Formwork (CPF) fabric, such as the Type II CPF system by Du Pont (trade name Zemdrain MD2), is to control the permeability of the formwork and significantly enhance the durability of the concrete. See the CIRIA Report C511 (2000).

CPF works by allowing the pressure of the concrete to force the surplus pore water and any air into the fabric, and drain away. There is an almost complete elimination in blowholes on the face, an increase of 30 per cent in face strength, increased freeze/thaw resistance, reduction in sorptivity, and many other beneficial factors for durability. A benefit being that, unlike silane treatment, a concrete surface using CPF is instantly recognizable, thus ensuring that the face has been improved for durability.

Tests have confirmed that the use of Zemdrain makes good concrete ten times less permeable than similar mixes cast against conventional forms. It increases the surface strength by up to 30 per cent, hence giving stronger 'covercrete' in the critical surface zone of the concrete, faster striking times, and improved abrasion qualities against scour etc.

CPF also has the benefit of providing an almost blowhole-free surface, even on complex sloping faces of formwork, also saving finishing costs. This property is particularly advantageous on inwardly sloping surfaces where both blowholes and sand runs can be a problem with conventional formwork on the upper face. This is a solution to the concreting of complex seawall shapes. It also provides significant benefits against carbonation *and* chloride penetration, both items affecting the long-term durability of concrete.

Use of CPF gives a slightly matt and patterned surface to the concrete which is very distinguished, thus giving specifiers a visual confidence that the surface has been cast against CPF, unlike certain colourless spray applied improvement materials (e.g. silane).

CPF has been successfully used in the UK on wall, soffit and column formwork, and with a range of backing materials such as plywood, panel systems and steel forms, and has also been used on complex double curvature hoppers. Its material cost is generally offset by the use of inexpensive backing materials, no release agents, elimination of finishing costs, and in many cases, permits a lower grade of concrete to be used, yet achieving the specified face strength for durability.

## 20.3.9 Design of wall formwork

### *Double faced formwork*

A typical example of double faced wall formwork is shown in Figure 20.7. The principle of wall formwork design is straightforward: You follow the force exerted by the concrete pressure from the face through to suitable restraints, knowing that the forces will balance.

- Establish the design concrete pressure for the formwork. The pressure acting on a face always *acts* at right angles.
- Calculate the forces, the force being the pressure multiplied by the area acted upon.
- Check the safe span of the face contact material or plywood (**ONE**). This is usually from tables plotting pressure against span. This obviously then gives the maximum spacing of the next level of supports – usually horizontal walings.

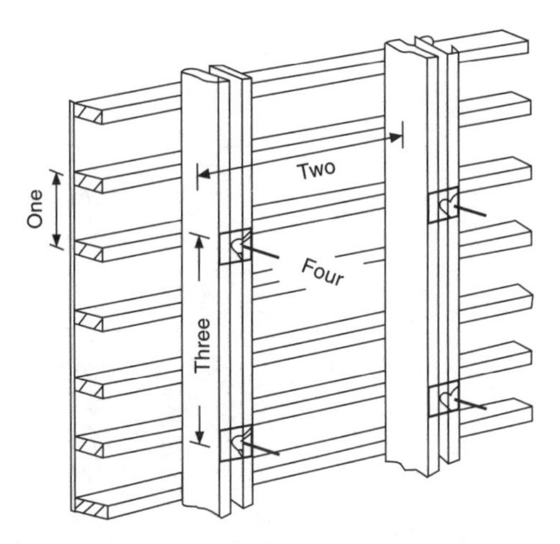


**Figure 20.7** Wall formwork.

- Check the safe span of the walings (**TWO**). This is determined by the location of the next level of support, usually vertical soldiers. The walings may be timber or aluminium. You can check the timber waling size required from standard tables, or by using continuous beam analysis programs – or by using, say, Appendix B of the guide (Concrete Society, 1995).
- Check the tie rod positions. These ties (see above) connect to the opposite face of the wall, and provide the restraint in tension to avoid the walls 'bursting' apart under the pressure. As the pressure reduces near the top of the wall, fewer ties will be required.
- The soldiers have a central slot to allow tie rods to be passed through. The tie rod locations, already determined, dictate the span of the soldier (**THREE**). Several computer programs will assist checking the design of the soldiers.
- Check the capacity of the tie rods to take the loads from the soldiers (**FOUR**).

Safety will be paramount – placing tie rods, platforms for placing concrete, and the CHSW Regulations (1996) need careful consideration. All working platforms will be a **minimum 600 mm wide** complete with toe boards, intermediate guardrails etc.

The centres of the soldiers and the ties might be predetermined by the permanent works designer, for example a particular surface finish with striations will dictate the only acceptable locations of ties, i.e. in the recesses, hence the soldier spacings are established! With such high forces in the formwork system, the design should take careful account of deflections and the final tolerances. Remember that the deflection limit of 1/270th on members is applied to the plywood, the walings and then the soldiers individually – it is not cumulative. In certain situations, kickerless construction can be adopted. See Bennett (1988) and also the Qi video on Formwork.

A check will need to be made on the stability of the formwork, and any propping designed for a minimum factor of safety on overturning of **1.2** when considering the most adverse conditions (Figure 20.8). The three checks used are:

1. Maximum wind plus overturning from a nominal load on any working platforms.
2. Working wind on day of concreting, plus full load on the platforms. The working wind represents the upper operating limit for the plant on site.
3. A minimum stability force representing 10 per cent of the self-weight of both faces of formwork, considered to act at 3/4 height up the form.

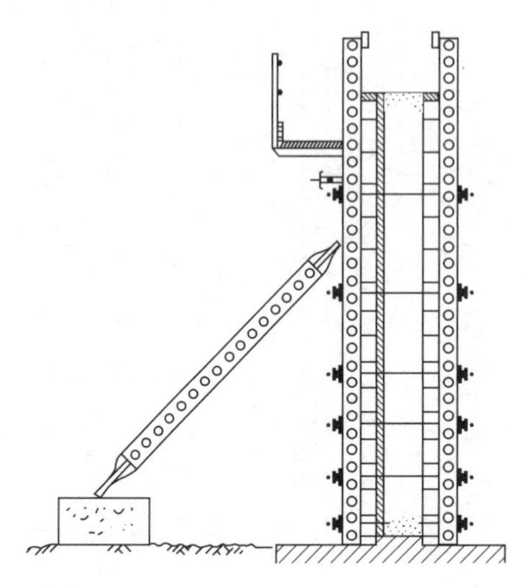

**Figure 20.8** Propping of wall formwork.

## 20.3.10 Soffit formwork

The design of soffit formwork is slightly different from that for walls, the loads are usually much lighter and the deflections more critical – as the spans can be larger. A typical example is shown in Figure 20.9. The principles of design are the same, the loads are followed through from the face to the supporting falsework.

There are many different decking systems to soffits, from traditional plywood with timber bearers, to aluminium bearers, to those with voided slabs, such as trough or waffle floors. Guidance on trough and waffle floors is given in the Concrete Society Report 42 (1992). Some proprietary systems actually have quick strip arrangements allowing the expensive formwork face components to be struck early, normally at a minimum of 5 N/mm$^2$ to avoid damage, yet leaving the main slab supported until approval to strike is received.

The time at which soffit forms can be removed, and the exact procedure for striking the forms, requires detailed consideration. If not, there is a real risk of damaging the permanent works and possibly initiating an accident! See section 20.5.2 below.

The applied loads to the soffit formwork will include the weight of the structure being supported, plus an allowance of 1.5 kN/m$^2$ as a construction operations load to cater for operatives placing the concrete, plus, of course, the self-weight of the forms and any edge forms.

All slabs will have edge forms, stop ends and/or construction joints. The design of these is often left to 'site' to complete. See also the use of Hy-Rib stop ends in section 20.3.6. Particular care is necessary if stop ends are greater than 400 mm depth (See Figure 20.9).

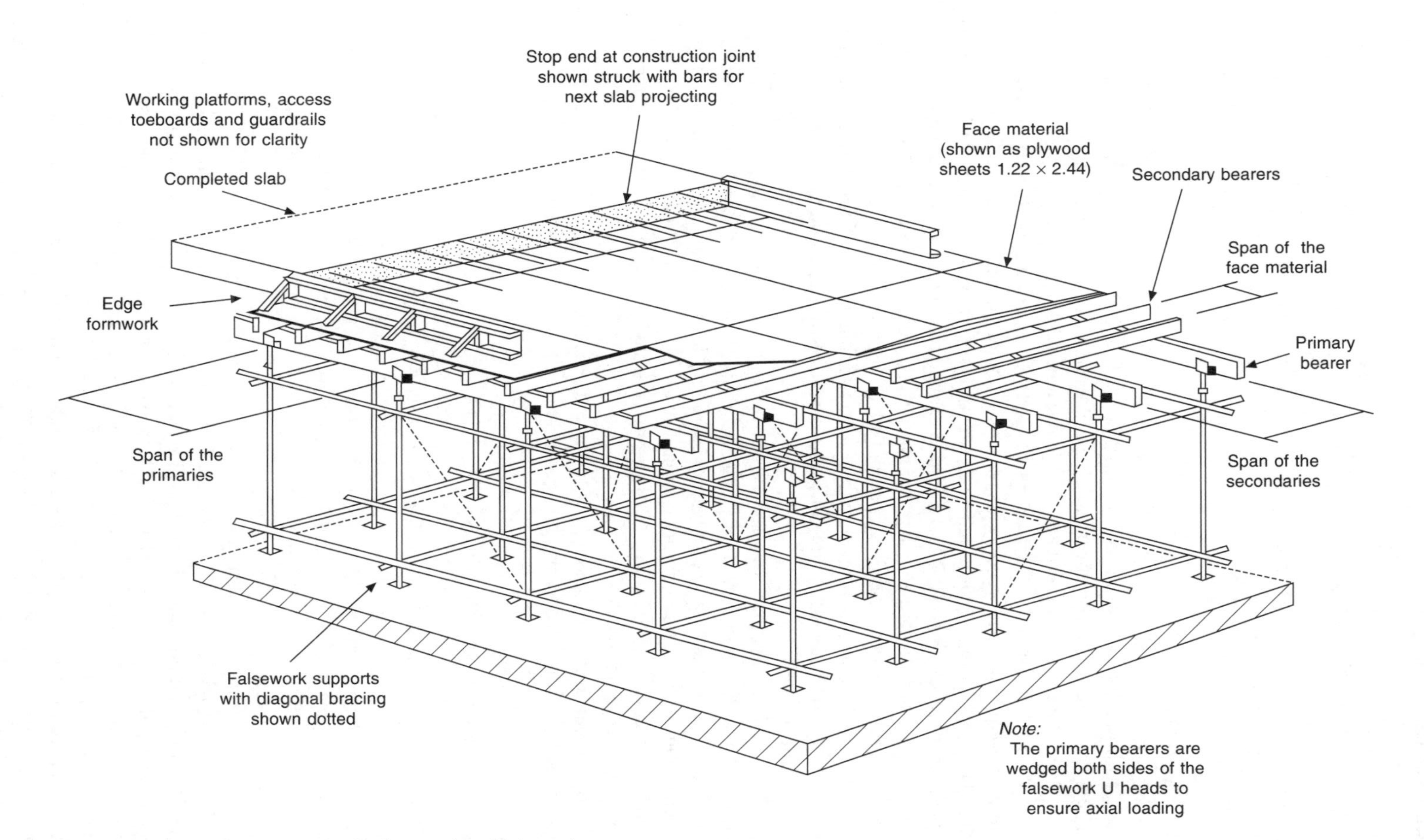


**Figure 20.9** A typical arrangement of soffit formwork and falsework.

## 20.4 Falsework

**Definition: Falsework**: Any temporary structure used to support a permanent structure while it is not self-supporting.

### 20.4.1 General

The responsibility for falsework safety and use is often left entirely to the contracting organization, yet its effect on the final shape, location and any residual stress in the permanent structure can be significant. The Government Advisory Report on Falsework, as long ago as 1976 (Bragg, 1975), recommended tighter control on the procedures for falsework, subsequently reinforced by the recommendations in the first BS code of practice in 1982. The quality of erection and understanding of stability of falsework has not improved; recent research (HSE 2001b) commissioned took measurements on many sites and showed little, if any, improvement in accuracy of erection, even using modern falsework systems. Of greater concern was the apparent lack of appreciation of the need for stability and creation of node points in falsework structures. The final report CRR394 recommended simple rules of thumb to assist site staff and designers to appreciate the necessity for lateral restraint for stability.

The Construction (Health, Safety and Welfare) Regulations (1996) require all working platforms to be a minimum of 600 mm wide, be regularly inspected, and have guardrails fitted so that there is no unprotected gap exceeding 470 mm, and that all falsework be stable and erected under competent supervision. The Construction (Design and Management) Regulations, updated in 2002 (HSE 2001a), imposes duties at Reg. 13 on *all* designers, including falsework designers.

A European performance standard for falsework prEN12812 (BSI, 1997) is at present being prepared in limit state terms. It is based on the German standard DIN 4421.

### 20.4.2 Contents of BS 5975

The UK code of practice on falsework is BS 5975:1996. It recognizes that falsework has a different design philosophy to that used in the permanent works; it's loaded for a short time; rarely is it an item in the 'Bill of Quantities'; it generally comprises re-usable components; has the unique structural requirement to be de-stressed under load (to get it out!); it is not tied down and therefore relies on its own weight for stability; and is usually stressed to 90 per cent of its safe capacity.

BS 5975 is a permissible stress code with sufficient information, such as steel stresses from BS 449, contained in the appendices. It has also got a very comprehensive wind loading section based on CP3 Chapter 5. The technical part of the code accepts either a fully designed solution or a standard solution using tables. It includes two tables of the standard solution for simple slabs and beams with the limits necessary to control their use.

The code recommends that *all* sites should appoint a ***falsework coordinator*** to control the work. The principal activities listed in the code (Clause 2.5.2.2) are not an onerous task – they simply set down in writing the actions that a responsible engineer should be carrying out any rate! To assist this list of items, **a design brief** is suggested to ensure that the temporary works designer has available *all* the relevant information prior to starting the design.

Obviously checking of the falsework during erection and prior to use is necessary, but there are two very important principal activities of the falsework coordinator suggested:

(m) after a final check, issue formal permission to load if this check proves satisfactory.
(n) when it has been confirmed that the permanent structure has attained adequate strength, issue formal permission to dismantle the falsework.

## 20.4.3 Materials used in falsework

- *Steel scaffold tube* – In the UK the specification for 48.3 mm OD scaffold tube, BS EN 39: 2001 specifies galvanized tube and gives two wall thicknesses as 3.2 mm known as Type 3; and 4 mm known as Type 4. The 4 mm tube is the preferred tube in the UK. There remains in use quantities of a slightly weaker (about 5 per cent) tube manufactured to an earlier BS 1139 specification from 1982.

Obviously, the Type 3 tube has a lower load capacity than the Type 4 tube. The two types of tube should *not* be mixed.

- *Couplers for steel tube* – The specification for couplers is BS 1139: (1991). The safe slip load for right angle couplers is 6.3 kN but reduces to 5.3 kN when using swivel couplers. Note also that sleeve couplers only have a capacity of 3 kN in tension, and internal joint pins have *no* tension capacity. (The European code for fittings is EN 74.)
- *Adjustable steel props* – BS EN 1065 published in November 1998 classifies 32 different types of prop in five classes (A to E) by length and states characteristic[1] values, but for only one load case, with 10 mm eccentricity. The traditional props made in the UK, to the earlier BS 4074, have five types of prop (known as Nos 0, 1, 2, 3, & 4) their properties and two tables of safe working load stated in BS 5975: 1996 for:
  1 Axially loaded and maximum up to $1^1/_2°$ out of plumb.
  2 Allowing the load 25 mm eccentric and maximum $1^1/_2°$ out of plumb.
- *Proprietary systems* – There are many different types of systems on the market; many for hire as well as for purchase. Always refer to the suppliers' recommendations for load capacity etc. The load capacity is not always obvious because although the vertical standards will often be 48.3 mm OD, to suit the scaffold couplers, they may have different wall thicknesses and/or be of different grades of steel. They are often made of lighter materials to reduce transportation weight and to ease erection and dismantling.

## 20.4.4 Loadings

- *Self-weight* The formwork soffit is usually taken as 0.5 kN/m$^2$, maximum is 0.85 kN/m$^2$.

---

[1]The characteristic strength is normally set at the 95 per cent confidence level, i.e. 95 per cent tested would fail above the characteristic strength, and therefore 5 per cent below! Note characteristic strength is definitely *not* working strength.

The falsework weight itself is very critical to the design. These structures are rarely tied to the ground and their self-weight is often the only load preventing them from overturning under the maximum wind loads.

- *Imposed loads* Obviously the weight of the structure, i.e. *the permanent work*, is the imposed vertical load. It can be just vertical (static and impact) or it can have some non-vertical components, such as from the pressure of concrete on other works.
  *Note:* Permanent works designers should make available to the constructor the design imposed loading for the permanent work – including finishes etc. This is necessary so that the striking times and striking procedures can be established safely (see section 20.5.1). This is vital on multistorey work.
- *Construction operations loads* (c.o.l.) To allow for the operatives placing the concrete/steelwork an allowance is made as a c.o.l. of 1.5 kN/m$^2$ – Service Class 2. (Note that this represents only 55 mm of excess concrete!) A reduced c.o.l. imposed load is allowed for inspection and access, such as underneath the soffit, at 0.75 kN/m$^2$ – Service Class 1 loading.
- *Environmental loads* The wind loading method stated in BS 5975 is based on a 3 s gust at a height of 10 m likely to occur once in 50 years! Generally on most falsework structures, placing concrete will not occur when the wind speed exceeds the operating limits set for the plant. This is usually at a Beaufort Scale 6 and corresponds to a wind speed of 18 m/s. This is known as the **working wind.**
- *Specified loads* Investigations during the 1970s into the cause of falsework collapses identified that one of the principal causes was the absence of stability in members and the structure as a whole. For this reason the code (Clause 6.4.4.1) recommends that *all* falsework be designed for a horizontal force applied at the top of the falsework, as the greater of either a specified $2^1/_2$ per cent of the vertically applied loads or all known horizontal loads plus 1 per cent of the vertical load as an erection tolerance load (Cl 6.3.1.3.2) to cater for workmanship.

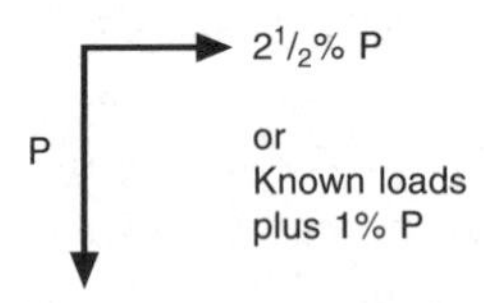


*Note:* This specified load could, in some cases, be restrained by connecting direct to the permanent works at the top of the falsework, such as at column tops – but only if the permanent work is strong enough to withstand the loads!

## 20.4.5 Stability of falsework

The Bragg Report (1975) identified that failures occurred not because there was insufficient bracing but because there was *no* bracing! This prompted considerable interest in the industry at the time and was finally incorporated into BS 5975 at Clause 6.4.1, which recommends that for *every* falsework structure, three design checks are required, namely:

1 Structural strength of members and connections
2 Lateral stability
3 Overall stability

Put very simply,

THINK VERTICAL
THINK HORIZONTAL and then
THINK HORIZONTAL AGAIN

## *Check 1 – Structural strength of members and connections*

Obviously the designer will check for bending, shear, deflection, (occasionally torsion!) and for stability of individual members and connections, such as welded or bolted joints.

The effective lengths of members in compression will need particular attention, especially long struts where the slenderness ratio is often the limiting criteria. Generally for all falsework with scaffold tube and fittings, the node points/couplers are considered as pin joints provided that the couplers are fitted within 150 mm of the node points (Clause 6.7.2 and Figure 20.10).

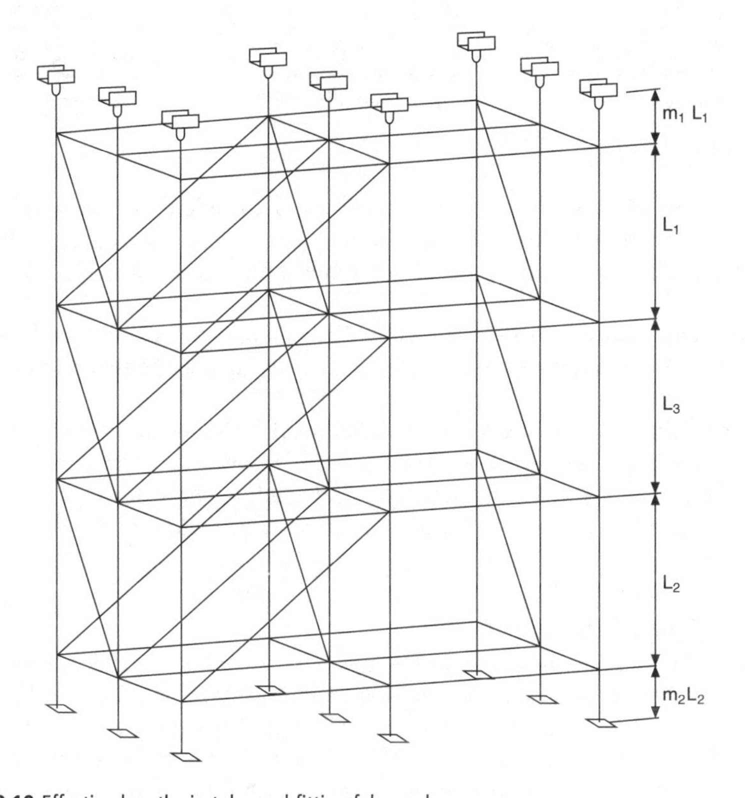


**Figure 20.10** Effective lengths in tube and fitting falsework.

Many of the proprietary falsework systems have patented joints which provide some moment restraint, and these can reduce the effective length factor below unity, thus increasing their load. Follow the supplier's recommendations. Particularly look out for unrestrained cantilever extensions on falsework at the head and bases. Often this will be the jack extension(s) and they have a crucial effect on the effective length to be considered of the adjacent lift of falsework.

The effective length ($l$) is considered to be that of the adjacent length of standard *plus* twice the length of the unrestrained cantilever (see Figure 20.10).

$$l = L + 2mL$$

### *Check 2 – Lateral stability*

The lateral stability needs to be considered at different stages, namely during erection when exposed to the full wind, see (1) in Figure 20.11, when concreting but unlikely to have the full wind but the working wind, see (2), or be considered with the lower bound as the specified load, see (3).

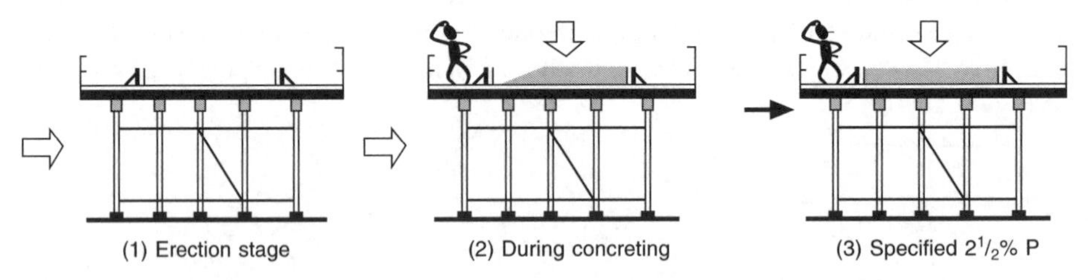


**Figure 20.11** Lateral stability.

The minimum specified load, explained above, is considered to act at the point of contact between the vertical load and the falsework. In building, the majority of falsework schemes completed by suppliers will assume the third case, and design for $2^1/_2$ per cent P.

The above cases illustrate the falsework as freestanding with stability by bracing; but it is also possible to provide restraint to falsework by tying it at the top to the permanent works.

Lateral stability also applies to the webs of steel beams at reaction points and at concentrated loading positions. BS 5975 Annex K.1 states that steelwork at *all* load transfer points requires **WEB stiffeners** unless calculations show that such stiffeners are not required! In other words you always put in web stiffeners *unless you can prove otherwise*.

### *Check 3 – Overall stability*

Tall slender structures and those exposed to high winds will have a tendency to be blown over. This third check requires that the **minimum factor of safety is 1.2** against overturning (BS 5975 Clause 6.4.5.1). As falsework is generally a gravity structure, relying on its own weight for stability, the most onerous condition is often just after the soffit formwork is fixed, and just before the reinforcement is fixed, i.e. with little restoring forces; once the concrete is placed there will be ample vertical load to resist overturning. If the falsework is unstable, then use holding-down bolts or kentledge.

## 20.4.6 Continuity of members

The structure will have many redundancies, and the distribution of load into the verticals from the soffit formwork bearers will also be very random. The face contact material, the secondary bearers, and often the primary members in the forkheads, will be continuous over several supports giving rise to increased reactions at internal supports from their elastic reactions.

The support reactions of beams change when they are continuous over the supports, such as a long scaffold tube or timber beam. The following examples assume that the distributed load on each span is 10 kN, and that the spans between the supports are equal:

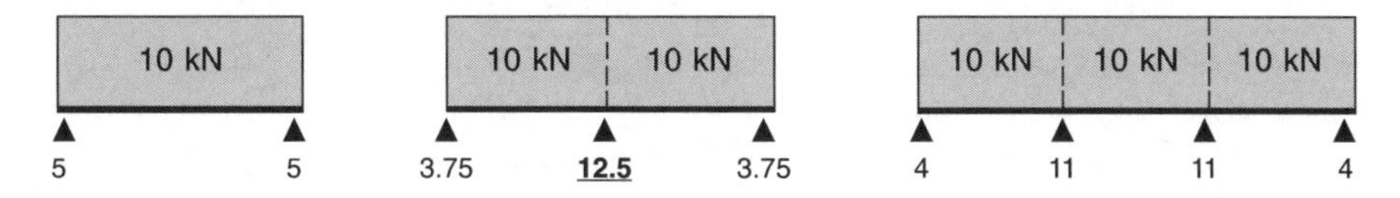


A worst case is a single beam continuous over two spans with three supports giving a central reaction of the static load × 1.25 for continuity, i.e. a staggering 25 per cent increase in load!

BS 5975 at Clause 6.4.3.1 accepts that in the case of falsework comprising random bearers, you design by working out the simple load on the area supported by the standard and **add 10 per cent** for continuity (the 10 per cent rule) but certain cases, e.g. over two spans, a more precise calculation may be justified. Certain proprietary systems incorporate simply supported beams and this 'rule of thumb' is also not applicable.

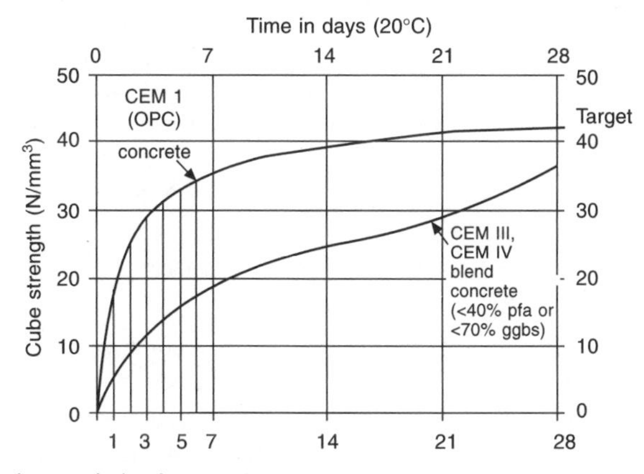


**Figure 20.12** Typical strength development in C30/37 concrete.

## 20.5 Striking formwork and falsework

The time at which striking is permitted is usually related to the strength of the concrete. The determination and assessment of the compressive strength of concrete is important, many factors are involved, such as the water/cement ratio, the blends in the cement (e.g. pfa and ggbs). The rate of strength development for concrete containing either pfa or ggbs will be different to that with CEM 1 concrete. A typical comparison of the strength development for a C30/37concrete is shown in Figure 20.12).

Efficient use of the forms will significantly alter the economics of the contract. It is for this reason that the main change in formwork in recent years has been the development of labour-saving equipment, such as the increase in use of aluminium table form soffit support systems. Savings in equipment and manpower costs are as a direct result of faster construction time, hence the importance of early age striking and correct selection of equipment.

The following striking recommendations are taken from the CIRIA Report R136 (1995).

## 20.5.1 Criteria for striking wall and column formwork

The main criterion for the removal of formwork from walls and columns is a requirement to have achieved sufficient strength to avoid mechanical damage during striking. If the form face has a feature then the draw and size of feature, as well as the direction of strike, can affect the striking time. The following striking times criteria related to the DTp classification of surface finish are now the industry norm. They are also stated in detail in Concrete Society (1995) at section 5.2.5. The criteria are given in Table 20.3.

**Table 20.3** Striking vertical forms – walls, beam sides, columns etc.

| Finish description | Striking criteria |
|---|---|
| Basic, plain or rough finish<br>DTp Classes F1 and F2 | Either: A minimum period equivalent to:<br>8 hours at 20°C for unsealed plywood<br>6 hours at 20°C for impervious forms.<br>or Strike next morning if the mean air temperature is above 10°C overnight. |
| Fair, fine smooth quality finish<br>DTp Classes F3, F4 and F5 | Minimum *in-situ* cube strength of 2 N/mm$^2$ |

Reference should also be made to the contract specification for any limitations on time of striking of vertical wall and column formwork. Tables in specifications may have different parameters; times presented in hours or in days, temperature related to surface temperature of the concrete, while others to the mean of the minimum and maximum daily air temperature.

When striking formwork to slender walls care is necessary to ensure that there is sufficient concrete strength at time of striking to withstand any wind loads. In sloping sections of walls, columns etc., the concrete has to support itself on striking and a minimum structural strength will be needed to be achieved *before* striking can be approved.

## 20.5.2 Criteria for striking soffit formwork – beams and slabs

The time and procedure by which beam and soffit formwork can be struck should be carefully controlled and should be carried out in accordance with the requirements of the contract specification and drawings. The following striking criteria for soffits are stated in more detail in Concrete Society (1995) at section 5.3.6. The criteria are given in Table 20.4.

**Table 20.4** Striking soffit forms – slabs and beams etc.

| Finish description | Striking criteria |
|---|---|
| Basic, plain rough finish or quality<br><br>DTp all classes | Either: use specifications, codes of practice or tables (e.g. CIRIA R136) or assess the concrete strength at the time of striking knowing the maturity, the concrete mix etc. |

The main consideration in striking soffits is the gain in strength of the concrete to ensure that the member, when released, can support its own weight together with any imposed construction operations loads. Additional considerations are the elimination of frost and mechanical damage, reduction in thermal shock, and limiting excessive deflections. The BCA Best Practice Guide No. 1 (2000) recommends the LOK test based on four results as the means of reliably understanding the strength.

A recommendation of the Formwork Guide is that '**the minimum concrete strength at time of striking be specified**', after taking into account the stage of construction. Calculation to justify a lower value of time and/or strength at striking should be approved by the permanent works designer. The temporary works designer will need to assess the proportion of loading on the structure at time of striking to the permanent works designer's design working load.

The load on the structure at the time of striking will include the formwork self-weight, the mass of the concrete, plus a minimum construction operations load for access of 0.75 kN/m$^2$. Where there are several other levels of construction additional loads may also be required to be supported. In most cases, striking can be considered when the *in-situ* concrete has obtained a characteristic strength at least proportional to the ratio of the loads multiplied by the grade of concrete.

Subsequent to the *Formwork Guide* method outlined above, a faster method of striking flat slabs in buildings, with flat slabs up to 350 mm thickness, was to consider the cracking of the slab at early age to the cracking expected by the designer at 28 days in the design. Establishing the cracking factor and the loading factors using this method gives a lower required equivalent concrete cube strength in order to strike the slab. This method was first detailed in BRE report BR 394 in May 2000 and has been introduced in several BCA Best Practice Guides. The authoritative practical guidance to this method is detailed in the CONSTRUCT *Guide to flat slab formwork and falsework* (2003).

## 20.5.3 Backpropping slabs in multistorey construction

In multistorey construction the ratio of design service load on a slab to the applied construction loads needs careful assessment. When the supporting slab immediately below the level under construction has not gained ful maturity, construction loads will require to be supported through several levels. This is known as backpropping. Ideally the supporting slab should have been struck and allowed to take up its instantaneous deflected shape. Refer to the *Guide to Flat Slab Formwork and Falsework* for a fuller explanation.

Note that repropping can be used at lower levels. Figure 20.13 indicates *two* levels of backpropping.

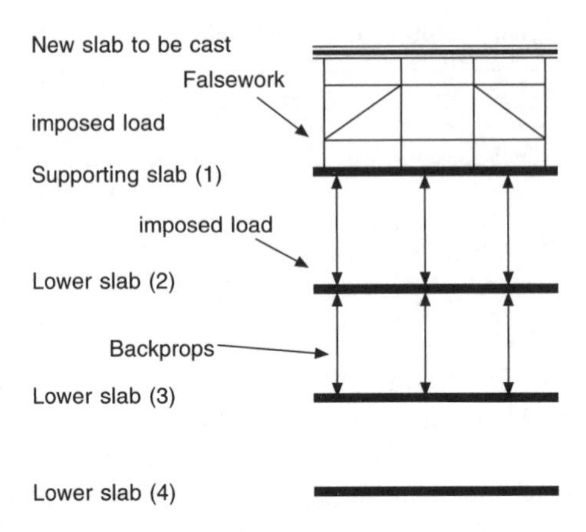


**Figure 20.13** Typical backpropping: two levels.

Research has shown that due to the elasticity of the aluminium shores used as backpropping and considering the stiffness of the floor, possibly only 33 per cent of the load from the new slab being cast is actually transmitted through the supporting floor to the backpropping, compared to over 66 per cent conventionally assumed. This means that the supporting slab will nearly always require to be checked for structural capacity. The Excel spreadsheet (Concrete Structures Group, 2003) is a useful tool to highlight the need for checks on the supporting slab, generally requiring the full 28-day strength to be achieved before the new slab can be cast. This will affect programming. Backpropping 'to the ground' is shown *not* to relieve the load on the supporting slab.

## References

Bennett, D.F.H. (1988) Kickerless construction, *BCA Guide*, BCA 47.023, London.

Bragg, S.L. (1975) *Final report of the Advisory Committee on Falsework*. Her Majesty's Stationery Office, London, June.

British Cement Association (1993) *Concrete on Site Series – No. 3 Formwork*, Ref 45.203. Autumn, Crowthorne.

British Cement Association (2000) *Early age strength assessment of concrete on site*, Best Practice Guide No. 1 for *in-situ* concrete frame buildings, BCA Ref. 97.503, April.

British Standards Institution, BS 1139: Part 2 Section 2.1 : 1991, Specification for couplers for steel tube, London. Corrigendum No.1 October 1998.

British Standards Institution, BS 5975: 1996: Code of Practice for Falsework, London.

British Standards Institution (1997), Draft prEN 12812 Falsework – Performance requirements and general design, Draft for Public Comment 97/102975DC, London, April. (When implemented will partially supersede BS 5975:1996.)

British Standards Institution, BS EN 39: 2001, Loose steel tubes for tube and coupler scaffolds – Technical delivery conditions, London, August.

British Standards Institution, BS 5268: Part 2: 2002: Structural use of timber, Code of Practice for permissible stress design, materials and workmanship, London.

British Standards Institution, BS EN 13377: 2002, Prefabricated timber formwork beams – requirements, classification and assessment, London.

Building Research Establishment, (2000) *National Structural Concrete Specification for Building Structures*, CONSTRUCT, Crowthorne, July.

Concrete Society (1992) *Trough and Waffle Floors*, Technical Report 42, Slough, June.
Concrete Society (1995), *Formwork – a guide to good practice*, 2nd edn, Special Publication CS030, The Concrete Society, Slough, June.
Concrete Society (1999) *Plain Formed Concrete Finishes – Illustrated Examples*, Concrete Society Technical Report No. 52, Slough, March.
Construction Industry Research and Information Association (1995) *Formwork striking times – criteria, prediction and methods of assessment*, CIRIA Report R136, London, August.
Construction Industry Research and Information Association (2000) *Controlled permeability formwork*, CIRIA Report C511, London, March.
Construction Industry Research and Information Association (2001) *Permanent formwork in construction*, CIRIA Report C558, London, May.
Concrete Structures Group (2003) *Guide to Flat Slab Formwork and Falsework*, Construct report, Crowthorne, Report CS140, June.
Harrison, T.A. and Clear, C. (1985) *Concrete pressure on formwork*, Construction Industry Research and Information Association, Report 108, London, September.
Health and Safety Executive (2001a) *Managing Health and Safety in Construction*, Approved code of practice, Publication HSG 224, HSE Books, Sudbury, December. (Includes all the Construction (Design and Management) Regulations.)
Health and Safety Executive (2001b), Investigation into aspects of falsework, HSE Contract Research Report 394/2001, December 2001. Available on web www.hse.gov.uk/research/crr_htm/2001/crr01394.htm (2.5 Mb).
Monks, W. (1980) Appearance Matters Series No. 1. *Visual concrete: Design and production*, British Cement Association Publication No. 47.101, Slough.
Monks, W. (1981) Appearance Matters Series No. 3. *The control of blemishes in concrete*, British Cement Association, Publication No. 47.103, Slough.
Qi Training, *Good Formwork Practice*, VHS Video, Qi Training, Swindon, Nov. 1989.
Statutory Instruments 1996, No. 1592, Health and Safety – The Construction (Health, Safety and Welfare) Regulations 1996, Her Majesty's Stationery Office, London, 2 September 1996.

# Part 7

# Precast concrete

# 21

# Precast concrete structural elements

*John Richardson*

## 21.1 Summary

The benefits obtained from the adoption of precast concrete components in building and civil engineering construction are identified. Much of the material goes towards satisfying the measures recommended by the Egan Report, 'Rethinking Construction' i.e.:

- Committed leadership, a focus on the customer, integrated processes and teams, a quality-driven agenda and committed people
- Product development, project implementation, partnering the supply chain and production of elements

Methods of production of elements in works and on the construction site are examined from the viewpoint of the experienced technologist and described, with reference to organization of throughput and achievement of quality. Works equipment and services are described as well as mould design and construction, reinforcement location, casting procedures, handling techniques, quality assurance procedures, and troubleshooting.

Production is examined in respect of normally reinforced and prestressed elements including: building frame elements, cladding, wall and floor units, bridge beams and segmental components, culvert and tunnel segments, retaining walls, piles, sea-defence elements, railway sleepers and other special products.

While a concrete technologist working within the precast industry will have encountered certain of the content of this chapter, coverage is particularly aimed at those who, working in other sectors of the construction industry, require knowledge of the topic. The chapter provides a 'survival kit' such that the technologist from those other areas can make a useful contribution when working with precast concrete designers and producers and in problem solving at works or site.

## 21.2 Structural precast concrete

Precast concrete is concrete which is cast in one place for use elsewhere and is a mobile material. The largest part of precast production is carried out in the works of specialist suppliers although in some instances, due to economic and geographical factors, scale of product or difficulty of access the elements are cast on or adjacent to the construction site. The distance between the place of manufacture and point of use varies immensely, for example distances as little as several metres at a site where the scale of the elements prohibits transport from an established works or where a replacement bridge is cast adjacent to an existing bridge which is beyond repair and is then jacked into position. Much of the product of specialist suppliers is transported more than 100 kilometres in the UK and there are instances where products have been transported hundreds and even thousands of miles when exported from the country of origin.

While this chapter focuses on production and installation of structural precast concrete elements, with particular reference to elements produced in a supplier's factory, the principles and methods of production at site are similar.

The commercial supplier's factory generally includes facilities for manufacturing a range of products of similar nature according to the firm's specialization, frame elements, cladding, bridge beams etc. The equipment is thus often highly mechanized including high-speed batching and distribution of concrete, provision of intensive means of mechanical handling and sophisticated plant for accelerated curing of product.

Typically, site production facilities take the form of production units dedicated to a particular product, bridge elements and piers, piles, tunnel segments etc., possibly employing some of the site facilities such as the batching plant for concrete production and site cranage for product handling. Production in both situations increasingly depends on outside supply of key items such as reinforcing steel, moulds and in some cases concrete supply from a supplier of readymixed concrete.

Wherever the production is carried out, the basic skills of management, supervision and the operatives employed are essentially similar. Differences in discipline between precasting and *in-situ* concrete work arise due to the repetitious nature of much of precast concrete and the opportunities in precast production for mechanization and controls both of outputs and quality.

## 21.3 The advantages to be achieved by employing precast concrete

The engineer, designer or contractor selecting the option to precast concrete elements takes advantage of a number of benefits inherent in the precast process. In many instances preferred dimension schemes exist, with elements of tried and tested designs available from a catalogue. Elements as diverse as massive prestressed beams and simple stair units can be ordered from a precaster specializing in that particular type of production. By this route, lead times between order and delivery are dramatically reduced as is overall contract time.

Precast producers can provide design services staffed by specialists in the precast discipline with a knowledge of the capabilities of the material. While building and civil engineering contractors are dependent to a large extent on a mobile labour force, the precast producer has a static production team, many of whom will have had years of experience with the particular processes of precasting. Major savings in time result from manufacture of structural elements apart from the series of events which determine overall duration of the construction, known by planning engineers as the 'critical path'. By employing precast manufacturers, clients avail themselves of the use of the producer's technology, skilled labour force, buildings, machinery and land:

- Laboratory facilities capable of the required control tests, many being certified for specific testing in accordance with National Standards. Provisions usually exist in works for even the largest elements to be tested before delivery, enabling validation of innovative design
- Labour experienced in the special processes employed in production.
- Equipment with capability suited to specific types of production such as stressing beds with appropriate capacity, moulds and machinery dedicated to particular products. Sophisticated automated batching plant, concrete handling and moulding machines are also employed ensuring high rates of output and short lead times between order and delivery.
- Buildings designed for particular processes and to facilitate controlled curing of cast elements.
- Land for quantity stockholding, such that elements can be delivered on demand in the correct sequence for erection.

A contractor may elect to produce elements of considerable size, or units required in very large numbers, on the construction site. Examples of such production are piles, bridge segments and tunnel segments. The principles of precast production outlined elsewhere in this chapter apply in the same way whether the work is carried out in works or on-site.

## 21.3.1 Advantages of precast concrete in terms of control and innovation

Pre-manufacture, taking work from site and into factories, is a way to ensuring higher-quality product of greater accuracy than can be obtained under the rigorous conditions often prevailing at the construction site. With the tendency towards greater and earlier cooperation between the designer and producer opportunities are presented for innovative solutions to constructional problems.

Designers can capitalize on the experience of producers, gained as it often is from a number of previous contracts of similar nature. Alternative solutions to the problems faced by the designer are offered such as, the use of alternative materials, alterations to element details to improve access for fixings and service installation and incorporation of features which improve the weathering characteristics of the units. Other instances of a supplier's input to the design of buildings and structures are local redesign to resolve structural problems such as local high stresses by the introduction of composite construction or, for example, incorporation of fibre reinforcement to permit the use of thinner concrete sections.

High-quality finishes achieved direct from the mould eliminate the need for interior decoration and ensure low maintenance costs.

As the size of contracts increases many are carried out employing schemes which the larger precast manufacturers' support, where design and construction activities are carried out in close cooperation, often concurrently, based upon:

- Clear definition of the client's brief at the outset.
- Agreement of a detailed specification including quality of all critical elements prior to start of manufacture.
- Strict quality control procedures.
- Encouragement and acceptance by the client and designer of suggestions from the producer regarding innovation and buildability.

Early cooperation between designer and producer as in the current 'partnering' schemes being fostered by major producers is changing the contract process from the historic conflict situation to one of cooperation and trust. Examples are quoted where this early cooperation between all parties to the contract has resulted in a change of location of activities and greater prefabrication. The expensive process of fixing bracketry to a steel frame, frequently carried out under site conditions, is performed in the works of the supplier of the steel frame with consequent saving in time and cost. Windows are also often fixed into panels at the precast works. Other instances of early cooperation include redesign as suggested by the supplier to combine a number of small units into a single, large unit reducing the amount of joints in the building skin and, importantly, site crane handling time.

Recent large-scale contracts, and contracts where energy efficiency has been a paramount consideration, have utilized 'hybrid' constructions comprising, for example, reinforced concrete frames, some incorporating post-tensioned beams, prestressed hollow core flooring as well as precast concrete stairs and stair enclosures. These measures combined with pumping techniques for placing *in-situ* concrete have proved to save design and construction time as well as reducing construction costs in concrete and reinforcement compared with 'traditional' construction methods.

Whole-life costing of precast product for civil engineering contracts such as bridges indicates that low initial construction costs are achieved by economies in formwork, falsework and access scaffolding. In building construction there is the added early provision of enclosed working conditions for following trades. In both cases low maintenance costs provide a competitive margin when compared with similar *in-situ* concrete or steel-based constructions (see Figures 21.1 and 21.2).

Precasting provides opportunities for the site constructor to overcome skills shortages by taking advantage of the skills employed in the factory.

The regular cycle of deliveries of precast elements whether works or site produced act as a regulator in speeding the overall construction cycle.

**Figure 21.1** The Flintshire Bridge: setting up reinforcing steel around inner mould face during site production of bridge tower segments. (*Courtesy* Flintshire CC and Norwest Holst Construction Ltd.)

## 21.4 Principles of precasting

### 21.4.1 The construction sequence

*In-situ* construction imposes many restraints, certain of which generate time penalties. Building work generally starts on foundations and progresses through the supporting structure floor by floor to the roof and final enclosure. Erection and dismantling of falsework and formwork for concrete feature heavily in the construction progress. Precasting can reduce these requirements considerably, saving contract time and providing the client with an earlier return on investment.

In precast building construction, with foundations established, a frame construction can proceed with the simplest of access scaffolding, column, beam and floor panels being erected with bracing and ties and except in special cases without formwork or falsework. The provision of precast stair and landing units and precast lift enclosures and walls where required translates the frame construction process into a straightforward mechanical handling proposition. The progress of the frame provides early access for following trades and service installation.

Where architectural cladding is a feature of the construction this can, if the programme so demands, be erected floor by floor as the construction proceeds. In recent very tall

**Figure 21.2** The Flintshire Bridge: Installing the first tower section onto shims on prepared foundation. (*Courtesy* Flintshire CC and Norwest Holst Construction Ltd.)

framed constructions the panels have been delivered in stages to selected points on upper floors then transported by fork lifts for erection from inside the structure.

In the case of *in-situ* construction in civil engineering works, substantial falsework, formwork and access arrangements are required, for example under the span sections of a bridge. Much of these activities are avoided when, given crane access, precast beams are employed in short and medium spans. In large-span situations where segmental construction techniques are employed the major temporary works expenditure will be that of providing the erection frame or traveller or the launching nose and jacking equipment.

Kurt Billig (1955) set out further details of the possibilities offered by precasting. His numerous books, which will interest a serious student, describe a wide range of element types and production processes, many of which, innovative in his day, are currently as viable as they were at the time he described them.

Some products lend themselves to the precasting process. The building and civil engineering industries, have focused on fast-track construction techniques during the past two decades. Precasters have always been aware of the benefits of fast-track construction.

## 21.4.2 Establishing standards

In the case of standard products manufacturers' stocks and samples serve to provide a true representation of the standards a client can expect in response to his order. In the following pages are recommendations that the technologist may observe in assisting with the selection of supplier. In the process of initiating production of bespoke products to be manufactured for a contract, early establishment of standards and agreement over the interpretation of specification are essential to success of the project. Samples may have formed a part of the quotation for supply, or a part of a similar contract may have been quoted in the enquiry. The most important step is to ensure that the designer and the precaster have the same aims as to product standards as regards accuracy and finish. Case studies later in this chapter indicate some of the problems encountered in this matter of establishing standards.

Samples should ideally be full size, where possible cast from the moulds similar to those to be employed in production, cast and finished by operatives in conditions similar to those in which the contract will be produced. Small samples cause problems, are misleading and frequently only act as a testimony to the sample makers' skills. Samples should include as many features of the projected design, such as returns, openings, finishes and such detail as possible. At the time of establishing approval of samples, arrangements should be made for specimens to be viewed from the distance at which they will be viewed in their final position. Sample elements should be suspended from a crane such that the play of light replicates the normal changes at site. In this way the designer will be clearly informed of the interpretation that the supplier has put upon the specification and at this stage modifications can be made to benefit both them and the producer.

An extension of this process where a mock-up of specimen elements from the early production have been assembled on a temporary foundation at works and site has proved to be of value.

As an example of the benefit to all parties to a contract obtained from such a trial, as many as twenty modifications were made following a trial erection and inspection at activity number 50 of a series of some hundreds of activities in the eventual path to completion of the structure. These changes to the initial detail of components resulted in improved appearance and weatherproofing details, proved the connection design and enabled handling techniques to be tested.

### 21.4.3 Drawings and controls

Skilled production control, the sequencing of casting operations and the dissemination of up-to-date information is a key function in efficient precasting operations. Personnel responsible for coordinating the production process, the supply of services and maintenance of supply of moulds and fittings must have a sound appreciation of the methods employed in the various sections of the works such as mouldmaking, fitters shop and steel shop. Good liaison with purchasing and dispatch departments can ensure a steady supply of the materials into and out of the works.

Perhaps the most critical of the activities of production control is the distribution of drawings and schedules, closely allied with the monitoring of the production in response to those schedules. An examination of the flow of information in the form of drawings within a typical works will underline the importance of this activity.

### 21.4.4 Organizing the precast operation

The production facilities of the majority of precast concrete manufacturers, described by Richardson (1991), are organized on similar lines and comprise design/drawing office, production control department, quality department and reinforcement, fitters, mould and casting shops. Finishing, storage, loading and transport are also important to the production.

As well as in-house design of bespoke work and standard product the drawing office staff are responsible for liaising with architects, engineers and designers regarding contract details, and for issuing drawings to the production control department. A works engineer liaises with works management over provision of production equipment and special arrangements for handling product.

The production control department is responsible for ensuring a flow of information on product detail to the mould shop or mould supplier(s), fitters shop, reinforcement shop and casting department and quality control department. Some of these departments may have an issue of more than one drawing each and there will be supporting schedules. A supervisor or foreman reporting to the works manager will control each production centre. The whole operation will be carried out in accordance with approved procedures, monitored by quality personnel and facilitated by access to laboratory and test facilities.

### 21.4.5 Drawing and documentation control

A works may have ten contracts under way at any one time, this is of course dependent upon the scale of production, the type and scale of the product. Some works will have more than ten contracts in the course of progress through the various departments or sections. For the purposes of explanation the drawings issued to works for ten contracts are discussed here.

Considering that there may be of the order of twenty drawings, for a contract the number of drawings likely to be in circulation within various departments of a works is of the order of 120 drawings per contract, in most instances many more. Each drawing is critical in terms of accuracy of product. The precaster may well have up to ten other contracts in various stages of production running concurrently. Just one out-of-date drawing or a drawing with unauthorized alterations can cause problems in the course of production if not in the final product. The situation develops as revisions and alterations are made to design and further elements are brought into production as contracts progress.

Analysis of a recent major stadium contract showed that the precast producer, while servicing several other clients with structural elements and flooring, processed a contract involving 604 supports and raker beams, 7696 terracing units and 1368 miscellaneous units such as stair flights. The concrete content of these items coupled with associated flooring and hollow core planks totalled more than 10 000 $m^3$ of concrete. While in a stadium the elements fall into rationalized sets, the control of drawings, issue of amendments and recovery of superseded drawings would obviously be a demanding task. A small non-conformity within the system would have caused substantial upsets to both production and the eventual site erection process.

## 21.4.6 Design of mouldwork and the way up of casting

Accurate moulds are the key to accurate elements. No amount of work subsequent to the casting operation can improve the face condition, or accuracy of product from an unsatisfactory mould. The traditional ability to produce moulds capable of casting product to satisfy demanding specifications has been the foundation of the majority of key precast concrete producers. Historically moulds were mainly produced in works though today there are many highly capable specialist suppliers of moulds to the industry.

A critical decision in the process of designing methods and moulds for precast production is that of the way up, or orientation of the unit within the mould. The precaster has the option of casting elements within the mould oriented in the manner which will yield the best results in terms of quality of product. The decision as to the way up of casting depends on a number of factors:

- Whether the accuracy of the product can be improved by a change of aspect.
- The attitude which best suits the concrete placement methods and equipment.
- The suitability of the method to the means of applying vibratory effort.
- In the case of a geometric element, which faces are best generated from the mould.
- Which faces provide opportunities for transfer casting of inclusions such as facing materials, connections and projecting steel reinforcement.
- Reinforcement characteristics, whether additional steel will be required to facilitate element handling from the chosen aspect of casting.
- Whether the casting operation can be carried out by simply skilled operatives or mechanized.
- Whether mould alterations to cast other types of element within the set would be simplified.

In the case of elements with complicated geometrical form or with several types of finish, after discussion with the designer the precaster may pre-produce parts of the element for later incorporation in to the final casting operation.

The way up of casting (apart from prestressed elements where the tendon pattern determines that the element cannot be rotated) is a major decision that has to be made by the production engineer. While there is no one correct way, not all ways are equally valid and one way will provide the best solution to meeting specification in terms of accuracy, sound compaction and finish.

Where, for example, stair elements are cast as drawn moulds require a supporting substructure, also the casting operation becomes timing critical due to the need to control surge of material under the influence of vibration. If the elements are cast inverted, non-slip treads can be transfer-cast into the elements and placement and compaction as well as the finishing of the soffit are simplified. Proprietary manufacturers often cast stair flights on edge, soffit to soffit about a central divider, inclusions such as treads and pocket formers being transfer-cast from the profiled mould sides.

## 21.4.7 Daily make

In order to operate profitably the precaster manufacturing medium-sized product usually aims to work on a daily casting cycle. Large elements with complicated reinforcement or cladding elements with included finishes may be produced on a two- to three-day cycle.

Works operations, provision of services, investment in moulds, concrete supply and handling equipment are geared to constant turnover in optimum time. Controlled working conditions, coupled with accelerated curing techniques, are an essential part of the precast, regime, permit shift working and the turnround of normal reinforced product in four to six hours should exceptional circumstances so require.

In the case of bespoke manufacture for a specific contract the number of moulds provided for a particular type of element will be determined on the basis of the production period divided by the casting cycle time. There will be an allowance of time for mould alterations to accommodate variations in the elements to be cast from the mould. Alterations to suit modifications to the unit have to be skillfully carried out, the sequence of casting being arranged so that units critical to programme are cast before exceptions, the whole production being planned to meet stockholding and delivery requirements.

## 21.4.8 Mechanical handling

The manufacture of precast concrete is essentially a mechanical handling proposition. Examination of the events in a medium-sized works producing a mix of normally reinforced concrete elements, reveals the extent of the dependence of all involved on the handling capabilities of the available equipment:

> Assuming 300 tonnes of product per day. At the start of each day there will be 300 tonnes of product to be stripped from moulds and casting beds, and this operation may involve the handling of 100 tonnes of moulds and mould components. The cast product will be handled from the factory bays to the stockyard, some via a finishing bay involving, say, a further 450 tonnes of handling. During the day, steel reinforcement has to be distributed, say, 50 tonnes of handling about the works. With reinforcement and inclusions set up the moulds are re-installed, say, 100 tonnes. Next, 300 tonnes of concrete are to be distributed to fill moulds as they become available. Meanwhile 300 tonnes of product has to be loaded for delivery. At this stage 1600 tonnes gross of handling operations have been required to achieve the 300 tonne output.

While this simple calculation has not taken into account the handling of aggregate supplies onto the works or the ancillary handling of moulds and casting equipment within the works it clearly illustrates the importance of the selection of efficient equipment appropriate to the product.

Typically concrete is handled in works by forklift in medium-quantity production. Where high outputs are involved crane and skip or bullet transporters are used.

Forklift or side loaders as well as gantry cranes, tower cranes and straddle loaders are employed in product handling. To reduce crane handling, where possible moulds are designed to be self-stripping or, as in the case of battery moulds, to include dedicated gantries.

## 21.4.9 Clear shed principle

A principle which affects outputs from a manufacturing plant, whether the factory of a commercial supplier or the workplace set-up at the construction site, is that of the clear shed or shop. It is essential that the teams making product see gainful work ahead of them, that previous product can be or is being cleared to enable mould stripping, element handling and the whole sequence of manufacture of units to continue without delays. In order to achieve this the balance of resources, cranes, hoists, forklifts or whatever the means of handling employed must be suited to the production. There has also to be a supply of reinforcement, fittings, connections and embedments to hand, to provide continuity of work.

When stages in the work are reached that demand in-process checks or inspections then the appropriate personnel must be available, particularly in the latter stages of assembly when the pre-concreting inspection is required.

## 21.4.10 Accelerated curing

In many instances the key to achievement of the required casting cycle to enable programmes to be met rests with the accelerated curing methods available. These can range from primitive arrangements such as the provision of free steam to the latest efficient methods employing electrical current via the reinforcement or in heating elements under the casting beds.

It is essential that, whatever the curing arrangement, the process should be properly controlled with regard to rate of heating, maximum temperatures applied, and reached within the concrete, and the rate of cooling.

Precasters use concrete mixes designed to generate early strength for demoulding reinforced concrete units and for the early transfer of force into prestressed units. Cement contents are such that only gentle heat is required to promote the exothermic temperature of the cement and maintain it for the required period. It is critical that the heat should be uniformly applied and that the essential humid environment is maintained.

Generally control specimens are stored under covers in the same condition as the unit and these are as to determine the suitability of the element for handling or transfer of force.

### 21.4.11 Condition of stock

Successful completion of contracts depends upon the ability of the supplier to provide elements of the right standards of quality and accuracy at the right time in the contract. This will only be possible if the stock at the point of precasting is finished complete to specification and ready for delivery. The abilities of the precaster can often be judged by the state of the stockyard and conditions of storage of the elements.

Elements requiring rework must be clearly identified and procedures established for the work required to bring them up to specification. The work of the finisher, whether repairing damaged surfaces, replacing displaced tiles or making a small repair to handling damage, is extremely demanding and, on occasion, time consuming. The work of the finisher is as important as that of those making product and should be programmed accordingly.

In factories producing certain products, mainly precast, prestressed concrete flooring and blocks, and particularly in production centres with high standards for product quality, the amount of substandard stock may build with time. In such circumstances and with client approval, reclaimed aggregate from crushed concrete may be incorporated in to new product. The crushing and cleaning process is carefully monitored and the reclaimed material used as a limited percentage of the total aggregate content. At least one major manufacturer with multiple production centres is known to use a mobile plant which is moved from factory to factory as the stock situation demands.

## 21.5 Moulds for precast concrete

Decisions as to type and number of moulds to be employed in the production process are critical to the economic outcome of the process.

Each mould will be designed to cast a set, or part of a set of elements, provision being made in manufacture to accommodate variations in the overall profile, outline and section of the set. These aspects of production are examined by Richardson (1973).

Where there are ‘runs’ of 30 to 50 elements in a set moulds used will generally be constructed of timber and ply, allowance being made in the costing for a degree of refurbishment of the mould faces as casting proceeds. Ply-faced, timber-framed moulds are also used where inclusions such as brick, tile and similar materials are set into matrices of fillets on the sides and soffit of the mould. Where extreme accuracy is required as in the case of cladding elements thirty uses can be achieved although greater numbers of uses from a cladding mould would generally dictate provision of steel moulds. The provision of several soffit forms and one set of sides complete with stop ends can make economies in cost of moulds. Element geometry permitting, the sides can be used to cast several elements on the series of bases in the course of a working shift.

In casting sets of 50 to 100 elements, including those having an exposed aggregate or featured face, steel moulds may be used. Depending on the thickness of the face material, and the substance of the backing members, steel moulds are capable of a hundred or more uses, again with a degree of refurbishment.

Where there are small sets, particularly where there are frequent variations in section and dimension, frame members and bridge beams are cast from ply-faced timber moulds.

Bridge beams, standard proprietary products such as tunnel lining units and rationalized elements for frames are produced in specially constructed steel moulds used and maintained as items of plant. Extremely large elements for civil engineering cast on-site are generally cast from standard, proprietary formwork or purpose-made forms.

The materials used in *in-situ* formwork, described elsewhere in this book, are all used in some form of precast production supplemented in many instances by materials which cope with the intricate details or complicated forms that precasters are required to produce. Materials employed to meet specific requirements of the precasting process have included:

- Casting polyurathane is employed as a mould liner to produce textures at the face of concrete and to form recesses and pockets
- Glass-reinforced cement has been used as a mould-facing material
- Concrete, used as a mould material. The mass of this material coupled with its characteristic strength and durability is utilized in pile and in tunnel segment manufacture as well as being used as features and soffit members for large panel casting.

By taking castings from master elements the mould maker can rapidly breed concrete moulds of extreme accuracy and stability capable of providing several hundred uses with a degree of refurbishment. Typical applications for concrete moulds are in casting tunnel elements, retaining wall elements, arch bridge units and pile beds.

All moulds require a release agent or mould oil to ensure ease of stripping without damage to the mould or the product. Currently the trend is towards the use of vegetable oil-based release agents. These have been thoroughly researched in Europe and Scandinavia and meet health and safety requirements in that they are non-toxic, biodegradable, safe to handle and easily sprayed or spread. They are also non-inflammable and safe to transport.

Elements are generally demoulded by removal of sides and stop ends prior to lifting from the soffit or base mould. Demoulding of cladding elements with brick faces or exposed aggregate faces of a complex nature is often carried out by turning the mould and concrete element complete through 180 degrees and depositing the element onto bearers for later turning and finishing. Tilting tables facilitate stripping of slender elements.

## 21.6 Casting techniques

### 21.6.1 Concrete supply

Distribution, placement and compaction methods employed in precast production are quite different from those of *in-situ* construction. Precasting is carried on in a static factory and thus more highly mechanized techniques are employed. Concrete batched in a central plant has to be distributed at a high rate. The demand can be as much as 2 $m^3$ per minute per plant throughout the working shift and this can only be achieved using automated batching and mixing equipment and distributing concrete by 'bullet' or monorail skips discharging to receiving hoppers or machines.

In some works concrete is called-up by the line supervisor from consoles adjacent to the casting beds, and these instruct the automated battery plant which mix from a standard menu of authorized mixes is required for the work in hand.

Poker vibrators within the concrete or external vibrators attached to the moulds are used to achieve compaction in works. High-frequency vibration is employed and to ensure effective action care is taken to securely attach external vibrators onto, or into, the fabric of the moulds.

## 21.6.2 Static machines

Repetitive outputs of reinforced elements up to 4.5 tonnes mass can be produced by dedicated machines. The concrete will be supplied from a computer-controlled batching and mixing plant matched to the throughput of elements. The concrete mix is discharged into hoppers feeding a placing device set above the making station. With a reinforcing cage installed the concrete is spread, levelled, and compacted into a mould that is secured by clamps to a vibrating table, an integral part of the machine. The frequency and amplitude of vibration can be adjusted to suit the mass of the mould and the contained concrete. Where the design of the element permits the next operation in the cycle is that of rotating the mould and demoulding element onto a pallet for conveyance to a curing tunnel. The mould is then cleaned, oiled and re-used. Where the design of the element is such that it cannot be immediately demoulded the element is cured in the mould prior to passing to a demoulding station where the cleaning and oiling process is carried out. The machine may be fitted with a self-stacking transporter.

## 21.6.3 Mobile casting machines

In producing sleepers and linear products such as piles on the long line bed, use is made of mobile casting machines. These comprise a receiving hopper for the concrete, spreaders, and equipment that transmits vibratory effort through clamps to the moulds on the bed. Only final trowelling is required to produce the specified finish to the exposed face.

## 21.6.4 Casting in individual moulds

Elements such as frame components and cladding panels as well as bridge parapet units are cast in individual moulds. The materials used in mould construction are described later in this chapter. The moulds are carefully established on level foundations, and frequently include mountings which promote the vibratory effort. Production is of the nature of jobbing production where teams each specialize in one part of the process work on each of a number of moulds in turn, carrying out assembly, steelfixing, casting, finishing and subsequent stripping, cleaning and oiling of moulds. Mouldmakers and fitters carry out alterations to type between the casting operations.

## 21.6.5 Tilting tables

These are steel frames on which moulds for panels can be mounted. After the element is cast and cured the frames can be raised from horizontal into the near vertical position by crane, or hydraulic ram, to facilitate stripping. This permits flat casting of items such as cladding with exotic aggregate veneers and sandwich elements in the face-down position also avoiding bending stresses in the stripping process. The mould is usually designed such that the element is restrained yet free to slide as the tilting action is carried out, thus assisting the demoulding by breaking the bond at the mould/concrete interface. The steel frames may support timber or composite moulds or, for series production, may form an integral part of the mould.

A development of the tilting table permits production of sandwich and insulated wall panels. One leaf of the element, incorporating wall tie steel, is cast and secured to a tilting table. The second leaf is cast and prior to the concrete stiffening the first unit is lowered into position, leaving the required cavity between the leaves. With sufficient strength achieved in the secondary casting the whole cavity element is demoulded by tilting the mould.

## 21.6.6 Gang casting

Gang moulds provide an economic solution where large numbers of similar elements of simple section such as piles or rectangular section beams and columns are to be cast. Within a gang mould elements are set side by side, or end to end with mould divider generating the section. Where sufficient draw can be allowed in the design these dividers may be a fixture, where sections are rectangular then the dividers are arranged so that they can be with drawn to allow elements removal. Gang casting techniques are frequently employed on long and short line prestressing beds. Beam, lintel and sleeper production provide examples of gang casting.

Moulds set gang fashion facilitate casting and finishing operations by machine. Stability of moulds and accuracy of section are improved where set in pairs, gang fashion, about a central fixed divider.

## 21.6.7 Battery and cell unit casting

Battery casting is in essence an extension of gang casting although used for different sections of element. Wall elements are cast side by side in mould assemblies that comprise a series of dividers suspended on rails to form cells or moulds. Batteries may comprise twenty or more cells and 'continuous' batteries may comprise 100 or more cells.

Materials used in battery mould construction are heavy steel plate for single plate dividers or lighter plate mounted on a steel framing where the dividers are composite and include heating elements. Stop ends and soffit members are fabricated, often from structural steel sections.

The mould dividers are spaced by soffit and stop end members which determine the wall height and profile. Economy in reinforcement is possible as elements are cast and handled vertically reducing turning and handling stresses. The dividers complement a fixed back former that is attached to the mould carcass. Chairs from the soffit of the mould locate reinforcement. Lifting bolts, which form site connections, are suspended from jigs at the top of the cell. Inclusions such as switch boxes, service conduit and so on are attached to the inner divider in each cell for the operatives working within the cell. Attachments are made to the back leaf only, spacers being used to position fittings that are required on the outer face. As each cell is set up, working from the back or fixed leaf outwards, it is completed by closing up the next divider which in its turn becomes the face to which fixings are made. With the complete battery set up the closing side is jacked into position and tie bars are installed to resist the concrete pressures developing during casting. Operatives working on a walkway at the top of the battery carry out concrete placement. As the mould dividers are suspended from the mould frame further economies are achieved as the main craneage in the production shop is relieved of activities of mould handling.

Elements are stripped from the mould successively as each divider is removed providing access to the fixings of inclusions.

'Chevron' casting is a variation of battery casting where L-shaped wall elements are nested is employed for lift enclosures. Alternatively an L-shaped wall may be cast in pairs about a central box or core. Currently prison cells and utility units for apartment blocks are cast using steel side forms set upon a soffit about a central collapsible box. Moulds similar to the tunnel forms used in *in-situ* construction have been used in the Americas for casting room-sized elements for condominiums.

## 21.6.8 Long and short line casting

Long line casting is employed in the production of prestressed concrete elements such as bridge beams, sleepers, piles and prestressed linear building elements.

Casting beds of up to 200 metres or more long are set between gravity abutments where posts cast into massive foundation blocks serve to support beams between which the tendons are stressed. In the case of short line casting of elements such as lintels, the bed incorporates side beams, designed to contain the forces generated in the stressing operation, thus avoiding the need for heavy abutments. In each case one end of the bed becomes the fixed anchorage end while the other incorporates either a stressing beam to which all the tendons are attached or a facility for tensioning single tendons. The stressing or jacking operation is carried out using calibrated equipment to achieve the specified force in the tendons, gauge reading and tendon extension being carefully monitored and recorded. It is important that the characteristics of the steel are known as these are used in calculating the extension required to achieve the specified force in the tendons.

In series production comprising large sets of similar elements, complete lines of mould sides are dedicated to a particular product. When there are modifications within a set of elements one or several pairs of mould sides will be moved progressively along the line as the concrete in each element achieves demoulding strength. In both instances once the whole line of moulds has been filled the bed will be finally covered with an insulating blanket to aid the curing process. Timing of the curing cycle and uniform application of heat are critical in the achievement of consistent cambers in slender elements. Any local differences, such as might be caused by lack of covering or a draught from an open door overnight, can dramatically effect the gain of strength locally in the concrete, lack of strength leading to local excessive camber. Compressive strength is also critical to the bond between the concrete and tendons and build-up of prestressing force within the elements at the time of transfer of force, and the cutting of elements to length.

When cube specimens indicate that the concrete has achieved the specified transfer strength the prestressing force is transferred into the concrete by releasing the stressing beam or cutting the tendons. This detensioning of the bed and transfer of force has to be carried out in a controlled fashion to avoid damage to the elements.

Beams, piles and similar discrete elements cast on a long line bed may be separated by stop-ends. Heavy divider or distribution plates are used where differing patterns of tendons are required in adjacent elements along the bed.

Long line casting beds form the basis of the highly productive extrusion and slipforming processes employed in the production of prestressed hollow floor products, some solid walls and sandwich wall panels, beams and lintels. These methods employ a casting machine which travels along the bed dispensing and forming concrete from a self-contained hopper. Various configurations of core and side formers are employed to impart a massive compactive effort achieving a cohesive strength in the concrete capable of retaining the formed section. The process eliminates the need for continuous side forms or stop ends, although openings and checkouts are formed manually as the machine proceeds along the bed. The elements are cut to the required length by either wet forming prior to transfer of force or sawing after transfer, the saw cutting both concrete and tendons. Not all continuous casting employs prestressed tendons; in some cases, particularly in sandwich wall construction, long lengths of high-yield steel form the reinforcement.

If extreme stresses are to be avoided at the ends of beams where the moment due to self-weight is less, and in order to enhance the shear capacity of a beam, either debonding or deflection of tendons is employed. Debonding techniques involve the systematic sleeving of tendons at the end of beams using plastic tubes sealed to prevent the ingress of paste during the casting process. This prevents bond forming between tendon and concrete over the distance of the sleeve.

In deflected strand production hold-down fittings are installed at strong points in the mould bed, typically at the quarter points of the beam, to restrain the tendons. Once the straight tendons have been stressed jacks apply an upward force to tendons which are to be deflected, at or adjacent to the ends of the unit thus achieving the required 'catenery' pattern. In the deflected strand process a full bond is achieved throughout the length of each tendon.

## 21.6.9 Post-tensioned beam production

Post-tensioned techniques are employed in a number of precast situations, particularly where large elements are in production, the precast works providing the controlled environment essential to quality results. Post-tensioning yields economies in both reinforcement and concrete particularly in extremely large or deep elements.

Casting of linear beams is carried out in individual moulds, plastic ducts for the stressing tendons being supported in catenery formation by grillages attached to the reinforcement or by dowel bars passed through the mould. The anchorages, aligned with the ducts, and surrounded by heavy reinforcement are attached to formers on the stop ends. These are blocks forming recesses in the cast element which are, after the post-tensioning operation, filled with concrete to protect the ends of the tendons and seal the ends of the ducts.

In the case of deep, slender sections such as beams particular attention is required in the casting operation to ensure that concrete does not bridge between the ducts and the mould, causing voids. In this stage of production the ducting is liable to sideways and upwards displacement unless securely fastened in position. Also joints in the ducting are particularly vulnerable to ingress of paste.

Stressing is carried out after initial curing and upon achieving the specified strength for the post-tensioning operation. Jacks, powered by hydraulic pumps, tension the tendons, either individually or in groups. Calibrated gauges and extension measurements monitor the force in the tendons. With the specified force achieved, the tendons are secured at the anchorages, either by wedging action or, in some systems, by screw collars.

On completion of the stressing operation and after confirmation from the designer that the forces in the beam are as specified the ducts are filled using a colloidal grout pumped in under pressure. It is extremely important that the ducts are completely filled with grout to avoid corrosion. Subsequently the tendons are cropped and the anchor recesses filled.

## 21.6.10 Tilt-up construction

The process of casting wall panels flat on the ground floor slab of a building then tilting them into their final location in the structure is identified by Southcote (1998) as being essentially a precast process. The system provides economy in moulds and simplicity of production activities combined with a reduction in transport costs while still permitting production of attractive concrete elements with varied finishes.

Mould sides are set out on a suitably prepared ground slab, reinforcement, opening formers, embedments and lifting fittings are installed and concrete is cast upon a bond breaker applied to the slab. On completion of the curing operation the elements are handled using spreader bars and bonds designed to limit the handling stresses within the newly cast concrete. In the case of extremely large panels temporary stiffening members may be attached to the slab at the time of lifting. Installation into, or onto, prepared foundations in the ground slab and stitching between units completes the operation.

Walls cast in this fashion may include brick and stone-faced panels. As well as exposed aggregate panels and sandwich panels, featured and striated finishes can all be produced. Panel sizes are typically 6 metres long and up to 2 storeys high.

Internal walls and gable end walls omitted for crane access may be cast on a clear area of the slab for later location. Stack casting has also been employed for the production of such panels. This is a technique where elements are cast one upon another in a stack. Bond breaker is applied to the top of each succeeding cast, the side forms being raised between each operation and supported from the previous casting.

## 21.6.11 Casting massive elements for civil engineering

### *Caissons*

Where elements such as caissons are produced the contractor often employs the techniques of *in-situ* construction in the precasting process. Caissons of 6500 tonnes and more have been cast using slipform techniques and 58 000 tonne dock entrance structures have been precast using climbing formwork then sunk into position.

To avoid thermal cracking, in casting such heavy sections it is imperative that temperature gradients are controlled. Typical specification limits are:

- Maximum concrete temperature at delivery 20°C.
- The maximum difference of 20°C between the centre and the surface of any section with a maximum gradient of 15°C in any 1 metre distance.
- The difference in temperature between newly placed concrete and any adjacent previously placed concrete not to exceed 10°C.
- The maximum peak concrete temperature in an element to be 70°C.

These criteria will vary with the locality in which the operation is carried out, in hot climates ice-making plant and water chillers may be required. Careful regulation of insulation during the curing period is also needed to maintain the specified curing regime.

### *Immersed tube construction*

Tunnel sections of immense proportions, typically 100 metres long by 25 metres wide and 10 metres high, incorporating a central spine wall and weighing 30 000 tonnes, are cast in dry docks or dewatered basins. The completed units are floated out. The construction methods are those of the *in-situ* constructor, employing proprietary formwork components or specially constructed forms on travellers in the casting process. Sections of the structure are cast on a concrete base, and once the ends of the section are sealed, a bond-breaking membrane permits the segments to be floated out into position then settled onto the river or sea bed.

The large quantities and heavy sections of concrete in these structures demands careful temperature controls. These range from simple precautions, such as shading aggregate stocks from the sun and spraying with cold water to injecting liquid nitrogen into the vehicles used to transport concrete from the plant to the point of placing.

## 21.7 Curing

For many years precast producers manufacturing large structural elements simply stacked product in their yards without consideration of the need for systematic curing. In the factories product was covered and kept moist by spraying then to be outshopped into whatever were the prevailing weather conditions. Now with the accent on speed of production, combined with durability of product, curing has assumed immense importance. The precaster working in controlled working environment has every opportunity for a scientific approach to the topic.

The technology of accelerated curing in respect of *in-situ* work is discussed elsewhere in this book. Accelerated curing is of particular importance in the precast situation where rates of turnover of expensive equipment have to be sufficiently high to absorb overhead costs.

Curing methods employed in works tend towards those of accelerated curing using free steam or infrared heater elements under cloches, closed-circuit steam or hot water piped under and into moulds, electric curing by direct or indirect means and the use of chemical accelerators and admixtures. Concrete heated using hot water or steam injected into the mixer provides a good basis for any curing system, this being boosted, early in the curing cycle, by the heat of hydration of the cement. Thermostatic control is exercised by sensors installed in the bed mould and others embedded in the concrete elements. Results from these probes are recorded as part of quality control procedures.

It is essential that, whichever system is employed, the rates of heating, the period of maintenance of heat and the rate of cooling should be monitored and controlled. Failure to control temperature gradients within the concrete unit can result in curling, warping, and excessive cambers.

Heating processes, coupled with maintenance of humidity within the concrete and the surrounding environment during the curing period, promote gain of strength permitting early demoulding of reinforced elements and enhanced transfer times for prestressed product.

## 21.8 Testing and controls

It is a requirement of any well-organized laboratory or test facility that all equipment is properly calibrated and that calibration is traceable to National Standards. Of the considerable range of materials and product tests, the main methods and tests employed within works by producers of structural precast concrete elements are:

(a) Sampling fresh concrete
(b) Slump test
(c) The Vebe test
(d) Degree of compactability
(e) Flow table test
(f) Density
(g) Air content – pressure methods
These tests are covered by BS EN 12350: Parts 1 to 7 respectively.

Requirements on specimens and methods employed include:

(h) Shape, dimensions and other requirements for specimens and moulds
(j) Making and curing specimens for strength tests
(k) Compressive strength of test specimens
(l) Density of hardened concrete.
These are detailed in BS EN 12390: Parts 1, 2, 3 and 7 respectively.

Further tests and methods employed as controls in production and inspection include:

(m) Methods of accelerated curing of test cubes
These methods are covered by BS 1881: Part 112.
(n) Cored specimens – taking, examination and testing
(p) Non-destructive testing – rebound number
These methods are covered by BS EN 12504: Parts 1 and 2 respectively.
(r) Determination of ultrasonic pulse velocity
(s) Determination of pullout force
These methods are covered by BS 1881: Parts 207 and 203 respectively.
(t) The use of electromagnetic cover meters
(v) The use of radiography
(w) Determination of water absorbtion
These methods are covered by BS 1881: Part 201 Guide to the use of non-destructive methods of test for hardened concrete, also BS 1881: Parts 204, 205 and 208 respectively.

Additionally testing of finished product may be specified, load testing of large structural elements being carried out in special test rigs within the precast works or by simply loading elements set upon bearings by loading with test blocks or elements of known mass. Connections, cladding fixings, lifting fixings and proprietary equipment will also be tested according to relevant specification or standards.

Flooring producers employ load testing at a predetermined frequency, depending upon the throughput of product. The frequency of testing is determined by test results, being reduced as the number of elements meeting set criteria increases, and increased in the event of failure occurring under the reduced test regime.

Further tests as described elsewhere in these volumes may be specified to assess chloride content of fresh concrete, reactive alkali content of concrete and similar materials properties.

## 21.9 Frame components

Precast building frames consist of columns, beams, walls and flooring. Proprietary 'Preferred dimension schemes' exist. Structural/visual concrete components combine the structural elements of a construction with the visual envelope designed to satisfy the overall architectural concept. This form of precast concrete includes product such as Double-tee elements employed as cladding, also H-frame elements incorporating parapet walling. In each case these units provide structural support for floors and roof as well as cladding the structure. These have been designed for use in constructions such as offices, schools, factories and warehouses. The elements, designed to fit a standard grid, permit the use of standard moulds with only a small degree of modification. Elliot (1992) describes the use of rationalized systems in reducing the time between order and completion of the frame structure.

Non-standard frames, manufactured for a particular application, involve all the processes of design and approval and establishment of standards plus time for special mould manufacture and installation.

Frame construction provides examples of all the benefits of precasting. Speed of construction results from the availability of pre-produced elements in accordance with programme. The elements will be of the specified standards of quality and accuracy. Proprietary systems will have been proved in previous construction and exceptional designs can have been tested in trial assemblies at the point of production. Precast frame construction provides early access for following trades and ease of access for service installation.

### 21.9.1 Structural visual components

Much of bespoke frame production tends towards structural/visual or architecturally finished concrete elements. 'H-frame' construction used in combination with double tee elements has been used extensively in frames for car parks and similar structures. Many impressive buildings have been constructed with the main column and beams of the structure combined in single precast elements forming the building enclosure.

In response to the findings of research, major producers are promoting systems employing structural concrete in composite 'Hybrid' form (Figure 21.3). Columns and main beams are cast *in-situ* incorporating the precast floor elements. As well as presenting architectural finishes, full use can be made of the physical properties of the precast units as diaphragms to stiffen the structure, such floor units also form insulating and heat-retaining components of the structure. The method also results in reduced numbers of elements with savings in cost of transport and erection.

**Figure 21.3** Paddington Station, London: Structural steel combined with precast and *in-situ* concrete, joints first bolted then welded, an elegant example of hybrid construction. (*Courtesy* Railtrack, Taylor Woodrow Construction and Trent Concrete Ltd.)

Other examples of hybrid construction have included the use of structural steel columns, encased in the precast works in concrete to produce an architectural finish. These columns have been used in conjunction with precast floor elements, connections being made by bolting and subsequent welding prior to the placements of stitching and topping concrete. In this way the directness of connection ensured stability throughout the erection process (see Figure 21.4).

**Figure 21.4** Toyota (GB) Ltd, Head Office: Precast concrete two-storey columns and self-finished floor elements combine with *in-situ* concrete beams and floor topping to enhance the buildability of a cost-effective structure. (*Courtesy* Takaneka (UK) Ltd and Trent Concrete Ltd.)

## 21.9.2 Flooring

Prestressed floor elements are mainly produced by extrusion or slipforming processes. Casting machines that comprise a receiving hopper for concrete set above short-mould and core-forming members travel along the prestressing bed. Vibratory equipment compacts the concrete which has early cohesive properties. As the concrete, in which are formed voids of various configurations, clears the machine it is self-supporting. To avoid displacement of the tendons during the compaction process the prestressing wires are held in position by guides within the machine. The moulds and core formers are mounted in cassettes which are changed to facilitate casting of different concrete sections.

Openings and check-outs for columns and services are formed while the concrete is in the fresh state. In some systems the elements are wet shaped to length, with others, once the product has been cured and the prestressing force transferred, the concrete is sawn into elements of the required length.

Flooring products include hollow core elements, solid prestressed slab elements, prestressed planks as well as beam elements for beam and block floor construction.

## 21.9.3 Connections

Design of concrete frame structures requires that the completed structure performs as a monolith, just as does an *in-situ* reinforced concrete structure. The range of connections employed to satisfy this requirement include welded connections, bolted connections, cast connections employing cement or resin-based materials, and combinations of these methods.

The proof of a connection system is service performance coupled with simplicity of jigging during production and ease of assembly during erection. A key requirement is speed of achieving initial fixity while elements are suspended from the handling equipment. Welding and bolting meet this requirement although these connections have usually to be enclosed in some form of fire protection as a secondary operation.

Some producers employ patented connections, developed and tested for construction of their proprietary frames. These generally comprise bolted or welded assemblies. Welded assemblies comprise pairs, or sets of plates anchored into the concrete, or billets of steel passed through the concrete which provide bearings for steel inclusions in the mating element. Welding at site has to be carried out by certificated welders.

Bolting connections replicate the connections in structural steelwork, and the heavier connections employ rolled steel sections. When the frame has been plumbed and levelled the joints are completed with *in-situ* concrete.

Columns can be set into pockets in foundation blocks. Alternatively cast-in baseplates welded to the main steel reinforcement are fixed to bolts cast into the bases. Swaged connections and screwed or wedged connectors are used for connecting reinforcement in column-to-column joints in some systems, the complete joints of columns and bearing beams being made by formed concrete as a secondary operation.

Column-to-beam connections may be by bearing angles anchored into or bolted onto the column face or by billets of steel cast into the column or passed through pockets cast in the column.

Wall elements within the frame are generally set upon levelling bolts, horizontal joints being made by dry packing with concrete mix (Figures 21.5–21.7). Vertical joints such as at columns and walls and between walls are made using 'stitches' of concrete cast around tie steel with joggles formed in the wall ends or the columns. It is essential that the mixes and methods of compaction of concrete in stitches and joints should be carefully specified, and not left to site operatives as has been the case in past constructions.

**Figure 21.5** Gatwick Airport: Acoustic wall. Setting up the core formers and reinforcement for the waveform wall. Architectural quality concrete, countercast to ensure fit. (*Courtesy* Gatwick Airport Ltd (BAA plc), the Pavement Team (AMECJV) and C.V. Buchan.)

## 21.10 Cladding

### 21.10.1 Surface finishes

A variety of surface finishes and treatments are available to the designer. These finishes which are examined in depth by contributors, (Taylor, 1992) include:

- Brick, slate, tile and stone facings that are transfer-cast from the mould into the concrete element.
- Exotic aggregate facings that are exposed by tooling, abrasive blasting and water jetting and acid etching, as well as the use of surface retarders that are painted onto the moulds.

**Figure 21.6** Gatwick Airport: Acoustic wall. Stripping a freshly cast element from a purpose-made steel mould. (*Courtesy* Gatwick Airport Ltd (BAA plc), The Pavement Team (AMECJV) and C.V. Buchan.)

- Featured faces cast from formers and ribs secured to the mould face.
- Reconstructed stone used as a veneer or structurally throughout the element.

**Figure 21.7** Gatwick Airport: Acoustic wall. Erection in progress, preparing to thread an element onto continuity steel, the operation to be completed by grouting after positioning the eighth tier element. (*Courtesy* Gatwick Airport Ltd ( BAA plc), The Pavement Team (AMECJV) and C.V. Buchan.)

## 21.10.2 Brick, tile and stone-faced panels

In these instances the producer has the advantage discussed earlier that the orientation of the element within the mould can be arranged as best suits the incorporation of the face materials. Brick, stone and similar facings can be set into matrices on the mould and transfer cast into the unit. The profiled back of brick and stone facings enhance the bond between facing and concrete. Stone and slate are secured by the insertion of stainless steel dowels at the interface with the structural concrete. The pins are set in neoprene. This, coupled with a debonding agent, permits differential movements between the facing and concrete. In each case pointing between the facings can be carried out in the mould before concrete placement or as a secondary operation in works or at site.

The size of individual bricks and tiles varies considerably and allowances have to be made for this in the set out or the formation of the matrix. Brick and tile-faced panels cast in this manner exhibit a degree of regularity which differs from orthodox brickwork which is laid to level. Quoins and returns serve to mask these differences, as do articulated faces.

### 21.10.3 Exposed aggregate, textured and featured finishes

Face-down casting permits economic production of exposed aggregate faced with veneers of expensive exotic aggregates. Panels comprising mixes of less expensive materials, and materials where the exposed aggregate is expressed on more than one face, are often cast with a through-mix of the aggregate to be exposed. In either case use of surface retarders can produce the required degree of exposure of aggregate.

Lacquer-type retarders, that can be applied to either horizontal or vertical faces, dry on the mould yet react with the concrete. The material is available graded to provide differing depths of exposure when brushed from the face.

A recent development in exposed aggregate cladding elements employs paper liners on which screen prints of enlarged photographs have been made using a retarding agent. In the course of casting, the retarder is thus transferred to the concrete as a series of small dots. The use of different coloured aggregates in the element combines with the retarding action to produce panels with copies of the photographic illustrations on the panel face.

Featured, textured and striated faces are produced by inclusions in the mould, and these may be steel, timber or the now popular elastomeric liners. The liners can be obtained in standard patterns or, should the designer wish, special liners can be produced from featured moulds. After casting, the liners are peeled from the face of the concrete. Used with care the liners can yield fifty or more uses.

Abrasive blasting and water jetting are employed to provide varying depths of exposure. These techniques are also employed to remove a fine film of material from smooth surfaces to reduce efflorescence. These secondary operations are carried out economically between 1 and 3 days of casting according to the detail.

Acid etching is a traditional means of lightly exposing aggregate as well as cleaning up faces produced by other means. The procedures have been improved by the introduction of materials in gel form rather than the hazardous brush application of a 10 per cent solution of acid previously employed. Gel-type materials permit a delay of up to 5 days from demoulding to treatment. After a set period of about 30 minutes from application water jetting ensures a consistent degree of exposure. The material can also be used to remove cement contamination from other types of decorative finish.

Textured cladding surfaces have the advantage of being relatively vandal-proof compared with alternative materials.

### 21.10.4 Reconstructed stone

Reconstructed stone finishes are popular with designers, and are formulated to resemble the appearance of natural stone such as those from Portland and Bath. The architectural possibilities of the material are examined by Dawson (1995). In production an earth dry mix is usually employed for smaller elements and a plastic mix for veneered and solid cast panels. Earth dry casting involves mechanical compaction of a facing veneer that is subsequently backed by the structural concrete. Wet cast plastic mixes are used for large panels the unit being cast solid with larger crushed stone aggregate than that employed in the earth dry mixes.

The basic design criteria of reconstructed stone which should comply with BS 1217 are as follows:

- Minimum 28-day compressive strength 35 N/mm$^2$
- Minimum cement content 400 kg/m$^3$
- Minimum cover to all reinforcing steel 40 mm
- Surface absorption (ISAT permeability test upper limits); 10-minute test, 0.50 Ml/m$^2$/s; 1 hour test, 0.20 Ml/m$^2$/s
- If the finishing mix is different from the facing mix, the difference in cement content between facing and backing mixes should not be more than 80 kg/m$^3$
- The cover to reinforcement measured from the face of the panel should not be less than 40 mm provided that facing and backing mixes form a monolithic panel
- Tolerances should comply with CP 297 Clause 4.1.3, or more demanding limits set by some producers.

## 21.10.5 Plain-formed concrete finishes

While generally less popular than the wide range of finishes described above, cladding elements may be required to exhibit surfaces 'As cast' or 'From the mould', sometimes with a limited permitted amount of face finishing. Elements of this nature demand extreme care in the production processes, mould making, reinforcement location concrete placement and curing to ensure that the finished product exhibits the required degree of consistency of appearance.

Uniformity of appearance, though frequently specified, is impossible to achieve in practice. Slight variance in the timing of operations, temperature, mix proportions, curing cycle and many other details can result in disappointing results. Quality production can be achieved provided the specifier takes note of the recommendations of selected manufacturers of high-quality product.

Section 21.19.1 later in this chapter underlines important aspects of the production of such finishes.

## 21.10.6 Size of elements

Site and works handling facilities govern the eventual size of precast cladding elements. Historically, one designer employed prestressing techniques to permit the production of cladding elements spanning six storeys of a structure. The norm for reinforced concrete cladding appears to be spandril panels of the order of 6–9 metres and cladding panels up to 6 metres long, one or two storeys high with a mass of about 20 tonnes. Practical considerations such as site access and transport restrictions govern the eventual panel size and weight.

## 21.10.7 Joints in cladding

Joints are such a critical part of cladding and much of precast construction that elements have been described as 'the concrete that fits between the joints!' Joints are required to look good, be weatherproof, redirect surface water yet allow the normal movements of the concrete.

Many joint arrangements have been tried some extremely complicated configurations with washboard features and incorporating gaskets set in grooves as well as elastomeric sealants and baffles.

Current practice utilizes simple panel edge detail avoiding complicated edge profiles which are liable to be damaged in erection. Square edges at joints provide a sound seating for sealants which are almost universally employed. Joint widths have to be of sufficient size to ensure that the sealant when installed can accommodate thermal movements of the concrete as well as shrinkage and creep, without the sealant/concrete bond failing. With adequate joint width this arrangement allows good access for sealant insertion and inspection. The concrete surfaces at the joint are carefully primed in order to improve adhesion of sealer. Two seals are employed in the face of the joint with an impregnated foam seal at the inner face. Seals are usually one part, moisture curing, low-modulus silicone rubber, gunned against a foam backing strip.

## 21.10.8 Cladding fixings

This area of construction is one where the supplier's inputs on design of elements, possibly redesign of the original architect's detail, can provide immense economies through reduction of numbers of elements and simplification of fixing detail.

Fixing systems combine hangers for instant attachment to the frame thus early release of craneage. These are supplemented by restraint fixings which cater for plumbing and lining while the panels are so suspended permitting normal movement of the concrete in service. Stainless steel components avoid problems with corrosion although it is essential that the proprietary systems are used in their entirety to avoid local corrosion cells where, for example, sub-standard shims might be used for packing.

## 21.10.9 Weathering

A long-standing and definitive report on weathering of concrete surfaces is that prepared by Hawes. This identifies factors contributing to weathering in terms of water flow across and down surfaces, the speed of such flow combined with the angle of the surface determining the amount of deposits causing staining. Hawes describes means of channelling surface water about the surface and away from joints by the use of quirks and drips, pointing out that much of traditional stone detailing is effective when used in concrete design. Hawes also makes the point that concrete quality has considerable influence on weathering capabilities.

It is known that some surfaces are more durable than others, external surfaces of smooth concrete tending to dilapidate sooner than those with texture. Textured surfaces can be used to direct water about the surface and to mask joints in the building face.

Polished and ground stone surfaces, slate and tile promote the flow of water at the surface and become to some extent self-cleaning.

## 21.11 Bridge beams

Production of precast, prestressed bridge beams in the factory provides a guarantee of high-quality durable concrete elements with high standards of dimensional accuracy. As well as savings in falsework and formwork the high quality of finish obtained from casting under factory conditions ensures low maintenance costs. Suppliers of prestressed beams offer the civil engineer an immense variety of elements of tested and proven design for use in a range of situations. Products for use in bridge spans of 12–40 metres include beams of a variety of configurations for either closed or open soffit bridges. The Prestressed Concrete Association 'Y-beam' series provides an open soffit solution for medium- and long-span major road bridges. 'M-beams' provide a voided bridge deck (pseudo box) for medium- and long-span road bridge construction. 'Super Y-beams' cater for the longest spans in major motorway schemes, providing an economical solution for spans up to 40 metres. 'T-beams' are available for use in short- and medium-span bridges, car parks, jetties and heavily loaded structures, a rapidly erected alternative to a wholly *in-situ* slab. Other configurations include 'U-beams' which by forming a strong box section in combination with the *in-situ* slab cater for torsion forces in skew bridge construction. Edge beams are available for use with all these standard beams. Single beams of various configurations including double-tee beams are used as footbridges.

## 21.12 Double-tee beams

While the technology and methods described in the case of bridge beams apply equally to double-tee beams there are some aspects of the manufacture and use which differ. Double tee beams are versatile elements and can be used for other uses than in bridging. Examples of applications for double tee include:

- Footbridge construction
- Floors and roofs for buildings and car parks
- Walls for building enclosure
- Seating elements and ramps in grandstands and stadia.

As well as providing an instant working surface, double-tee beams used in floors and roofs allow early placement of topping concrete.

Double-tee beams are manufactured in individual moulds or on long line beds. The geometry of the elements is such that the moulds can be of one piece (incorporating soffits, sides and flanges). The normal hog or camber which develops as a prestressing force is transferred to the concrete and acts to break adhesion at the concrete/form interface so that demoulding becomes a simple matter of lifting the element from the mould.

Connections within the structure are made by welding plates at the bearing positions and at abutting flanges.

## 21.13 Sea-defence units

### 21.13.1 Dollos units

These concrete elements covered by patent are of the form of a 'Twisted H'. One leg of the H is set at 90° to the other. Generally site-produced the elements are cast from steel moulds at the rate of one per mould per shift. Economy in moulds is achieved by providing two bases per set of sides, the sides being used twice per shift. Elements are cast with each leg of the H horizontal and the bar of the H vertical. In order to facilitate early stripping of the side members standing supports left in position support the uppermost leg during the curing process.

### 21.13.2 'Shed' units

These proprietary sea defence elements take the form of a concrete cube with a hollow centre. Each face is pierced so that when set into place the elements stifle the force of waves on a shore. The elements are cast in steel moulds the sides of which carry profiles to form the piercing in the faces. The central void is formed by an inflatable former which can be simply deflated for stripping purposes.

### 21.13.3 Walling units

Sea wall elements provide a smooth wave-breaking protection for the shoreline. Contracts generally require large numbers of elements and they are cast from steel or concrete moulds either cast on edge or reversed to accommodate the geometry. Precasting face down ensures a dense well-compacted durable surface.

## 21.14 Piles

Reinforced concrete piles are manufactured both at site and in the works of specialist suppliers. Long line casting techniques are employed in conjunction with casting machines. Piles are generally simply reinforced and concrete placement into the open mould face presents little difficulty.

Prestressed piles are cast on long line beds, shoes and caps, acting as distribution plates, locate the wire or strand. In the case of major civil engineering works such as power stations contractors have set up site casting facilities employing concrete moulds between gravity abutments. Readymixed concrete has been used to good advantage, the delivery vehicles discharging directly into the moulds.

In the case of prestressed pile production using concrete moulds the shortening of the free wire or strand between the elements at the time of transfer breaks the bond between element and mould. Where reinforced concrete piles are cast in concrete moulds, rigs comprising pumps and hydraulic jacks have been used in pairs at the lifting points to jack the piles sufficiently to break the bond at the concrete/mould interface.

## 21.15 Sleepers

Railway sleepers, or rail ties as they are known outside the UK, are ideally suited to linear production on the long line prestressing bed. Some sleepers are manufactured in other countries in individual moulds employing post-tensioning techniques but all manufacture in the UK is in gang formation six or eight elements wide on heated lines.

Accuracy required of sleepers is high, particularly the rail seatings and dimensions affecting rail gauge. The elements are cast underside up with the rail attachments inserted into accurately placed pockets in what becomes the soffit of the mould. Steel moulds are employed, fabricated within demanding tolerances. Each mould has identification marks which transfer cast an indelible mark into the elements to permit traceability of the time of placing and thus the batch of concrete. Importantly the mould can be traced in order to permit remedial measures in the event of inaccuracy. Quantities of concrete cast in each shift are high and placement and compaction are carried out by a casting machine.

A high percentage of the make is inspected and dimensions are gauged each shift. As the product is highly repetitive the gauging employs sophisticated equipment with results sent to data plotters and thence to the quality department records. A requirement of specification is that a percentage of elements from each line are tested. These tests are also automated and the results analysed against standard requirements then added to records. Curing using electrical systems is rigidly controlled by embedded thermostats.

While the Standard PrEN 13230 Concrete sleepers and bearers, parts 1 to 5 covers manufacture, curing, and test methods and provides a design guide, manufacture is carried out to more stringent requirements to meet the specification of the railway authority. In the UK the authority, Railtrack plc, requires compliance with RT/CE/S/030.

## 21.16 Structural elements for groundwork and support

### 21.16.1 Box culverts

Progress of construction work in the ground depends to a great deal on resolution of problems of support, drainage and other contingencies. In many instances the use of precast elements provides solutions to these problems.

It is always interesting to compare the activities involved in *in-situ* construction with those of the precast alternative. For example, in the construction of an underpass or sizeable culvert where a 'cut and cover' process has been decided upon the *in-situ* constructor has first to excavate to the base level, then place hardcore, blinding concrete, then install a membrane, if specified, prior to side forms reinforcement and stop end member pierced by continuity steel. The base can then be concreted. All this work is carried out in the prevailing weather conditions. Form fabrication, steel erection and closure of the forms precedes the wall concreting operation. After wall form striking the roof forms are installed and steelfixing and concreting operations are carried out. Subsequently this formwork is stripped. Of course, walls and roof formation may be combined where a tunnel type form is employed. Many handling and transporting operations are involved as operations are repeated bay by bay in the course of the construction.

Using box culverts, with hard base prepared, installation proceeds, elements being placed by crane from the delivery vehicle then lined and levelled. Final jointing is achieved by winching units to close the joints. Jacking techniques are employed in constructing underpasses and culverts, the elements being driven as the ground is excavated, as in tunnelling with a shield.

In production, box culverts are usually cast on end around a central collapsible form to facilitate concrete placement. External vibration is applied as the concrete is distributed evenly around the four sides of the unit. The whole mould is covered down for curing purposes. After stripping, the elements are turned into the service position for delivery (see Figure 21.8).

**Figure 21.8** Setting one of a batch of precast culvert elements onto a prepared foundation. Note the absence of falsework and formwork. (*Courtesy* Hanson Concrete Products Ltd.)

## 21.16.2 Retaining wall units and sound barriers

These elements are generally proprietary standard elements and can be ordered from a catalogue. Steel moulds are used in production, the elements being cast on edge for ease of placement and compaction. Sound barriers of similar section having exposed aggregate faces are cast face-down against a retarder. To facilitate handling at site, holes are formed through the upright limb of the element.

## 21.16.3 Tunnel segments

Tunnel segments are cast to extremely demanding levels of accuracy, because, in service, the physical form of the elements generates the final geometry of the tunnel. The moulds employed are of machined steel or of cast concrete, the latter being cast from accurately produced steel or composite masters. Some concrete moulds include heating elements to promote the hardening of the cast concrete. Concrete moulds are frequently assembled in gang form, and these gang moulds are circulated through a series of stations where casting, curing and demoulding take place.

## 21.17 Erection at site

Erection is generally carried out by small teams of specialists. These may be in the employ of the precast supplier although there are firms who undertake erection work on a sub-contract basis. Key operations are the unloading, handling and installation of elements, the lining and levelling of the components of a frame or the cladding, jointing or stitching and the eventual waterproofing of the structure.

### 21.17.1 Delivery

It is essential that programming of delivery matches the erection cycle, where possible elements being handled into position direct from the transport. Elements are usually delivered to site loaded such that they can be lifted directly into the structure without the need for turning. Large cladding panels may require to be loaded on inclined stillages on low-loading trailers to clear obstructions en route. Long units such as bridge beams and double-tee will be delivered on turntable vehicles with articulated trailers. Floor elements are generally close stacked on flat-bed vehicles and trailers. Loading is a skilful business as elements can be subject to stress when, for example, vehicle chassis flex.

Whether elements are manufactured off-site or in a casting yard detached from the main construction it is important that deliveries are phased in with the erection program. Site-stored elements are liable to be damaged and any double handling resulting from site stacking is an unnecessary expense.

### 21.17.2 Handling equipment

The lifting equipment and lifting attachments at site must replicate those employed in production. Spreader beams, lifting connections and tackle must be suited to the operation. In some instances specialized equipment has to be designed to cater for lack of headroom or a need to reach under previously erected components. Spreader beams allow for differing positions of lifting points permitting lifting attachment to be made on panels and beams at the designed points. Articulated beams allow for non-cyclic lifting points on slabs ensuring that elements seat uniformly onto bearings, avoiding cracking and damage due to high local stresses. In the event that elements are stored at site as a buffer against delays in delivery, proper racks and stillages are required to prevent damage.

### 21.17.3 Responsibilities

At site a full set of structural drawings must be available and the supervisor responsible for the erection must be informed on critical detail. Details such as acceptable locations for plant to be used in the erection process and limitations on construction loads to be placed on the structure in the course of the erection process must be known and observed. It is imperative that the structure should be properly braced, stayed and propped such that it is stable at all stages in the work. Whatever the basis of the erection contract it is essential that responsibility is established for the structural integrity of all fixings and connections. Where cladding is involved responsibility for weathertightness has to be established. Where firms of quality assured capability are employed there will be published procedures for all activities and persons responsible for those activities will be named.

## 21.18 Segmental casting

As bridge spans have increased, constructors have made greater use of precast concrete. The economies outlined earlier in this chapter, reduction in falsework and temporary works, the concentration of operations into a dedicated production facility with greater utilization of fewer skills have all tended to influence the decision to precast.

Initially large-span linear beams were cast on-site or in factories then installed by erector beams. Today it is common practice to cast beams in segments several carriageways wide, each often equal or greater, in mass than the linear counterpart. Elements are produced either at site or in a precast producer's factory.

Segmental casting techniques permit the constructor to rotate elements in the mould such that changes in plan or elevation profile can most readily be accommodated. For example, a segment of bridge or elevated motorway having a curved alignment would be cast deck uppermost, the plan geometry being catered for by the inclination of the stop end members of the mould. Segments for a balanced cantilever construction would be cast on end, permitting the elevation geometry to be varied by adjustment of the mould member forming the soffit of the segment. Elements cast in this fashion are supported at site by falsework while jointing or stitching concrete is cast, the spans being completed by post-tensioning.

### 21.18.1 Countercasting

Current trends in segmental construction are towards countercasting, sometimes called matchcasting. Segmental casting using countercasting techniques ensures immaculate joints between elements. Countercasting is carried out in two ways. Where casting machines are employed each element is cast using the adjacent element as a stop end. In casting the segments of a balanced cantilever bridge with its arched soffit, the inclination and height of the mould base within the machine is adjusted as each element is cast to generate the correct geometry. An alternative method of casting employs a continuous soffit on which succeeding elements are cast and a pair of mould sides are moved along the length of the soffit. In each cast the previous element forms one end profile and a stop end forms the other. In this way the faces provide an immaculate joint in the final erection. Reinforcing cages are assembled on a jig to ensure accuracy and maintenance of cover. In both cases ducts for stressing tendons are supported within the reinforcement cage which includes anchorages and additional reinforcement at the anchorage positions.

No remedial work is permitted on countercast faces as they have to match perfectly in the finished structure. The final joints between segments were generally made using epoxy resins although now, in some instances, the segments are jointed dry, the erection process being speeded by the omission of the epoxy application stage. In the latter case the tendons run within the cores, positioned by cast-in distribution plates.

As well as carriageway sections the countercast technique has, for reasons of accuracy and finish, been employed for bridge piers and towers, in these instances the elements are stack cast, each element upon the preceding element. In this instance variations in geometry can be made by lapping or adjusting the length and shape of the side forms.

Countercasting yields economies through localization of production, reduction in falsework and formwork requirements (although the casting machines employed in countercasting are expensive items of equipment). Handling is reduced to that of turning if necessary from the casting position then stacking and final transport to the erector beam.

### 21.18.2 Incremental launching

Incremental launching using precast segments provides further economies beyond those obtained from simple segmental construction. The technique involves the establishment of a production facility at one of the bridge abutments. This facility generally takes the form of a simple shelter set over the casting base. It includes a static bed mould with removable mould sides and core formers. The base is heated for curing purposes and is anchored by piles to resist lateral movement when jacking commences. This assembly is used to cast segments of 10–30 metres long, 400–1200 tonnes mass, each segment being the complete width of the carriageways. Reinforcement is preassembled in jigs and the concrete may be site mixed or supplied from a readymixed concrete supplier.

With the first segment cast, cured and stressed, a launching nose is attached and the assembly is then jacked forward. Succeeding segments are each cast onto the previously cast segment and connected by prestressing tendons. Jacking continues on a regular cycle, usually of 7 days or less. As the ribbon of segments passes over the bridge piers which are anchored against turning, forces generated by the jacks lift and thrust the assembly forward. In course of installation every section of the bridge is subject to deflection and contraflection as it passes over the piers.

Where the bridge alignment is mildly curved this is achieved by casting straight segments. The tops of the piers are inclined, effectively warping the concrete to the required line. Where the bridge is markedly curved in plan the segments are cast curved.

## 21.19 Troubleshooting precast concrete

The producer aims to achieve a zero defects regime but, however well organized the manufacturer and well skilled the production staff, a small percentage of defects and non-compliances are inevitable. Defects result from such details of manufacture as wrong interpretation of drawings and instructions, poor moulding, casting and curing techniques, early handling in the 'green' state. Unsatisfactory handling and storage of finished product in the works and on-site in course of final erection also contribute problems.

In the event of non-compliance the experienced technologist with knowledge of precast techniques, materials and method can make a major contribution to the processes of rectification and the prevention of ongoing problems.

Defects and failure to comply with specification fall into several categories, i.e. visual defects, structural damage and lack of accuracy of product. While defects in precast elements are mainly readily apparent the causes are generally not so easily recognized arising as they often do from the special processes employed in production which differ from those in *in-situ* working. Careful examination of a suspect element, the conditions under which it was produced and factors such as temperature, timing of operations and also the skills employed, are required to find solutions and to be able to specify effective remedial action. The technologist charged with finding solutions to systematic defects is advised to examine the whole sequence of events in the production of the elements, commencing with the design of moulds and continuing to the stages of final loading, delivery and erection. Test records and prior history of non-compliance should be examined to establish whether like defects have been recorded for these or similar elements in the past.

Due to the repetitive nature of much of precast production a problem may develop and be replicated several times depending upon the frequency of in-process checks and inspections. Early identification is the key to resolution of problems hence the importance of those in-process checks and especially the initial inspection of moulds when first set up and the post-stripping inspection of the first product from any mould or line. At this stage minor alterations to equipment and method coupled with useful feedback to the production team can bring products under control. Ongoing checks and inspections then serve to confirm the effectiveness of remedial actions and to ensure the maintenance of quality.

## 21.19.1 Visual defects

Visual defects are frequently the result of some defect in mould design. Moulds have to be designed to enable stripping of the element or removal of the mould parts without damage to either the mould or the cast element.

Mould faces, finishes and treatments must be selected to generate the required texture and detail at the concrete face while yielding economic numbers of re-uses in series production. Spalling, scabbing corner and edge damage, and particularly the tearing of edges of features, frequently indicate a wrong choice or poor application of parting agent or mould oils.

Local darkening, most noticeable at edges and corners as well as at joints in the form face, is generally caused by mould leakage. These can usually be avoided by provision of seals at all joints and the application of elastomeric sealants around openings accommodating inclusions and projecting steel.

Dark patches on smooth concrete faces and poor display of aggregates in exposed aggregate work result from local upsets in vibration where a thin mould facing vibrates at a different frequency from heavier or better supported areas of the mould. These patches indicate where the finer particles of the paste have been drawn to the surface. In the case of exposed aggregate faces the larger stones are impelled into the mass of concrete to be replaced by paste. Various forms of accelerometer can be used to identify areas of a mould where the mode of vibration is distinctly different from others. 'Tuning' of the mould by the incorporation of extra framing members behind timber or ply faces and attachment of additional patches of steel to thin metal sheathing can overcome these problems. To avoid such problems in exposed aggregate work, and in order to achieve a good display of what are often expensive materials, the mix should contain as much coarse aggregate as possible consistent with maintaining the workability required to achieve sound compaction.

A defect sometimes encountered in the face of precast elements is that of prominent straight lines of fine paste at the time of exposing aggregates by tooling or the use of surface retarders. The defect will be found to extend some distance into the mass of the unit. This defect results from poorly secured joints in the mould face which allow local differential movements between facing panes, propelling the larger aggregates from the face and replacing them with paste. Refixing the facings and sealing the joints provides the only solution to this problem.

On panels cast face down in the mould, areas of concentrated aggregate surrounded by lines of paste are caused by concrete being dumped into one place in the mould then spread by vibration. The type of skip employed should permit concrete to be spread evenly about the mould.

Where reconstructed stone elements have been produced employing a veneer backed by structural concrete, poorly timed operations can allow the structural concrete to bleed through to the face of the element.

Plucking of the arises to features and striations is generally a result of insufficient amounts of lead and draw.

Surfaces, whether achieved by use of retarders or by tooling, often exhibit variations in the degree of aggregate exposure. The top, bottom and edges are less well defined than the parts of the panel which can be easily accessed during the exposure process. This results from difficulty of reaching areas while handling a heavy tool, or the natural reluctance of the operative to reach or stoop to carry out the work. The edges of tooled work are often less well exposed as the operative is reluctant to tool near vulnerable edges. This is an access problem and the conditions in the workplace must be such that the operative can reach to work comfortably, particularly if the task is an ongoing one. Balance equipment for tooling such as is used on the automobile assembly line can also avoid operative fatigue.

Variations in exposure of aggregate and texture occurring where abrasive blasting or water jetting is employed to expose aggregate, often particularly apparent in circular or featured work, result from changes in the angle of incidence of the jet to the face affecting the energy applied. To avoid such problems free access is important and where possible a means of rotating circular elements.

## 21.19.2 Structural defects

Structural defects often result from displacement or misplacement of reinforcing steel. With the extreme amount of energy imparted by the vibratory equipment used in site or works production of precast elements particular care is required over selection of spacers. The wrong type of spacer, a chair-type spacer, for instance, used to locate steel from a vertical face, will rotate under the influence of vibration and loss of essential cover will result. A detail that is often overlooked is the provision of spacers at the end of main bars in beams and floor units.

Projecting steel and inclusions such as bearing plates and bracketry used to tie cladding to the frame or to provide connections between frame elements can, if not correctly jigged into position, move when the concrete is vibrated to project more or less than required by the design. Elements which get to site exhibiting incorrect location and projection of these items may prompt local *ad hoc* adjustments at site, welded attachments or the inclusion of shims which upset the design of the joint or connection, causing problems later in the life of the element or structure.

Steel in stair elements and balcony units requires careful location, particularly where elements are cast other than the way up in which they are set in the construction. Balconies are often cast soffit uppermost, causing confusion regarding the location of the main steel reinforcement.

Corner and bearing damage often occurs in early handling or in secondary handling operations between the casting location and final erection at site. It is essential, and a requirement of quality assurance procedures, that repairs which may involve reinstatement of steel or the insertion of additional reinforcement are subject to approval by a suitably qualified engineer. All such remedial work must also be subject to re-inspection before despatch for final assembly into the structure.

Normal shrinkage and creep movements of concrete, particularly at early age, often cause cracks. An apparently minimal crack identified in stack may in time develop further providing a route for the ingress of corrosive materials, leading to serious problems in the course of time. For this reason all cracks must be given serious consideration with a view to elimination of the cause. Shrinkage and settlement cracks are generally located at sharp changes in section such as at the corners of openings and at the junction between web and flange in I-section, tee-section and double-tee beams. This is another situation where engineering decisions are required regarding the integrity of the unit, although in the case of openings and through-holes early precautions in the provision of elastomeric gaskets in formers to avoid restraints can avoid difficulties. In production, revibration applied to I-beams etc., can overcome the problem of cracks resulting from plastic settlement such as at the web/flange junction in I-beams and double-tee beams.

Cracks occurring at the corners of door openings in battery cast elements can be avoided by providing a continuous ring reinforcing bar around the opening, the part of the bar across the threshold being cut away once the element is installed. The introduction of a small amount of prestress along the line of the lintel assists in avoiding cracks at the head of the opening.

Incorrect handing, or elements which are incorrectly marked as right- or left-handed can cause major problems at site. The worst situations here arise from confusion in the minds of the operative and frequently wrong interpretation of drawing instructions. Drawings which state 'Element Mk xx RH, similar in all respects to MkxxLh but. . .' and then proceed to list a number of differences can cause difficulties for all concerned in the mould making, setting-up and alteration processes.

Product storage and stacking methods can cause considerable problems. Bearers within stacks may cause local cracking when not placed correctly. No more than two bearers should be used for reinforced beams, placed at the fifth points, or at the points of support for cantilever elements. Prestressed elements must be stacked with bearers at or near the ends of the element. Slab and hollow core elements are stacked in direct contact after the surface of the lower element has been swept clean.

At the time of transfer of force damage can be caused by shortening of elements and movement along the stressing bed when mould features, or connections set into pockets in the mould base, cause restraints.

Connections and lifting attachments which are embedded in other than sound, well-compacted concrete are a hazard in all stages of handling and erection. Sockets or bracketry which are wriggled or vibrated into the fresh concrete become surrounded by paste rather than concrete and are suspect. All such fittings should be fastened to brackets or cleats at the time of setting up the mould and the complete concrete mix vibrated around them. Lifting attachments which are to resist other than axial forces must be trimmed by bars or contained within the reinforcing cage.

Early handling damage results from force applied in stripping the mould or the element. Ideally the stripping action should be translated into a shearing rather than a peeling action, taking advantage of the weight of element or that of the mould. With the mould anchored down force can be applied by the lifting equipment although unless this effort is applied uniformly to concyclic lifting attachments or via a balancing spreader system binding within the mould faces or corner cracking may occur. Highly featured elements will benefit from the application of a jack or wedge at the mould/concrete joint to allow air to infiltrate and overcome the adhesion between the two.

At site the technologist may encounter defects such as corrosion of reinforcement and spalling of concrete. Covermeter surveys will permit assessment of such problems. Remedial measures must be approved by a qualified engineer and may include removal and replacement of affected steel as well as ongoing protective measures. Sound compaction is essential in protecting reinforcement from corrosion, and ultrasonic testing provides a reliable measure of the degree of compaction in the hardened concrete.

Connections may exhibit corrosion where, for instance, dissimilar metals have set up corrosion cells and particularly where water has penetrated the joint. Such problems are often caused by the introduction of unauthorized washers or packing materials.

Cladding elements, while restrained, must be free to move due to thermal movements as well as normal shrinkage and long-term creep. Internal linings such as brick or block skins have been known to restrict movements and cause cracking.

## 21.19.3 Accuracy of product

Tolerances are subject to client specification for specific products and vary over the wide range of elements included in this chapter. Often manufacturers elect to set closer permissible deviation in manufacture to those of BS 8110 :1995 Structural use of concrete. In the case of cladding elements, guidance is provided by BS 8297 Design and installation of non-loadbearing precast concrete cladding. Sub-standard mould design and manufacture causes problems in the critical stages of erection. Inaccuracy may result from moulds which have been fabricated without regard to the pressures resulting from placement and compaction, and the result is bowing and poor alignment of faces, and wind (twist) or distortion. Moulds may pass checks and inspections in the unfilled state yet fail to comply in service.

Stop-end movement in the course of placement and compaction can cause problems of loss of bearing area. In-process checks and inspections should identify the cause of such inaccuracy early in the course of production permitting corrective action to be taken. Certain production methods tend to cause specific problems. For example, in gang and battery casting movement of mould dividers causing overthick elements in one cell will also generate adjacent elements thinner than specified.

Normal shrinkage and creep can result in bow and wind when elements are stacked out-of-plumb or on inclined bearers. Early identification of the problem and time in stock permitting, these can be corrected by reversing the inclination of the bearers, so inducing a reversal of the bow or wind.

Elements stacked in situations where the curing conditions on opposing sides vary considerably will result in bow setting in. A survey of the elements in stock may permit grading and reprogramming of deliveries to reduce the discrepancies between adjacent elements in the structure. A survey of varying camber in prestressed units can likewise reduce steps in floor and bridge soffits. Markedly differing cambers are, however, indicative of some variation in concrete quality, strength at transfer of force or in the curing conditions and warrant careful investigation in the production department.

Opening formers in beams and panels tend to be displaced, or rotate during the vibratory process, cores in wet cast hollow beams tend to float and in each instance positive ties are essential. Where formers and fittings are attached to the leaves of a battery mould the fixing has to withstand displacement by concrete dropped as much as 2 or 3 metres. The casting procedures should define the sequence of filling moulds. Service provisions such as conduit and pipe which lie horizontally within a cell or mould require support by bars passed through them or by attachment to the mould.

A human aspect of accuracy is that where operatives in a repetitive production operation, 'learn a mould'. Each day or each shift in repetitive work they put the same fitting into the same jigs or screw the same anchors back to the same predrilled holes. If one of the jigs or holes becomes lost or filled with paste, that fitting may well be omitted from subsequent units. In process checks and inspections are vital to eliminate such errors.

## 21.20 Conclusions

Many of the disciplines of precast concrete manufacture and erection set it aside from the normal practices of *in-situ* construction with concrete, for example the possibilities outlined in this chapter for casting outside the series of events which govern the overall duration of construction. Precast producers also have the opportunity to cast elements oriented within the mould as best suits achievement of specified finishes as well as casting elements out of sequence to suit the moulding technique employed. For optimum results, early input of the specialist knowledge of the precaster and possible modification of design are required to improve the performance of the elements in service and also speed the erection process at site.

Standard precast structural elements such as those for frame structures and bridges meet many of the requirements of the civil and structural engineer, reducing design time and lead times from contract placement to commencement at the construction site.

The concrete technologist with a background knowledge of concrete properties, mix design procedures and test procedures for fresh and hardened concrete has much to offer in early stages of design, establishment of standards, mix design, quality control and troubleshooting in contracts involving precast concrete.

## Case studies

The following case studies provide an illustration of the importance of early establishment of standards of accuracy and finish.

(1) A precast manufacturer produced a series of column casings for a prestige apartment block incorporating a restaurant. Work had progressed for several floors above the location of column casings which formed a colonnade directly outside a restaurant with views of a river scene. The clerk of works, and eventually the architect, expressed dissatisfaction with the exposed aggregate finish to the column casings.

An independent technologist was invited to give an opinion regarding whether the finishes met specification requirements which called for the casings, 4 metres high, circular in plan and of 1 metre diameter to exhibit a uniform degree of aggregate exposure. This was to be achieved by abrasive blasting, all to be in accordance with an approved sample.

On examination it was found that the casings cast in a solid through-mix containing Cornish Granite aggregate showed distinct vertical lines of unevenly exposed aggregate and pitting of the paste. Enquiries established that there was a sample of the finish which had been approved at the stage of placing the contract. When this sample was eventually located it became apparent that the finish on the casings was inferior to that of the sample, which took the form of a typical small flat sample-maker's model.

Abrasive blasting being extremely operator and material sensitive, it would have been difficult to achieve a similar degree of consistency of aggregate exposure on the circular work as could be achieved on the flat, small model. There has to be adequate aggregate in the mix to present a good display. The operative has to be able to apply a consistent amount of effort in spraying the abrasive medium normal to the surface of the element at the time his vision is partially obscured by dust in the process.

As rework would have reduced the concrete cover to the reinforcement, the casings were condemned. An unfortunate instance of failure to ensure that samples submitted represented the attainable finished result. More so because the authorities attention had initially been drawn to the casings because a small fillet forming a chase around one of the elements but not included in the sample had moved in course of casting.

(2) Another case that demonstrates the importance of establishing standards early in the course of a contract concerns cladding elements to an industrial building. Again the architect and client were dissatisfied with the appearance of board-marked precast concrete panels, stating that they did not comply with specification and did not match a specified sample panel on a building previously erected on the same site. Inspection confirmed that 100 white concrete panels, each 6 metres long by 2 metres high, to one elevation of the building, did exhibit the normal minor instances of variability to be expected of precast concrete. The architect quoted the specification which called for uniformity of colour and pointed out the specified sample on the adjacent building.

Some research revealed that the building containing the sample had in fact been cast *in-situ* and while the elevation displayed some differences in shadation, these were consistent with *in-situ* construction and were systematic, resulting from the joints and variations in the degree of compaction achieved in the course of lift upon lift construction.

At one stage in discussion it had been suggested that the panels might be painted although this would have resulted in ongoing maintenance costs. On reflection and hearing of the range of factors which govern the appearance of precast elements the architect accepted that while the panels were not as he originally intended, in view of the crispness of the joints between the precast elements and the excellent definition of the board-marked finish he was prepared to accept the work.

Samples must be realistic, of large size as possible and produced by similar means and in similar conditions to those in which the elements for the contract will be produced.

## References

Billig, K. (1955) *Precast Concrete*, McMillan, London.

Dawson, S. *Cast in Concrete*. (1955) Leicester, UK Architectural Cladding Association. BPCF.

Elliot, K.S. and Tovey, A.K. (1992) *Precast Concrete Frame Building – Design Guide*. British Cement Association, Crowthorne.

Hawes, F. (1981) *The Weathering of Concrete Buildings*. Report for Cembureau, British Cement Association, Crowthorne.

Richardson, J.G. (1973) *Precast Concrete Production*. Allied Book Publishers, Basingstoke.

Richardson, J.G. (1991) *Quality in Precast Concrete: Design–Production–Erection*. Longman, Harlow.

Southcote, M.F. and Tovey, A.K. (1998) *Tilt-up Concrete Buildings: Design and Construction.* British Cement Association, Crowthorne, UK.

Taylor, H.P.J. (ed.) (1992) *Precast Concrete Cladding*. Edward Arnold, London.

# PART 8

# Concrete roads

22

# Concrete roads and pavements

*Geoffrey Griffiths*

## 22.1 Introduction

Concrete is a particularly useful form of construction, which can be used in situations where pavements are subjected to considerable point loads, and aggressive environments; concrete pavements have a number of uses that make them beneficial when compared with alternative bituminous designs. They are specifically useful when:

- high point loads are expected
- diesel spillage or other chemical spills may attack alternative materials
- low subgrade strengths are expected
- high pavement temperatures will be experienced, e.g. in tropical regions
- heavy axle loads can be anticipated
- poorly skilled labour is the only possible labour force

The specific advantage of a concrete pavement is that it is a rigid stiff plate, applying a load to a wide area. Concrete pavements are also relatively easy to construct. Readily available concrete and steel can be quickly and easily assembled and laid in association with the main construction works of a large building project. Many examples exist of large projects, constructed in various forms of concrete; motorways, industrial hard standing areas, airport taxiways, runways and aprons as well as simple bus lay-bys, all successfully constructed in various forms of concrete.

In its simplest form a concrete pavement can consist of a domestic drive laid as a humble 100 mm mass concrete slab. The most sophisticated form is probably the continuous reinforced, exposed aggregate finish Whisper concrete (Charonnat *et al.*, 1989) used in some motorway projects. The successful application of concrete is beyond question; the form of construction is evident throughout civil and building engineering projects. This chapter is intended to provide a brief introduction to the different methods of construction and common applications, which may be found in our current environment.

## 22.2 Typical concrete pavement types

Three alternative styles of construction can be employed. Each system is appropriate to different applications; the three main types of construction are:

- jointed unreinforced concrete pavement (URC)
- jointed reinforced concrete pavement (JRC)
- continuously reinforced concrete pavement (CRC)

The systems are described and discussed separately in this section of the chapter.

### 22.2.1 Jointed unreinforced concrete pavement (URC) description

Jointed unreinforced concrete construction has been extensively used on major highway projects. The system is currently out of favour in UK highway schemes but is extensively used on general infrastructure projects and in other parts of the world. Figure 22.1 illustrates a typical industrial application. The pavement consists of a patchwork of concrete slabs joined together with dowel and tie bars or crack induced joints. Each slab

**Figure 22.1** A typical industrial lorry yard.

will consist of approximately square units. The detailing of the joint layout is crucial to the successful design, execution and operation of the pavement.

The system relies on the tensile capacity and flexural strength of the concrete to resist cracking and successfully carry a load. When the pavement is built, the size of the concrete panels is controlled by the shrinkage strain generated by the hardening process. As the concrete sets, gains strength and cools, shrinkage strains generate a tensile force in the pavement. The size of the concrete slab controls the magnitude of the force. If the tensile capacity of concrete is exceeded the slab cracks.

A number of rules govern pavement detailing. The most important feature is to ensure that the joints are detailed, designed and most important, spaced correctly. A pavement joint must be arranged to produce a patchwork of rough square panels, the longitudinal joints running in one direction and transfers joints arranged at 90 degrees. Joint spacing is controlled by standard practice and is a function of pavement thickness: thicker pavement slabs can have greater joint spacing. Table 22.1 (ACPA, 1992) details the accepted practice for joint spacing. It is noted that the recommended maximum ratio of longitudinal to transverse joint spacing is 1.25. Pavement joints may be constructed as dowelled or undowelled: current practice is to construct most pavements with dowelled tie bars. Removing the steel dowels reduces the efficiency of the joints and gives an increase in pavement thickness.

**Table 22.1** Joint spacing for mass concrete pavements (ACPA, 1992).

| Pavement thickness (mm) | Maximum recommended joint spacing (m) | |
|---|---|---|
| | Gravels and crushed rock | Limestone aggregate |
| 150 | 4.5 | 5.4 |
| 200 | 4.9 | 5.9 |
| 250 | 5.3 | 6.4 |
| 300 | 6.0 | 7.2 |

## 22.2.2 Typical applications for URC pavement

Several different types of mass concrete pavement are currently found in major construction projects.

### *22.2.2.1 Airfield pavements*

One of the most significant Western European applications of mass concrete pavements is currently the construction of airfield taxiways and aprons. Large, thick slipformed areas of concrete are constructed with sawn, crack induced joints. The pavements are built across a slipformed platform of typically 150 mm thick leanmix concrete. The leanmix acts as a support to the crack induced joints.

### *22.2.2.2 Industrial yards and hard standings*

Many large industrial sites are constructed in mass concrete. The method of construction is particularly useful for large lorry manoeuvring and turning areas. Typical industrial projects consist of 200 mm thick pavements in 4.5 m square grids with dowel and tie bar joints laid on to a 300 mm thick crushed rock sub-base.

#### 22.2.2.3 Example: Mumbai (Bombay) city streets

A surprising application for mass concrete pavements can be found in Mumbai, India. The city authority currently uses mass concrete as a standard pavement option for city streets. The city engineers have found that system to be more durable than alternative asphaltic concrete systems in the hot humid climate. Concrete slabs are constructed using labour-intensive techniques. Joints are sawn, cut, and sealed in the normal manner. A standard brush finish is used. Service crossings are accommodated by omitting concrete panels at regular intervals. Figure 22.2 illustrates a typical street layout.

**Figure 22.2** A Bombay street scene using mass concrete construction.

### 22.2.3 Jointed reinforced concrete pavement (JRC) description

Reinforced jointed concrete pavements are frequently constructed as a variation on the mass concrete theme. The reinforcing steel provides two functions; it controls cracking but also instils additional stiffness in the concrete slab. Reinforced concrete slabs are frequently used in place of mass concrete when

- workmanship and materials are suspect
- the pavement will be subjected to large unplanned differential settlement forces

The recognized design methods allow a significant reduction in the pavement thickness to allow for the reinforcement structural contribution when compared to URC. Reinforced systems are generally constructed with longer slabs than the equivalent mass concrete pavement. Slabs can be designed and constructed up to 20 m long. A frequently accepted slab length is 10 m.

Reinforced systems can be designed as either cracked or uncracked slabs. Reinforcement is typically placed in the centre of the slab but some designers use reinforcement on each face. The minimum thickness of a reinforced concrete slab is approximately 150 mm; which is fixed by the practical problems involved in providing adequate cover to the reinforcement. A distinct advantage of fixing the reinforcement in the centre of the slab is that the positive and negative moments are equally balanced thus allowing the slab to flex equally before cracking and failing. A disadvantage of reinforced slabs is that the wider joint spacing precludes the use of undowelled joints. The wider transverse joint spacing produces larger thermal strains resulting in increased joint movement when compared to mass concrete pavements. The movements can only be accommodated using dowel bars or tied joints. The movement is too large for untied joints.

## 22.2.4 Typical applications for JRC pavement

Jointed reinforced concrete is typically used when designers are not confident that their work force will construct a pavement correctly or a pavement may be subjected to high differential consolidation force.

### *22.2.4.1 Example: Bangkok city streets*

One of the best examples of the efficient use of reinforced concrete can be found in the city streets of Bangkok. The streets are subjected to a very aggressive trafficking regime within a very hot tropical climate. The city is also subjected to a number of subgrade problems that make the effective operation of bituminous pavements very problematic. The city is built across a very low-lying river delta; frequently subjecting the streets to severe flooding. Subgrade strengths are exceptionally low: a desiccated surface crust provides most of the strength at formation. Groundwater abstraction also produces massive differential consolidation problems. Poor workmanship is also a major additional problem. Each of these contributory factors produces an exceptionally difficult environment but the reinforced concrete pavements operate effectively despite the extreme environment. Pavements can be seen containing massive cracks, surface polishing and various forms of acid attack but the system generally functions to an acceptable level of service despite the fact that very little planned maintenance is carried out. Figure 22.3 illustrates a typical city street.

### *22.2.4.2 Western European industrial sites*

Many small industrial sites are built with reinforced concrete slabs if designers are not confident that a mass concrete alternative would be constructed correctly. If small areas of concrete are required it is frequently preferable to use a reinforced slab and ensure that an unskilled labour force will successfully complete the project without any problems.

## 22.2.5 Description of continuously reinforced concrete pavements CRCP

This form of construction has been developed from reinforced concrete pavements. Continuous reinforced concrete pavements are constructed as long slabs with longitudinal reinforcement fixed at the centre of the slab. Figure 22.4 illustrates a typical CRC pavement. The longitudinal reinforcement is intended to control shrinkage cracking. A nominal amount of transverse reinforcement is also provided; to hold the longitudinal reinforcement in place. Essentially a continuously reinforced concrete slab consists of a regular section

**Figure 22.3** Bangkok reinforced concrete pavement.

**Figure 22.4** CRC pavement.

of cracked square concrete plates connected together by the steel. The system is very similar to a mass and reinforced concrete; except that the cracks are formed in a random fashion and remain unsealed. A second feature of a continuous reinforced concrete system is that ground anchors are required at terminations. A CRC slab will move extensively under the influence of changing environmental temperatures. The ends of the slab are therefore anchored to prevent massive movement. If a continuously reinforced concrete slab is not provided with terminations a large bump or ripple will occur at the start of the bituminous material. Figure 22.5 illustrates a typical anchorage arrangement.

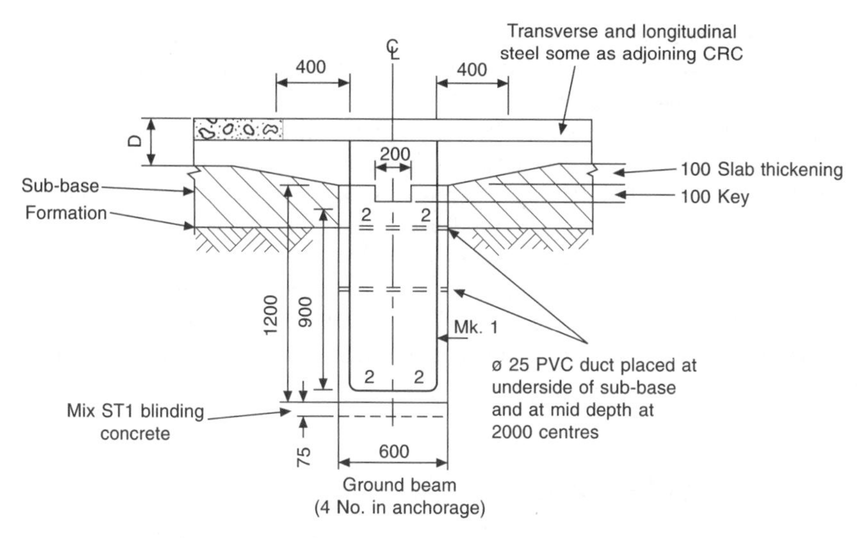


**Figure 22.5** CRC ground anchor, taken from the UK Highways Agency Standard Details.

Crack spacing is essential to the efficient operation of this type of pavement. Transverse cracks must be spaced between 0.9 m and 4 m centres; if the cracks are too closely spaced the blocks of concrete can fail in shear as punch-outs. Cracks can also be spaced too widely; if the cracks are spaced too far apart aggregate interlock is lost across the joint. Crack spacing is controlled by the longitudinal reinforcement content which is currently fixed at 0.6% of the section area.

The surface finish is a particular engineering problem associated with this type of construction. Several different types of finish are currently used. A patent form of construction known as Whisper concrete (Charon *et al.*, 1989) uses an exposed aggregate surface to provide a running surface. Figure 22.6 illustrates a section of Whisper concrete. The top

**Figure 22.6** Whisper concrete.

40 mm layer of concrete is constructed in a thin overlay of high quality, air entrained 10 mm aggregate concrete. The surface is then sprayed with a retarding agent immediately after the concrete is laid. The surface can then be removed by wire brush as the concrete sets, thus removing the cement paste from the coarse aggregate matrix. Whisper concrete is very popular in mainland Europe but has been used only on a number of small experimental projects in the USA and UK. The system requires expensive specialist aggregates, skills and equipment, thus making it uneconomic on most projects. Whisper concrete is currently banned in new construction on UK trunk roads.

A thin bituminous 35 mm wearing course is currently considered the most practical UK form of surface finish to this type of construction. The wearing course is held in place using a bituminous pad coat. The main concrete slab can then be constructed in non-air-entrained material.

### 22.2.6 CRCP applications

This form of construction may not be confined to highway projects. It is practical to construct any major pavement in a continuous reinforced option. The efficient application is restricted by the limited design methods. No real analytical technique is currently recognized. The highway design methods are basically empirical.

#### *22.2.6.1 Highway projects*

The system is found only on major motorways and trunk road schemes. This form of construction is usually the most economic type of pavement when a large aggregate source is available within the site area.

#### *22.2.6.2 Airfield runways*

Liverpool Airport runway was constructed using continuously reinforced techniques and is a notable example of a successfully completed project.

## 22.3 Manner of pavement failure

A number of different factors contribute to the failure; each issue needs to be considered separately in undertaking the pavement design.

### 22.3.1 Fatigue cracking of concrete

The first design issue to consider is the cracking of concrete. This form of failure occurs when the concrete strain exceeds the tensile capacity of the material. Fatigue cracking is easily recognized as either corner breaks or longitudinal cracks following the line of wheel loading. Figure 22.7 illustrates a typical failure.

**Figure 22.7** Fatigue cracking.

## 22.3.2 Loss of support

At high numbers of load repetitions, over 3000000 load applications, pavements with unbound sub base may fail as a result of undermined foundations. This type of failure is known as loss of support. Failure can be prevented using bound foundation materials.

## 22.3.3 Frost damage

Much debate has occurred around this issue. Some frost resistant high strength concretes can be produced but the technique is not accepted in the USA or UK. Frost damage is a major problem in concrete pavements. If the pavement is constructed in normal un-air-entrained concrete the surface can be quickly removed by the weathering action of frost. Figure 22.8 illustrates frost damage. The concrete must be air entrained. A number of researchers (BSI, 1997) have suggested that if concrete achieves a 50 N strength the material will not be susceptible to frost.

A typical standard of frost protected concrete will be achieved with a 5% plus or – 1.5% air content. The air-entraining agent acts has a cracking agent, reducing the size of any bubbles to a point when the formation of ice lenses within the pores will not cause damage to the concrete matrix. In a conventional concrete ice lenses are formed in the voids contained within the structure of the concrete. The ice is then able to crack the concrete thus resulting in the formation of surface scaling.

Adding air-entraining agents reduces the concrete strength by approximately 10%.

**Figure 22.8** Frost damaged concrete compared to non frost damaged concrete.

## 22.3.4 Surface abrasion

Surface abrasion is an important design consideration. The surface of a heavily trafficked pavement will quickly scrub and abrade away under the action of traffic if the concrete is of an inadequate strength. Figure 22.9 illustrates a good example of surface failure.

**Figure 22.9** Surface abrasion.

### 22.3.5 Skidding characteristics

Most national standards fail to specify a skidding requirement. The British highway specification is an exception and rigorously controls pavements' skidding characteristics. The standard specifies the polishing characteristics and aggregate abrasion values for the coarse aggregate. Different polished stone (PSV) and abrasion values (Highway Agency, 1999) are dictated for different highway characteristics. The same aggregate characteristics are used for both bituminous and concrete coarse aggregate. Other national standards fix skidding characteristics using the strength of the aggregate.

The UK standard uses a different surface texture for concrete roads. The surface texture is fixed at 1 mm plus or minus 0.25 mm.

## 22.4 Design methods

Two alternative approaches may be used to undertake the design of a pavement. The two styles of design are:

- the empirical approach
- the semi-empirical approach

Both methods relying on statistical calibration relationships to produce accurate designs.

### 22.4.1 The empirical approach

The simplest method of designing highway pavements is to use the statistical back analysis equations presented in Mayhew and Harding (1987). The method forms the basis of UK highway pavement design. Two alternative equations are presented in the document.

- the relationship for estimating mass concrete pavement design lives
- a second similar formula for estimating the design life of a reinforced concrete pavement

The relationships are then used to present design graphs for mass concrete, reinforced concrete, and continuously reinforced in Highways Agency pavements.

#### *22.4.1.1 Mass concrete pavement design after Mayhew and Harding (1987)*

$$\ln (L) = 5.094 \ln (H) + 3.466 \ln (S) + 0.4836 \ln (M) + 0.08718 \ln (F) - 40.78 \tag{22.1}$$

where the following terms are used:

$L$ = traffic loading in msa

$H$ = the intended pavement thickness in mm

$S$ = concrete 28-day mean compressive cube strength

$M$ = the estimated equivalent modulus of the pavement support platform in MPa, derived from core data extracted from the pavement foundation

$F$ = the percentage of failed bays

### 22.4.1.2 Reinforced concrete pavement design

$$\ln(L) = 4.786 \ln(H) + 1.418 \ln(R) + 3.171 \ln(S) + 0.3255 \ln(M) - 45.15$$

where $R$ = the amount of longitudinal reinforcement in $mm^2/m$.

The following key assumptions are used with the expressions and are used to generate the standard UK pavement designs:

- A standard 40 N concrete mix is assumed to produce concrete with a strength of 50 $N/mm^2$ at 28 days.
- The support platform is conservatively assumed to be a 270 MPa stiffness, which is represented by a 150 mm concrete leanmix over 350 mm of capping onto a 2% CBR formation.
- The equations are set to a 50% confidence.
- The design graphs (Highways Agency) also assume that a kerb or 1 m wide concrete strip, to prevent edge loading conditions occurring supports the pavement edge. The standard design graphs contain a reduction in thickness of approximately 10% when compared to the equations to allow for the edge beam stiffness.

## 22.4.2 The semi-empirical approach

A number of scientifically based rational design methods may be found in various publications. All these methods use modifications and adjustments to the Westergaard (1926) pavement stress/strain calculation techniques linked with a fatigue model to produce a design method. Some design methods use Westergaard (1926) corner loading conditions, some edge conditions and others internal loads. Figure 22.10 illustrates the standard Westergaard load cases. A summary of the most popular and common design methods is presented here:

1. AASHTO (1992) design method, using a corner loading condition
2. Modified AASHTO (1998) design method, using an internal loading condition
3. TR 550 (Chandler, 1982) design for heavy loads, Westergaard (1926) corner loading condition

These are the most important design methods, many other national highway design standards, airfield pavement design methods and a number of methods for designing port

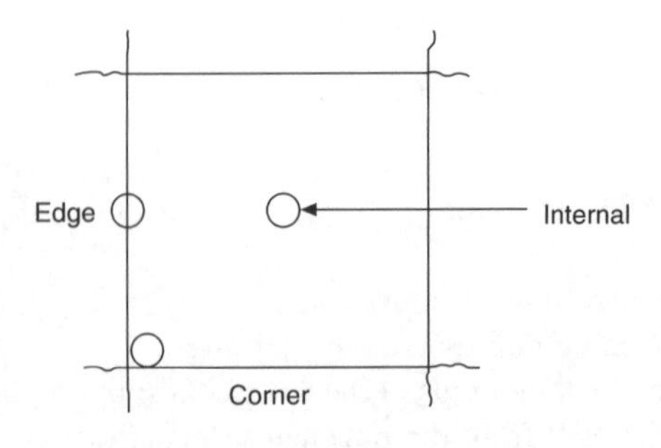


**Figure 22.10** Westergaard loadings.

facilities may be found. All of the design methods use variations, developments or adjustments on the basic Westergaard (1926) design equations.

The design methods combine concrete tensile fatigue models with the estimated pavement stress and Miners Rule (1945) to produce practical design methods. The most commonly used fatigue model is the Portland Cement Association method which assumes that a cement bound material has an infinite life if the magnitude of the stress impulse falls below 50% of the ultimate tensile strain.

#### 22.4.2.1 The AASHTO design method (AASHTO, 1992)

The AASHTO design method is the most commonly used pavement design technique in the world. The standard uses a derivation of the Spangler (1942) corner loading equation to produce mass concrete pavement designs. The original standard uses a graphical nomograph presentation derived from a complex equation to produce pavement thickness. The following variables may be adjusted in the design method:

- flexural strength
- subgrade strength
- drainage conditions
- design confidence level and probability of survival
- joint conditions and types
- subgrade support condition

The standard is written using Imperial units. Pavement designs produced following Mayhew and Harding (1987) and AASHTO (1992, 1998) give the same pavement thickness when comparable input data is used. The AASHTO (1992) system is far more comprehensive and flexible than the Mayhew and Harding (1987) relationship.

#### 22.4.2.2 The AASHTO relationship

An alternative, more complex design method was published by AASHTO in 1998. The method is a modified internal Westergaard loading condition assessment of pavement stress linked to climatic conditions and the results obtained from an extensive monitoring programme. The standard includes a set of catalogue designs that may be applied to most common pavement problems. The results obtained from all three design methods are comparable.

### 22.4.3 Continuous reinforced concrete pavements

Great difficulty is experienced in effectively designing continuous reinforced concrete pavements. Most of the common current design methods are based on modifications of either AASHTO (1992) or Mayhew and Harding (1987), both of which were originally produced for jointed reinforced concrete pavement design. A pressing need exists to review the existing design methods and produce a rational or semi-empirical design method truly focused on the specific problems associated with continuously reinforced concrete pavements. The design of CRC pavements must therefore be considered as more empirical than rational.

## 22.5 Joint design and detailing

It has already been noted that joint design is fundamental to the efficient operation of a concrete pavement. Joints may be designed as either dowelled or undowelled with crack induced or formed joints. A maximum traffic level of 3 msa is considered to be the limit for an undowelled construction. Various techniques are currently available to form joints. If joints are formed using crack-induced techniques the crack or saw cut must extend to a depth of between $^1/_4$ and a $^1/_3$ of the slab depth. Joints are classified in three different types, each of which forms a separate function in the pavement.

### 22.5.1 Transverse contraction joints

Transverse joints are formed at regular intervals in mass concrete pavements and jointed reinforced concrete pavements at points where hydration cracking might occur. This type of joint is used in jointed reinforced concrete and mass concrete pavements. Figure 22.11 illustrates a standard joint design. It is important to note that the dowel bars should be de-bonded to allow free movement.

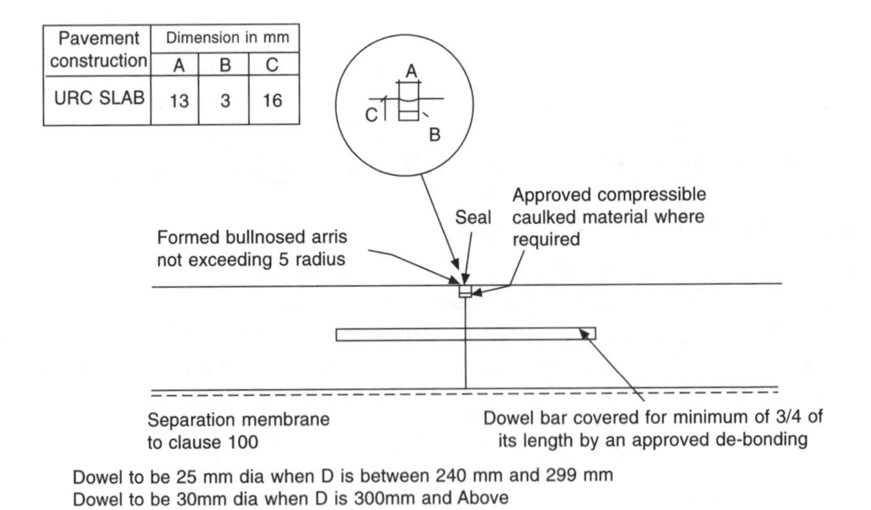


**Figure 22.11** Transverse joints, taken from the UK Highways Agency Standard Details.

### 22.5.2 Longitudinal joints

Longitudinal joints are required in all pavement types. The joint is needed to produce a regular crack between construction bays or at a point in CRC construction where the transverse reinforcement would induce an irregular crack. Longitudinal joints are generally constructed with tiebars. The tiebars are bonded to produce a positive connection between adjoining slabs. Figure 22.12 illustrates a typical arrangement.

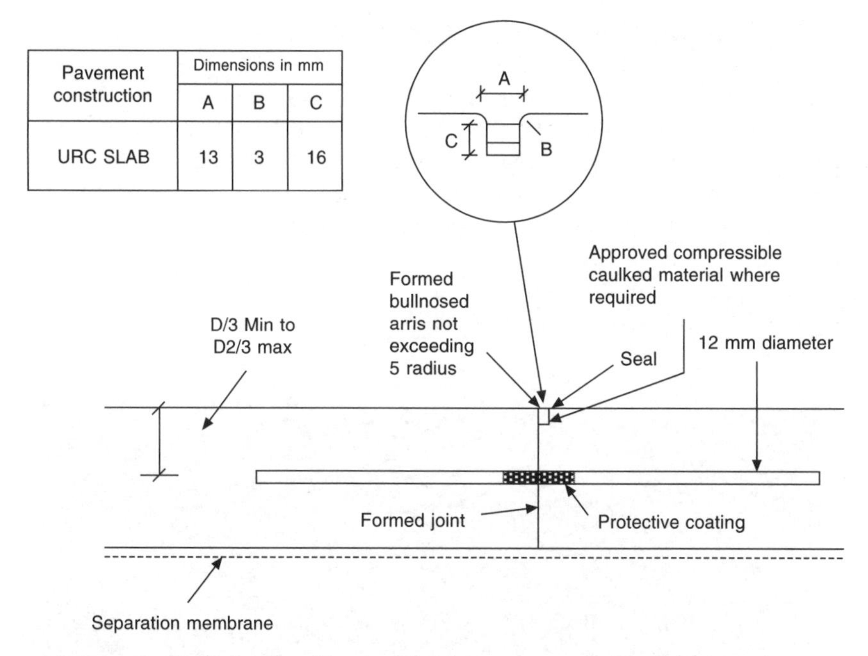

| Pavement construction | Dimensions in mm | | |
|---|---|---|---|
| | A | B | C |
| URC SLAB | 13 | 3 | 16 |



**Figure 22.12** Longitudinal joint taken from the UK Highways Agency Standard Details.

## 22.5.3 Expansion or isolation joints

Expansion joints and isolation joints provide similar functions. They are intended to allow movement between slabs and/or hard objects. Expansion joints are essential to reinforced concrete pavements but may be omitted in mass concrete crack induced design options. Manholes and piled foundations must be isolated from a concrete slab. Concrete moves extensively under the influence of changing environmental temperatures. Detailing manhole, gullies, and the other protrusions through the concrete slab is essential to the success of the pavement. Figure 22.13 describes a standard arrangement.

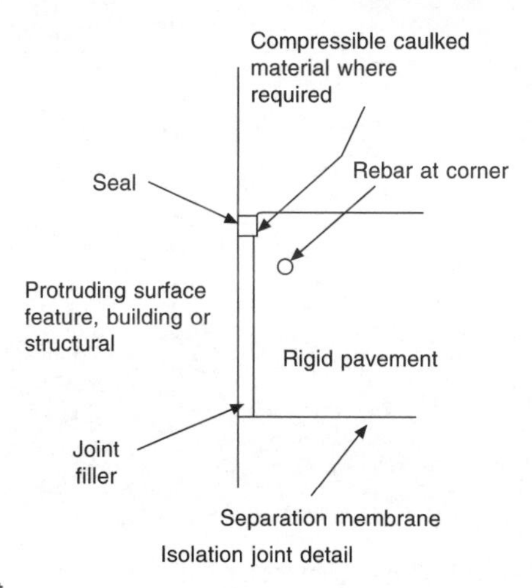


**Figure 22.13** Isolation joint.

## 22.5.4 Typical jointing related problems

The following construction problems are commonly encountered and illustrate the importance attached to good sound detailing. Many contractors fall foul to poor construction detailing problems. The following figures illustrate typical problems:

Figure 22.14 shows an incorrect joint location
Figure 22.15 shows a joint which is too shallow
Figure 22.16 shows poor detailing
Figure 22.17 shows good manhole detailing
Figure 22.18 shows badly debonded tie bars

**Figure 22.14** Joint cut too late and incorrect joint location.

**Figure 22.15** Joint cut too shallow.

## 22.6 Materials and concrete properties

High-quality concrete is important to the success of all concrete pavements. It is already noted from the design method that one of the most important issues influencing the bearing capacity of a concrete pavement is the strength of the concrete mix. Strength may be measured by either of the standard structural control techniques using compressive cubes or cylinders. Tests are normally conducted using standard 28-day cured cubes or cylinders. UK practice is to use cubes, European practice is to use cylinders, Americans use beams and cylinders. Standard structural relationships exist linking the two parameters.

### 22.6.1 Compressive cube strength and tensile strength relationships

Concrete slabs are basically brittle unreinforced members which rely on the tensile capacity of the main structural element, concrete, to provide bearing capacity and durability.

**Figure 22.16** Poor detailing.

**Figure 22.17** Good quality detailing.

Concrete tensile strength is therefore fundamental to the efficient operation of a pavement. Tensile strength may be measured using two alternative techniques. The standard method is to use the Modulus of Rupture ($R$) to define the bending tensile capacity of concrete. The test is conducted on prisms using a bending test. The British Standard is *BS 1881 Parts 109, 118 and 111* (BSI, 1983a, b, c). The test is a four-point bending test producing a tensile strain, which reflects the mass concrete beam strain capacity at failure.

An alternative technique known as the Brazilian Test or Indirect Tensile Strength test may also be used to control tensile strength. *BS 1881 Part 117* defines the test (BSI,

**Figure 22.18** Badly bonded tie bars.

1983d). Again standard relationships exist linking indirect tensile strength to the Modulus of Rupture. Figure 22.19 illustrates the different techniques for testing tensile strength.

A number of researchers have identified relationships linking tensile capacity to compressive cube strengths. Different relationships may be identified for gravel aggregates and crushed rock aggregates. The most commonly used relationships are those identified by Croney and Croney (1991):

For gravel aggregate — Modulus of Rupture = $0.49\ (f_{cu})^{0.55}$
For crushed rock aggregate — Modulus of Rupture = $0.36\ (f_{cu})^{0.70}$

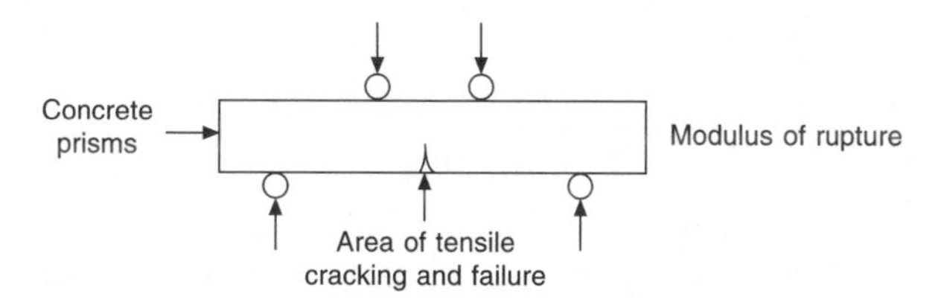


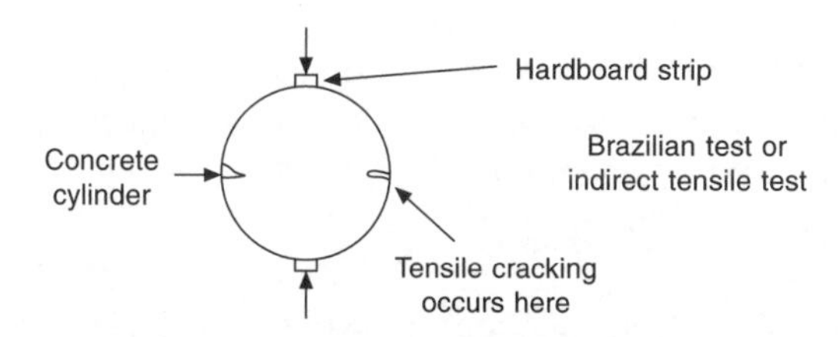


**Figure 22.19** Tensile bending test, Modulus of rupture and the Indirect/Tensile test or Brazilian Test.

Similar relationships are also available for Indirect Tensile Strength and compressive Cylinder Strength.

### 22.6.2 Slip membranes

Slip membranes are important to the efficient performance of all types of pavement. A de-bonding layer is needed underneath a concrete pavement to prevent interaction between the base foundation and the slab but also to allow differential movement to occur. Plastic sheeting is the commonest method of providing a slip membrane. The sheeting is intended to:

- prevent loss of fines into the sub-base
- form a separation layer between the concrete and the sub-base
- reduce the risk of cracking between the joints during early shrinkage

If sawn crack induced joints are used plastic sheeting is not the most appropriate form of membrane. The plastic sheeting can allow too much free movement. If joints are sawn crack induced they may not break correctly at each joint. A bituminous sprayed layer is a more effective method of de-bonding providing partial separation thus producing cracks at every regular joint.

## 22.7 Construction methods

A number of different techniques are used throughout the world for constructing concrete pavements. Each is appropriate to different environments, economies and labour forces. It is important to recognize that slip forming is a useful technique in Western Europe, on large construction projects, but will not work on small building projects or third world schemes.

### 22.7.1 Traditional techniques

The labour-intensive method of construction is to erect wooden shutters along the line of each longitudinal joint and pour concrete into one continuous strip. Intermediate contraction joints are formed by either joint sawing, into the hardened concrete or dropping patent plastic crack inducers into the hardening pavement surface can be used to form intermediate contraction joints. The disadvantages of this method are:

- the technique is slow and labour-intensive
- erecting shutters and laying concrete by hand is a skilled activity
- joint sawing must be undertaken after the concrete has hardened but before any shrinkage has taken place. The timing of the joint sawing is critical

The advantages are:

- the technique is flexible and requires no major capital expenditure
- difficult pavement shapes can be worked around without problems

## 22.7.2 Slipforming

Modern slipforming plant can produce an economical fast, well constructed alternative to hand-laying techniques. Slipforming machines have been employed on the construction of recent UK airfield projects. A typical slipformed pavement system will include the following construction:

- a stabilized formation layer, to allow good support to the slipformer
- a sprayed bituminous slip membrane, in place of a plastic membrane

Mechanical dowel bar placing techniques can be employed in association with slip formed paving machines for mass concrete pavements. Slipforming is the most effective method of construction for continuously reinforced construction. Figure 22.20 illustrates a typical slipform system.

**Figure 22.20** Slipforming a CRCP pavement, picture courtesy of Wirtgen slipform manufacturers.

# 22.8 Maintenance issues and repair of pavements

The maintenance and repair of concrete pavement is an important area of specialist engineering application. A summary of the subject is presented in the current Highways Agency concrete pavement maintenance manual published in 2001. The first essential step in undertaking a repair is to identify the cause of failure. When the precise nature of failure is understood, it is then possible to move on and identify suitable repair techniques. The standard contains a set of flow charts to classify and evaluate contributory factors producing a defect.

The standard divides maintenance techniques into three different categories:

- emergency repairs
- medium-term repairs
- long-term repairs

Defects are classified into the various different types:

- jointing problems
- surface abrasion, spalling or cobweb cracking problems
- general forms of cracking including corner breaks, longitudinal cracks or shrinkage cracking
- joints stepping, slab rocking and loss of support through pumping
- surface irregularities, pop outs or general durability problems

The severity of defect must be understood and evaluated in the design process but the following techniques may be used to remedy problems.

## 22.8.1 Emergency repairs

### *22.8.1.1 Bay replacement*

Immediately after a concrete bay has failed the first line of defence is to knockout the bay between the existing joint and replace it with a bituminous inlay. This technique is considered as a very short-term emergency measure. The continuity of the concrete slab is lost as the weak bituminous material is placed in the gap left by the removal of the slab. Progressive chipping away and failure of the remaining adjoining slabs will quickly follow if the bituminous material is not replaced with concrete at the earliest opportunity.

## 22.8.2 Medium-term repairs

### *22.8.2.1 Cracks*

Any cracks identified within a pavement with a width greater than 0.5 mm should be repaired using some form of bituminous seal. If the cracks are not sealed detritus, dust and dirt will quickly fall down into the gap forcing open the joint.

Wide cracks, which have width greater than 1 mm, are considered to have a structural implication and must be repaired. Crack stitching techniques can be used to repair wide cracks but these techniques are undesirable and technically difficult. A stitch repair technique must also involve some form of pressure grouting. If a joint contains detritus or moisture, grouting techniques are unlikely to be successful.

### *22.8.2.2 Surface dressing*

Concrete pavement surfaces may be restored on a temporary basis by a simple application of a conventional bituminous surface dressing. This system is used to repair skidding problems caused by erosion of the pavement surface or aggregate polishing problems.

#### *22.8.2.3 Pressure grouting to restore support*
A number of patent repair techniques are available which use pressure grouting to fill voids under the pavement. The systems are of mixed success and require highly skilled operators. These types of repair should only be attempted in exceptional circumstances.

### 22.8.3 Long-term repairs

#### *22.8.3.1 Full depth bay replacement*
The most effective long-term repair is to replace completely badly cracked concrete bays. Bay replacement should only be attempted when it is clear that the majority of the concrete pavement is sound and likely to give many years further service.

#### *22.8.3.2 Joint repairs*
Many techniques exist which can give exceptionally good long-term service. Joint repairs require very careful design and detailing. If long-term joint repairs are considered appropriate it is recommended that section 8 of the repair manual (Burkes Green and Partners, 2001) is carefully examined.

## 22.9 New techniques

A number of research techniques are currently being investigated which will further develop the technology of concrete pavements.

### 22.9.1 Fibre-reinforced concrete

A number of researchers are examining the possibilities of using steel fibres as enhancing material to mass concrete pavements. High tensile steel fibres are extensively used in internal industrial floor slabs. The techniques are applied to heavily loaded industrial slabs. Steel fibres are added at the rate of approximately 0.4% to a cement bound roadbase material. The fibres enhance the durability of the mix, increasing the failure ductility. The point when initial cracking occurs remains unaltered but the ultimate strength of the mix is increased substantially before failure occurs.

### 22.9.2 Thin overlays and white topping

A number of interesting experiments are currently being conducted involving the use of continuously reinforced overlays to existing bituminous pavements. Heavily rutted bituminous material can be removed and a thin 150 mm CRCP system installed as a replacement. The system is particularly effective for heavily trafficked motorway slow lanes and may become a standard future repair technique.

## 22.10 Conclusions

This chapter is intended as a brief introduction to the basic concepts of cement bound pavements. The views expressed here are the personal opinions of the author and should not be considered as design standards or definitive guidance. If serious design work is considered appropriate, based on advice given in this chapter, it is recommended that the primary sources of information quoted as references are used for detailed design.

## References

Charonnat Y., Lefebvre J. and Sainton A. (1989) Optimization of non-skid characteristics in construction of concrete pavements, 4th International Conference on Concrete Pavement Design and Rehabilitation, April, Purdue University.

American Concrete Pavement Association (1992) Proper Use of Isolation and Expansion Joints in Concrete Pavements, ISA400.0IP.

BSI (1997) BS 5328 Part 1 1997, Guide to specifying concrete.

The Highways Agency (1999) *Design Manual for Roads and Bridges*, Vol. 7, HD 36/99 Surfacing Materials for New and Maintenance Construction.

Mayhew, H.C. and Harding, H.M. (1987) Research Report 87, Thickness design of concrete roads, Transport and Road Research Laboratory, ISSN 0266–5247.

The Highways Agency *Design Manual for Roads and Bridges*, Vol. 7, HD 26/01 Pavement Design.

Westergaard, H.M. (1926) Stresses in concrete pavements computed by theoretical analysis. *Public Roads*, Vol. **7**, No 2. Also *Proceedings of the 5th Annual Meeting of the Highway Research Council*, Washington DC 1926 as Computation of Stresses in Concrete Pavements.

American Association of State Highway and Transportation Officials (1992), *AASHTO Guide for Design of Pavement Structures*, ISBN: 1-56051-055-2.

American Association of State Highway and Transportation Officials (1998) *AASHTO Guide for Design of Pavement Structures Part II – Rigid Pavement Design and Rigid Pavement Joint Design*. ISBN 1-56051-078-1

Chandler, J.W.E. (1982) Technical Report 550, Design of floors on ground. Cement and Concrete Association.

Miner, M.A. Cumulative damage in fatigue, *ASME Transactions*, **67**, A159.

Spangler, M.G. (1942), Stresses in the Corner Region of Concrete Pavements, Bulletin 157. Engineering Experiment Station, Iowa State College, Ames.

Ioannides, A.M., Thompson, M.R. and Barenberg, E.J. (1985), Westergaard solutions reconsidered. *Transportation Research Record*, **1043**.

Croney, D. and Croney, P. (1991) *The Design and Performance of Road Pavements,* second edition. McGraw Hill International, ISBN: 0-07-707408-4.

BSI (1983a) BS 1881 Parts 109, Method for making test beams from fresh concrete.

BSI (1983b) BS 1881, Part 111, Method of normal curing of test specimens (20ºC).

BSI (1983c) BS 1881 Part 118, Method for determination of flexural strength.

BSI (1983d) BS 1881 Part 117, Method for determination of tensile splitting strength.

Burkes Green and Partners (2001) *Concrete Pavement Maintenance Manual.*

23

# Cement-bound materials (CBM)

*David York*

## 23.1 Introduction

Cement-bound material (CBM) may be defined as a granular material or soil, mixed with cement and compacted at or about optimum moisture content by external vibration. Main applications are in paving construction but it has been successfully used as a mass foundation material.

The term covers a very wide range of materials from products whose strength is measured in terms of Californian Bearing Ratio (CBR) (BS 1377: 1990 Part 4) to products which have flexural strengths of 5 N/mm$^2$ or more. The former would be appropriate for use as a capping layer, while in some cases the latter could be the actual pavement surface, e.g. container terminal or log-handling yard.

The strengths in Table 23.1 are indicative only. The individual categories can be assessed as follows.

### 23.1.1 Cement-stabilized soil (CSS)

This material is often produced *in-situ* as a means of improving subgrade to provide a capping layer. In the UK the SHW Clause 616 deals with this work and a CEN standard for soil cement is being drafted. With cohesive soils, it is usually necessary to pre-treat with lime. This breaks down clays to a friable condition, increasing the surface area

**Table 23.1** The whole family

| Very weak (CBR) | | Range of CBMs ⟷ | | Very strong Flexural strength 5 N/mm$^2$ |
|---|---|---|---|---|
| Cement-stabilized soil | Soil cement CBM1 | Cement-bound granular materials CBGM | Lean concrete CBM 3,4,5 | Roller-compacted concrete |
| 15–30% CBR | 1 to 5 N/mm$^2$ at 7 days' compressive strength | 4 to 10 N/mm$^2$ at 7 days | 7 to 20 N/mm$^2$ at 7 days | Over 20 N/mm$^2$ at 7 days' compressive strength |
| | | Increasing quality of raw material required → | | |

which can be coated with cement. Such layers, when properly compacted, will have very low permeability. They also need a very competent sealing membrane to prevent moisture loss and 'cracking' or 'block forming' occurring with associated loss of performance.

## 23.1.2 Soil cement (SC)

This may be produced '*in-situ*' or 'in-plant'. The material is useful as a sub-base/capping or even a road-base for very lightly trafficked roads. The raw material will be granular and ideally not of very uniform grading but generally unprocessed. A low organic content is also desirable. The Specification for Highway Works Clauses 1035 and 1036 (CBM1) and a future CEN standard for Soil Cement are appropriate.

Examples of ideal materials include 'as-dug' sands and gravels, quarry wastes, recycled materials like crushed concrete, incinerator bottom ash, asphalt planings and so on.

## 23.1.3 Cement-bound granular materials (CBGM)

At this point, the raw material needs to be more controlled. A fairly broad grading envelope and a minimum raw material hardness requirement will normally be required. CBGM can be used as a sub-base or even road-base material for lightly trafficked roads. The UK's *Specification for Highway Works* (SHW) Clauses 1075 and 1037 (HMSO, 1998b) and the impending CEN standard for CBGM apply.

## 23.1.4 Lean concrete (LC)

This is generally a road-base material for flexible composite pavements but is also used extensively as a sub-base for rigid constructions. The requirement for aggregate is that which would be required for any lower-grade concrete. The grading envelope is narrower than that for CBGM. In the UK Specification for Highway Works Lean Concrete is now covered by the Clauses for CBM 3, 4 and 5 Clause 1038 plus general Clause 1035. The CEN standard for this material will be the CBGM document.

## 23.1.5 Roller-compacted concrete (RCC)

This gives a performance equivalent to pavement quality concrete (PQC). It can be specified in terms of flexural, indirect tensile or compressive strength. At this time there is no UK specification for RCC. However, the Specification for Highway Works Clauses relevant to CBM 3, 4 and 5 can be used as a foundation excluding compressive strength and inserting the desired strength of RCC. A CEN standard is being produced while in the USA the 2 Corps of Engineers has a specification for this material.

The aggregate required is of a high quality and the grading needs to be tightly controlled. Often RCC mixtures incorporate pulverized-fuel ash (pfa). This pozzolanic material can be used to reduce the OP cement content. This reduces heat generated by the mixture and gives long-term strength gain. The amount of pfa can be tuned to fill all voids in the compacted layer. This ensures good durability of the layer.

When being considered as the running surface, the designer should remember that texturing the surface may be required in order to give adequate skidding resistance. The technique used for 'whisper concrete' can be employed (see Chapter 22 of this volume by Geoffrey Griffiths). This involves spraying the compacted surface with retarder and after a few hours mechanically brushing the surface to remove mortar to give a predetermined texture depth. A curing agent should then be applied. Table 23.2 shows how these grades of material are currently specified in the UK in terms of strength.

**Table 23.2** Strength classes for various CBMs

| | |
|---|---|
| Cement-stabilized soil | CBR or 15% or 30% min is typical |
| Soil cement (CBM1) | Compressive strength of 4.5 N/mm$^2$ at 7 days (mean of 5 specimens)<br>This is the SHW requirement but varies from nation to nation<br>The Dutch, for instance, have a class of 1 N/mm$^2$ at 7 days |
| CBGM (CBM2) | Compressive strength of 7 N/mm$^2$ at 7 days (mean of 5 specimens)<br>As above, different nations have different strength classes |
| Lean concrete (CBM3, 4 and 5) | Compressive strengths at 7 days of<br>10 N/mm$^2$ – CBM3<br>15 N/mm$^2$ – CBM4<br>20 N/mm$^2$ – CBM5<br>Again different requirements abroad |
| Roller-compacted concrete | In the UK there is no specification currently. However, overseas 28-day compressive strengths of 40 N/mm$^2$ are achieved with flexural strengths of 4 or 5 N/mm$^2$ specified. |

## 23.1.6 Modes of failure

During normal use, CBM pavements are subjected to a combination of traffic-induced and thermally induced stresses. A properly designed, specified and constructed CBM will give a very long life indeed. Failure of designers/specifiers to predict the real traffic and thermal conditions can lead to premature failure. However, CBM performance is also very sensitive to workmanship.

Poorly mixed, poorly compacted, or poorly cured CBM will undoubtedly lead to early failure. Unfortunately, in the 1960s and early 1970s, too many of these bad practices took place and lean concrete was unfairly given a bad name. More recently, better mixing equipment and more professional control resulting from greater understanding have led to clients now getting the benefits of CBM bases.

Occasional failures of CBM pavements have been attributed to 'reflective cracking' through overlying bituminous layers. (This occurs occasionally and is a result of thermal movement in the CBM which with time creates stresses which aged bitumen is unable to withstand.) This may be true for moisture-sensitive subgrades when CBM cracks in an uncontrolled manner through thermal movement. However, there are now technologies for inducing frequent, very fine transverse cracks which make this kind of failure a thing of the past. The stresses induced in the bituminous material due to thermal movement are consequently minimized.

There are also very occasional failures in compression where rupture occurs. Invariably, this is where CBM has been laid in cold conditions (i.e. the materials are thermally contracted) and day joints are not cut vertically. In such circumstances, during hot periods, the CBM thermally expands and non-vertical joint faces 'slide' over each other leading to one side of the joint lifting, inevitably causing damage to the CBM and overlying layers. Vertical day joints usually prevent this type of occurrence.

## 23.2 Methods of specifying CBM

### 23.2.1 Compressive or tensile strength

Previously, only compressive strength or CBR has been referred to as a means of classifying performance. Compressive strength at 7 days is the normal means of specifying CBM in the UK. This is a convenient means of controlling quality and checking contract compliance and is therefore very useful.

However, from a design point of view, compressive strength alone is of limited value. UK designs for roads are contained in the Highways Agency Vol. 7, *Design Manual for Road and Bridge Works* (HMSO, 1998a). This is really a catalogue of options. The designer only needs to know the traffic loading which the road will take during its life and then use the charts in Vol. 7 to determine the thickness of individual pavement layers.

Some other nations have similar systems, but one notable exception is France. Here the designer selects the aggregate source, thoroughly tests CBM mixtures incorporating the aggregate and determines the tensile strength and modulus of elasticity ($E$) at 28 days. These values are used to determine the thickness of layer required to fulfil pavement life.

The higher-grade CBMs can also be classified by flexural strength. Flexural strength, tensile strength and $E$ can be used in analytical design methods which optimizes the thickness of a layer by truly exploring an aggregates potential.

The catalogue designs based on compressive strength tend to be inefficient in terms of aggregate usage, as they take no account of aggregate performance. As an example, a CBM produced using rounded sand and gravel may have the same compressive strength as a CBM produced using crushed limestone. However, their flexural strengths are likely to be very different; the crushed rock will be higher.

## 23.2.2 Designing CBM (the mix)

In designing the mixture, there are a number of basic considerations:

- The specification to be met, e.g. CBM1 or CBM5. This will impose the limitation on raw material source and set the strength requirements.
- If one of the lower categories is specified is the aggregate variable? If it is reasonably consistent, life is much easier than if the aggregate is extremely variable. In the latter instance more than one mix design may be appropriate. As an example on a road scheme, the earthworks may expose a formation in some areas as a fine sand, where in others it is a sand/gravel mixture. Clearly in such circumstances two designs would be advised.
- From that point the object is to determine:
  - The cement content which gives a sensible margin of safety on strength.
  - The optimum moisture content for the mixture using the selected compaction regime, e.g. proctor or vibrating hammer.

Typically a mix design procedure may involve mixing the aggregate with the 'guesstimated' amount of cement and adding water to obtain optimum moisture content (omc) with the specified compaction, e.g. Proctor, vibrating hammer (Figure 23.1).

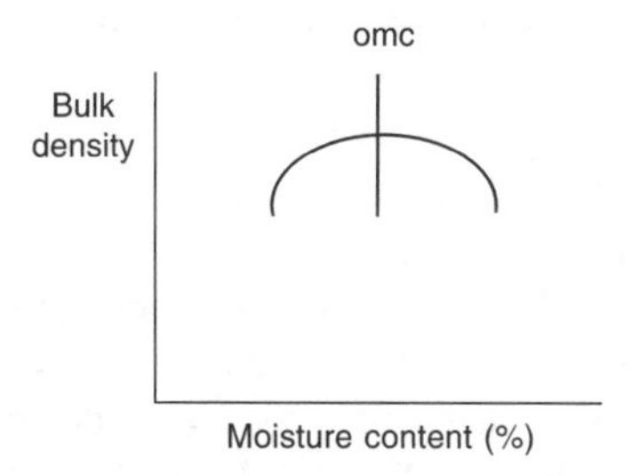


**Figure 23.1** Determining optimum moisture content (omc).

A number of strength test specimens (cubes or cylinders) may be made from the mixture. The more results available, the more robust the mix design will be.

At least two further cement contents should also be used to manufacture strength test specimens. Once all results are available then the cement content for the work may be chosen (Figure 23.2).

A typical design may then be expressed as follows:

| Material: | CBM 1 | |
|---|---|---|
| Aggregate | Sand and gravel dug on-site | 2078 kg/m$^3$ |
| Cement | Red Square ex Manham Works (42.5N grade) | 125 kg/m$^3$ |
| Total water | | 133 kg/m$^3$ |
| | | 2336 kg/m$^3$ |

In undertaking pavement design, the designer must appreciate the relationships between (i) mix design – laboratory strength; (ii) specified strength and (iii) field strength. Specified

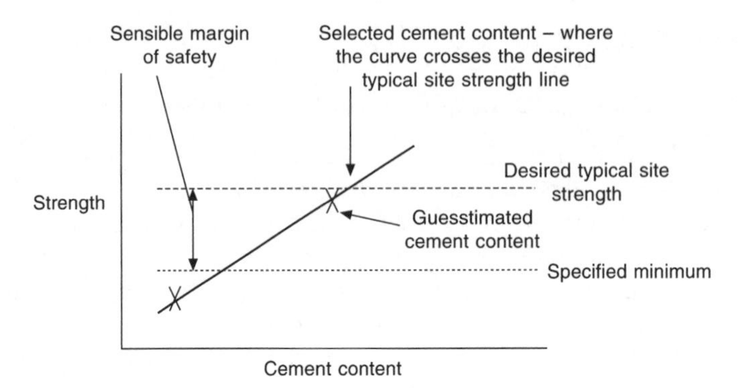


**Figure 23.2** Determining cement content of the mixture.

strength could be minimum or characteristic. The compaction regime to be used will also have a bearing on how to design and specify the pavement layer and the materials.

Typically, for a compacted layer, a 1 per cent variation in density has a 10 per cent impact on compressive strength. So for regimes which use Standard Proctor compaction (BS 1377: 1990 Part 4) where 100 per cent compaction is achievable in the field, the pavement designer can specify a mix of, say, 10 N/mm$^2$ at 7 days and know the field strength will be similar.

However for regimes as employed by the UK Highways Agency which employ Vibrating Hammer (BS 1377: 1990 Part 4) (effectively a refusal density), field density can be down to a specified 95 per cent min. As the performance of the CBM is so sensitive to density, field strength will typically be 50 per cent of laboratory test specimens (fully compacted) if field density is at 95 per cent of specimen density. The designer must therefore make this minimum field strength the basis of his pavement design.

The 28-day modulus of elasticity ($E$) of CBMs ranges from, say, 5 N/mm$^2$ soil cement up to, say, 40 N/mm$^2$ for some roller-compacted concretes.

## 23.2.3 Producing CBM

Some CBM can be produced *in-situ*, in fact with clayey soils it may be the only way to mix. Such material is generally low in the pavement structure as a capping, though some sub-base is produced this way. However, the control of production is superior when mixed 'in plant' so for higher-quality CBM, plant mix is normally employed. Continuous mixers rather than batch mixers are ideally suited to this kind of work where the 'high energy' mixers ensure good distribution of the cement into the mixture (Figure 23.3). Modern plants can employ sophisticated surveillance and recording to monitor the plant's performance. In addition, electronic weighing equipment can work to very high tolerances.

The *Mix in Place* technique involves preparing the aggregate in its final position. The area is covered by a carpet of cement at the appropriate rate/m$^2$. This is rotovated into the soil or aggregate by purpose-built equipment, which adds water at the same time. The resultant mixture is trimmed to level and rolled to specification.

*Mix in Plant* means stockpiling the soil or aggregates adjacent to the mixing plant. The

**Figure 23.3** Mixing plant pugmill.

CBM produced through the plant is loaded into tippers for delivery directly to the laying area.

## 23.2.4 Laying CBM

The CBM can be laid by grader or other 'bladed' machine, but specifications sometimes require the material to be laid through a paving machine. This undoubtedly produces a better job particularly from a surface tolerance point of view. There is sophisticated equipment available to enable pavers to produce tight surface tolerances. These include slope and level automatic controls. Various sytems are available:

- string lines with sensors
- laser guided
- geo stationary positioning

Paving machines can be fitted with 'high-density' screeds which impose greater compactive effort than a conventional vibrating/tamping screed. With some mixtures a high-density screed can eliminate the need for rolling, leaving a very flat surface (see Figure 23.4). Occasionally CBM is laid through a slipform paving machine when longer schemes require substantial daily quantities to be laid in excess of, say, 2500 tonnes/day.

One extremely important recent development in the UK is incorporating 'controlled cracking' into pavement design and practice, particularly for CBM road-base as a means of reducing the risk of reflective cracking and with the potential to permit reduced thickness of bituminous overlay.

A number of full-scale trials (TRL Report 289) took place around the country incorporating various methods of inducing cracks in the laid material during the mid-1990s. It is now accepted that 3 m spacing between transverse cracks is effective in producing fine cracks, which exhibit good load transfer capability, retaining in excess of

**Figure 23.4** Paver.

95 per cent of mid-point stiffness. Uncontrolled cracks can be at wide spacing with typically 50 per cent of mid-point stiffness. Clearly surfacing thickness reduction can be justified if controlled cracking is employed.

One very easy method of inducing cracks is to use a vibrating plate with a fin welded to the base (see Figure 23.5) to form grooves about two-thirds the depth of the uncompacted layer at 3 m centres transversely on single carriageways with longitudinal ones at lane dividing points on wider carriageways. On large hardstandings, a 3 m grid may be formed.

**Figure 23.5** Vibrating plate with fin'.

The grooves are formed directly behind the paving machine and the inside is sprayed with bitumen emulsion. Rolling and subsequent operations continue as normal. When the hardened layer goes into tension (initially usually through thermal movement), the films of bitumen are weaker in tension than the CBM and hence cracks are induced at the intended positions. There are other techniques available, mostly formed at the 'wet' stage. There is also an option to crack the hardened CBM when it is a few days' old, using an 'Arrows breaker' (see Figure 23.6). These machines employ a heavy blade, raised and dropped along the intended crack position. The resultant cracks perform extremely well, but the technique is relatively expensive.

**Figure 23.6** A breaker with a heavy blade which could be used to induce cracks in hardened CBM.

## 23.2.5 Compacting CBM

It is crucially important to achieve the highest density in CBM laid in a pavement. The *in-situ* strength of the CBM will be directly proportional to the *in-situ* density. The traffic life of the pavement will be also in proportion. A simple rule-of-thumb relationship exists as shown in Figure 23.7. A 50 per cent loss in strength for 5 per cent loss in density! Therefore it is essential to have well-selected rollers and a carefully monitored rolling regime.

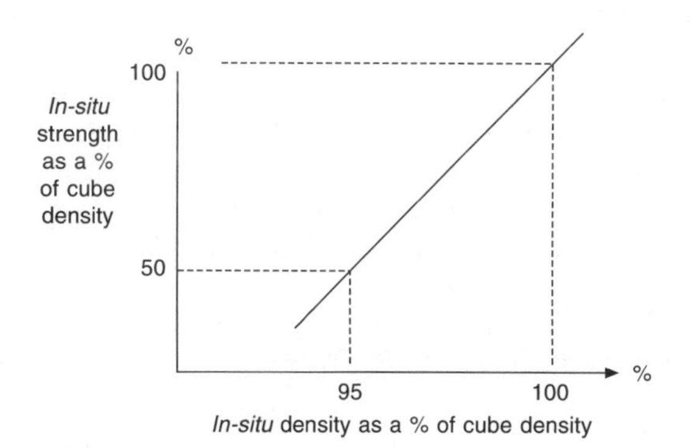


**Figure 23.7** The sensitivity of strength to density.

The UK *Specification for Highways Works* (SHW) specifies vibrating hammer compaction for cubical CBM specimens. *In-situ* density from the sample location must be a minimum of 95 per cent of the cube density. From the relationship above it can be deduced that field strength will be between 50 per cent and 100 per cent of cube strength.

It is normal to expect mean field density of 96–98 per cent. It is also good practice to design mixtures which will give, say, a 30 per cent safety margin over the specified strength. The combination of *in-situ* density and strength safety margin means that field strength will be approaching specified strength. Other nations use Proctor or Modified Proctor (BS 1377: 1990 Part 4) compaction on test specimens.

A very effective combination of rollers is a 10.5-tonne single-drum self-propelled earthworks compactor followed by either a 2.5-tonne tandem vibrating roller or a pneumatic tyred roller (PTR) of a minimum 7.5-tonne ballasted weight. The earthworks compactor will give the necessary centrifugal force during vibrating to compact at the base of the layer (up to 350 mm with sand and gravel aggregate). The wide drum (about 2.3 m) leaves an even plane surface unlike tandem rollers where drums 'wobble' as they meet varying resistance in the CBM, leaving a less even surface.

After the deep compaction, a smaller vibrating or PTR is used to roll out any creases or lines in the surface. The PTR is particularly useful on 'clean' materials, which do not bind together easily. The surface of such material is continually disturbed by vibrating rollers whereas the kneading action of the non-vibrating PTR presses the particles together.

A good way to set up the compaction regime right at the start of the works is to undertake a compaction trial. This may be incorporated into the works, provided specification requirements are met. The recommended method would be as follows:

1 Lay 40 linear metres or so of CBM (either paver or blade).
2 Have the earthworks compactor make two passes on the layer, one up to the paver and back again.
3 Use a nuclear density meter (NDM) (SHW Clause 104) to check the *in-situ* density.
4 Repeat steps 2 and 3 until there is no significant gain in density. This will usually be six passes of the roller ± two.
5 At this stage, use the NDM to produce a density profile through the layer by checking density at 25 mm increments of depth, right through the layer.
6 Assess the density profile to see whether:
   (a) homogenous impaction is achieved – if so, so use the tandem only to tidy up the surface prior to curing;
   (b) there is variable compaction. If so, does the layer have low compaction at the top or at the bottom? If it is low at the top, use the tandem roller with and without vibration and check again. If the layer has significantly lower density at the bottom of the layer, the main compactor is probably not heavy enough or the amplitude of vibration is not large enough. It is also a good idea to reduce the rolling speed right down to a slow walking pace.

Once the regime is established, training operatives and robust supervision is important to ensure proper compaction of the whole of the layer.

## 23.2.6 Curing CBM

The curing period for CBM is normally 7 days. However, when the programme requires, it is possible to add extra cement and crush test specimens say at 3 or 4 days. If the 7-day strength has been achieved at this early age then the CBM may be trafficked. In such circumstances it is useful in cooler weather to cure the test specimen adjacent to the pavement, rather than in the laboratory-curing tank, as strength gain is related to temperature.

The laid CBM is cured by applying a thin membrane to the surface. This could be a bituminous emulsion or aluminized resin. Such material is sprayed onto the CBM at specified rates/$m^2$. An alternative is to use polythene sheeting though this technique is expensive and cumbersome. It can also be embarrassing in very windy conditions! The curing membrane should be applied within two hours of the CBM being mixed. In order to do this the laying, compacting, checking of levels etc. needs to be very efficiently handled. In warmer climes, retarding agents are used to maintain a useful working life in the CBM mixture.

In two-layer work, it may also be appropriate to cure the first layer with water sprays, ensuring that at no time does the surface dry out.

## 23.2.7 Testing CBM

It is important to recognize the distinction between production control and specification testing. In order to ensure the correct mixture, production control must include testing of the raw materials, and a well-controlled process. If detailed records are produced by the mixing plant, all the better. Specification testing provides data for acceptance of rejection of work.

Production control is a management tool for ensuring that the mixture is produced to the quality required by the designer. The elements of production control will cover the type and maintenance of mixing plant, testing of raw materials, surveillance of the process, records of mixing, testing of the final product and so on. Some companies operate under BS EN ISO 9000 quality schemes, while European standards may incorporate a model factory production control annex which producers will be encouraged to use.

Production control is a positive step to ensure the product is right first time. This is quite distinct from specification compliance testing which is only a means of accepting or rejecting the mixed material. The *in-situ* density and either compressive or tensile strength of hardened specimens are the normal test requirements for specification compliance. Both are equally important.

Other testing may include a check on the particle size distribution of the mixture, modulus of elasticity of hardened test specimens or indeed assessing the stiffness of the *in-situ* layer by, say, a falling-weight deflectometer (*Design Manual for Roads and Bridge Works*, Volume 7, Section 3, Part 2, Chapter 5) (HMSO, 1998a).

The *Specification for Highway Works* (HMSO, 1998b) requires tests on the aggregate, plus compressive strength testing of hardened cubic specimens and *in-situ* density testing of fresh laid CBM using a nuclear density meter. This must be done within 4 hours of the mixture being produced. However, most contractors use the gauge during the rolling of CBM to ensure specification compliance.

Other *in-situ* density tests are available on hardened CBM. These include a 'sand-replacement test' and a 'water balloon test' where the CBM is excavated using a hammer and chisel to form a roughly cylindrical hole. The recovered CBM is weighed. The volume of the hole is measured using a reference sand or water balloon. A simple calculation yields the *in-situ* density.

## 23.3 Summary

The term 'cement-bound material' covers a wide range of possibilities from a treatment of clay soil subgrades, to a tightly controlled high-strength roller-compacted concrete. Suitable secondary and by-products can be used as aggregate.

For sub-grade treatment and capping/sub-base production, *in-situ* mixing may be employed. Mix in plant can be employed across the range apart from the clayey soil treatments.

Compaction is by rolling and it is essential to obtain the maximum possible density in order for maximum field strength to be gained and hence the design life to be fulfilled.

Laying and compacting plant is readily available although for larger schemes more sophisticated equipment and techniques may be employed.

If properly designed and constructed, the combination of thermally induced and traffic-induced stresses will not be exceeded and the CBM can have a life tending toward permanency.

## References

Croney and Croney (1998) *Design and Performance of Road Pavements.* McGraw-Hill, London.

Ellis, S.J., Megan, M.A. and Wilde, L.A. *Construction of full-scale trials to evaluate the performance of induced cracked CBM Roadbase.* TRL Report 289. Crowthorne, UK.

HMSO (1998a) *The Design Manual for Road and Bridge Works.* The Stationery Office, London.

HMSO (1998b) *The Specification for Highways Works.* The Stationery Office, London.

Williams, R.I.T. (1998) *Cement Treated Pavements – Materials, Design and Construction.* Elsevier Applied Sciences Publishers, London and New York.

# Part 9

# Industrial floors

# 24

# Concrete floors

*Neil Williamson*

## 24.1 Introduction

### 24.1.1 Floor slab function

Simply, a floor slab functions to distribute without deformation or cracking, the loads applied to it to the weaker sub-grade below, in the case of a ground bearing slab, or to the piles supporting it if designed as a suspended ground slab, and to provide a suitable wearing surface upon which the operations in the facility may be carried out efficiently and safely. Different specific properties are required by different industries, and even by competing operators in the same industry. Typical requirements of facility operators and owners are shown in Figure. 24.1.

Operators and owners require that their floor slab should:

- Be capable of supporting applied loads without cracking or deforming.
- Have a minimum number of exposed joints.
- Have maintenance free joints that do not impede MHE operating speeds.
- Be dust-free (i.e. highly abrasion resistant).
- Have tolerances appropriate to the materials handling system to be used.
- Be smooth and easy to clean, but not slippery.
- Be flexible enough to accommodate possible future changes in operating systems.
- Contribute to a safe, pleasant working environment.

**Figure 24.1** Principal requirements of facility operators and owners for floor slabs.

### 24.1.2 Concrete floor slab design and construction: key issues

The key issues to consider when designing and constructing an industrial floor slab to meet the above requirements are given in Figure 24.2.

1. An appropriate specification is the essential starting point.
2. Materials selected must be compatible with the requirements of the specification.
3. Site conditions must be managed to enable construction within a suitable environment.
4. Quality Control procedures must be agreed in advance and adhered to.
5. Operator skill is a key element – and difficult (but not impossible) to account for.
6. Supervision should be experienced in the type of installation being carried out.
7. Clear responsibilities will help to ensure everyone works together.

**Figure 24.2** Key issues in the design and construction of an industrial floor slab.

It is wise to start our understanding of the design and construction of industrial concrete floors by considering the special requirements of the concrete to be placed in these floors. We shall return to the other key issues later in this chapter.

## 24.2 Concrete for industrial floors

The single most important ingredient in the successful construction of a concrete ground floor is an appropriate concrete mix. Put simply, with poor concrete even the most skilled crews can struggle to produce an acceptable floor, whereas with a properly specified and controlled concrete mix, relatively inexperienced crews can produce very good results. The specification, selection, production and quality control of concrete for use in ground floor slabs is a process of compromise, balancing a number of conflicting properties and requirements. As a minimum, the concrete specification and production process should take into account the following:

- the environmental conditions at the time of placing and finishing the concrete
- the degree of weatherproofing of the slab from sun and wind
- the method of placement and levelling of the concrete
- the resources available for placing and finishing the concrete
- if (and when) a topping is to be applied to the concrete
- whether any admixtures or steel fibres (and type) are to be added at site
- any special floor flatness requirements

### 24.2.1 Mix properties

Having taken the above factors into account, the specifier, designer and producer of the concrete should enable a mix to be delivered to site that has the properties given in Figure 24.3.

Concrete in industrial floor slabs should:

- Have good finishability and bleed characteristics.
- Be suitable for achievement of the specified tolerances.
- Possess low shrinkage characteristics.
- Have workability and mix characteristics appropriate to the method of placement to be used.
- Set at the desired rate, consistent with the resources available.
- Achieve the specified strength at the specified time.

**Figure 24.3** Industrial concrete floors – desired concrete mix properties.

### *24.2.1.1 Finishability*

Concrete described as having good finishability principally means a mix with good cohesion, not deficient in fines, which can be floated and trowelled without undue working of the surface. Difficulties in adequately closing the surface of the concrete result in increased permeability and poor aesthetics. Finishability is principally a property of the mortar that is brought to the surface during the placing and finishing operations. The water, fine aggregates, cement, and admixtures all influence finishability. In addition, toppings applied to the plastic concrete such as dry-shakes may have a considerable impact. Sometimes quite harsh and difficult to finish concrete can be transformed by the application of a dry-shake topping during the finishing operations. Under no circumstances should cement alone be added to the surface in order to make finishing easier.

Finishability cannot be measured in the same way as workability or setting characteristics, and is a property that is only adequately determined by the finishers themselves. Experienced finishers can pinpoint adjustments necessary to improve this property. Finishability is not something that can be easily tested for in the laboratory. Thus, prior to commencing work on critical flooring projects, trial mixes should be delivered to site and placed in the same way and under the same conditions as the main works. Feedback from the finishers at this time should be taken full account of by the concrete supplier.

Adequate water is necessary to lubricate the cement and aggregate particles. The amount of water required for this lubrication is in excess of that required for hydration of the cement. It can also be more (or less) than that required for the specified workability. Very low workability concrete cannot only be very difficult to place, but it is usually even more difficult to finish. Plasticizers may be used to enhance workability, but (by design) do not add any free water to the mix, hence will not overcome the potential finishing problems. Minimum free water contents of around 160 litres per cubic metre are typically required to achieve a dense, close finish. The majority of UK readymixed concrete with the strength specified below 35 N/mm$^2$, containing a water-reducing agent and with a slump of 75 mm or less, would contain less water than this.

It is clearly good practice to minimize the amount of water in the mix in order to help to minimize shrinkage, and other important properties. However, if the surface of concrete is not properly finished, wear resistance, permeability (i.e. durability), and colour consistency are compromised. Thus there is a trade-off required in the mix specification and design process due to these (apparently) conflicting requirements. The issue is further complicated in that the water needed at the surface is 'free water', which reduces from the moment that it is added to the mix, through hydration, evaporation, absorption by the aggregate and so on.

The bleed characteristics of the mix have a great effect on the finishability of the mix. However, some of the procedures described later for improving the bleed characteristics of a mix will not necessarily improve finishability.

Finishability can also relate to the setting, or stiffening, characteristics of the mix. There is a window of finishablity during which the necessary finishing procedures must take place. Clearly, if a mix sets faster than the resources available can cope with, defects in the finish will occur, most commonly a poorly closed surface, an uneven finish, and blackening from trowelling too late, particularly if the mix is inherently difficult to finish. However, if this window can be extended, for example by using a blended cement or cooling the aggregates, it will give the finishers a longer period to work the concrete, and perhaps also enable them to bring more mortar to the surface.

Other measures known to improve finishability include:

- improving the weatherproofing of the pour area
- changing the type of fine aggregate (or blending two fine aggregates)
- use a finishing aid admixture
- changing the cement source
- the use of mist sprays or fine sprinkling of the surface with water
- the use of an evaporation retardant
- the use of a dry-shake topping

The use of a rounded fine aggregate may help to improve finishability. Less water (and hence cement paste) is required to encapsulate the rounded fine aggregate particles, and they are more easily worked by the floating operation. Angular fines can be more difficult to move, needing more water to lubricate their larger surface area, and if they remain poorly orientated it can be difficult to close the spaces between the particles with the cement paste.

Admixtures known as finishing aids work essentially by making the mortar at the surface 'slicker', in conjunction with the water added for mixing and workability.

Changing the cement source has been known to improve finishability. There is no clear evidence on why this should be the case, but it is most likely due to the varying particle sizes and shapes resulting from minor variations in the grinding process.

The use of a fine mist, or 'fog' spray is an excellent way of enhancing the finishability of concrete when evaporation rates are high. These are very fine mists of water from a pressure hose sprayed into the air over, but not directly onto, the concrete. However, the best solution is to ensure that all the necessary steps are taken to provide a placing environment and concrete mix such that the application of any water to the surface is unnecessary.

Evaporation retardants (otherwise known as monomolecular films) are applied to the surface immediately the concrete is struck off to slow down the rate of evaporation from the plastic concrete in a similar way to that of a curing membrane applied to hardened concrete. The film produced is immediately broken up by subsequent power-floating operations and does not reform unless a further application is made. In extreme conditions this procedure may be carried out between successive passes of power-floating. There is no detrimental effect on the hardened concrete.

The simplest way to improve finishability is to increase the cement content. It is also common in the United Kingdom for a minimum cement content of 325 kg per cubic metre to be specified for finishability and durability purposes (Ref TR34). In principle, the higher the cement content the easier the slab is to finish. However, sometimes this ease of trowelling and closing of the mortar can inadvertently lead to delamination by enabling the premature sealing of the surface. Equally, higher cement contents, even at low water to cement ratios, are not recommended due to the increased shrinkage potential of the paste. Crazing is also more common with cement-rich mixes, and the shorter finishing window resulting from the higher cement content may effectively compromise the finishability of the mix, particularly in elevated temperatures. If the concrete does not require a particular cement content for strength, bleed control or other properties, it would be more appropriate to apply a well graded dry-shake topping or use a finishing aid to enhance finishability. The application of a dry-shake topping will result in an increased cement and fine aggregate content at the surface, without compromising the shrinkage and other properties of the base concrete mix.

## Bleeding

The bleed characteristic of the concrete mix is a critical element in the production of a high quality slab with good abrasion resistance.

Bleeding is the accumulation of water added during the mixing process at the surface of the fresh concrete. It is simply the settlement of the heaviest components in the mix to the bottom of the slab, through the forces of gravity, with the lightest (the water) coming to the surface. Excessive bleeding occurs when the mix is not sufficiently cohesive to prevent this vertical movement of the constituents of the mix. In most instances of floor construction a small amount of bleeding is desirable, particularly when a dry-shake topping is to be applied to the base concrete.

The most important factors in determining the amount and rate of bleeding are the overall grading of the aggregates and the amount of free water in the mix (Figure 24.4). The following can also contribute to the severity of bleeding in concrete for ground floors

- cement type and particle size
- the method of placement
- the type and rate of addition of any admixture used
- the presence of an impermeable barrier (such as the slip membrane) below the slab

| Sand <1.18 mm, % by weight of total aggregates | Minimum fines <0.30 mm, including cement per m$^3$ of concrete | Fineness modulus of sand (FM)* |
|---|---|---|
| 26–32% | 450 kg | 2.65–2.80 |
| *FM here is the sum of the accumulated percentage retained on 2.38 mm to 0.15 mm sieves. | | |

**Figure 24.4** Recommended fine aggregate proportions for non-bleeding concrete (20 mm maximum aggregate size).

Bleeding is also a time-dependent phenomenon in that it cannot take place once the concrete has set, thus any factor that can affect stiffening times will influence the overall bleed characteristics of the mix.

Some bleeding of the concrete may be desirable, but a continuous layer of water a few millimetres deep is likely to be a serious problem. The concrete finishing process should never commence until all of the bleed water has evaporated or been removed from the surface. If bleed water is present and an attempt is made to finish the slab, a weak surface will occur as a result of the high water to cement ratio present. However, the concrete may still be bleeding even if there is no standing water on the surface. This may happen when the rate of evaporation is the same as, or slightly greater than the rate of bleed. This may be the result of particularly low humidity and high temperatures, or more usually, due to wind blowing across the slab.

Whilst many of the recommendations below have merit to solve the specific problem of excessive bleeding, caution should be exercised with all of these steps (Figure 24.5). For example, switching to rapid hardening cement may not be prudent in summer if the higher early strengths are accompanied by faster stiffing rates, and, again, although air-entrained concrete can often be easier to finish, an air-entraining agent should not be used when the slabs are to be power-floated.

- Use more cement
- Switch from ordinary Portland cement to rapid-hardening cement (not usually available in UK)
- Use PFA as a filler
- Use more fine aggregate, especially at the lower end of the sieve range. Increase the amount of fines passing the no 100 sieve (0.150 mm)
- Use less water and/or use a water reducing agent
- Use entrained air
- Use an accelerator

**Figure 24.5** Measures to be considered to reduce bleeding.

### *24.2.1.2 Suitability for achieving specified tolerances*

Generally, the same factors influencing mix design and production that ensure good finishability apply to enabling the achievement the tolerances required. In addition to the effects of plastic settlement, the workability of the concrete mix can influence the level tolerances achieved, and the quantity of fines and bleed characteristics of the mix can have a profound effect on flatness.

Inconsistency is also a major issue influencing the achievement of specified tolerances, and is addressed separately in a following section.

Very wet concrete with a high rate of bleed affects the initial screeding or strike off operation. In extreme cases the bleed water itself is being levelled, leaving a slightly 'uneven' surface below. With floors measured to an accuracy of 0.1 mm, even a small amount of bleed water on the surface at the time of screeding (strike-off) and re-straightening is influential. The effect of plastic settlement is self-evident. Good cohesion of the mix is thus essential.

Low slump concretes are difficult to handle and slow to place. They may also be hard to strike-off to the required tolerances, even when using a vibrating beam. Very low slump concretes tend also to produce an inherent waviness in the surface of the concrete, at right angles to the orientation of the strike-off beam.

For flatter floors, in general, a longer period for working the concrete is desirable. This enables repeated re-straightening of the slab surface using straightedges. Floors made with fast setting concrete tend to be less flat than those using slower setting mixes, all other factors being equal.

Mixes for Superflat floors also tend to require a higher fines content, or at least a slightly higher concentration of 'fat' at the surface. This enables the 'cut and fill' re-straightening of the surface profile, necessary to achieve the very onerous tolerances required for these floors, to be readily undertaken. Equally, of course, very high fines contents need to be avoided to prevent creation of a weak surface and a high shrinkage concrete. In many cases it is better to achieve the increased fines content at the surface by use of a 'dry-shake' topping.

### *24.2.1.3 Shrinkage characteristics*

While in theory it may be possible to produce a concrete containing exactly the correct amount of water for the full hydration process to take place and nothing more, such a mix would almost certainly be impossible to place and finish. Thus all commercially produced concrete shrinks (contracts), even those made using shrinkage-compensating cements. The performance of shrinkage compensating cements is outside the scope of this chapter.

**Thermal contraction** In addition to shrinkage of the concrete resulting from the drying out of the surplus mixing water, concrete floor slabs may also have to accommodate thermal contraction. Thermal contraction in floor slabs generally occurs early in the slab's life (normally within 24–72 hours), primarily as the concrete cools after heating during cement hydration.

**Drying shrinkage** Drying shrinkage takes longer to manifest itself than thermal contraction. Low shrinkage also tends to result in reduced curl at joints. Thus the long-term problems associated with curled joints, such as edge damage, poor local flatness, and cracking are also reduced, or eliminated.

Drying shrinkage primarily occurs as the water used for workability and finishability in excess of that necessary for hydration leaves the concrete through microscopic pores in the cementitious matrix. This pulls the walls of the pores inward creating a shortening effect on the concrete. Ultimately the combined shortening induces a stress into the slab due to the restraint to movement from friction with the sub-base, any intrusion through or into the slab, or in the case of a heavily reinforced slab, the reinforcement steel. Thus the two most important elements determining the ultimate shrinkage of the concrete are the quantities of water and cement present in the mix.

The aggregates are also important in that they influence the amount of free water that is added to the mix, and that they restrain the shrinkage of the cement paste. Aggregate size and shape also influences shrinkage. As illustrated by comparing the values given in Figures 24.3 and 24.6, larger aggregates require less fine material for a homogeneous mix than smaller maximum aggregate sizes. Mixes containing larger aggregates thus require less water for a given level of workability. Similarly, because of their total smaller surface area in a given volume, rounded aggregates demand less water than irregular angular

| Sand < 1.18 mm (No 16 sieve, USA) % by weight of total aggregates | Minimum fines < 0.30 mm (No 50 sieve) including cement per m$^3$ of concrete |
|---|---|
| 18–26 | 380 |

**Figure 24.6** Suggested fine aggregate portions for flooring concrete when using 40 mm maximum coarse aggregate size

materials. The maximum aggregate size used should not be greater than 1/3 of the slab's thickness, however.

The following summarizes the principal methods of minimizing the drying shrinkage potential of a concrete mix:

- avoid aggregates with high drying shrinkage values
- use the largest possible maximum aggregate size
- reduce the amount of free water in the mix
- use less cement

All of the above must be undertaken consistent with the requirements for finishability.

### 24.2.1.4 Workability and other mix characteristics appropriate to method of placement

While the long-term benefits of a low shrinkage concrete are clear, floor-layers do not want a concrete that while possessing very low shrinkage characteristics is impossible to place, compact, and finish properly.

The method of placement determines workability and influences setting characteristics of the mix. Lower slump concrete may be used for direct truck discharge and screeding with a vibrating beam. Slumps in the order of 100–150 mm are required for levelling using a Laser Screed (see Figure 24.7), and slightly higher than this for hand-screeded large pour installations. Similarly, if the concrete is to be pumped, the workability (and fines content) may have to be adjusted. Ideally this should be done without a detrimental effect on the shrinkage characteristics of the mix. Due account should be taken of the method of compaction when selecting the level of workability. Generally it would not be appropriate for the floor designer to specify the level of workability required, since this is so dependent on the actual construction procedure.

The concrete should not lose its workability too quickly, particularly if it is to be pumped a long distance. Delays to progress as a result of difficulty in pumping or levelling

**Figure 24.7** Laser screeded large area pour, using steel-fibre reinforced concrete.

the concrete after discharge are expensive to the flooring contractor and may result in defects in the finished floor. Adding water at the point of discharge will not necessarily help to increase the period of workability, and may worsen any bleed present. Where an increase in workability at the point of discharge is required a superplasticizer with neutral shrinkage characteristics should be used. Care should be exercised in the selection of such admixtures, as not all of these extend the concrete workability window with no, or only a marginal effect, on the shrinkage of the concrete.

With laser screed operations in particular, the rate of loss of workability should also be closely related to the rate of pouring, so as to eliminate the risk of 'cold joints' or difficulties in screeding. Mixes therefore may need modification in summer, particularly where access or placement is likely to be tricky.

**Note on fibre concrete**

Consideration should be given to the effect of the various types of fibres on the workability of the concrete. A small reduction in measured workability is to be expected when adding polypropylene fibres. This varies slightly between the two main types of fibre. When the concrete is being vibrated, or a superplasticizer added it may be prudent to merely accept this slight loss of workability. The option is to increase the amount of water in the mix, and in order to maintain the water:cement ratio, to increase the cement content. This is often done by readymixed concrete producers, but should normally be avoided if possible. A greater reduction in slump is noted after the addition of steel fibres, perhaps 25 mm or more, depending on the type and quantity of fibre used. This needs to be accounted for in the mix design, but again the use of an appropriate superplasticizer will mean that no significant changes need be made to water or cement contents.

#### *24.2.1.5 Setting characteristics*

Both instances of setting either too quickly or too slowly can contribute to problems.

As discussed earlier, concrete that sets too fast frequently causes problems. The flooring contractor may simply not have the resources to start the various stages of the finishing operation at the right time, especially where a large area is being poured. Inadequate finishing can lead to a poorly closed, discoloured surface, and there is the risk of the finishers adding excess water to the surface to try to 'save the floor'.

Where dry-shake and other monolithic toppings are being incorporated into the slab, the slab may set too quickly relative to the rate of bleed and there is thus inadequate water at the surface to hydrate the topping, and to bond this to the base concrete.

In extreme cases of fast setting concrete there is a risk of cold joints in the slab. This can occur particularly with large pour floors where the 'face' may be 40 or 50 metres long, and is exposed for some time. The use of fog sprays or evaporation retardants may be useful in these circumstances.

If the concrete sets too slowly, this may result in premature finishing, with the consequent increase in the risk of delamination, and if the finishers do access the slab too early, they may also cause disruption to floor flatness. During the winter months, the concrete may also be susceptible to frost damage, and the longer the floor remains unfinished and uncured, the greater the risk of damage from the elements.

From a commercial point of view, the floor-laying contractor does not want to have to pay men to wait around for the concrete to stiffen before power floating can commence. Further, late finishing increases lighting, and in some cases heating, costs.

The worst situation for a floor finisher is when the set of the concrete is variable. This is addressed in the section on production issues, below.

#### *24.2.1.6 Achieve specified strength*

Because the majority of floor slabs are judged by their wearing surface performance, an over-emphasis on the strength of the concrete may be detrimental to the long-term durability of the floor. As we have already seen, high cement contents that may be required to achieve even only moderately high strength concrete can be detrimental to other more important properties of the concrete. Similarly, using admixtures to achieve higher strengths at lower cement contents may result in a concrete with inadequate water and or paste to enable a good finish to be achieved. All too often the 28-day compressive strength of the concrete used in floor slabs is reached after just 7 days. This usually results from unnecessarily high cement content. Few (if any) floor slabs fail because the concrete strength was below that accounted for in the design. However, shrinkage cracking and excessive joint opening often cause major disruption.

The designer may also want to exploit the creep characteristics of the concrete in its early days. Creep reduces as the strength of the concrete increases (and the modulus of elasticity changes). Thus this characteristic is best achieved in slower maturing concrete. Creep is particularly beneficial when large areas are poured incorporating a single layer of mesh and regular sawn joints.

In most circumstances we need to avoid concrete that has a very slow rate of gain of strength, in particular, tensile strength. Because of the high surface area to volume ratio of floor slabs, the concrete must gain tensile strength at a rate comparable to the increase in stresses in the concrete resulting from early age drying and thermal movements. Again, this is particularly relevant to large area pours and other systems where saw cutting of control joints within 24 hours or so is specified, or where steel fibre concrete is used. If the concrete gains strength too slowly compared to the increase in tensile stresses in the slab there is a risk of cracking of the slab before saw cutting takes place or sufficient bond is achieved with the steel fibres or bar reinforcement (depending on the design), particularly at re-entrant corners. This is particularly the case in warmer weather when higher percentages of the non-PC component in blended cements are used.

In addition, if the rate of gain of strength of the concrete in the early days is too slow, programmes can be disrupted, or the floor layer's resources severely stretched.

In some instances an early gain of strength and drying out of the slab may be beneficial, such as when a rigid floor covering is to be applied to the slab with adhesive. Additionally, in these cases, low shrinkage is also important to prevent reflective cracking in the covering. A low free water content will also lead to a faster slowing down of the rate of moisture migration from the slab, thus minimizing the delay before commencement of laying of the covering. Proprietary mixes containing micro-silica, and high-range water reducing agents are often promoted to provide this performance.

## 24.3 Production issues: consistency

Consistency of the concrete mix is probably the most important of all of the properties of the concrete. The two principal areas of concern to the floor layer are set and bleed consistency.

Factors that can affect uniformity of the concrete include:

- moisture content of the aggregates
- amount of free water added to the mix
- variations in cement contents or the percentages of blended cement
- admixtures – variations in quantity or dispersion in the mix
- variations in truck mixer efficiency, particularly where the mixing water is added into the truck
- variations in travelling times from production plant to site, particularly in hot weather
- change of aggregate or cement source during pour
- delays in placing
- areas of shade and sun on the slab surface

The vast majority of floor finishers would rather have a relatively poorly designed mix that is consistent, than a good 'theoretical' mix that varies greatly in setting or bleeding characteristics from load to load. Even if the concrete is relatively fast setting, but consistent, and provided that there are adequate equipment resources, the finishers will usually produce a first class job. They admit to being unlikely to be able to do so, on a consistent basis, with very variable concrete.

Not only must the mix be uniform throughout the load, but also from load to load. If this is not achieved the following defects can result:

- poor tolerances at the interface between 'soft' and 'hard' loads at the time of finishing
- marks, e.g. 'smears', on the floor surface where machines pass from soft to hard loads
- scratching and tearing of the surface where machines pass over already finished areas
- poor colour consistency, particularly with white and other light coloured dry-shakes
- pinholes and small unfinished areas along the interface between 'wet' and 'dry' areas
- delamination in slower setting areas, due to unintentional premature finishing.

Even the best finishers may be unable to cope in extreme circumstances, and the consequences for all involved can be severe, particularly for the flooring contractor, both financially and due to loss of reputation. One of the worst cases is when a large pour is being undertaken and one of the loads away from the edge of the slab sets faster than the loads surrounding it. In the first instance the finishers may not be aware of this, because they are some metres away, and even if it is possible to judge from the perimeter of the pour, how are they to reach this load?

In addition, difficulties can be encountered with striking off the concrete, slowing down the speed of installation, in extreme cases risking cold joints in the floor. The slow placement in turn can affect the concrete suppliers, as a result of delays to their trucks at site.

The supply rate must be balanced against the following:

- resources of the floor-layer
- depth of the slab (faster rate required for a deeper slab)
- method of placement – and capacity of pumps or conveyors etc.
- dimensions of the pour (area, length of open face etc.)
- weather conditions
- anticipated setting times of the mix

Clearly by importing aggregates, cements and so on from other locations across the country, and by only using the most modern mixer plants, most concrete suppliers could theoretically meet the foregoing requirements. In many cases this is not practical. However, by appreciating the most important factors, and taking the necessary steps to balance the conflicting requirements there need be little, if any, cost increase over standard mixes. Minimizing the cement content, for example, is likely to contribute to a reduction in cost. The real challenge on a day-to-day basis is to consistently produce concrete at a price at which the benefits of the designed mix outweigh any cost differential that may result from ensuring that the properties meet the requirements of the floor-layers.

## 24.4 Floor construction

### 24.4.1 Methods

The majority of industrial concrete floors are now installed using large pour methods. Up to 4000 square metres per day can be installed using such systems, with typical pour sizes in the order of 1500 to 2000 square metres. In particular, in the UK widespread use is made of the Somero Laser Screed to assist with the placing and striking off of the concrete (see Figure 24.7). Manually screeded large pour techniques are more common on the continent of Europe. Other methods include the more traditional long-strip system, when the concrete is placed in long bays of limited width. In the UK, this method is generally confined to small projects or where onerous tolerances are required, such as in the case of 'Superflat' floors.

Superplasticizers may be used with all forms of construction. The Laser Screed imparts some compaction into the concrete, but cannot cope with very stiff concrete. Stiffer mixes are sometimes used in long-strip construction, and vibrating beams or poker vibrators used to compact the concrete.

With many 'Laser Screed' pours any dry-shake topping specified is applied immediately after initial strike-off. Clearly, if the concrete then bleeds badly, difficulties can be encountered with removing the water, or reducing the w/c ratio at the surface. Similarly, adequately closing the floor can be difficult if the concrete does not bleed at all, or sets faster than anticipated.

Almost all industrial floors are 'directly finished' using power-floats. Smaller walk-behind models are used for long-strip and specialist floor finishes, and ride-on machines used to cover the large areas required with large pour installations (see Figure 24.8). Final power-trowelling is followed by curing, typically with a spray applied membrane.

### 24.4.2 The finishing process

Finishing of the floor normally commences when an imprint of not more than 3 mm can be left in the surface of the stiffening concrete. Many finishers now leave floors much later than this when using very large heavy machines, which cover the floor very quickly. There is some risk involved in this, both by reducing the length of the 'finishing window', and also by increasing the risk of delamination. The initial process, called floating (or breaking) the floor is undertaken using float pads laid flat in contact with the surface of the concrete, or with large diameter discs (pans) fitted to the power-float arms. The initial

**Figure 24.8** Ride-on power-floating and trowelling of a large area pour to polished finish.

floating operation depresses the coarse aggregate by drawing 'fat' to the surface, opens the surface to permit any trapped bleed water to escape, and allows the slab to dry out. Used properly, pans can also assist in removing minor irregularities in the floor surface, particularly when used in conjunction with a highway straight edge. Further floating continues to produce a smoother but 'open' paste at the surface. Monolithic toppings or dry-shakes may also be applied early in this process. As the surface tightens and dries out the trowelling process starts. Finishing blades are much smaller than float pads, and are used to smooth and close the surface. As the concrete stiffens and dries further, the angle of the blades with the floor surface is increased (often along with the speed of rotation of the blades), resulting in densifying of the surface. It is this process that produces the glass like reflection found on a well-finished industrial concrete floor.

## 24.4.3 Planning

### *24.4.3.1 Concrete*

Prior to construction of any floor slab, but particularly a large pour floor, it is essential that all of the factors affecting the concrete suitability and consistency are discussed, solutions agreed where necessary, and the control regime agreed. In the past many concrete suppliers have been unaware of the special needs of the floor-layer in terms of concrete consistency. In some cases the standards for sampling and testing of the concrete simply do not take these requirements into consideration.

#### 24.4.3.2 Concrete production and delivery

In order that the concrete has the intended properties at the point of discharge, wherever possible the mix specification and procurement process should take into account the following factors:

- travelling time between plant and site, and any potential delays
- anticipated ambient temperatures at time of production and delivery
- whether the concrete is pre-mixed prior to discharge into the truck, or mixed solely in the drum of the truck mixer
- any specific issues concerning truck mixer access to the point of discharge
- method and location of the addition of any admixtures used

Particular consideration should be given to mixing procedures at 'dry-batch' plants. That is, plants where the concrete is mixed in the truck-mixer. For example, specific procedures should be agreed to ensure uniformity of addition of admixtures (these should ideally be pre-mixed with the batching water before discharge into the truck in two stages, at the beginning and middle of the load).

Where appropriate, trial mixes should be commissioned well in advance of the installation.

Before the main work starts it is useful (and is essential on a very large project) to carry out a trial pour by using the method proposed for the main slab. This may be in an office or amenity block, for example. Following this it may be necessary to agree final adjustments to the mix (and methods of placement).

#### 24.4.3.3 Preparation

As the date for the installation draws near it is necessary to monitor progress of preparation, and to set up a pre-pour meeting. Such a meeting is VITAL to the success of any floor construction project. Attendees at the pre-pour meeting should include:

- the floor-layer
- main contractor
- concrete supplier
- floor surveyor (where appropriate)
- client's representative
- suppliers, e.g. admixtures, toppings

At this meeting the following should be addressed:

- clarify quality control procedures to be used, and responsibilities
- confirm resources required are available
- instigate any necessary training, including Health & Safety issues
- likely weather conditions

## 24.4.4 Supervision and workmanship

It is important that the following are carefully monitored both during and prior to installation of the floor slab:

- Building environment
  - weathered building envelope
  - temperatures
  - other construction activities
- Preparation
  - sub-base – condition, level
  - slip membrane, reinforcement
  - joint details – workmanship
- Resources
  - labour and plant
  - stand-by equipment
- Concrete supplier
  - materials – availability, notify any changes
  - truck availability and condition
  - contingency plans – stand-by plants etc.

During the pour the concrete supplies need to be checked for consistency, the amount of any admixtures and water added, and where appropriate that the correct quantity of steel fibres has been incorporated. In addition great attention has to be paid to the placement and levelling operation. This involves the floor placement team, and any other supervisors, constantly monitoring:

- workability
- level and flatness
- edges/special details
- temperatures, both air and concrete

Checks should also be made on floor flatness during the finishing operation, particularly the hand finishing at edges, and that the curing membrane is properly applied.

Within 24 h of completion of the first pour, level/flatness, the overall finish quality and the attention to detail should be checked. Following this, and as required, it may be necessary to:

- adjust concrete mixes
- vary finishing/placing techniques
- revise the pour layout or the sizes of pours

## 24.5 Design and loading

The principal elements in floor design are as follows:

- Structural design – thickness
- Shrinkage crack control
  - * joint layout
  - * joint details
  - * reinforcement
  - * other measures – e.g. Shrinkage reducing admixture
- Wearing surface performance

## 24.5.1 Structural design

### *24.5.1.1 Floor loading*

The frequently quoted u.d.l. specifications are inadequate for the purpose of the structural design of ground floor slabs. Such specifications are unrepresentative of the actual loading patterns to which slabs in warehouse facilities are subjected. Only in rare cases of block-stacking is the u.d.l. condition relevant, and even here the aisles are unloaded, thus the aisle width itself may become a more critical factor in the design.

BRE Paper IP19/87 addressed this issue by defining non-dedicated warehouse floor loading in terms of four classes. Each of the proposed classes had typical critical loadings specified, depending on floor use. Specifying floor loading requirements in terms of a class which has a set of parameters associated with it is very useful as all parties can be confident (from a loading point of view) that the floor will be fit for purpose. Today, this classification remains the safest, most cost-effective, means of specifying speculative warehousing.

When an industrial floor is to be designed for a known set of loadings, the specification should include both the magnitude and distribution of the point-loads, any restrictions to be placed on the proximity of joints in relation to these loads and the frequency of any dynamic loads such as fork-lift truck movements.

For example in a warehouse, the loading from the pallet racking can be determined from the number of levels of racking, the number of pallets per set of beams (usually 2 or 3), and the maximum pallet loads. Only in exceptional circumstances is the self-weight of the racking significant. Point loads from pallet racking are typically up to around 12 tonnes per leg, but can reach 30 tonnes in stacker crane served facilities.

Care should be taken when considering the point loading from mezzanine floors. Very often mezzanines are used in facilities where there are low stack heights hence the slab may not be designed for particularly high point loads elsewhere. If the point loads are particularly high compared to those computed from the rest of the systems found in the facility, it may be appropriate to investigate the advantages and disadvantages of using discrete foundations.

Mezzanine floors may often be associated with shelving systems located below the decking. The floor designer must take this combination of loads into account.

#### Base plates

As a rule the base plates for pallet racking and shelving end frames are quite thin, typically of around 4 mm in thickness. The plates also have holes for fixing – normally the base plates are designed primarily as a member to fix the frames to the floor. As a result, these base plates are relatively flexible, and cannot be considered to assist in spreading the load from the end frame leg. Therefore they do not serve to effectively relieve the flexural stresses in the concrete.

In practice, the base plates need to be very rigid to spread the load adequately. Even those used for mezzanine posts are rarely thicker than 16 mm (typically 200 × 200 mm. in area). In some instances upgrading the plan size and thickness (and the stiffness) of the base plates to mezzanine floors may be effective in reducing slab depth, or ensuring that flexural stresses are within the capacity of an existing floor slab. To be worthwhile the base plates may have to be designed of a size large enough such that they would then project unacceptably into trafficked areas, hence this technique is rarely used to enhance the plan area and stiffness of pallet racking base plates. Very occasionally beams at ground floor level have been used to spread loads below pallet racking when uncertainty exists with the load carrying capacity of an existing slab. Provided that the floor is flat enough to permit full contact with the underside of the beam, this considerably enhances the effective loaded area.

Other than in special cases where the effective load area is increased as described above, a loaded area of 100 × 100 mm is normally assumed for shelving and racking, and 200 × 200 mm for mezzanine floor posts when designing floors to carry the point loads. Refer to BCA ITN11 for more information.

### *24.5.1.2 Loading configurations*

When calculating loads from pallet racking, the position of the bottom pallet must be determined. This is either on the floor or on a beam a few inches above the floor. In the latter case all of the loading is transferred onto the end frames. Where the racking is five levels high the difference in magnitude of point loads between a pallet on floor and pallet on bottom beam configuration is 25%. A bottom beam configuration occurs in very narrow aisle facilities, due to the presence of the truck guide rail. However, this situation can be found in many types of pallet racking installations and it is important to determine the position of the bottom pallet prior to designing the slab (Figure 24.9).

Depending on the position of joints and so on, the combination of loading from the front leg of the racking system and an adjacent forklift truck needs to be considered. It is not usually the worst case, particularly with standard pallet racking. Normally the back-to-back legs of a standard pallet racking system combine to produce the worst case for structural design purposes.

Generally the wheel loads from forklift trucks are not as high as loads from pallet racking systems. Dynamic loading across joints may be a serviceability issue, as dynamic loads from the trucks can exceed twice the static wheel (point) loading. The design for load transfer should take these into account. Factors of safety should take into account high repetitions of dynamic loading across joints. It may appear that a relatively thin slab section is appropriate to modest point loadings from racking, or because the facility may be blocked stacked. However, the designer should check the edge loading (or corner condition, if relevant) from the transient forklift truck, against the static load capacity of the slab section. The more load repetitions that are expected the higher the factor of safety should be. Joint failure is far more likely than structural failure of the slab.

Contact pressures are not usually considered in floor design calculations. Even for hard polyurethane wheeled trucks, these pressures are unlikely to exceed 15 $N/mm^2$. Manufacturers of pneumatic tyred trucks sometimes quote contact pressures, but in any case these can be misleading, as the contact area while the truck is in motion is likely to vary. It is recommended that contact pressures and the relatively small horizontal loads resulting from friction during braking, be ignored by the floor designer except in extreme circumstances. These effects are more relevant to the abrasion resistance characteristics of the slab than from a structural loading point of view.

Equipment used in factories can be very heavy, and in many cases it is better to design separate foundations for this. However, if there is the possibility that the layout may be reconfigured at a later date the floor must be designed accordingly. In many cases potential settlement may be more of an issue than the ability of the slab to accommodate the load.

**Figure 24.9** Typical pallet racking system – wide aisle (Note: no bottom beam in this system).

Take, for example, a printing press weighing 60 tonnes. This may have say six or eight fixing points to the slab. The individual point loads are thus not dissimilar to a moderately high drive-through racking system. It would probably be prudent to take the location of the heaviest loading as edge and corner conditions and design such a slab as described using one of the standard references referred to later.

Stresses from internal walls built off the slabs can be calculated on the basis of line loads. When these are properly assessed it will be noted that slab thickenings are rarely justified.

### 24.5.1.3 General principles

If an industrial floor slab is to be durable then random cracks are to be eliminated since under trafficking these will deteriorate, possibly ultimately rendering the surface unserviceable. Cracks result from tensile stresses exceeding the tensile capacity of the concrete. In ground slabs one of the causes of these stresses is slab flexure under load. In other flexural applications the high compressive strength of concrete is exploited and it is reinforced with steel to provide the necessary tensile strength. In order for this reinforcement to contribute to the load carrying capacity of the composite section, the concrete in the tension zone has to crack. This is clearly not desirable in floor slabs. Non-suspended ground slabs should therefore be designed such that the flexural stresses generated under and around loads do not exceed a factored proportion of the tensile strength of the concrete, unless the slab is specified to include steel fibre reinforcement.

A ground bearing industrial floor is part of a system that comprises sub-grade, sub-base, the slip membrane (if specified) and the concrete slab itself. The performance of the sub-base and sub-grade is critical to the slab's performance.

The slip membrane may also act as a gas membrane or damp proof membrane (dpm) in certain circumstances, and in the case of cold stores the sub-base is replaced by the insulation panels. In the case of an unbonded overlay slab, the sub-base is replaced by an existing concrete slab. Similar requirements for tolerances and uniformity of support apply – in fact they may need to be stricter since many overlay slabs are thinner than slabs-on-ground.

It is essential to include a requirement for the sub-base not to deform under truck mixers during the construction phase. In some instances this may result in the concrete being placed by dumper or pump. If the truck mixers are allowed to travel over the slab it is important that no rutting takes place. Any depression caused by the wheels from the truck mixers will have an adjacent 'hump' and the disturbance will cause not only an increase in slab restraint but also loss of section. As a result the load carrying capacity and resistance to shrinkage stresses of the slab will be locally reduced.

### 24.5.1.4 Ground bearing slabs

For many years it has been accepted that the work of Westergaard provided a sound basis for ground slab design. His original theoretical work has been modified by experiment, and all of this was brought together in the authoritative Technical Report 550, published in June 1982. In 1988 a more concise document, by Chandler and Neal, BCA Interim Technical Note 11, developed these design principles further, following detailed research into the basic parameters determining loading types and the material characteristics of both the concrete and sub-grade. This latter document contains useful general load tables, and effective guidance on assessment and enhancement of the modulus of sub-grade reaction to use in the slab design. The influence of the sub-grade on ground bearing slab design is taken as that of an elastic medium. Tensile stresses induced in the slab by point loads are, however, quite insensitive to variations in the modulus of sub-grade reaction. See Figure 24.10.

CBR test results alone should be considered as too unreliable for slab design, as they reflect only a comparatively shallow stress bulb, and hence do not indicate to what extent the sub-grade is stressed at depth. An adequately defined site investigation (SI) is therefore necessary, particularly in the case of high-bay warehouses with materials handling systems sensitive to floor tolerances which can be disturbed by even very small degrees of

| Sub-grade classification | Typical modulus of sub-grade reaction $K$ (MN/m$^3$) | Slab thickness (mm) |
|---|---|---|
| Very poor | 13 | 235 |
| Poor | 27 | 230 |
| Good | 54 | 225 |
| Enhanced | 82 | 220 |
| Note: Based on characteristic concrete strength of 40 N/mm$^2$. | | |

**Figure 24.10** Variation of slab thickness with modulus of sub-grade reaction ($K$) (loading based on BRE IP 19/87 load class: heavy).

consolidation. Further, as loadings on industrial floors have increased, some ground bearing slabs located on plastic soils, designed in accordance with the recognized documents, but based on inadequate soils information, have suffered tensile cracking longitudinally in the aisles as a result of differential consolidation between these unloaded areas and the heavily loaded sections beneath the racks. There is no authoritative guidance to this design problem at this time, but the importance of a site investigation and the assessment of soil plasticity and likely total and differential settlements cannot be over-emphasized. It may be necessary to employ a specialist soils engineer as part of the design team.

Where consolidation of plastic soils is determined to be a potential problem following an assessment of the SI and the actual loading pattern anticipated, a suspended slab is the only effective solution. However, where the ground can be improved using techniques such as vibro-replacement or dynamic compaction, and an adequate granular sub-base is provided (typically 300–450 mm, compacted in 150 mm layers), the slab may be considered to be ground bearing and designed in accordance with the principles for such a slab.

Ground floor slabs are still found with fabric specified in the top and bottom. This is fundamentally wrong, unless the slab has been designed as suspended between piles or beams. The practice of incorporating relatively light fabric to 'span any soft spots' is not consistent with any recognized method of design. Recent research at the University of Greenwich and RMCS (unpublished) has suggested that a layer of mesh reinforcement will increase the load carrying capacity of the slab. However, this does not necessarily result in a theoretically crack-free slab.

Post-tensioned slabs provide an excellent technical solution for both suspended and ground bearing slabs. In ground bearing post-tensioned slabs, random shrinkage cracks can be eliminated, and thinner slabs can result from the elimination of critical corner and edge loading conditions. Similar benefits apply to suspended ground slabs, and here post-tensioned slabs are often much simpler to construct and are frequently shown to be more economic than 'traditional' reinforced concrete design. Post-tensioned slabs have been successfully constructed worldwide for many years, and design is relatively straightforward. As no authoritative guide exists in the UK, it is recommended that reference be made to ACI 360.R-92. This also contains guidance on design of slabs with shrinkage compensating cement.

### *24.5.1.5 Suspended ground floor slabs*

Suspended slabs are usually designed as flat slabs spanning between pile heads, with or without drops, as per BS 8110, and occasionally as simply supported between ground beams. The latter solution is preferred, unless steel fibre reinforcement is being considered. This allows for stress relief joints to be detailed over the beams, and by constructing the beam/slab interface so that no connection occurs there is no inhibition to free movement of the slab. As a result, there is significantly less risk of random shrinkage cracking occurring than with flat slab design where considerable restraint is imparted into the slab by the piles or 'drops'. Structurally the beams can also double as portal ties.

Some designers consider flat slabs to be more economic. For simplicity, flat slabs are usually designed based on a u.d.l., which does not reflect the true loading pattern, and unexpectedly severe cracking sometimes occurs. To help prevent this, steel must be provided continuously in the top of the slab, and should be detailed to control crack width as suggested by Simpson in Concrete Advisory Service Data Sheet Number 5 (The Concrete Society, Slough, August 1993). When detailing such slabs, several small diameter bars at close centres are preferred to a lesser number of larger diameter bars at a wider spacing. Problems can also occur as a result of plastic settlement where inadequate cover to the top steel is specified, particularly if the settlement is of a differential nature where 'drops' are detailed. A minimum of 40 mm cover is required, and 50 mm is preferred by most floor layers. Wire guidance systems are often specified in high bay warehouse facilities, and in these instances a greater depth to the reinforcement may be required.

### 24.5.1.6 Fibre reinforced slabs

The second edition of Concrete Society Technical Report 34, published in1994, describes the main properties of fibres used in concrete floor slabs. From a design point of view the concepts of 'toughness' and 'equivalent flexural strength' are introduced for steel fibre concrete slabs. This toughness, or ductility, can improve the load bearing capacity of floors considerably when compared to plain concrete. Steel fibre manufacturers offer warranted designs, based upon test work (mostly outside of the UK). Many of the designs are based upon the work of Meyerhof, and guidance on the use of this method is given in the Appendix to TR34:1994. TR34 is currently under review, but until further authoritative guidance is available, this method should only be used for steel fibre reinforced concrete. Typical dosages of steel fibres for ground bearing slabs are in the range of 25–40 kg/m$^3$. Steel fibre concrete slabs utilizing high dosage rates of fibres (45–50 kg/m$^3$), with or without additional bar reinforcement over the pile heads, are also designed as suspended floors.

### 24.5.1.7 Properties and performance of steel fibres

The mechanical properties of steel fibre reinforced concrete are influenced by a number of factors, including:

- type of fibre
- length to diameter ratio (aspect ratio)
- quantity of fibres in the mix
- strength of the concrete matrix
- distribution of the fibres in the mix

Steel fibres particularly influence the following properties of the concrete:

- flexural strength
- shear strength

- fatigue failure strength
- impact resistance

The fibres add strength to the concrete by transferring stresses from the concrete matrix by:

1. interfacial shear (i.e. the bond), and/or
2. interlock between the fibres and the concrete where the fibre surface is deformed or where the fibres have end anchorage.

In simplistic terms, the stresses are shared by the fibres and the concrete matrix until the matrix cracks, when the total stresses are transferred to the fibres. The randomly distributed fibres act to intercept microcracks preventing them joining together to form larger cracks. As the loading increases the fibres spanning the crack provide a residue load carrying ability. Ultimately the composite section fails, usually by fibre pull-out.

The resistance to pull-out is normally referred to as the fibre efficiency. This and the weight of fibres per cubic metre are the most important factors, aside from the concrete matrix itself, governing the performance of the composite Steel Fibre Reinforced Concrete. The usual mode of failure in the composite mix is by fibre pull-out, rather than failure of the fibres themselves. Increased resistance to pull-out is thus searched for by the fibre manufacturers by adopting various deformed fibre shapes, and using enhanced end anchorage. Some failures may involve a combination of fibre pull-out and tensile fracture of the fibres, but failures by the exclusive breaking in tension of the fibres are rare. This is a benefit in as much that pull-out failure tends to be more gradual and ductile when compared with the potential 'sudden' breaking of the fibres in tension.

There is a much greater effect on the flexural strength of concrete when fibres are incorporated than there is with direct tension or compressive strength. There are usually two flexural strengths given for a steel fibre reinforced concrete:

- first crack flexural strength
- modulus of rupture (ultimate flexural strength)

Flexural strength of fibre reinforced concrete is usually based on testing beams under third point loading in accordance with the Japanese standard JSCE-SF4. The standard beam size is 150 × 150 × 450 mm. (US procedures for 4″ × 4″ × 14″ specimens are described in ACI 544.2R and ASTM C 1018).

Ultimate flexural strength generally increases in relation to fibre volume and the aspect ratio of the fibres. There is also a great variation in the post first crack performance of the fibres depending on their shape, cross-section and whether or not they have enhanced end anchorage. The deformation characteristics of the steel fibre reinforced concrete between first crack and ultimate flexural strength are what we have previously referred to as its 'toughness'.

An assessment of the 'toughness' (sometimes referred to as ductility) of steel fibre reinforced concrete is useful to enable the design of slabs made with this concrete to take advantage of this property. The equivalent flexural strength ($f_{e,3}$) of steel fibre reinforced concrete can be obtained either from ASTM C 1018, or JSCE-F4. The equivalent flexural ratio ($R_{e,3}$) is the ratio of equivalent flexural strength to the flexural strength of concrete multiplied by 100.

While it is acknowledged that the equivalent flexural ratio is mainly dependent on fibre type and dosage, some caution should be exercised in assessing laboratory test data, particularly if this research was carried out abroad. As ultimate failure is usually as a result of pull-out of the fibres from the matrix, the improvements in strength are also related to the properties of the concrete matrix in terms of bond strength and so on. This is influenced by a number of factors, including water cement ratio, cement type and so on. This may be particularly important if the slab is loaded early or, for example, if the rate of gain of strength is slow but shrinkage of the matrix is high. In these circumstances pull out may occur at much lower loads than laboratory tests would suggest.

The maximum aggregate size also affects the ultimate flexural strength obtained by the fibre reinforced concrete. For a given fibre content, the strength normally decreases with larger maximum aggregate size, and/or the proportion of coarse aggregate in the mix. To some extent this is the reverse of the case with plain concrete, where larger angular aggregates tend to produce higher flexural strength concrete for a given cement content. However, for any given aggregate size or cement content, within the range of aggregate size used in ground floor slabs, steel fibre reinforced concrete will always have a higher ultimate strength than plain concrete.

There is considerable laboratory data to suggest that steel fibres will increase the shear capacity of concrete. For slabs on ground this advantage is of minimal value. Similarly, although research at the University of Greenwich has clearly demonstrated that the punching shear loads applied to concrete slabs on ground are considerably higher with steel-fibre concrete than with either plain or mesh reinforced slabs, punching shear failure is extremely rare in ground floor slabs in service. However, this property is particularly useful when considering ground supported slabs on piles.

### 24.5.1.8 Polypropylene fibres

Polypropylene fibres do not contribute to the load carrying capacity of slabs. They have a useful contribution to make in the plastic concrete state, but from the structural design point of view slabs containing these fibres are normally designed as for plain concrete. The very recent introduction of structural synthetic fibres is an interesting development that may prove an alternative solution where rusting of steel fibres is an issue to the owner.

### 24.5.1.9 Concrete compressive strength

Although the design of ground bearing slabs, whether plain or steel fibre reinforced, is based upon flexural strength this is rarely tested routinely at site. It is therefore commonplace to specify and monitor compressive strength of the concrete, using established relationships between these two related properties. Design of industrial slabs should be limited to the range of 30–40 N/mm$^2$ for the design, or characteristic compressive strength of the concrete. In many cases 35 N/mm$^2$ provides a good compromise, as cement content (and the potential for shrinkage) is not as high as with 40 N/mm$^2$ mixes, yet this strength offers reasonably economic solutions. The choice of design strength should be based upon consideration of all the required characteristics of the floor, not simply the structural design. In any case, due to the properties of modern cements, and the very rapid construction programmes routinely required, no factoring for 90-day strength of the mix should be allowed.

### 24.5.1.10 Portal ties

When designing slabs it is still sometimes practice to tie the slab into the portal frame in order to accommodate the horizontal thrust from the legs. This is to be discouraged, and if necessary separate ties under the slab designed. Since ground bearing slabs fail under flexure it is clear that any additional tensile stress will reduce the concrete's ability to withstand the loads applied to it. By good fortune, the designers' details and site practice do not always result in fully tying the portal frame, and induced stresses tend to be significant only at the perimeter of the slab, where it is usually relatively lightly loaded. However, even in these instances, cracking often occurs as a result of the ties restraining the natural drying shrinkage of the concrete.

## 24.5.2 Design and workmanship

### *24.5.2.1 Floor joints*

It is widely acknowledged that joints provide the greatest source of problems in industrial floors, particularly warehouses. Careful planning of joint location and clean detailing are essential to minimize the risk of damage to the potentially vulnerable edges of these, particularly in respect of construction joints. It is also important to understand what the intended purpose of each particular type of joint is.

Joints are provided in ground bearing slabs for two principal reasons. First as construction joints, the number and location of which are related to the method adopted and the tolerances specified, and secondly, for the purpose of limiting tensile stresses in the concrete resulting from thermal and drying shrinkage movements and hence to eliminate random cracking, collectively known as control joints. It has been accepted for many years that expansion joints are not required in slabs on ground constructed within a weathered building envelope.

For suspended slabs not designed as simply supported, the Concrete Advisory Service advocates the pouring of these in sections as large as practical. Control joints are not applicable, shrinkage crack control is achieved by appropriate reinforcement design and detailing. Joints in simply supported suspended slabs may be detailed as if ground bearing, subject to construction or control joints perpendicular to the direction of span being located directly over the support beams, and the reinforcement detailed accordingly.

### *24.5.2.2 Construction joints*

Ideally, construction joints should be eliminated altogether wherever possible. Provided that tolerances are not too onerous, areas up to 4000 m$^2$, or more, can be poured continuously in one day using large pour construction techniques. Clearly, this enables many facilities to be installed free of construction joints, and dramatically reduces the numbers required in many others. With a minimum of construction joints it is much easier to detail these in low traffic areas or, if this is not possible, to allow for incorporating high performance details without contributing significantly to the overall cost of the floor. In warehouse floors with onerous tolerances, in order to meet these requirements more frequent joints are required. In these instances the longitudinal construction joints should be located beneath the racking system, the ideal location for these joints being mid-way between the front and back legs of the racks, with the transverse joints located at the mid-points of the racking entry bays. Consideration should always be given to specifying a large pour floor in the 'free movement' areas outside of the racked area.

Particular care with both workmanship and detailing should be exercised at 'free joints', i.e. where the dowels are de-bonded. By design, these joints will open up over time, and particularly in the case of large pour floors where the spacing may be very large, the final width can be considerable. Joints formed using very straight square edged formwork, sawn 3 mm wide after some of the initial movement has taken place, have proved successful for 'long-strip' or similar pours. For large pour floors the use of steel angle reinforcement or left in place metal forms at these free joint edges is now considered essential at the 'day-work' joints and at perimeter doorways subject to fork lift truck traffic (see Figure 24.11). Provided that care is taken to ensure adequate anchorage and that the angles do not twist or rotate during concrete placement, steel angles continue to provide an excellent detail. As a good practice, this detail should be adopted in all trafficked perimeter locations, whatever the method of construction.

**Figure 24.11** Typical armoured joint detail.

Tied construction joints, as the name implies, should not open up. In large pour systems they should be located mid-way between sawn joints installed at their usual spacing. Where a mix of tied and free joints are used, as is often the case with 'traditional' long-strip construction, clearly it makes sense to adjust the joint layout to try to ensure only tied joints occur in the heaviest trafficked areas.

### *24.5.2.3 Control joints*

Control, or 'contraction' joints, should preferably be formed by saw cutting with a diamond tipped blade within 24 hours of concrete placement. The use of plastic inserts, or similar methods of forming grooves in the wet concrete, should be discouraged in industrial floors, because the insertion of these involves disturbing the surface, and consequently creates unnecessary difficulties in obtaining fine tolerances. Further, saw cutting at the appropriate time provides a much neater joint. Although the initial cost of saw cutting is higher, particularly in areas of flint or similar hard aggregates, the long-term costs of disruption in the event of joint break-down of a poorly inserted strip are clearly many times the initial additional outlay. Under no circumstances should top and bottom crack inducers be specified. There is no guarantee that these will align vertically, and in all probability, a crack propagating from the bottom inducer will occur close to, but not coinciding with, the top one.

Saw cutting must itself be carefully timed. This must not commence too early or a ragged joint that will break down easily under traffic will occur. If it is left too late the stresses resulting from thermal shrinkage will cause cracks to occur ahead of the cutting. It is also important to commence saw cutting at locations of maximum stress, such as internal corners.

The spacing of control joints should be determined such that in an unreinforced (or nominally reinforced) slab, uncontrolled cracks do not occur. This requires spacing such that the tensile stresses resulting from restraint of the natural shrinkage movement of the slab do not exceed the tensile capacity of the concrete. In respect of long-term drying shrinkage, due account should also be taken of the loads imposed upon the slabs (both from the point of view of induced flexural stresses, and the increase in frictional forces), and other external restraints, such as portal ties.

### *24.5.2.4 Isolation joints*

The slab should be isolated from the main structure and any other intrusions through or partially into it. Failure to adequately isolate the floor will cause a considerable increase in local stresses in the concrete, resulting in shrinkage cracks occurring. A typical isolation joint detail for an internal column is shown in Figure 24.12.

Diamond shaped column surrounds result in unnecessarily large 'infills', and it is often difficult to align form-work and saw cuts to the points of these. Unless it is necessary

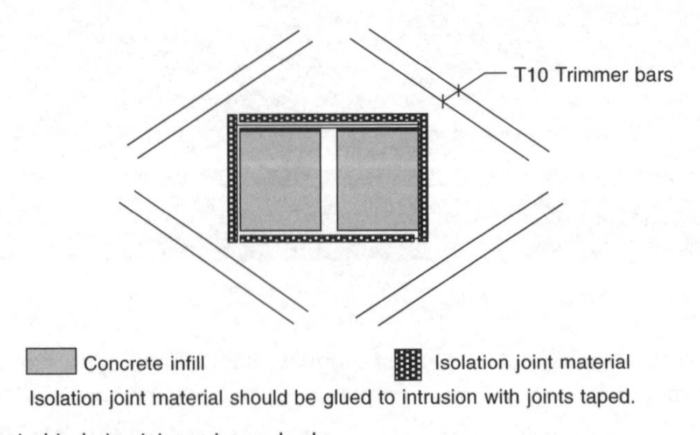


**Figure 24.12** Typical isolation joint at internal column.

to encase steelwork independently of the slab for other reasons, it is preferable to use the detail shown in Figure 24.12. This detail is particularly suitable where onerous tolerances or coloured finishes have been specified. Corners of square intrusions should always be trimmed with a minimum of two T12 bars 900 mm long. Circular column surrounds eliminate the requirement for these bars, but as they take up a larger area of the floor slab they may become subject to traffic with the resultant implications for maintenance of the joints. However, the use of circular surrounds eliminates the need for the trimmer bars.

The isolation joint material must be sufficiently compressible, and is usually manufactured from expanded polyethylene. 'Compressible boards' are not suitable for effective isolation.

### 24.5.2.5 Joint sealing

The sealing of joints subject to warehouse traffic needs to be done with a material that will provide support to the edges of the joints. Unfortunately, most of the materials that have this property also have a very low strain capacity. Therefore sealing should be left as late as possible to enable much of the drying shrinkage to have taken place. However, modern fast-track programmes often result in there being only a few weeks between laying of the slab and hand-over to client, insufficient for any significant drying shrinkage to have taken place in a well cured slab. If the joints are sealed at this time it is likely that re-sealing will be required at a later date. If they are left unsealed all but the narrowest of joints may suffer damage, thus requiring some repairs prior to sealing after a period of 12 months or so. The use of flexible sealants in narrow joints subject to fork-lift traffic is not cost-effective. These provide little edge support and are often 'plucked out' by the rubber tyres of the trucks. Clients need to be educated into the need for the two stage 'hard' sealing process, clearly only the aisles and free-movement areas of warehouses needing to be re-sealed, at, for example, the end of the defects liability period. A recent development is the use of a pre-compressed joint strip to support the edges of the joint until the second stage sealing with the semi-rigid material can be undertaken. This strip is installed immediately after saw cutting (see Figure 24.13).

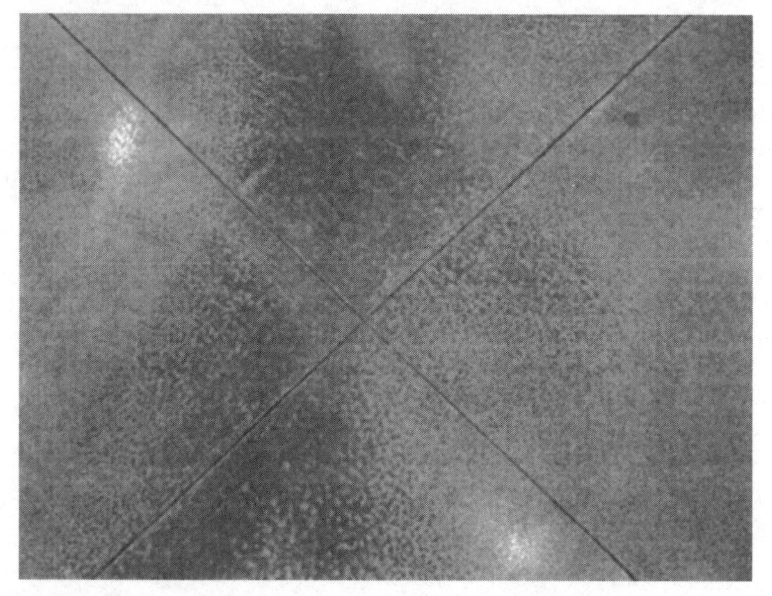

**Figure 24.13** Joint strip inserted into joint immediately after curing and saw cutting.

Due account must be taken of the proposed traffic at the time of slab design. For example if small steel wheels are to be used in the facility even narrow well formed and sealed joints will suffer degradation over time. Either the client should be persuaded to change the tyre material, or a jointless floor slab will need to be specified. Various alternative solutions for 'jointless floors' are now available including the use of shrinkage compensating cement, post-tensioning and steel fibres (see Design section of this chapter).

## 24.6 Floor toppings

BS 8204-2:1999 now refers to these as wearing screeds. Compressive strength is often used to predict properties such as wear resistance. This is not always appropriate, since other factors such as finishing technique, curing regime and bleed characteristics can have an equal or greater influence on the abrasion resistance achieved. High compressive (or flexural) strengths are also uneconomic, due to the amount of cement required, where they are used to achieve high standards of abrasion resistance.

'Dry-shake' toppings provide a cost-effective solution to achieving high abrasion resistance. They permit the use of a relatively low strength base concrete (usually in the range of 30–40 N/mm$^2$) with economic cement contents, to provide the required structural performance, and result in a very dense, high strength monolithic wearing course when applied correctly. Dry-shake toppings consist of a blend of cement and selected natural or synthetic aggregates (typically in the ratio of 1:2), and may also contain pigments, plasticizers, micro-silica, polypropylene fibres and other trace ingredients. Dry-shake finishes should be applied by specialist contractors.

A further enhancement in long-term performance is achieved by the use of 'wet-on-wet' monolithic toppings, using essentially the same components as found in dry-shakes, but these are pre-mixed with a prescribed quantity of water, pumped into place and bonded to the base concrete during the finishing operation. These may be 8–10 mm thick, depending on the system used.

A few specialists bond a thin topping to the base concrete some days or weeks later, using specially developed bonding agents and low shrinkage mixes to produce very flat, extremely hard-wearing floors. These toppings often incorporate steel fibres.

Suitable finishing specifications and good finishing techniques result in high abrasion resistance of the floor surface and ensure a smooth, highly reflective finish. Reflectivity, a function of surface density and the quality of the final trowelling operation, not only helps to provide an environment that is pleasing to the facility's employees, but also offers real savings in terms of reduced lighting costs and enhanced productivity. Similarly, a smooth surface, free of defects, is easier to clean (and hence reduces costs), and enhances the corporate image of the facility operator.

## 24.7 Floor specification

### 24.7.1 Tolerances

If joints cause the greatest practical problems in ground floor slabs, then tolerances undoubtedly cause the greatest number of arguments and disputes. These either result from an ill-defined specification, poor workmanship, unsuitable materials and laying conditions, differing interpretation of survey results, inaccurate or non-representative surveys, or a combination of all of these.

Where there are no racks, or no indication of the materials handling system to be used is given, tolerances should be specified that allow some form of large pour system to be adopted, with a view to eliminating as many construction joints as possible. Floors to speculative warehouse developments should not therefore be specified to flatness tolerances that demand narrow strip construction. The alternative is a floor with a large number of potentially troublesome joints, which in all probability will remain exposed for all of the slab's working life. Also, at the time of construction of the slabs in such units, traffic paths (by definition) are not known. In order to achieve and measure the most onerous tolerances, the wheel track path of the fork-lift trucks etc. must be known prior to the commencement of floor laying.

An attempt has been made in Technical Report 34 and its subsequent Supplement (1997) to address the most fundamental of the issues, namely that of producing a specification that is acceptable to the materials handling equipment suppliers, is clear and unambiguous, achievable in practice, and can be measured relatively straightforwardly. However, disputes continue. For defined movement floors, Table 7.1 of Technical Report 34 should be used as it is written (see Figure 24.14). If this is done, warehouse floor slabs specified and constructed – and checked for compliance – in accordance with the tolerances shown will be fit for the purpose described.

| Category | Allowable limits | | | | | | | |
|---|---|---|---|---|---|---|---|---|
| | Property I | | Property II | | Property III | | | |
| | | | | | Wheel track up to 1.5 m | | Wheel track over 1.5 m | |
| | A | B | A | B | A | B | A | B |
| Superflat | 0.75 | 1 | 1 | 1.5 | 1.5 | 2.5 | 2 | 3 |
| Category 1 | 1.5 | 2.5 | 2.5 | 3.5 | 2.5 | 3.5 | 3 | 4.5 |
| Category 2 | 2.5 | 4 | 3.25 | 5 | 3.5 | 5 | 4 | 6 |

All allowable limits shown in millimetres
Column A = '95% property limit', Column B = '100% property limit'
The pour will be considered satisfactory when:
(a) not more than 5% of the measurements exceed the particular property limit in column A
(b) none of the measurements exceed the particular property limit in column B

| Diagrams showing measurements taken for Table 7.1 Properties | | |
|---|---|---|
| Property I | Property II | Property III |
| 300 mm | 300 mm 300 mm | Wheel track |
| Difference in elevation over 300 mm | Difference in slope over 600 mm | Difference in elevation across wheel track |

**Figure 24.14** Extract from concrete society technical report (2nd edition, 1994). Table 7.1 'Allowable values of the properties of flatness for defined movement areas'.

The free movement area tolerances of TR34, second edition, caused some controversy among the floor laying contractors. As a result a supplement to TR34 was published in 1997 providing new limits for these (see Figure 24.15). Both the Defined and Free Movement Specifications are again under review at the current time, with revised guidance likely to be published during 2003.

In order to minimize disputes, materials handling equipment supplier(s) and concrete flooring specialists should be consulted before the tolerance specification is decided upon. In addition prior to work starting on site complete clarity needs to be established on how the floor is to be measured, and what is to be done if areas are found to be out of tolerance (this procedure is required by BS 8204-2 1999).

Warehouse floor slabs should always be checked for compliance as work proceeds. Electronic floor measurement devices are now available from a number of specialist floor survey companies to simplify this essential procedure.

| Floor classification | Maximum permissible limits | | | | |
|---|---|---|---|---|---|
| | Property II | | Property IV | | |
| | A | B | C | D | E |
| | 97% | 100% | 90% | 97% | 100% |
| FM1 | 2.5 mm | 4 | 3 | 4.5 | 7 |
| FM2 | 3.5 | 5.5 | 6 | 8 | 12 |
| FM3 | 5 | 7.5 | 8 | 10 | 15 |

*Note 1: Property II*
The floor area being measured shall be considered satisfactory when, for the floor classification specified:
(a) Not more than 3% of the total number of measurements for property II exceed the limit in column A (97% limit).
(b) None of the measurements exceed the limit in column B (100% limit).
*Note 2: Property IV*
The floor area being measured shall be considered satisfactory when, for the floor classification specified:
(a) Not more than 10% of the total number of measurements for property IV exceed the limit in column C (90% limit).
(b) Not more than 3% of the total number of measurements exceed the limit in column *D* (97% limit).
None of the measurements exceed the limit in column E (100% limit).

**Figure 24.15** Extract from supplement to Technical Report 34 (1997), Revised Table 7.2R: allowable values for the properties of flatness and levelness for free movement areas of floors.

## 24.7.2 Wearing surface performance

In addition to cracking, poor joint performance and inappropriate tolerances, some floor slabs fail to meet the expectations of the client in terms of abrasion resistance, slip resistance, or appearance.

### *24.7.2.1 Abrasion resistance*

High abrasion resistance is important to eliminate wear of the slab surface. This wear leads to dust in the atmosphere (and the consequential contamination of products), and in severe cases to disruption of the materials handling system.

Abrasion resistance of concrete floor slabs is influenced by many factors. Extensive research work has been undertaken at Aston University by Kettle and Sadegzadeh, and by Chaplin and others, into the various influencing factors. A class system for abrasion resistance was first defined in BS 8204; Part 2 1987. However, BS 8204-2 1987 gave confusing guidance to the specifier, particularly as the concrete mixes specified for the various classes were often uneconomic, and unsuitable for use with modern construction methods. The adoption of 'wear-in' values in TR34:1994, subsequently incorporated in BS 8204-2 1999 is therefore of significant importance (see Figure 24.16). This enables *in-situ* abrasion resistance to be measured using the portable accelerated abrasion testing apparatus. Developed jointly by Aston University and the C&CA, this portable apparatus measures the wearing of the surface under rolling steel wheels. The writer recommends that the measurement should be taken on a section of the floor cured using polythene or water, in order for the effect of any resin curing membrane not to distort the results.

| BS 8204 Class | Duty | Description | Maximum depth of wear (mm) |
|---|---|---|---|
| Special | Severe abrasion and impact | Very heavy duty, engineering workshops etc. | 0.05 |
| AR1 | Very high abrasion steel wheel traffic and impact | Heavy duty industrial workshops, special commercial, etc. | 0.1 |
| AR2 | High abrasion steel or hard plastic wheel traffic | Medium duty industrial and commercial | 0.2 |
| AR3 | Moderate abrasion rubber tyre traffic | Light duty industrial and commercial | 0.4 |

**Figure 24.16** Wear-in values applicable for BS 8204 (Part 2) classes of floor slabs – simplified.

### *24.7.2.2 Slip resistance*

This is measured using the TRL-developed apparatus. This involves swinging a pendulum with a rubber 'sole' fitted at the end across the floor surface, and measuring the resistance of the floor to its travel. From this a Slip Resistance Value (SRV), approximating to the co-efficient of friction is calculated. BS 8204-2: 1999 includes suggested values and a description of the apparatus.

### *24.7.2.3 Appearance*

This is clearly a subjective area, and depends greatly on the use of the facility. However, one area of controversy at the present time is the number of steel fibres that may be expected to be found in the surface of a steel-fibre reinforced slab. The acceptable incidence of fibres, and the method of sampling should be agreed prior to work starting on site. Guidance is also available in a Belgian Standard NIT204.

### *24.7.2.4 Other measurements*

Sometimes electrical conductivity is checked to BS 2050:1978 (the standard for anti-static resin flooring). Bonded screeds may be checked for soundness using the BRE screed tester.

## 24.8 Defect identification and remedial measures

Industrial floors are routinely subjected to unrelenting punishment throughout their working lives. Thus maintaining and repairing these floors is an area of considerable importance. Unfortunately there is little authoritative guidance to professional advising end-user clients.

Floor repairs can be not only expensive to execute, but cause disruption to the efficient running of a facility while they are being carried out. Thus the consequences of inadequate performance of such repairs are, at best, further disruption resulting in another increase in operational costs. It is therefore crucial that the repairs are correctly specified and carried out diligently.

Whatever the defect experienced or the maintenance items to be attended to, it is important to first establish the cause of the problem. This, in turn, will considerably influence the choice of remedy. In addition, it is useful to consider methods of prevention of the occurrence of the same situation in the future. Thus all flooring problems should be addressed in the same way:

- identify the cause(s)
- specify the remedies
- consider means of prevention of future reoccurrences

Defects in new construction all too readily occur as a result of some failing in the specification or construction (or both) phases. Figure 24.17 summarizes some of the most common defects, contributing factors and causes.

| Defect | Principal causes and key issues |
|---|---|
| Delamination | Closing surface too early. Lack of bleed-water (dry-shakes): Set of Concrete, Weathering, Workmanship |
| Dusting | High water: cement ratio at surface. Poor curing: Specification, Bleed of Concrete, Workmanship |
| Plastic shrinkage cracks | Rate of evaporation exceeds rate of bleed: Bleed of Concrete, Weathering |
| Finishing defects | Poor concrete finishability. Contamination. Poor timing: Concrete Materials Selection, Workmanship, Supervision |
| Poor tolerances | Incorrect procedures. Mistiming of finishing: Workmanship, Concrete, Weathering |
| Crazing | Poor curing, excess fines at surface: Specification, Weathering, Supervision |
| Cracks in hardened concrete | Drying & thermal shrinkage, inadequate design: Concrete shrinkage, Specification, design |
| Curling, joint damage | High shrinkage, poor curing, poor detailing: Specification, Supervision, Workmanship |

**Figure 24.17** Defects in new floor construction – causes and critical issues.

## 24.8.1 Common flooring defects

Problems relating to out of tolerance floor surfaces are not considered here, the reader is referred to Concrete Society TR34:1994 and the 1997 Supplement, for further information. Such problems are usually apparent prior to occupation by the end-user or soon after, whereas the other problems listed above may not be apparent on day one, but often manifest themselves over time with use of the floor.

Other 'defects' such as crazing, discoloration and other finishing defects are primarily of an aesthetic nature. These defects do not generally affect long-term wearing surface wearing performance. However, where a coloured concrete floor is specified, particularly in a retail environment, the occurrence of such defects is a critical issue. Unfortunately, only resurfacing, replacement or over-coating of the aesthetically unacceptable slab usually provides an acceptable remedy in these circumstances.

## 24.8.2 Joint spalling

The direct cause of the joint spalling is dynamic loading at the joints. The indirect cause is poorly detailed or constructed joints that are unable to accommodate these loads.

Spalling may result from any one or more of a number of reasons. The principal reasons and remedies are:

**Excessive joint opening** This may occur at either designed free movement joints or at sawn induced joints. It is usual for these joints to be initially sealed with a relatively soft elastomeric joint filler soon after construction. Such materials give little support to the edges, and hence the edges are soon degraded by the pounding from small wheeled pallet-movers and the like. Fast track programmes rarely allow adequate time for the concrete to have exhibited the majority of its shrinkage before hand-over. Thus any joints in the floor will be at risk.

Where damage has occurred it is necessary to repair the joint, prior to resealing. This involves cutting out the damaged section of joint, to a square edge, typically to a depth of 15 mm, applying a primer and making good with an epoxy mortar. It is usually easier to repair the full width of the joint for saw cuts or narrower construction joints, and to repair each face individually with wider joints. All repairs should be finished smooth and flush with the existing floor profile. In most cases full width repair is followed by reintroduction of the saw cut or joint. However, if the slab is over two years old, and not subjected to excessive temperature variation, alternate sawn joints may be left uncut. Upon completion of the repair, a semi-rigid (usually epoxy or polyurethane-based) joint sealant is used to provide support in the future.

**Curling** This may occur with or without excessive joint opening. If this occurs it is first necessary to stabilize the joint by pressure grouting or similar, prior to undertaking the edge repair. In addition it will almost certainly be necessary to grind a small width of the floor either side of the joint to remove the vulnerable 'peak' at the joint caused by the curling.

**Incorrectly positioned joint details** This includes misplaced dowels, and (less commonly nowadays) inaccurately placed crack inducing strips. The result of such errors is often a crack a few centimetres away from the location of the formed joint. This results in a strip of loose concrete that very quickly breaks down under traffic. In some cases it is possible to resin inject the crack and re-cut the joint to an adequate depth. If the slab is some years old, it may be practical to inject the crack and repair the damage as described above, without re-introducing the joint. Where the slab is still liable to volumetric change, in extreme cases it may be necessary to cut out the damaged area and replace this full depth, thus creating two dowelled joints, one either side of the original location.

**Left in place forms** Some left-in-place forms, particularly those used to construct shear keys to transfer vertical loading, can cause problems over time as the concrete shrinks away and a section of the keyway becomes unsupported. A crack (or number of cracks) occurs parallel to the joint on the female side of the joint as a result of dynamic loading on what is effectively a cantilever. Support needs to be introduced into the joint – by grouting in extreme cases, a check made for curling, and the damaged section repaired flush with the surface of the left-in-place form, prior to sealing the joint as described above. If left unattended to, the metal form itself may become damaged. This will result in the necessity either to cut out the metal strip at the surface and make good as described above, reintroducing the joint with a saw cut if possible, or to cut out a section of the joint full depth. In other instances the temporary bolts holding the halves of the left in place form are not removed, thus preventing opening of the joint and causing cracks and/or spalling behind the metal edge of the form.

### 24.8.3 Cracks

Cracks may be caused by:

- plastic shrinkage
- thermal movements
- drying shrinkage
- loading above slab capacity
- a combination of shrinkage and loading conditions

In general, all shrinkage cracks may be treated in the same two ways, depending whether or not they are still 'live' cracks, i.e. whether the slab is still moving as a result of ongoing volumetric change. If such cracks are less than 0.5 mm wide they are usually more of an aesthetic issue rather than of concern from the point of view of slab performance. However, wider cracks can quickly break down under the pounding from the hard narrow wheels of pallet trucks and the like.

If the shrinkage crack is still 'live' it is not prudent to repair this with a normal resin injection system. It may be useful to seal the surface with a medium rigidity joint sealant, in order to prevent further damage, until a permanent repair can be undertaken. If the crack is reasonably straight it is also possible to insert a preformed joint insert of the appropriate width to undertake the same function – this being much easier to remove at the time of permanent repair. In addition low viscosity acrylic copolymer emulsions may be successful in repairing narrow cracks that are expected to continue to move. These materials have an excellent bond with dry concrete, while providing good flexibility.

If the shrinkage cracks are no longer 'live', there are two alternative methods of repair.

The cracks may be injected with a resin of suitable viscosity, depending on crack width. Specialists should normally be employed for this work, although there are 'off-the-shelf' kits available for small volumes of repairs. Where damage has occurred at the surface, the edges of the joint should be cut out square and made good with an epoxy resin mortar. Alternatively, again particularly if the joint is wide, but straight, the ravelled top edge of the crack may be routed (usually 3–5 mm wide) to a depth of 10–15 mm. This 'rectangular' section then acts as a reservoir for a low viscosity resin, which is allowed to penetrate into the crack, the reservoir being constantly 'topped-up'. When the crack has been fully 'sealed' the top section is made good with a suitable resin or mortar that will provide abrasion resistance and support to the edges. This is then sanded smooth to complete the job.

Structural load-related cracks may also be repaired using injection systems, but first it is necessary to determine the precise cause of the crack and to address the situation that has led to the excess stress being induced in the floor. Without careful assessment of the nature of the problem, solely repairing the crack will result in another crack forming very close by as the stress is redistributed elsewhere.

## 24.8.4 Delamination

Delamination primarily results from closing the surface of the concrete too early, i.e. before bleeding has finished or where there is a risk of air being trapped below the densified surface. In the case of dry-shake toppings, it may also result from a lack of bleed water being available to adequately hydrate the topping, thus resulting in a poor bond at the interface between the base concrete and the topping. There are four principal methods of repair:

- resin injection into the void caused by the delamination
- removal of delaminated section and reinstatement with a thin bonded repair
- removal of delaminated area and part of base concrete and reinstatement with a HD screed
- slab removal and replacement, full depth (severe cases only).

## 24.8.5 Pop outs

The most common cause of pop-outs is contamination of the fine or coarse aggregate. A typical source of such contamination is lignite in former coal-mining areas.

The solution involves establishing the extent of the problem, confirming that the slab's abrasion resistance is unaffected, and simply drilling out the contamination and filling the resultant holes with resin, colour matched to the floor finish. In extreme cases, it may be necessary to apply a high-build resin coating for aesthetic reasons, or where abrasion or impact resistance is impaired to consider removal of the affected area of the slab.

## 24.8.6 Dusting

Dusting results from the wear of the surface under traffic, i.e. poor abrasion resistance. Such floors may be renovated by:

- grinding and sealing with a penetrating sealer
- preparing and applying a high-build resin coating
- preparing and applying a resin 'self-levelliing' product
- preparing and applying a thin, bonded cementitious overlay
- grinding followed by a spray applied thin surfacing
- applying a proprietary sealing system

The selection of the solution will depend on the use of the facility and the soundness of the base concrete.

# Part 10

# Reinforced and prestressed concrete

# 25

# Reinforced and prestressed concrete

*Ban Seng Choo*

## 25.1 Objectives

The main objective in this chapter is to explain the basic principles of limit state design relating to reinforced and prestressed concrete structural elements to concrete technologists who need to be aware of the use of concrete as a practical structural material. It is not the intention to set out the relevant design theory or to explain in detail the behaviour of structural elements or to illustrate practical design to code requirements through worked examples. There are many textbooks that cover the detailed theory and/or provide useful worked examples to which the reader is directed for practical design purposes (Mosley *et al.*, 1996, 1999; MacGinley and Choo, 1990; Kong and Evans, 1987; Beckett and Alexandrou, 1997).

The topics which will be covered are:

- Serviceability and ultimate limit states
- Use of partial safety factors
- Functions of the principal members of a structure
- Nature of the different types of loading which can occur
- Load transfer through a structure to the foundations
- Behaviour of a reinforced concrete beam under load
- Assumptions made in the design of a reinforced concrete beam
- Principles of prestressing concrete
- Methods of prestressing and application of pre- and post-prestressing

- Reinforcements for reinforced and prestressed concrete
- Loss of prestressing force
- Load testing of simple structures

The principles described herein apply to the design of buildings and other structures constructed in reinforced and prestressed concrete.

## 25.2 Principles of limit state design

### 25.2.1 Introduction

The aim of design is to achieve an acceptable probability that the structure will perform satisfactorily during its life. It must be able to safely resist the structural actions (loads) acting on it, not deform excessively and have adequate durability and resistance to the effects of chemical and physical attacks. Early approaches to structural design included the permissible stress method and the load factor method.

### 25.2.2 Permissible stress method

In the permissible stress method, the ultimate strengths of the material were divided by a safety factor to provide permissible design stresses which were considered to be within the elastic range. An elastic method of analysis is then used to determine the stresses which result from the working loads acting on the structure. If the stresses are less than the permissible values, the structure is considered to be safe. This is a simple and useful approach but, as it is based on elastic stress behaviour, it is not really applicable to a semi-plastic material such as concrete.

### 25.2.3 Load factor method

In the load factor method, the ultimate strengths of the materials and working loads, multiplied by a load factor, are used in the calculations. The method does not take into account the variability of the material strengths and loads actions.

### 25.2.4 Limit state method

The limit state method of structural design was developed to overcome the limitations of the above methods. It uses the concept of probability and is based on the application of statistical methods to the variations which occur in practice on the loading and other structural actions and as well as in the strengths of the materials. The method recognizes that it is not possible to make a structure completely safe. It is only possible to reduce the probability of failure to an acceptably low level.

In Europe, the Eurocodes (BSI, 2002) are being adopted for use. Initially, they would serve as an alternative to the different rules and standards in force in the various member states of the European Union but would eventually replace them. They comprise a group of standards for the structural and geotechnical design of buildings and civil engineering works. Eurocode 1 (EC1) defines the basis of design and actions on structures (BSI, 1992). Eurocode 2 (EC2) Design of Concrete Structures relates to the design of buildings and other civil engineering structures to be constructed in plain, reinforced and prestressed concrete. EC2 is written in several parts and its clauses are set out as principles and application rules. Part 1: 'General rules and rules for buildings' has been written such that its principles will generally apply to all parts when they have been developed whilst the specific rules have been worked out for building structures only. The further parts to EC2 will complement and adapt Part 1 as appropriate and will provide rules for the design of particular types of structures. By complying with EC2 the structural engineer will have satisfied the requirements of the Construction Products Directive in respect of mechanical resistance.

Principles in Eurocodes are distinguished by the prefix 'P' and application rules are indented. Principles are general statements and definitions for which there is no alternative. Application rules are generally accepted methods, which follow the principles and satisfy their requirements. Alternative methods which follow the principles are permitted. Thus the procedures in current national codes are by and large likely to be acceptable as the principles are in general similar. Some numerical values in the Eurocodes are given within a 'box' and are referred to as boxed values. Boxed values are given only as indications and each member state is required to fix the boxed values applicable within its jurisdiction. The boxed values will be found in National Application Documents (NAD) to be produced by each member state.

In the Eurocodes, limit states are defined as states beyond which the structure no longer satisfies the design performance requirements. In common with most limit state design codes, the Eurocodes distinguish between ultimate limit states and serviceability limit states.

Ultimate limit states are those associated with collapses or with other similar forms of structural failure. States prior to structural collapse are also treated as ultimate limit states. Situations which should also be considered at ultimate limit states are those where the structure or a part of it may fail due to:

- Fatigue or other time-dependent effects
- Loss of equilibrium
- Excessive deformation, loss of stability or transformation into a mechanism

The main concern of the structural engineer when designing for ultimate limit states is the safety of the structure, its contents and the safety of the people in it.

Serviceability limit states relate to conditions beyond which specified service requirements for a structure or a structural element are no longer met. The serviceability requirements concern the

- functioning of the construction works or parts of them
- comfort of the people using the structure
- appearance of the structure

The Eurocodes require the designer to distinguish, where relevant, between reversible and irreversible serviceability limit states. Serviceability limit states which should be considered include:

- Deformations or displacements which affect the appearance or the effective use of the structure or cause damage to the finishes and non-structural elements
- Vibrations which cause damage to the structure or which limit its functional effectiveness
- Damage, including cracking, likely to adversely affect the appearance, durability or the function of the structure
- Observable damage caused by fatigue or other time-dependent effects.

There are a number of other publications which support the designer of concrete structures to EC2, to which the reader is directed (British Cement Association, 1993; Ove Arup *et al.*, 1994; Beeby and Narayanan, 1995; Betonvereniging *et al.*, 1997).

## 25.3 Structural elements

Generally, the design process starts with the client specifying the requirements for his building or structure. This is usually done by or through an architect resulting in a layout and arrangement of the building or structure. The structural engineer then determines the structural system and form to meet the client's requirements. Other forms of contract such as those by consortiums to design and build are also used. Regardless of the contractual approach adopted, material properties and construction methods have to be considered in deciding the most economical solution. After the structural form and arrangement have been finalized, the design process is as follows:

- Idealization into load-bearing frames and elements for structural analysis element design.
- Estimation of the loads and structural actions that the structure will be subjected.
- Conduct structural analysis to determine the moments, shears and forces for element design.
- Conduct structural element design, i.e. to determine the element shape size and reinforcement arrangement.
- Generate detail drawings for construction purposes.

The purpose of the process is to verify that the limit states are not exceeded when design values are used in the structural and load models which have been set up to represent the various design situations and load cases for the relevant ultimate and serviceability limit states. In order to be able to develop a structural model, it is necessary for the structural engineer to break down a building or structure into the main structural members, namely:

- Slabs – horizontal plate elements carrying lateral loads
- Beams – members (usually horizontal) carrying lateral loads. These are typically of rectangular, tee or ell cross-section
- Columns – members (usually vertical) carrying primarily axial loads but are generally subjected to axial loads and moments
- Walls – vertical plate elements carrying vertical, lateral or in-plane loads
- Bases and foundations – pads or strips supported directly on the ground which spread the loads from the columns or walls so that they can be supported by the ground without excessive settlement. Alternatively the bases may be supported on piles.

The elements listed above are shown in Figure 25.1 which shows a typical cast *in-situ* concrete building structure, in which the elements are rigidly connected to form a monolithic frame. Alternatively the elements may be precast and are connected on site in a rigid manner or as simply supported or pinned connections.

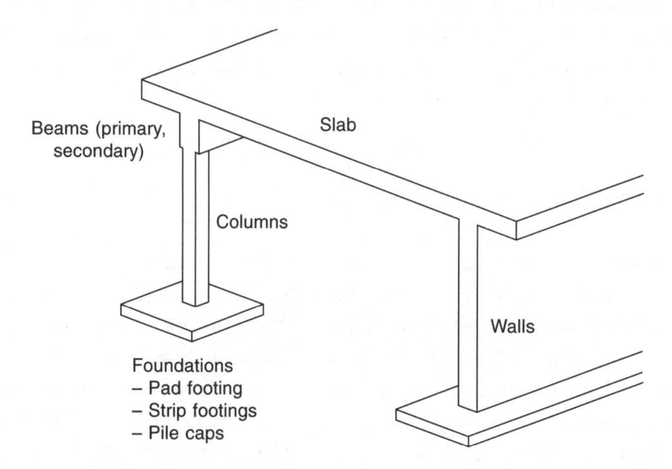


**Figure 25.1** Structural elements.

## 25.4 Design values

### 25.4.1 Actions

Vertical gravity loads acting on a structure are generally carried by floor slabs. The loads on the slabs are then transferred to columns or walls through beams which support the slabs. The columns and walls then transfer the loads to the foundation bases or piles, as can be visualized from Figure 26.1. It is important that the designer is able to visualize the paths through which the loads acting on the structure are transferred to the foundations.

The Eurocodes refer to 'actions' on structures and define these as either

- a direct action (i.e. a force or load applied to the structure) or
- an indirect action (i.e. an imposed or constrained deformation or imposed acceleration caused by temperature changes, uneven settlement or earthquakes etc.

Structural actions may be classified by their variation in time, namely as permanent, variable or accidental. Permanent loads are those with constant magnitudes throughout the structure's life and these include the self-weight of the structure and architectural components such as claddings, partitions, equipment and machinery which are permanent fixtures. Permanent actions can usually be determined quite accurately. However, for preliminary design, it is necessary to estimate the probable sizes and self-weights of the structural concrete elements to be designed. Variable actions such as wind loading and loads imposed due to human occupancy are transient and are not constant in magnitude. Accidental actions are those of short duration and are unlikely to occur with a significant magnitude during the design working life of the structure. However, it should be noted that an accidental action could be expected to cause severe consequences unless special measures are taken.

Actions may also be classified by their spatial variations. Actions such as the self-weight of structural elements are fixed but movable imposed loads, wind and snow loads are free actions. Further, actions may be classified by their nature and/or the structural response. Static actions are those which do not cause significant acceleration of the structure or member. Dynamic actions can cause significant acceleration of the structure or member. In many cases the effects of dynamic actions may be calculated from quasi-static actions by increasing the magnitude of the static actions or by the use of an equivalent static action.

## 25.4.2 Materials

As indicated above, limit state design is based on the application of statistical methods to the variations that occur in practice in the strengths of the materials used in the construction and also in the loads acting on the structure. Concrete strength serves as a good example of the variations in values that occur in practice. Tests on cubes or cylinders of 'identical' concrete gives values which may have a coefficient of variation of up to 10 per cent. Similarly, the variations occur with the yield strengths of reinforcing steel, albeit with lower coefficient of variation. Hence it is not practicable to specify that the concrete should have a certain precise concrete strength or indeed that the reinforcement should have a particular yield or proof stress for the purposes of the design of reinforced and prestressed concrete. The material strengths on which design is based are the values of strength below which results are unlikely to fall. These values are called characteristic strengths. In statistical terms, it is taken as the value below which it is unlikely that more than 5 per cent of the strength results will fall, namely

$$f_k = f_m - 1.64s$$

where $f_k$ = characteristic strength
$f_m$ = mean strength
$s$ = standard deviation.

Thus for a specified characteristic strength, the higher the value of the standard deviation of the strength values, a correspondingly higher value of the mean strength of the concrete will be required to achieve the specified characteristic strength.

## 25.4.3 Partial safety factors

In general, limit state design makes use of the concept of design strengths and design loads, obtained by applying partial safety factors and other factors such as the combination factors to characteristic or representative values of strengths or loads. In exceptional cases, it may be appropriate to determine design values directly. However, the values should be chosen to correspond to at least the same degree of reliability implied in the Eurocodes. In these calculations, it is assumed that the load and strength variables are normally distributed – see Volume 4, Chapter 10 on Statistics.

Partial safety factors are used to take account of:

- possible unfavourable deviations of the characteristic values

- possible inaccurate modelling of the characteristic values
- uncertainties in the assessment of the effects of actions, geometric properties and resistance model

Combination factors are used when actions are combined to take account of the reduced probability of simultaneous occurrence of the most unfavourable values of several independent actions. Combination values may be used for the verification of ultimate limit states and irreversible limit states.

In particular, design strength values are obtained by dividing the characteristic strength value by a partial safety factor $\gamma_m$ appropriate to that material and the particular limit state, that is:

$$\text{Design strength} = \text{characteristic strength}/\gamma_m$$

Ideally, the characteristic values of an action are the values with an accepted probability of not being exceeded during the life of the structure and are determined from the mean and standard deviation as for the characteristic strength described above. However, due to the lack of statistical data, it is not yet possible to express direct load actions in this manner. In practice, the so-called characteristic loads are the values which are designated as such. These are generally the actual service loads that the structure is designed to carry and can be thought of as the maximum loads which are not to be exceeded during the life of the structure. The Eurocodes considers the characteristic value of an action to be its main representative value. In statistical terms, the characteristic loads have a 95 per cent probability of not being exceeded. The Eurocodes also refer to mean, upper or lower characteristic action values or nominal action values (for cases where a statistical distribution is not known).

The variability of permanent actions is in general small, in which case a single characteristic value is sufficient for structural calculation purposes. However, if the variability of permanent actions is not small, upper and lower characteristic values will have to be used appropriately.

The situation is more complex for variable actions. In general, it is the maximum variable value which is critical and so the upper value (which produces unfavourable effects) is appropriate and can be obtained from the expression

$$G_k = G_m + 1.64s$$

where $G_k$ = characteristic load
$G_m$ = mean load
$s$ = standard deviation.

However, in certain situations, for example, where stability is being considered, it may be more appropriate to use the lower minimum characteristic load value given by

$$G_k = G_m - 1.64s$$

The structural design engineer must also take account of load actions which are caused by accidents, water, currents, wavers and tides, where relevant.

Having obtained the characteristic loads, the design loads are obtained by multiplying the characteristic loads by partial safety factors for the appropriate loading conditions. Partial safety factors range in magnitude from 1.0 to 1.4 depending on:

- the load combination, that is,
  - permanent and variable
  - permanent and wind
  - permanent and variable and wind
- ultimate or serviceability limit state
- loading is adverse or beneficial for the loading case.

The full implementation of the verification rules required by the Eurocodes can be fairly involved, however, it should be noted that the use of a simplified verification approach is permitted, where appropriate. To determine the behaviour of the concrete structure, EC2 permits the use of elastic analysis without redistribution or with limited redistribution as well as plastic and non-linear methods of analysis. By using the elastic method of analysis, the designer assumes that the structure behaves in an ideal linear elastic manner, that is, that all structural deformations are proportional to the loads acting on the structure or structural elements. At relatively low load levels, it may be assumed that concrete structures behave in a linear elastic manner. However, at realistic load levels, concrete as a material does not behave in the ideal elastic manner assumed. Concrete as a material exhibits non-linear characteristics, that is, the load deformation profile of concrete is not linear. And so the non-linear plastic method of analysis should be used for ultimate limit state (that is, strength) design while the other methods are suitable for both serviceability and relevant ultimate limit state calculations.

## 25.5 Load testing of simple structures

Although design entirely based on testing is not the approach used in structural engineering, it is recognized as a reasonable approach in situations where some of the basic values such as the strength of the material are unknown or where the engineer is unsure of the structural behaviour and/or whether the usual approaches to modelling are appropriate. Where load testing is employed to determine the design strength, it should be conducted in a manner that allows for the uncertainties covered by partial safety factors in conventional design described above. It is necessary to establish the influences on behaviour so that characteristic and design responses can be determined. Care should also be exercised when interpreting data from tests on structural elements which are smaller than prototype size as scaling and size effects should be accounted for.

Testing is also undertaken to appraise existing structures or to establish/verify values for use in design. The test method and procedure to be adopted will depend on the overall objectives of the test programme and the information required, together with the criteria for judging the acceptability of the data obtained should be specified.

## 25.6 Behaviour of reinforced concrete beams

Reinforced concrete is a composite material which uses steel or non-ferrous reinforcing material to compensate for the relatively low tensile strength and brittle characteristics of plain concrete. Steel reinforcement is generally used in construction. The use of alternative non-ferrous reinforcement is described in the following chapter. Improvement in tensile strength may also be achieved by embedding fibres into the concrete (see Chapter 6 of this volume). Reinforcing concrete with steel bars produces a ductile composite structural material that can be easily formed into complex shapes and is strong in both tension and compression.

A plain concrete beam will not be able to carry significant lateral loading as shown in Figure 25.2(a) because of its relatively low tensile strength (around 10 per cent of its compressive strength). The bending moments generated by the lateral forces will cause tension in the lower portions of the beam. As concrete has a low tensile strength in comparison to its compressive strength, the beam will crack as shown. However, by placing reinforcing steel in the tensile region of the beam, the resulting bond between the steel and concrete enables the transfer of tensile forces from the concrete to the steel (see Figure 25.2(b)). The beam acts in a composite manner with the concrete in the upper portion resisting the compressive forces and the steel resisting the tensile forces. Thus, one of the main ideas in the design of reinforced concrete beams is to place tensile reinforcement where needed. Design is conducted by considering axial equilibrium of the compressive and the tensile forces in the reinforced concrete section which forms a couple which is in equilibrium with the externally applied bending moments. The regions in tension in the cantilever and continuous beams shown in Figure 25.3 will need to be reinforced.

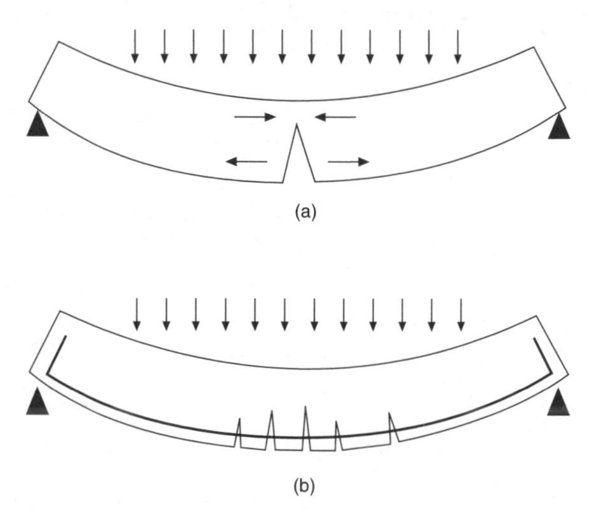


**Figure 25.2** (a) Plain and (b) reinforced concrete beams.

It should be noted that wherever tension occurs in concrete, it is likely that cracking will take place. This, however, should not affect the ability of the reinforced beam to carry the imposed loading and its overall structural safety, provided the cracks widths are small enough to continue to prevent the embedded steel from corrosion. Codes of practice for the design of reinforced concrete structures specify the necessary cover, concrete quality and detailing requirements such as spacing, minimum and maximum allowable steel content (BS EN 1992, Eurocode 2) to ensure that corrosion and other chemical reactions which can affect the durability of the concrete structure are minimized.

The load-deflection response of the unreinforced plain concrete beam shown in Figure 25.2(a) is more or less linear elastic until the concrete cracks and the beam fails by

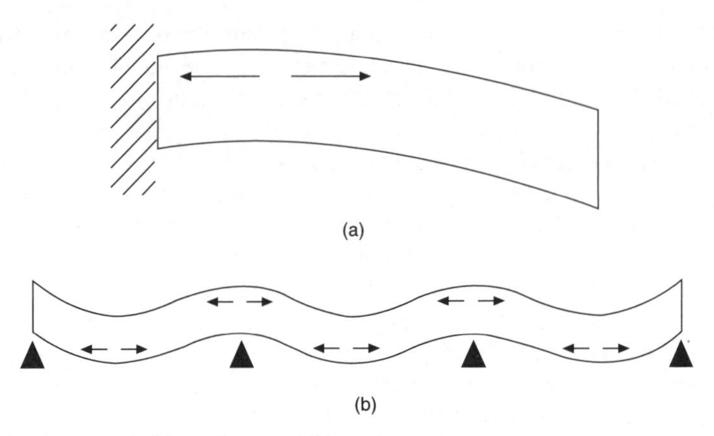


**Figure 25.3** Tension zones in (a) cantilever and (b) continuous beams.

collapsing suddenly without any warning. This behaviour is described as brittle failure as it is sudden and there is little deformation before collapse occurs. However, if a small amount of reinforcement is provided, the behaviour is similar to that of the unreinforced beam, before cracking occurs. With the development of some flexural cracking the stiffness of the beam reduces as indicated by the reduced slope of the load deflection curve in Figure 25.4.

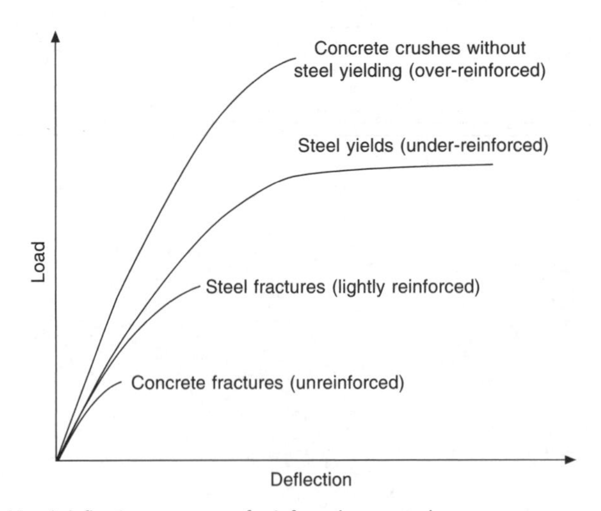


**Figure 25.4** Typical load-deflection response of reinforced concrete beams.

As the load is increased, more cracks form and existing cracks extend towards the compression zone of the beam. The load-deflection curve remains fairly linear until the reinforcement yields in the mid-span region of the beam. Due to the yielding of the reinforcement, the beam loses stiffness and deflections increases significantly with a small increase in load. Because this behaviour is similar to that of a hinge or pin, the region where there is yielding of the reinforcement is referred to as a plastic hinge. As the beam deflects, the crack widths also increase and extend towards the compression region of the beam, thus reducing the depth of the compression zone in the top region of the beam. This in turn, causes the compressive stresses in the concrete to increase until the concrete fails in compression. Because failure of the beam starts with yielding of the reinforcement, the response of the beam is considered to be ductile and the considerable cracking and deformation which occur give warning to the impending failure, a desirable mode of failure. It should, however, be noted that this ductile behaviour only occurs when the amount of reinforcement is below a critical level that permits the reinforcement to yield before crushing of the concrete occurs. Reinforced concrete beams with a reinforcement area equal to or less than this critical value are known as under-reinforced beams and the strength (moment of resistance) or load carrying capacity of such a reinforced concrete beam is approximately proportional to the area of reinforcement. However, if no compression reinforcement is provided, the failure mode of an over-reinforced beam (i.e. a beam with reinforcement area greater than the critical value) is different (non-ductile).

The initial behaviour of an over-reinforced beam is similar to that of an under-reinforced beam except that the beam is generally stiffer and the reduction of stiffness which occurs with cracking is less noticeable. As loading increases, the beam continues to carry the loading until failure occurs suddenly without warning. This non-ductile failure mode is undesirable and it indicates that the steel reinforcement is not fully utilized, i.e. yielding of the steel has yet to occur.

The ductility of both under- and over-reinforced beams can be improved by adding compression reinforcement and links which provide confinement to the concrete in the compression zone. For a more detailed discussion of the subject (see Volume 2, Chapter 6 on the strength and deformation of concrete under static loading). In over-reinforced beams, the load-carrying capacity is only significantly increased when the compression reinforcement increases the strength in the compression region such that the strength of the tensile reinforcement is allowed to develop fully by yielding. In addition to generating bending moments, lateral loads on beams also generate shear forces. Links, a common form of shear reinforcement, in beams also serve to resist the shear forces.

A considerable amount of research has been carried out on the shear behaviour of concrete beams. First, consider members without shear reinforcement. This is of limited relevance for beams where some form of shear reinforcement is usually provided. However, it is of major importance for slabs where it is very inconvenient to provide shear reinforcement. It is of particular relevance for punching shear situations where local shear failure occurs around a concentrated load on a slab, for example in the region around a column in a flat slab.

The major variables governing the shear strength of members without shear reinforcement are the concrete strength, the depth of the member and the reinforcement ratio. EC2 reasonably assumes that the shear strength is directly proportional to the tensile strength of the concrete. It also assumes that the shear strength increases with increase in the reinforcement ratio. EC2 uses a bilinear relationship to model this, up to a maximum value of 2 per cent of reinforcement. The absolute section depth of the member also has a significant influence on the shear strength beyond the influence which can be expected from normal geometrical scaling, that is, there is a size effect which results in shallow members having a higher shear strength. Most codes of practice, including the EC2, allow for this scale effect.

The generally accepted model for the prediction of the effects of shear links reinforcement is the 'truss' model in which the top and bottom truss members are the concrete in the compression zone and the tensile steel respectively. The members connecting the top and bottom members are represented by the steel tension link members and the 'virtual concrete struts'. EC2 expressions for the design of shear links are based on this model. It permits two approaches; the 'standard' and the 'variable truss angle' methods. The standard method uses a fixed truss angle of 45° and assumes that the shear reinforcement is required only to carry the excess shear force beyond the shear capacity of the member section without shear reinforcement. On the other hand, the variable angle method assumes that all the shear force is carried by the shear reinforcement and is considered to be a more rigorous of the two approaches and also to be more economical.

In designing reinforced concrete beams to codes of practice, it is also necessary to ensure that other detailing requirements such as bond requirements, minimum and maximum steel areas etc. are satisfied.

## 25.7 Behaviour of prestressed concrete beams

Prestressed concrete is treated in EC2 as part of reinforced concrete, a wide group of materials ranging from normal reinforced concrete through partially prestressed to fully prestressed concrete structures. Eugene Freyssinet is generally considered to be the 'father' of prestressed concrete. His interest in the material and experimental work from the early 1900s led him to believe that prestressing is a practical approach with high-quality concrete and high-strength steel. It was only after the Second World War that prestressed concrete started developing as a practical construction material. The formation of the Federation Internationale de la Précontrainte or FIP in 1952 helped spread and popularize the concept of prestressed concrete.

The methods given in EC2 are applicable to the full range of concrete structures. It should be noted that EC2 part 1 covers the design of prestressed concrete members with fully bonded internal tendons. Specific rules for external and unbonded tendons will be covered in other parts of EC2.

The principle employed in the design of prestressed concrete is that the internal stresses generated by given external loading are counteracted to a desired degree by the application of prestressing forces. This is to take full advantage of the relatively high compressive strength of concrete. Due to externally applied lateral loads on a beam, internal tensile and compressive stresses are generated to form an internal couple in equilibrium with the externally applied moment. At low loads or in beam members constructed in materials of similar tensile and compressive strengths the internally generated tensile and compressive stresses are of similar magnitudes (depending on shape and cross-section configuration). However, because of the relatively low tensile strength of concrete, the beam will not be able to resist the tensile stresses generated. In normally reinforced beams, this tensile weakness is remedied by reinforcing the tensile region of the beam. In prestressed concrete beams, the tensile weakness is remedied by applying a longitudinal compressive force creating internal compressive stresses which balances to a desired degree the tensile internal stresses. By so doing, the resulting internal stresses are mainly in compression. Depending on the relative magnitudes of the internal compressive stresses caused by the prestressing force versus the internal tensile stresses generated due to the applied lateral loads, the tensile stresses may be minimized or totally eliminated.

The prestressing force is usually applied internally via steel cables or bars which are fixed or locked at the beam ends. The prestress can be applied to the beam before or after the concrete has hardened. In the former, the approach is referred to as applying a pre-tension force as illustrated in Figure 25.5. The tension in the prestressing cables is only released into the beam after the concrete has hardened. In the latter approach, the alignment of the prestressing cables is fixed by the provision of ducts placed in the concrete and the cables are only tensioned after the concrete has sufficiently hardened as illustrated in Figure 25.6.

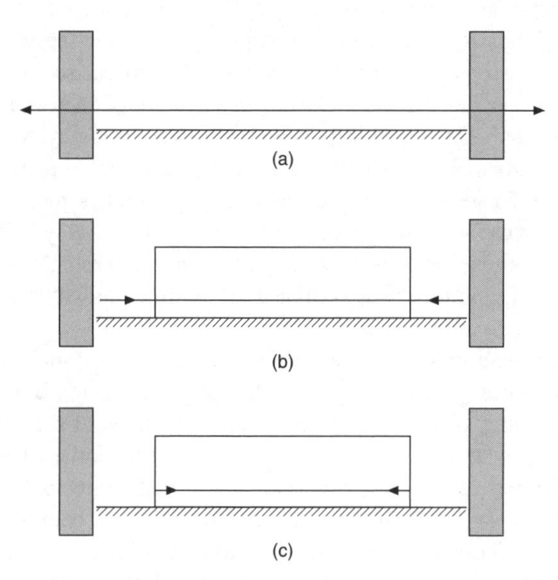


**Figure 25.5** Pre-tensioned prestressed beam.

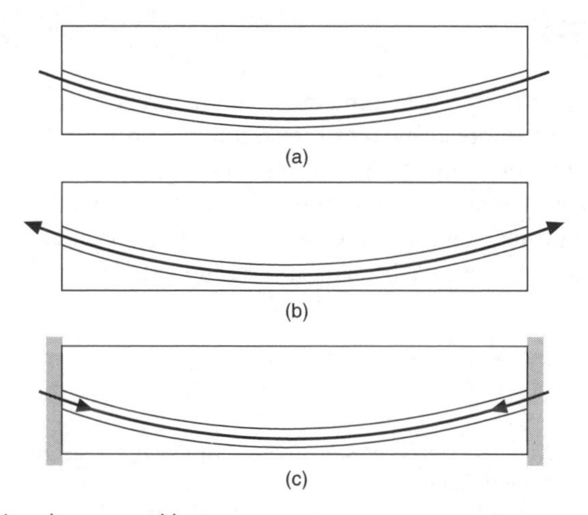


**Figure 25.6** Post-tensioned prestressed beam.

A major difference in the two approaches is that the alignment of the pretensioned cables are usually straight while post-tensioned cables can take non-linear profiles. However, if required, multiple linear alignments can be achieved in the pre-tensioned arrangement.

The pre-tensioned method is ideally suited to precast production of prestressed elements in which a length of prestressing cable is pre-tensioned with a series of moulds placed along its length to contain the concrete. There are a number of proprietary systems for tensioning concrete elements.

Whichever system is used for prestressing, it should be noted that there will be some losses in the prestressing force which is initially applied or transferred to the unit. The reasons for this loss range from the properties of concrete such as elasticity, creep and shrinkage characteristics to friction in the anchorage and ducting system used. Losses which occur after transfer can be of the order of 20 per cent of the force achieved at transfer. Losses which occur before or during transfer are of the order of 5–10 per cent depending on the system employed in pre- and post-tensioning systems.

There are applications for which the pre- and post-tensioning methods have been found to be suitable. In general, the pre-tensioning method is more suitable for small cross-section units that cannot easily accommodate the comparatively bulky post-tensioning cables and anchorage ends. Pre-tensioning is particularly suitable and economical for the mass production of large numbers of similar units such as railway sleepers, floor joists, beams, poles, piles etc.

The post-tensioning approach is more versatile and is able to make more efficient use of the prestressing forces. The losses are smaller and by curving the cables upwards at supports, the prestressing system adds to the shear resistance of the elements. However, in the post-tensioning system, it is necessary to have ducting and anchorages to effect the method. This additional cost is not economical in small units but in large units, the proportional increase is small. Post-tensioned prestressing has been successfully used in buildings, bridges, offshore oil platforms and other engineering structures. For many structures, prestressing can provide economical and aesthetically attractive concrete solutions.

## 25.8 Summary

The aim of this chapter has been to explain in simple terms the basic principles of limit state design relating to reinforced and prestressed concrete structural elements as a practical structural material without setting out the relevant design theory in detail. The ideas behind the limit state approach to the design of concrete structures, the basic behaviour and principles of reinforced and prestressed concrete have been outlined. The two usual ways of prestressing concrete have been described and uses for each have been indicated.

## References


Beckett, D. and Alexandrou, A. (1997) *Introduction to Eurocode 2: Design of Concrete Structures (including seismic actions)*, E&FN Spon, London.

Beeby, A.W. and Narayanan, R.S. (1995) *Designers' Handbook to Eurocode 2 part 1.1:Design of Concrete Structure*, Thomas Telford, London.

Betonvereniging, The Concrete Society, Deutscher Beton-Verein (1997) *Design Aids for Eurocode 2*, E&FN Spon, London.

British Cement Association (1993) *Concise Eurocode for the Design of Concrete Buildings*, British Cement Association.

British Standards Institution, BS EN 1992, Eurocode 1: Actions on Structures, publication date to be announced.

British Standards Institution, BS EN 1992, Eurocode 2: Design of Concrete Structures.

British Standards Institution, BS EN 1990: 2002, Eurocodes – Basis of Structural Design, BSI, London.

Kong, F.K. and Evans, R.H. (1987) *Reinforced and Prestressed Concrete*, 3rd edn, Chapman & Hall, London.

MacGinley, T.J. and Choo, B.S. (1990) *Reinforced Concrete: Design Theory and Examples*, 2nd edn, E&FN Spon, London.

Mosley, W.H., Bungey, J.H. and Hulse, R. (1999) *Reinforced Concrete Design*, 5th edn, Palgrave Macmillan, London.

Mosley, W.H., Hulse, R. and Bungey, J.H. (1996) *Reinforced Concrete Design to Eurocode 2,* Palgrave Macmillan, London.

Ove Arup, S.B. Tietz and British Cement Association (1994) *Worked Examples for the Design of Concrete Buildings*, British Cement Association.

# PART 11

# Alternative reinforcement for concrete

# 26

# Alternative reinforcement for concrete

*John L. Clarke*

## 26.1 Aims and objectives

When correctly specified and constructed, reinforced concrete is a cost-effective and durable construction material. However, the specification and/or use of inappropriate materials can lead to poor durability, particularly in severe exposure conditions and with poor levels of workmanship. The major problem is the corrosion of the embedded steel reinforcement. The principal approach for achieving durability is the selection of appropriate concrete materials and mixes, which are covered extensively elsewhere in these volumes. Another approach, and one that is becoming more widespread, is the use of alternative reinforcement materials that will be more durable. These include:

- coated reinforcement (either galvanized or fusion-bonded epoxy)
- stainless steel
- fibre composites

This chapter briefly describes the first two materials, considering their properties and giving examples of their use in practice. The chapter concentrates on fibre composite materials, generally known as FRPs (fibre reinforced polymers), which consist of glass, carbon or aramid fibres combined with an appropriate resin to form a solid. The topics covered include:

- properties of fibres and resins
- manufacturing processes and properties of composites
- research and development
- advantages and disadvantages
- amended design rules for concrete reinforced with fibre composites
- Health and safety considerations
- applications and demonstration projects

The use of fibre composite reinforcement is still very much in its infancy but it is predicted that its use in specialist applications will grow rapidly. This chapter provides the necessary introduction to the materials.

## 26.2 Other types of reinforcement

### 26.2.1 Galvanized steel reinforcement

Galvanized steel reinforcing bars have been successfully used in several countries over the past 50 years (Australia, Bermuda, Netherlands, Italy, the UK, and the USA) and consumption is increasing. The main advantages of galvanized steel are:

- it delays the initiation of corrosion and cracking
- it has very good performance in carbonated concrete
- it tolerates higher chloride migration levels than uncoated steel
- it provides protection to the steel during storage
- it has longer life in cracked carbonated concrete than uncoated bar.

Hot-dipped galvanized steel is produced by dipping clean and fluxed steel into a bath of molten zinc. The layer formed on the surface of the steel usually consists of a thin outer coating of pure zinc on a series of layers of zinc/iron alloys with increasing iron content.

The performance of galvanized steel in concrete as reported in the literature (Andrade *et al.*, 1995) is contradictory. Although it has been used successfully in practice, laboratory studies suggest that its performance would not be cost-effective. The factors behind this divergence of views, currently the object of discussion, are:

- the pH of the cement paste
- the bond between the reinforcing bars and the concrete
- chromate passivation of the galvanized steel
- the structure and thickness of the zinc coating
- the resistance of the zinc coating to corrosion induced by chloride ions

Zinc is passive in most cement pastes as the pH of uncarbonated cement pastes is 12–13.5. A passive layer would be formed when $pH < 13.3$, the upper limit for passivation, due to the formation of a layer of calcium hydroxyzincate, inhibiting further corrosion. The passivating process results in an homogenous zinc depletion of about 10 μm. A more protective film is produced from pure zinc than from an iron–zinc alloy. It is recommended that an external pure zinc layer of at least 10 μm and a total galvanized layer of at least 80–85 μm are needed to provide suitable protection when embedded in concrete.

In concrete made from a cement with exceptionally high soluble alkali, film formation could be inhibited during the setting period and corrosion of the zinc in the hardened concrete will depend on the environment (humidity, chloride penetration).

During the formation of the passive layer, hydrogen is evolved. Although the evolution of hydrogen raises the spectre of embrittlement, the reinforcing bars normally used in construction are not susceptible to hydrogen embrittlement. Similarly, the hydrogen evolved during the pickling process (pickling is part of the preparation of the surface prior to the application of the zinc, using a weak acid) before galvanizing does not cause a problem. Galvanizing is not generally recommended for steels with a tensile strength above 700–800 $N/mm^2$ , i.e. not for prestressing steels here the risk of hydrogen embrittlement is more severe than for unstressed reinforcement.

Several reports (Andrade *et al.*, 1995) compare the reduction in bond strength of galvanized and uncoated steels, both plain and deformed. Reduction in bond is attributed to the formation of hydrogen bubbles at the interface between the bar and the concrete. It has been suggested (Andrade *et al.*, 1995) that this can be overcome by adding chromate to the concrete mix or giving the bars a chromate passivation treatment. On the other hand, the zincates produced – which are less expansive and more soluble than iron corrosion products in the cement environment – could diffuse into the pores of the concrete and make the concrete more dense locally, increasing the bond strength above what would be expected for uncoated bar.

In practical terms, most construction is carried out with deformed bar and it is probable that the evolution of hydrogen will not affect the bond strength of galvanized deformed steel reinforcement. However, the use of a passivation agent is still debated. The most effective is a chromate but its use as a concrete admixture raises a number of serious environmental and health questions on-site and would certainly be rejected by cement manufacturers and contractors. It would be more appropriate to use chromated bars as, in the first instance, it would restrict the amount of chromate used and ensure it was where it was needed. It would furthermore provide additional corrosion protection before use and ensure that poor storage would not lead to white rust on the reinforcing bar.

Zinc coatings remain passive in carbonated concrete and the rate of corrosion is much lower than for uncoated steel. This makes galvanized steel reinforcement ideal for use in concrete which is at risk from carbonation.

As regards corrosion resistance in chloride-contaminated concrete, the distinction has to be made between cast-in chloride and that which penetrates from the outside. Cast-in chloride may attack the zinc coating before and during the formation of the passive calcium hydroxyzincate whereas chlorides penetrating from the outside will find the passive layer already formed and so may be less dangerous.

Though zinc can be depassivated and attacked in the presence of chloride ions, the tolerance of galvanized steel to chloride is higher than that of uncoated steel. Galvanizing protects the steel against chloride ingress because it is more tolerant to chloride, requiring a higher concentration for depassivation and it corrodes more slowly in chloride-contaminated conditions.

## 26.2.2 Fusion-bonded epoxy-coated steel reinforcement

Fusion-bonded epoxy-coated reinforcement (FBECR) has been developed over the past 25 years to combat the huge economic cost of deterioration of reinforced concrete structures caused by reinforcement corrosion. The reputation of the first FBECR was doubtful, with many failures reported in the USA. The critical processes in the manufacture of FBECR are now better understood.

There are ASTM (ASTM A775M-01 Standard specification for epoxy-coated reinforcing steel bars) specifications for FBECR. It has been recognized that, with good manufacturing techniques, the quality and performance of the FBECR can be improved significantly. It is also recommended that no cracking should be allowed on any part of the coating when a bar is subjected to the specified bend test (ASTM A755).

The chemical resistance of epoxy coating to alkalis is very good, with powder coatings being slightly better than liquid-applied coatings (Andrade *et al.*, 1995).

Suitably formulated epoxy coatings exhibit the necessary mechanical properties such as good adhesion, formability, impact resistance and abrasion resistance. As most bars are coated in straight lengths, they must be capable of being bent without rupture of the epoxy coating. To do this, the epoxy coating relies on its flexibility and its adhesion to the steel. The flexibility of the coating depends upon its formulation and its thickness: the thinner the layer, the more flexible the coating but if it is too thin its ability to protect against corrosion falls. For good corrosion protection and adequate flexibility, a coating thickness of 180–300 μm is recommended (ASTM A775M-01).

The abrasion resistance of epoxy coatings is usually good and wear resistance is slightly better with powder coatings than with liquid-applied coatings (Andrade *et al.*, 1995).

The adhesion of epoxy powder coatings to steel is in most cases good but pre-treatment of the steel is important. The best adhesion is obtained from steel that has been blast cleaned. Some patterns are more difficult to clean than others. The adhesion of an epoxy coating to an uncleaned or inadequately cleaned bar is poorer than a clean bar as the epoxy coating adheres to the contaminants rather than the steel and produces poor protection.

One important structural requirement that may be affected by the presence of the epoxy coating is the bond strength between the steel and the concrete, normally determined by pull-out tests. Epoxy coatings can cause a certain degree of slippage between the coated bar and the embedding concrete. An acceptable level of bond strength is generally considered to be 80 per cent (Andrade *et al.*, 1995) of that of an equivalent uncoated bar.

Reduction of bond strength is slightly less important for deformed bar as the deformations transmit the load from the bar to the concrete. However, the geometry of the deformation can be very important and affect the pull-out strength of epoxy-coated bar (Swamy, 1988).

The epoxy coatings used today to protect reinforcing steel contain no corrosion-inhibiting pigments but act solely as a barrier against the environment. No epoxy coating is completely impermeable to oxygen and moisture, but diffusion can be reduced if the coating is as dense (free from pores) as possible. If aggressive substances can reach the exposed steel because of damage to the coating, corrosion will be concentrated at these points and the cross-section will be reduced at these points. To control the spread of corrosion beneath the film, some manufacturers of epoxy-coated bars now pre-treat the bars before the epoxy coating is applied.

## 26.2.3 Stainless steel reinforcement

### General

Stainless steels are iron-based alloys containing at least 10.5 per cent chromium whose corrosion resistance increases with alloying metal content. The alloys are commonly identified by their micro structure: martensitic, ferritic, austenitic and duplex (with a microstructure of ferrite and austenite). Although these terms relate to the microstructure of the alloys, they also define the ranges of chemical composition as the microstructures of these alloys tend only to be stable in specific composition ranges. Interest in the use of stainless steels as concrete reinforcement is attributable to their increased resistance to corrosion particularly in chloride-containing media. Of the wide range of possible alloys, only a few have been investigated (Concrete Society, 1998) for their suitability as reinforcement for concrete.

### Ferritic stainless steels

Tests have shown (Concrete Society, 1998) that the threshold chloride content of concrete for corrosion of ferritic stainless steels is higher than for plain carbon steel. However, the susceptibility of ferritic stainless steels to corrosion with increasing chloride content suggests that, unless the level of chloride contamination can be contained below the threshold level (which has yet to be defined but is likely to be below 1.9 per cent chloride ion content with respect to the cement content), the use of ferritic stainless steels as reinforcement should be limited. At present there are no standards for ferritic stainless steel reinforcing steel.

### Austenitic stainless steels

Long-term studies of austenitic stainless steels have shown (Concrete Society, 1998) that they are resistant to corrosion in chloride-contaminated concrete. The critical chloride content has not been determined but is likely to be above 3.2 per cent chloride ion with respect to the cement content.

Work carried out on welded stainless steels has shown (Concrete Society, 1998) that it is essential to remove all oxide and scale produced during welding. If oxide remains on the surface, the steel is at risk from corrosion in chloride-contaminated concrete. Any blueing created by grinding austenitic stainless steel would create a similar corrosion hazard.

To ensure the best possible performance, the surface condition of austenitic stainless steel is therefore important: certainly it should be oxide-free.

At present there are only Standards for types 304 and 316 stainless steel reinforcing bars. Table 26.1 shows the chemical compositions, which are specified in BS 6744: Austenitic stainless steel bars for the reinforcement of concrete.

The use of the 316 grade austenitic steel is recommended in chloride-contaminated environments, as the molybdenum provides better corrosion resistance.

(Since the publication of BS 6744 in 1986 there have been many developments which now mean that updating is required to cover the latest chemical and physical properties of the stainless steels available. A BSI committee is currently reviewing this standard and it is proposed that the BS 6744 classification system should no longer be used. It will be replaced by the system used in BS EN 10088–1: Stainless steels. Part 1. List of stainless steels. Thus the commonly used names of 304 and 316 will be replaced.)

**Table 26.1** Chemical composition of austenitic steels complying with BS 6744

| Grade | C | Si | Mn | P | S | Cr | Mo | Ni |
|---|---|---|---|---|---|---|---|---|
| 304S31 | 0.07 | 1.0 | 2.0 | 0.045 | 0.030 | 17.0–19.0 | – | 8.0–11.0 |
| 316S33 | 0.07 max | 1.0 max | 2.0 max | 0.045 max | 0.030 max | 16.5–18.5 | 2.5–3.0 | 11.0–14.0 |

#### *Duplex stainless steels*

Laboratory studies (Concrete Society, 1998) have shown that duplex stainless steels can be more corrosion-resistant than austenitic type stainless steels. There are currently no Standards for duplex stainless steel reinforcing bars.

#### *Reinforcing bar clad in stainless steel*

Provided the cut ends are well protected, reinforcing bar clad in austenitic stainless steel performs as well as solid austenitic stainless steel. When the cost of manufacture of such steel became too high, production ceased. However, new manufacturing techniques have been developed and clad bars are now available again. There are currently no Standards for reinforcing bars clad with stainless steel.

## 26.3 General introduction to fibre composites

### 26.3.1 Background

FRP materials have been used for many years in the aerospace and automotive industries, where their high strength and low weight have shown distinct advantages over traditional materials such as aluminium and steel. They are slowly being adopted by the construction industry either as construction materials in their own right or for use in conjunction with traditional materials such as concrete. This section aims to give a brief introduction to FRP materials and to give indicative values for the properties appropriate to those materials likely to be used in connection with concrete bridges.

### 26.3.2 Materials

Currently, the most suitable fibres are glass, carbon or aramid. Each is a family of fibre types and not a particular one. Typical values for the properties of the materials are given in Table 26.2. The fibres all have a linear elastic response up to ultimate load, with no significant yielding. These values should only be taken as indicative; actual values should be obtained from the particular manufacturer.

The fibres are used in various forms, either in the form of ropes or fabric materials or, more generally, combined with a suitable resin to form a composite as described later. In each case the strength and stiffness will be lower than the anticipated value from Table 26.2, for a composite they will be in the region of 65 per cent.

The common forms in which fibres are used may be summarized as follows:

- for embedded reinforcement: composite rods or grids

**Table 26.2** Typical fibre properties

| Fibre | Tensile strength ($N/mm^2$) | Modulus of elasticity ($kN/mm^2$) | Elongation (%) | Specific density |
|---|---|---|---|---|
| Carbon – high strength | 3430–4900 | 230–240 | 1.5–2.1 | 1.8 |
| Carbon – high modulus | 2940–4600 | 390–640 | 0.45–1.2 | 1.8–2.1 |
| Aramid – high strength and high modulus | 3200–3600 | 124–130 | 2.4 | 1.44 |
| Glass | 3500 | 75 | 4.7 | 2.6 |

- for prestressing: fibre ropes or composite rods
- for external repair and strengthening: sheet material consisting of pure fibres or fibres pre-impregnated with part-cured resins (known as prepreg materials), composite plate material or preformed composite shells
- for permanent formwork: preformed composite shells

The main advantages of fibre composites are that they are lighter and stronger than steel and, with the correct resin and fibre combination, should prove to be more durable. However, there are possible disadvantages which include the lack of any yield at the ultimate load. This will be considered further in the section on design.

There is a wide choice of resins available, many of which, though not all, are suitable for forming composites. (The Draft Canadian Highways Bridge Design Code (Bakht, *et al.*, 1996) specifically prohibits the use of polyester resins for embedded FRP material.) The choice will depend on the required durability, the manufacturing process and the cost. Thermosetting resins are generally used but they have the major disadvantage that, once they have fully cured, the composites cannot be bent to form hooks, bends and similar shapes. One possible alternative is the use of suitable thermoplastic resins, which are now being developed. With these resins the composite components can be warmed and bent into the required shapes. On cooling the full properties of the resins are restored. However, there is likely to be distortion of the fibres in the region of the bend, which will lead to a reduction of the strength locally.

## 26.3.3 Manufacturing processes

The most widely used manufacturing process for forming composite rods is pultrusion. The fibres, which are supplied in the form of continuous rovings, are drawn off in a carefully controlled pattern through a resin bath which impregnates the fibre bundle. They are then pulled through a die which consolidates the fibre–resin combination and forms the required shape. The die is heated which sets and cures the resin allowing the completed composite to be drawn off by suitable reciprocating clamps or a tension device. The process enables a high proportion of fibres to be incorporated into the cross-section and hence relatively high strength and stiffness are achieved. However, the sections have a smooth surface, which provides insufficient bond if they are to be used as reinforcement in concrete. Hence, a secondary process, such as overwinding with additional fibres, is required to improve the bond.

There are variations to the process, such as the use of a braided rope, which is then impregnated with resin. The resulting profile provides a good bond with the concrete.

As indicated above, thermosetting resins are generally used at present. Once formed these cannot be bent into the range of shapes currently used by the concrete industry, and hence different manufacturing processes are required to form specials. Filament winding, in which resin-impregnated fibres are wound round a mandrel of the required shape, has been used to manufacture shear links. Other manufacturing processes, such as filament arranging, are being developed for more complex shapes.

Fibre composite two-dimensional reinforcement grids, and even three-dimensional grids, are made by a number of different patented processes.

## 26.3.4 Short-term properties

The physical properties of a composite will depend on the type and percentage of fibres used. Typically a pultruded composite would have about 65 per cent of fibre by volume. Thus with glass the ultimate strength might be 1200 $N/mm^2$ rising to 2000 $N/mm^2$ for carbon. The elastic modulus will be about 40 $kN/mm^2$ for glass fibre composites and may be 150 $kN/mm^2$ for carbon fibre composites. As composites are not currently manufactured to a common standard, their properties will vary from one manufacturer to another. Thus all design must be on the basis of the actual properties, as supplied by the appropriate manufacturer.

## 26.3.5 Long-term properties

Creep rupture, or stress rupture as it is often known, is the process by which a material with a permanent high load applied to it will creep to failure. This will be particularly important for prestressed structures and for reinforced structures with a high permanent load. There is a reasonable amount of data to confirm the performance for a few years under stress but there is still a degree of uncertainty about the exact form of the long-term response. Thus for design purposes, relatively large factors of safety are currently proposed (The Institution of Structural Engineers, 1999) to allow for the uncertainty, which may be revised as more test data become available.

The phenomenon of creep rupture does not appear to have any effect on the short-term strength. Thus an FRP element that has been loaded to a significant level for a period of time will retain its initial short-term strength. Similarly the stiffness of FRP is largely unaffected by permanent load (Institution of Structural Engineers, 1999).

The effect of corrosive actions on the fibre is complicated and varies from fibre to fibre. Some corrosive elements attack the internal bonds, breaking the long-chain molecules into short lengths, reducing the strength but not the stiffness. Other factors attack the fibre from the outside, physically removing some of the fibre, which will change both the strength and the effective stiffness. The effect of the resin is to shield the fibres from chemical attack. The durability of the resin itself, and its permeability to aggressive substances, will affect the durability of the FRP bar. Because of the lack of research data, relatively large safety factors have to be applied to the measured short-term properties. Future testing will allow these factors to be modified and will indicate the most suitable combinations of resins and fibres to resist attack.

### 26.3.6 Health and safety

Loose fibres on the surface of the FRP bars may cause irritation and a few people may suffer an allergic reaction to the resin matrix. Hence, during assembly of the reinforcement cage sensible precautions should be taken, such as wearing gloves. At all times, current Health and Safety Regulations, as indicated by the manufacturer of the FRP material, should be followed. When cutting the material, suitable dust extraction should be used.

Once the FRP material is embedded into the concrete there will be no problems as far as health and safety are concerned. Where adhesives are used, it should always be used in accordance with the manufacturer's recommendations.

## 26.4 FRP internal reinforcement

### 26.4.1 Review of materials and manufacturing processes

A number of manufacturers make FRP reinforcement for concrete, some of which are given in Table 26.3. Carbon and glass are the most common fibres, though aramid is also used. Rods are made by pultrusion, as described in section 26.3.3, with a secondary process to form a surface with adequate bond properties. This may consist of removing the smooth outer layer of resin, overwinding with additional fibres or adding a sand layer to form a mechanical key.

**Table 26.3** Details of some available types of FRP reinforcement

| Manufacturer | Country | Trade name | Fibre type | Comments |
|---|---|---|---|---|
| Fibreforce | UK | Eurocrete bar | Glass or carbon | |
| Hughes Brothers | USA | Aslan | Glass | |
| International Grating | USA | Kodiak | Glass | |
| Mitsubishi | Japan | Leadline | Carbon | Prestressing material |
| Nefcom | Japan | Nefmac | Carbon, glass or hybrid | Grid material |
| Tokyo Rope | Japan | CFCC | Carbon | Prestressing material |

The Japanese NEFMAC is a grid material made by a special process which forms successive layers of composite material in the two directions, which are then compressed together (Sugita, 1993).

The above are all with thermosetting resins, and hence cannot be bent once the resin has fully cured. Some attempts have been made to bend partly cured elements, though this is not generally satisfactory as it leads to displacement of the fibres around the bend. Thus various techniques are being considered for forming shapes such as shear links and hooks. These include filament winding, to form a cylinder or box. These can be cut to form appropriate closed shapes. An alternative approach is filament placing, in which the resin-impregnated fibres are wound round pins to give the required shape.

## 26.4.2 Advantages and disadvantages

### *Advantages*

The main advantage of FRP reinforcement should be improved durability, provided an appropriate combination of fibre and resin has been used. Thus FRP reinforcement would be most beneficial in highly corrosive environments, such as bridges in marine environments or those subjected to de-icing salts. The improved durability should lead to the possibility of a lower specification for the concrete and lower covers. The rods have high strength, probably twice that of normal high-yield steel, leading in some situations to lower reinforcement percentages. FRP would be particularly beneficial for precast members in which the reinforcement is required mainly for handling, transport and erection. Here the long-term behaviour of the material would not be an issue; unlike steel, it would not cause any damage to the concrete if it were to degrade.

### *Disadvantages*

One of the disadvantages of FRP reinforcement is its relatively low stiffness compared to steel; for glass it may be 25 per cent, for carbon 75 per cent or more. This will result in increased deflections and crack widths. In addition, the bond stress may be lower, again increasing crack widths. However, the latter will only be important from the point of view of aesthetics and not durability.

A major practical disadvantage at present is the difficulty of forming shear links, hooks etc. All such items will have to be factory made as the current bars cannot be bent on site. This may change with the introduction of thermoset resins, which can be bent once warmed, though the properties at the bend may be seriously affected.

## 26.4.3 Summary of research

### *Brief overview of research*

Throughout the world there are programmes of work developing the use of FRP in concrete structures. In Japan, North America and in Europe, considerable effort is being put into the development of embedded FRP reinforcement for concrete. The main areas being investigated around the world are:

- selection of suitable resins and fibres
- development of appropriate manufacturing techniques
- investigations to determine the durability of FRP rods exposed to aggressive environments either directly or embedded in concrete
- determination of the structural behaviour through testing and analytical techniques
- economic and feasibility studies
- development of case studies of trial structures and components
- development of suitable design guidance

### *Tests on reinforced concrete members*

Much of the early work was concerned with the bond between FRP reinforcement and concrete. Over the last 10 years or so there have been many programmes of tests on simple beams, considering both the flexural and the shear behaviour. Limited work has been carried out on frames, columns and slabs. The work has been reported in a number of major international conferences (Nanni and Dolan, 1993; Taerwe, 1995; El-Badry, 1996; Japan Concrete Institute, 1997; Burgoyne, 2001).

### *Durability*

One very important aspect of the use of FRP rods embedded in concrete is their durability. Surprisingly, this was not considered to any great extent in any of the early research programmes worldwide. It is important to note that carbons and aramids are inherently more durable in an alkaline environment than the standard E-glass. AR-glass (Alkali Resistant glass) has been used for some time for GRC (glass reinforced cement) and is used in some composite rods. However, many resins degrade in the highly alkaline concrete environment. The Draft Canadian Highways Bridge Design Code specifically prohibits the use of polyester resins for embedded FRP material (Canadian Standards Association, 1996). Thus the manufacturers' claims of long life need to be demonstrated.

Currently, trials are being carried out on the resins and fibres in isolation and in the form of the composite, both in a range of artificial aggressive environments and embedded in concrete, with a view to developing the necessary confidence in the long-term properties of the materials. The accelerated laboratory testing is being backed up by data from specimens on exposure sites. An important aspect in the development of new materials is the construction of demonstration structures, as outlined below. While they may not be economic, because the materials themselves are not yet fully understood and the design approaches are not fully developed, they give valuable experience of practical construction aspects and an indication of the long-term performance. They thus develop the necessary confidence in the new materials.

## 26.4.4 Design guidance

Modified design rules for use with FRP reinforcement have been developed for BS 8110: Structural use of concrete, Part 1, Code of practice for design and construction (1997): Part 2, Code of practice for special circumstances (1985) for buildings and BS 5400: Steel, concrete and composite bridges, Part 4, Code of practice for design of concrete bridges (1990) for bridges (Institution of Structural Engineers, 1999). The proposals are broadly in line with the codes being developed in Japan (Japanese Ministry of Construction, 1995; Japan Society of Civil Engineers, 1997) and in North America (Canadian Standards Association, 1996; Bakht *et al.*, 1996; ACI, 2001).

For detailed design guidance, reference should be made to the Institution of Structural Engineers document. The following significant differences when designing with FRP reinforcement are summarized as follows:

- *Analysis* Because FRP materials have a straight-line response to ultimate, with no yielding, it is appropriate to use only elastic methods of analysis. For design purposes it should be assumed that no redistribution of the elastic bending moments and shear forces will take place.

- *$\gamma$ factors* The effective strength and the effective stiffness of embedded reinforcement may change with time, due to alkali attack, depending on the types of resin and fibre used in the composite. Appropriate factors of safety applied to the short-term values will be required to take account of the changes.
- *Durability* The quality of the concrete will be governed mainly by strength considerations and the cover by the aggregate size and the size of the reinforcing bar. Design crack widths will be controlled by aesthetic considerations and, possibly, watertightness of the structure.
- *Flexure* The basic principles are unchanged. The design equations and design charts given in the Codes are not appropriate as they assume yielding of the reinforcement at ultimate. Because of the relatively low stiffness of FRP, it is likely that failure will occur by compression of the concrete and not by reaching the ultimate capacity of the tensile reinforcement.
- *Shear* The shear capacity of the concrete cross-section should be calculated on the basis of an equivalent area of steel, transformed on the basis of the modular ratio. The strain in FRP shear reinforcement should be limited.
- *Serviceability* Because of the lower stiffness, deflections and crack widths may become dominant design criteria. However, as indicated above, with no limitation on crack width required from the point of view of durability, aesthetics will be the only criterion. Thus the current rules could be relaxed considerably.
- *Columns* The strength of bars in compression should be ignored and the column designed on the basis of the concrete area alone. (There would appear to be considerable scope for using hoop reinforcement in columns, providing containment to the concrete and hence increasing its load-carrying capacity.)
- *Bond* Because there are no agreed Standards, it will be necessary to determine the ultimate bond stress for the particular material and then use this in design with an appropriate partial safety factor.
- *Fire* FRP reinforcement is generally not recommended for structures for which fire is a significant design consideration.
- *Detailing* Reinforcement cages should be assembled with non-metallic ties. Lapping of reinforcement should be satisfactory, but the relatively low bond strength may lead to uneconomic overlaps. Once formed, FRP reinforcement cannot be bent to form shear links etc. Thus the designer will be required to work not only to standard shape codes but also to fixed dimensions.
- *Construction* There should be no significant problems during the assembly of the reinforcement cage. Any cut ends of bars, or damaged areas, should be sealed with a suitable resin. During the casting of the concrete the reinforcement will have a tendency to float, because of its low density, and allowance must be made for this during fixing.

Care must be taken with the storage of FRP material prior to use, to avoid damage to the resin matrix. In particular, reinforcement should be protected from the effects of prolonged exposure to UV light, which can lead to the degradation of some resins.

## 26.4.5 Applications

### *Introduction*

This section concentrates on the use of FRP reinforcement materials in bridges but also considers a number of other applications, chiefly marine and coastal, that are relevant because chloride ingress is the prime cause of corrosion of the embedded steel.

### *Bridges*

The first footbridge in Britain, and probably in Europe, using glass FRP reinforcement was built at Chalgrove in Oxfordshire in 1995 (Clarke *et al.*, 1998). The bridge is a simple slab, 1.5 m wide and 5 m long reinforced on both faces with glass FRP bars. The structure was load tested and then monitored for about a year. Shortly after, a second footbridge was built in Oxfordshire using the concept of Supercover (Arya, 1996). This is a conventionally reinforced structure with additional cover to protect the bottom steel. To control cracking, the additional cover is reinforced with a layer of glass FRP bars.

In Norway, the Oppegard footbridge was built in 1997 on a golf course near Oslo (Grostad *et al.*, 1997; Haugerud and Mathisen, 1997). It has a span of about 10 m and consists of twin-arched beams, reinforced with glass FRP bars and stirrups made from a glass FRP with a thermoplastic matrix, which could be bent once warmed. The bridge had horizontal prestressed ties containing aramid tendons.

In Denmark, carbon FRP was used for the unstressed reinforcement and shear links for the 90 m long concrete footbridge at Herning in Denmark (Christoffersen *et al.*, 1999). The same FRP material was used for the longitudinal and transverse prestressing cables and for the cable stays.

In the USA, glass FRP bars were used in a bridge in Arkansas in 1994, though the actual application is unclear. The McKinleyville Bridge in West Virginia is a 54 m long, three-span continuous structure which was completed in 1996 (Thippeswamy *et al.*,1998). It carries two lanes of traffic. The 230 mm thick concrete deck spans between the main steel girders which are 1.5 m apart. Two types of glass FRP reinforcing bars were used. The structure was load tested on completion in 1996.

In Canada the five-span Taylor Bridge in Headingly, Manitoba, has a total length of 165 m. A number of the 1830 mm deep precast beams were reinforced with carbon FRP both for the shear links and also for longitudinal reinforcement (Rizkalla *et al.*, 1998). The links projected from the tops of the beams to provide longitudinal shear reinforcement. Two different types of carbon FRP material were used. The beams were prestressed using the same materials. Carbon FRP was also used for the reinforcement of part of the deck slab and glass FRP rods were used to reinforce the safety barrier. A large number of sensors were built into the structure and are being monitored remotely. The bridge was opened in October 1997.

In North America, and particularly Canada, there is a growing interest in the use of steel-free bridge decks, though this would appear to be generally for steel–concrete composite bridges. The concrete deck spanning between the main girders is designed using the principle of compressive membrane action (Cole, 1998). Hence, tension reinforcement is not required but some nominal anti-crack reinforcement is provided. This may be short, chopped fibres or else FRP bars, as were used for the Chatham Bridge in Ontario (Hearn, 1998) or FRP grid material, as was used in the Joffre Bridge in Sherbrooke, Canada (Benmokrane, 1997).

FRP grid material has been proposed for bridge parapets in Ontario (Maheu, 1994).

### *Other applications*

Apart from bridges, fibre composite reinforcement has been used in a number of different marine and coastal applications. Descriptions are given in the proceedings of various conferences (Nanni and Dolan, 1993; Taerwe, 1995; El-Badry, 1996; Japan Concrete Institute, 1997; Burgoyne, 2001). A significant marine application was a replacement fender support beam for a jetty in the Middle East (Grostad *et al.*, 1997) using glass FRP bars and thermoplastic links. Glass FRP bars were recently used in the sprayed concrete repair of a sea wall in West Palm Beach, Florida, as nominal anti-crack reinforcement.

In the USA, Japan and France glass fibre composite bars have been used in a number of applications in which stray electrical currents in steel reinforcement would be a problem. To date these have been mainly under sensitive electronic equipment, such as in hospitals or military installations. As indicated above, FRP reinforcement and prestressing tendons have been used for the support beams for experimental maglev systems in Japan. A number of other applications are being actively considered.

## 26.5 FRP prestressing strand

### 26.5.1 Review of materials and manufacturing processes

A number of different tendon systems have been developed, using carbon, glass or aramid fibres, as shown in Table 26.4. This is not an exhaustive list as there are many other Japanese systems which have not yet been used elsewhere.

**Table 26.4** Details of some available types of prestressing systems

| Trade name (manufacturer) | Country of origin/use | Fibre type |
|---|---|---|
| Arapree (Nippon Aramid plus Italian Company) | Netherlands, Italy, Japan, Portugal | Aramid |
| CFCC (Tokyo Rope) | Japan, USA, Canada, Germany | Carbon |
| Fibra (Mitsui/Shinko Wire) | Japan | Braided aramid |
| Leadline (Mitsubishi) | Japan, USA | Carbon |
| Parafil (Linear Composites) | UK, Norway | Aramid rope |
| Technora (Teijin) | Japan | Aramid |

It should be noted that one system, Parafil, consists of parallel Kevlar fibres without any resin. Thus it is not strictly speaking a composite but for the purposes of this report it has still been classed as an FRP. The remaining materials are all produced by variations of the pultrusion process, as described earlier. Their properties will depend on the particular amount and fibre used, but will be in line with those given in section 26.3.4.

Standard barrel and wedge type anchors as used for steel strands are not appropriate for fibre composites as they cut into the surface layer, damaging the material and leading to premature failure. Hence manufacturers have had to develop special anchorage systems for composite tendons. These are generally stainless steel barrels into which the tendons are bonded using specially formulated adhesive grouts. It is likely that the anchorages will have to be installed on the ends of the tendons under factory conditions, to ensure proper workmanship and a fully cured resin. The anchors can then be proof tested, to, say, 10 per cent above design ultimate load, before the tendon is installed in the structure. The design of the end-block, and of any parts of the structure through which the tendon has to pass, must be such as to allow the passage of the anchorage and not just of the tendon.

In addition, some manufacturers have developed wedge systems which grip the individual strands without cutting into them. These can obviously be fitted to the ends of the strands once installed in the structure, as with steel strands.

As indicated above, Parafil is the only system in which the fibres are used in isolation, rather than in the form of a composite. Anchorage is by means of a wedging system, using a tapered barrel and an internal spike drawn into the middle of the rope (Burgoyne, 1993).

## 26.5.2 Advantages and disadvantages

### *Advantages*

The major advantage of FRP prestressing tendons is their improved durability. They do not require the traditional protection provided by grout to bonded steel tendons; in fact, bonding is difficult for some of the systems because of the smooth surface texture. Thus they are ideally suited for use as unbonded tendons, either internal or external. Their light weight will make installation easier than with equivalent steel tendons.

### *Disadvantages*

From the point of view of construction, one disadvantage will be that the end anchorages of FRP tendons will generally have to be installed in the factory, making installation more difficult than with simple steel strand. Appropriate openings will have to be formed throughout to allow the complete anchor to pass through. In addition, their low elastic modulus will require large extensions during the stressing process.

As with all external tendons, appropriate protection against accidental damage must be provided. In addition, the possibility of vandalism must be considered with FRP tendons as some can be easily cut. In some situations, fire might be an additional hazard. Where necessary, protection against ultraviolet light should be provided; advice should be sought from the manufacturer.

## 26.5.3 Summary of research

A large amount of development work has been carried out around the world on FRP prestressing tendons (Burgoyne, 1993; Wolff and Meisseler, 1993; Gerritse, 1993), looking at the development of appropriate anchorage systems as well as the behaviour of prestressed concrete beams.

A number of studies have looked at the behaviour of beams in flexure (e.g. Yonekura *et al.*, 1996). Because the tendons are generally unbonded and because the elastic modulus of FRP will be relatively low, the ultimate strength is dictated by the concrete failing in compression.

Because of the lack of any yield, it has been suggested that local high strains in a bonded tendon at a crack might lead to sudden, brittle failure. Hence work has been carried out at Cambridge University (Burgoyne, 1997) on partially bonded tendons. This leads to a more ductile behaviour.

Little experimental work has been carried out on the shear behaviour of beams with FRP tendons, though Yonekura *et al.*, (1993) tested a number of I-beams prestressed with either carbon or aramid tendons.

Limited studies have looked at the thermal behaviour of pretensioned elements. The coefficient of thermal expansion along an FRP bar is controlled by the fibres and will be significantly lower than that of concrete. Transversely, it will be due largely to the resin and will be significantly higher. Cracking of some elements prestressed with an aramid FRP has been attributed to this high transverse expansion (Gerritse, 1993).

## 26.5.4 Design guidance

In principle there should be few changes required when designing with fibre composite tendons rather than steel ones. One significant factor that has to be taken into account, however, is the phenomenon of stress rupture, that is, the fact that a tendon stressed to a certain level will creep to failure. This will limit the applied stress in FRP tendons. The draft Canadian Bridge Design Code (Bakht *et al.*, 1996) gives the following maximum permissible jacking stresses as percentages of their ultimate strengths:

| | |
|---|---|
| Carbon fibre tendons | 65 |
| Glass fibre tendons | 55 |
| Aramid fibre tendons | 40 |

Depending on the type of tendon being used (see Table 26.3), the member may be pre-tensioned or post-tensioned. Many of the tendons are smooth, making grouting difficult if not impossible. If grouting is used it should be resin based rather than cement based. The member may be designed in accordance with the current rules for bonded tendons (using BS 5400 Part 4) or unbonded tendons (using BD 58/94: Highways Agency, 1994) as appropriate. When designing with bonded tendons, and clauses that assume that the tendon has yielded should be ignored as they are not applicable to FRP which is elastic to failure.

In determining losses, creep will be controlled by the fibres in the tendons and not the resins. The relaxation in the tendons will be offset by the lower elastic modulus leading to losses that are similar to those for steel. In addition, with grouted anchorages as described above, there is likely to be some loss due to creep in the resin grout. Appropriate information will have to be obtained from the manufacturer.

The clauses in BS 5400 for the design of end blocks are intended for bonded prestressing tendons. When using unbonded FRP tendons it will be appropriate to use the slightly more detailed clauses in BS 8110, which cover both bonded and unbonded tendons.

### 26.5.5 Applications

Prestressing tendons have probably been the widest use of fibre composites in concrete structures to date, though this is likely to change rapidly. A number of bridges have been built worldwide, generally with conventional steel for the unstressed reinforcement.

In Germany and Austria a total of five road bridges and footbridges have been built prestressed with glass-fibre composite tendons, using the Polystal system. The first highway bridge, the Ulenbergstrasse Bridge in Dusseldorf, which has two spans each of about 20 m, was opened to traffic in 1986 (Wolff and Meisseler, 1993). The bridge has been monitored and load tested periodically. However, it is not clear whether the system is still commercially available. Also in Germany, carbon fibre tendons have been used for one bridge (Zoch *et al.*, 1991).

In Spain, aramid FRP tendons were used for a cantilevered roadway in Spain (Casas and Aparicio, 1992).

As mentioned earlier, carbon FRP tendons were used for the 90 m long cable-stayed concrete footbridge at Herning in Denmark (Christoffersen *et al.*, 1999). Carbon FRP was also to be used for the unstressed reinforcement and for the stays.

Carbon fibre tendons are proposed for the three-span Dintelhaven Bridge near Rotterdam (Hordijk, 1998) which will have a total length of about 370 m. Short lengths of the proposed cables will undergo long-term trials before the bridge is constructed.

In Japan the emphasis of development has been on carbon or aramid composites. At least 10 bridges have been built to date (Noritake *et al.*, 1993; Tsuji *et al.*, 1993). These take a variety of forms. There are a number of road bridges, ranging from 7 m single-span up to four-span with a total length of about 80 m. A number of foot and cycle bridges have been built, the longest having a clear span of about 75 m. In addition, there have been a number of unconventional footbridges built; there is a 55 m long floating structure and a stressed ribbon bridge with a clear span of about 45 m. A further application has been for maglev structures.

In North America part of one bridge in South Dakota has been stressed with glass and carbon tendons (Iyer, 1993) and a bridge in Calgary contains carbon fibre composite strands (Anon, 1993). A number of other bridges are currently being planned.

In Canada the five-span Taylor Bridge in Headingly, Manitoba, has a total length of 165 m. A number of the 1830 mm deep precast beams were prestressed with carbon FRP tendons (Rizkalla *et al.*, 1998). Two different types of carbon FRP material were used. Carbon FRP was also used for the reinforcement of the beams and for part of the deck, see section 26.3.5. A large number of sensors were built into the structure and are being monitored remotely. The bridge was opened in October 1997.

To date no concrete bridge has been built in the UK with FRP tendons, though they have been used for a prototype stressed masonry footbridge.

## 26.6 Summary

This chapter has described alternative forms of reinforcing bars and prestressing tendons for concrete, concentrating on fibre composite materials. It has demonstrated that FRP materials are at an advanced stage of development and have been used in a variety of applications around the world.

## References

ACI (2001) ACI 440. IR Guide to the design and construction of concrete reinforced with FRP bars. American Concrete Institute, Farmington Hills, USA.

Andrade, C. *et al.* (1995) Comité Euro-International du Beton, State of the Art Report, Coating Protection for Reinforcement, Thomas Telford, London.

Anon (1993) Carbon-fibre strands prestress Calgary span. *Engineering News Record*, 18 October, 21.

Arya, C. (1996) Supercover concrete. *FRP International*, **IV**, Issue 3, 4.

Bakht, B. *et al.* (1996) Design provisions for fibre reinforced structures in the Canadian highway bridge design code. In El-Badry, M.M. (ed.), *Advanced Composite Materials in Bridges and Structures*, Canadian Society for Civil Engineering, Montreal, pp. 391–406.

Benmokrane, B. (1997) NEFMAC for a highway bridge in Canada. *FRP International*, **V**, Issue 4, 5.

Burgoyne, C.J. (1993) Parafil ropes for prestressing tendons. In Clarke, J.L. (ed.), *Alternative Materials for the Reinforcement and Prestressing of Concrete*, Blackie Academic & Professional, Glasgow, pp. 102–126.

Burgoyne, C.J. (1997) Rational use of advanced composites in concrete. *Proceedings of the Third International Symposium on Non-Metallic (FRP) Reinforcement for Concrete Structures*, Japan Concrete Institute, Vol. 1, pp. 75–88.

Burgoyne, C.J. (2001) *Fibre-reinforced Plastics for Reinforced Concrete Structures*, Thomas Telford, London, (two volumes).

Canadian Standards Association, (1996) Canadian Highways Bridge Design Code, Section 16, *Fibre Reinforced Structures.*

Casas, J.R. and Aparicio, A.C. (1992) A full scale experiment on a prestressed concrete structure with high strength fibres; the North ring-road in Barcelona. In *Proceedings of FIP-XI International Congress*, Hamburg, T15.

Christoffersen, J., Hauge, L. and Bjerrum, J. (1999) Footbridge with carbon-fibre-reinforced polymers, Denmark. *Structural Engineering International*, **9**, No. 4, November, 254–256.

Clarke, J.L., Dill, M.J. and O. Regan, P. (1998) Site testing and monitoring of Fidgett Footbridge. In Virdi, K.S., Garas, F.K., Clarke, J.L. and Armer, G.S.T. (eds), *Structural Assessment – The Role of Large and Full-scale Testing*, E&FN Spon, London, pp. 29–35.

Cole, M. (1998) Arching action. *New Civil Engineer*, 28 May, 32–33.

Concrete Society, (1998) Guidance on the use of stainless steel, Technical Report 51, The Concrete Society, Crowthorne.

Gerritse, A. (1993) Aramid-based prestressing tendons. In Clarke, J.L. (ed.) *Alternative Materials for the Reinforcement and Prestressing of Concrete*, Blackie Academic & Professional, Glasgow, pp. 172–201.

Grostad, T., Haugerud, S.A., Mathisen, L.L. and Clarke, J.L. (1997) Case studies within Eurocrete – Fender in Qatar and bridge in Norway. In *Non-Metallic (FRP) Reinforcement for Concrete Structures*, Japan Concrete Institute, Sapporo, Japan, Vol. 1, pp. 657–664.

Haugerud, S.A. and Mathisen, L.L. (1997) The design and development of a novel FRP reinforced bridge. In *Proceedings of IABSE Conference on Composite Construction – Conventional and Innovative*, Innsbruck, Austria, pp. 765–770.

Hearn, N. (1998) Strength without steel, *Innovator; the Newsletter of ISIS Canada*, ISIS Canada, University of Manitoba, Winnipeg.

Highways Agency (1994) *Design Manual for Roads and Bridges*, Volume 1, *Highway Structures: Approval Procedures & General Design*: Section 3, General Design; Part 9, Design of bridges and concrete structures with external and unbonded prestressing.

Hordijk, D.A. (1998) A concrete balanced cantilever box girder bridge in the Netherlands with carbon fibre prestressing cables. In Stoelhurst, D. and den Boer, G.P.L. (eds), *Challenges for Concrete in the Next Millennium*. A.A. Balkema, Rotterdam & Brookfield, pp. 29–33.

Institution of Structural Engineers (1999) *Interim Guidance on the Design of Reinforced Concrete Structures using Fibre Composite Reinforcement.* The Institution of Structural Engineers, London, p. 116.

Iyer, S.L. (1993) Advanced composite demonstration bridge deck. In Nanni, A. and Dolan, C.W. (eds), *Fibre reinforced plastic reinforcement for concrete structures*, SP 138, American Concrete Institute, Detroit, 831.

Japan Society of Civil Engineers (1997) *Recommendations for Design and Construction of Concrete Structures using Continuous Fiber Reinforcing Materials*, Japan Society of Civil Engineers, Tokyo. Concrete Engineering Series 23.

Japanese Ministry of Construction, (1995) *Guidelines for structural design of FRP reinforced concrete building structures*, Japanese Ministry of Construction.

Maheu, J. (1994) NEFMAC barrier walls. *FRP International*, **II**, Issue 2, 7.

Noritake, K. *et al.* (1993) Practical applications of aramid FRP rods to prestressed concrete structures. In Nanni, A. and Dolan, C.W. (eds), *Fibre Reinforced Plastic Reinforcement for Concrete Structures*, SP 138, American Concrete Institute, Detroit, pp. 853–873.

Rizkalla, S. *et al.* (1998) The new generation. *Concrete International*, June, 35–38.

Sugita, M. (1993) NEFMAC grid type reinforcement. In Clarke, J.L. (ed.), *Alternative Materials for the Reinforcement and Prestressing of Concrete*, Blackie Academic & Professional, Glasgow, 55–82.

Thippeswamy, H.K., Franco, J.M. and GangaRao, H.V.S. (1998) FRP reinforcement in bridge deck. *Concrete International*, June, 47–50.

Tsuji, Y., Kanda, M. and Tamura, T. (1993) Applications of FRP materials to prestressed concrete bridges and other structures in Japan. *PCI Journal*, July–August, 50.

Wolff, R. and Meisseler, H.J. (1993) Glass fibre prestressing system. In Clarke, J.L. (ed.), *Alternative Materials for the Reinforcement and Prestressing of Concrete*, Blackie Academic & Professional, Glasgow, pp. 127–152.

Yonekura, A. *et al.* (1993) Flexural and shear behaviour of prestressed concrete beams using FRP rods as prestressing tendons. In Nanni, A. and Dolan, C.W. (eds), *Fibre Reinforced Plastic Reinforcement for Concrete Structures*, SP 138, American Concrete Institute, Detroit, pp. 525–548.

Zoch, P. *et al.* (1991) Carbon fibre composite cables: a new class of prestressing members. In *Proceedings of 70th Annual Convention of the Transportation Research Board*, Washington, DC.

## Further reading

Clarke, J.L. (ed.) (1993) *Alternative Materials for the Reinforcement and Prestressing of Concrete*, Blackie Academic & Professional, Glasgow.

There have been a number of international conferences in recent years which have reported on the wide range of research and development being carried out around the world and on current applications for FRP reinforcement and prestressing. Some are listed below.

Benmokrane, B. and Rahman, H. (eds) (1998) *Durability of Fibre Reinforced Polymer (FRP) Composites for Construction*, Department of Civil Engineering, University of Sherbrooke, Quebec, Canada, 706.

El-Badry, M.M. (ed.) (1996) *Advanced Composite Materials in Bridges and Structures*, Canadian Society for Civil Engineering, Montreal.

Japan Concrete Institute (1997) *Non-Metallic (FRP) Reinforcement for Concrete Structures*, Tokyo, Japan, pp. 728 and 813 (2 volumes).

Nanni, A. and Dolan, C.W. (eds) (1993) *Fibre-Reinforced – Plastic Reinforcement for Concrete Structures*, American Concrete Institute, Detroit, SP-138, 977.

Taerwe, L. (ed.) (1995) *Non-Metallic (FRP) Reinforcement for Concrete Structures*, E&FN Spon, London, p. 714.

# Index